# Essentials of Bacteriology and Mycology

# Essentials of Bacteriology and Mycology

Editor: Donald Vincent

Callisto Reference

www.callistoreference.com

**Callisto Reference,**
118-35 Queens Blvd., Suite 400,
Forest Hills, NY 11375, USA

Visit us on the World Wide Web at:
www.callistoreference.com



ISBN: 978-1-64116-168-8 (Hardback)



Printed in China.

**Cataloging-in-Publication Data**

Essentials of bacteriology and mycology / edited by Donald Vincent.
p. cm.
Includes bibliographical references and index.
ISBN 978-1-64116-168-8
1. Bacteriology. 2. Mycology. 3. Microbiology. I. Vincent, Donald.
QR74.8 .E87 2019
579.3--dc23

# Table of Contents

# Preface

Bacteriology refers to the in-depth study of bacteria. It delves into ecological, morphological and genetic aspects of bacteria. It is also concerned with the characterization and classification of bacterial species. It also studies the uses of bacteria for varied purposes. Mycology is the study of fungi, their properties and taxonomy. It also explores the use of fungi for commercial purposes such as in the sector of food, medicine, etc. This book aims to shed light on some of the unexplored aspects of bacteriology and mycology. It aims to present researches that have transformed these disciplines and aided their advancement. This book is appropriate for students seeking detailed information in these areas as well as for experts.

Significant researches are present in this book. Intensive efforts have been employed by authors to make this book an outstanding discourse. This book contains the enlightening chapters which have been written on the basis of significant researches done by the experts.

Finally, I would also like to thank all the members involved in this book for being a team and meeting all the deadlines for the submission of their respective works. I would also like to thank my friends and family for being supportive in my efforts.

**Editor**

# 1

# *Lactobacillus acidophilus* alleviates the inflammatory response to enterotoxigenic *Escherichia coli* K88 via inhibition of the NF-κB and p38 mitogen-activated protein kinase signaling pathways in piglets

Haihua Li, Lei Zhang, Longbin Chen, Qi Zhu, Wenjie Wang and Jiayun Qiao*

**Abstract**

**Background:** A newly isolated *L. acidophilus* strain has been reported to have potential anti-inflammatory activities against lipopolysaccharide (LPS) challenge in piglet, while the details of the related inflammatory responses are limited. Here we aimed to analysis the ability of *L. acidophilus* to regulate inflammatory responses and to elucidate the mechanisms involved in its anti-inflammatory activity.

**Results:** The ETEC (enterotoxigenic *Escherichia coli*) K88-induced up-regulations of IL-1β, IL-8 and TNF-α were obviously inhibited by *L. acidophilus* while IL-10 was significantly increased. Moreover, *L. acidophilus* down-regulated pattern recognition receptors TLR (Toll-like receptor) 2 and TLR4 expression in both spleen and mesenteric lymph nodes of ETEC-challenged piglets, in accompanied with the reduced phosphorylation levels of nuclear factor kappa B (NF-κB) p65 and mitogen-activated protein kinase (MAPK) p38 as well in spleen of ETEC-infected piglets. Furthermore, *L.acidophilus* significantly increased the expression of the negative regulators of TLRs signaling, including Tollip, IRAK-M, A20 and Bcl-3 in spleen of ETEC-challenged piglets.

**Conclusions:** Our findings suggested that *L. acidophilus* regulated inflammatory response to ETEC via impairing both NF-κB and MAPK signaling pathways in piglets.

**Keywords:** ETEC, Inflammatory response, Innate immunity, *L. acidophilus*, Mitogen-activated protein kinase, Nuclear factor kappa B, Toll-like receptor

## Background

Probiotics contribute to maintaining a healthy digestive system of the host via regulating both intestinal development and flora balance [1–3], some of which have been shown to enhance the host immune responses by regulating cytokine and chemokine production [2]. *L. acidophilus* strain is a well-characterized probiotic bacterium, which has been reported to improve the production performance of animals as well as enhance the immune responses [4–6]. TLRs are essential for triggering the innate immune response by sensing pathogen-associated molecular patterns (PAMPs) [2, 7], which activates nuclear factor kappa B (NF-κB) and mitogen-activated proteinkinase (MAPK) signaling pathway [8, 9].

In our previous study, a newly isolated *L. acidophilus* strain has shown potential protective activity against inflammatory response to lipopolysaccharide (LPS) in piglets [4], but the underlying molecular mechanism is still unknown. LPS present in the outer membranes of some Gram-negative pathogens, such as ETEC, which can trigger the production of proinflammatory mediators that may contribute to intestinal inflammation and consequent inflammatory damages during the infection [9–11]. *Escherichia coli* (*E. coli*) is one of the predominant species of facultative anaerobes in the pig gut and an opportunistic pathogen to the host. ETEC, as a group of

* Correspondence: qiaojy1979@126.com
Tianjin Institute of Animal Husbandry and Veterinary Medicine, Tianjin, China

pathogenic *E. coli*, can causes diarrheal disease in pigs [10, 12]. Therefore, here we use piglets infected with ETEC as experimental models to investigate the role of *L. acidophilus* in both NF-κB and MAPK signaling pathways, thus to determine the regulative ability of *L. acidophilus* on inflammatory responses and deepen the mechanisms involved in reduced inflammation during ETEC infection by *L. acidophilus.*

## Methods

### Bacterial strain

ETEC strain K88 from the China Veterinary Culture Collection Center was grown in Luria-Bertani (LB) broth containing 1 % tryptone and 0.5 % yeast extract (both from OXOID) plus 1 % NaCl, pH 7.0. After incubation at 37 °C with vigorous shaking overnight, bacteria were 1:100 diluted in fresh LB and grown for 2 h. The bacterial cells were harvested by centrifugation at 3000 × g for 10 min at 4 °C, washed in 0.9 % NaCl solution and resuspended in saline.

The *L. acidophilus* was grown in MRS (De Man, Rogosaand Sharpe) medium at 37 °C under anaerobic environment. Culture solution of the strain was centrifuged at 3000 × g for 10 min at 4 °C. Bacterial powder was acquired according to the treatment in a vacuum freeze-drying machine (Tofflon, Shanghai, China), and there are $5 \times 10^{10}$ CFU/g *Lactobacillus* in freeze-drying powder. Bacterial concentrations of both ETEC and *Lactobacilli* were determined in preliminary experiments by densitometry and confirmed by serial dilutions followed by CFU counts of ETEC on LB agar after 16-h incubation and the lactobacilli on MRS agar after 48-h incubation under anaerobic environment.

**Table 1** Ingredient and chemical composition of basal diets (% w/w, as-fed basis)

| Item | Amount |
|---|---|
| Corn, yellow | 63.20 |
| Soybean meal, 43 % CP (crude protein) | 19.00 |
| Whey powder | 4.80 |
| Fish meal, 65 % CP | 8.60 |
| Glucose | 1.00 |
| Acidifier | 0.30 |
| Calcium hydrogen phosphate | 0.60 |
| Limestone | 0.70 |
| Salt | 0.30 |
| L-Lys•HCL, 78 % Lys | 0.30 |
| DL-Met, 99 % Met | 0.10 |
| L-Thr, 98 % Thr | 0.10 |
| Vitamin and mineral premix[a] | 1.00 |
| Calculated composition | |
| DE (digestible energy), Mcal/kg | 3.25 |
| Lys, % | 1.39 |
| Met, % | 0.53 |
| Analyzed composition, | |
| Crude protein | 18.75 |
| Crude fat | 3.42 |
| Calcium | 0.88 |
| Total phosphorus | 0.71 |
| Crude fiber | 2.20 |

[a]Supplying a minimum per kilogram complete diet of: 12,500 IU Vitamin A; 1250 IU Vitamin D; 125 IU Vitamin E; 90 μg Vitamin $B_{12}$; 10 mg riboflavin; 48 mg pantothenic acid; 35 mg niacin; 4.5 mg folic acid; 0.25 mg biotin; 130 mg Fe; 180 mg Zn; 15 mg Cu; 30 mg Mn; 0.60 mg I and 0.25 mg Se

### Animals and experimental design

All pigs, which were purchased from Tianjin Nongfu Agriculture and Animal Husbandry Co. Ltd, used in this experiment were born naturally at full term (114 days of gestation). A total of 12 crossbred healthy female piglets (Duroc × Landrace × Yorkshiere) were reared by sows and weaned at 21 ± 2 days of age. After a 7-day period of adaptation, the pigs (5.34 ± 0.09 kg) were allotted to 1 of 4 dietary treatments (3 pigs per treatment). *L. acidophilus* were included in the diet by replacing the same amount of corn. The corn-soybean meal-fish meal basal diet (Table 1) was formulated to meet the National Research Council (NRC 2012) requirements for all nutrients.

The experiment was arranged as a 2 × 2 factorial arrangement of 2 diets (basal diet with or without *L. acidophilus*) and subsequently ETEC challenged (pigs challenged with ETEC or treated with sterile saline). (1) Control group (piglet fed the basal diet and receiving oral administration of 0.9 % NaCl solution); (2) ETEC group (piglet fed the basal diet and receiving oral administration of ETEC); (3) *L. acidophilus* group (piglets fed the basal diet supplemented with 0.2 % *L. acidophilus* powder and receiving oral administration of 0.9 % NaCl solution); (4) ETEC + *L. acidophilus* group (piglets fed the basal diet supplemented with 0.2 % *L. acidophilus* powder and receiving oral administration of ETEC).

Each pen was equipped with a feeder and a nipple water to allow piglets free access to feed and drinking water, and maintained at ambient temperature of 20 ~ 30 °C. All piglets had free access to the basal diet (Table 1) between 21 and 28 days of age for adapting to solid food, and pigs were received the four diets respectively at 28 days of age throughout the 14-d feeding trial. At 42 days of age, the challenged group was orally received *E. coli* K88 at $1 \times 10^9$ CFU/kg BW and the unchallenged group was orally received with the same amount of 0.9 % NaCl solution. The dosage of ETEC was chosen according to Li et al. [13]. ETEC ($1 \times 10^9$ CFU/mL) was diluted in sterile 0.9 % NaCl.

**Blood and tissue sample collections**

Three hours after the ETEC or saline treatment, blood samples (5 ml per piglet) of piglets were collected through precava. Serums were obtained by centrifugation at 3000 rpm and 4 °C for 20 min and stored at −20 °C before analysis. The spleen and MLNs samples were harvested by scraping with a glass slide, immediately frozen in liquid nitrogen, and then stored at −80 °C for further analysis. According to the previous reports, ETEC can induce acute inflammatory responses within 1–6 h following the oral challenge in piglets, which results in intestinal morphologic damage and the impairment of intestinal barrier function [14–16]. Therefore, the time point of 3 h following ETEC or saline treatment was chosen for experimental measurements.

**Detection of serum cytokine levels by ELISA**

The concentrations of IL-1β, IL-8, TNF-α and IL-10 in serums were measured using commercially available ELISA kits specific for porcine IL-1β, IL-8, TNF-α and IL-10 (R and D Systems, Minneapolis, MN) according to the manufacturer's instructions.

**RNA extraction and real-time RT-PCR**

Total RNA was extracted from tissue samples using the TRIzol reagent [TaKaRa Biotechnology (Dalian)] according to the manufacturer's guidelines, and 1 μg of RNA was used for cDNA synthesis using Moloney murine leukemia virus (MMLV; Promega) to examine expression of TLR2, TLR4, Tollip, IRAK-M, A20 and Bcl-3. Real-time PCR was performed using specific primers for TLR2 [7] (sense 5′-TCA TCT CCC AAA TCT GCG AAT-3′, antisense 5′-GGC TGA TGT TCT GAA TTG ACCTC-3′), TLR4 [7] (sense 5′-CCG TCA TTA GTG CGT CAG TTCT-3′, antisense 5′-TTG CAG CCC ACA AAA AGCA-3′), Tollip [9] (sense 5′-TAC CGT GGG CCG TCTCA-3′, antisense 5′-CCG TAG TTC TTC GCC AAC TTG-3′), IRAK-M [9] (sense 5′-TGG AGC AGC CTT GAA TCCTT-3′, antisense 5′-TGG ATA ACA CGT TTG GGA ATCTT-3′), A20 [9] (sense 5′-CCT CCC TGG AAA GCC AGAA-3′, antisense 5′-GTG CCA CAA GCT TCC TCA CTT-3′), Bcl-3 [9] (sense 5′-CGA CGC GGT GGA CAT TAAG-3′, antisense 5′-ACC ATG CTA AGG CTG TTG TTT TC-3′), or β-actin [9] (sense 5′-CAT CAC CAT CGG CAA CGA-3′, antisense 5′-GCG TAG AGG TCC TTC CTG ATGT-3′) and the Real Time SYBR master mix kit [TaKaRa Biotechnology (Dalian)] in accordance with the manufacturer's protocol for the ABI7500 real-time PCR system (Applied Biosystems, Life Technologies). Gene expression was normalized to β-actin (internal reference) and presented as relative fold change compared with control group. All samples were run in triplicate.

**Detection of both NF-κB and p38 MAPK activation by western blot**

The tissue samples (50–100 mg) were homogenized in 1 mL of NP-40 lysis buffer supplemented with protease and phosphatase inhibitors and centrifuged at 12,000 × g for 15 min at 4 °C. Tissue proteins were separated by 12 % sodium dodecyl sulfate-polyacrylamide gel electrophoresis (SDS-PAGE) and transferred onto the nitrocellulose membranes. After blocking with 3 % BSA in Tris-buffered saline including 0.1 % Tween-20 buffer, membranes were incubated with primary antibodies against NF-κBp65 (Cell Signaling), phospho-NF-κBp65 (Cell Signaling), p38 MAPK (Cell Signaling), phospho-p38 MAPK (Cell Signaling) or β-actin (Sigma-Aldrich). Membranes were then washed and incubated with appropriate secondary antibodies. Proteins were visualized using the ECL reagent according to the manufacturer's instructions. One sample from each treatment was run in duplicate in a gel and 12 samples from 4 treatments ($n = 3$) were run in 3 gels at one time to minimize the variations of gel to gel. Densitometric analysis was performed by Gel-Pro analyzer and normalized to β-actin.

**Statistical analysis**

Data were evaluated using one-way ANOVA (SAS Institute Inc., Cary, NC, 2002). The Tukey multiple comparison test was used to determine differences among the means of treated groups. A probability value of <0.05 was considered statistically significant.

## Results

***L. acidophilus* regulates ETEC-induced serum cytokine production in piglet**

In our previous study, we evaluated the anti-inflammatory effect of *L. acidophilus* in piglets challenged with LPS, and established the appropriate dose of *L. acidophilus* (0.2 %) for further studies on inflammatory responses [4]. Here we used the same dose of oral administration to evaluate anti-inflammatory function of *L. acidophilus* in piglets challenged with ETEC. We found that serum levels of IL-1β (Fig. 1a), IL-8 (Fig. 1b), TNF-α (Fig. 1c) and IL-10 (Fig. 1d) increased significantly in ETEC group compared with Control group. On the contrary, the proinflammatory cytokines which were increased in ETEC group, such as IL-1β (Fig. 1a), IL-8 (Fig. 1b) and TNF-α (Fig. 1c), were suppressed in ETEC + *L. acidophilus* group, while the significant increase of anti-inflammatory cytokine IL-10 was observed (Fig. 1d). Meanwhile, the uptake of 0.2 % *L. acidophilus* (*L. acidophilus* group) had no effects on the production of the indicated cytokines (Fig. 1a-d). These results indicated that we established an appropriate model to study the anti-inflammatory mechanism of *L. acidophilus*.

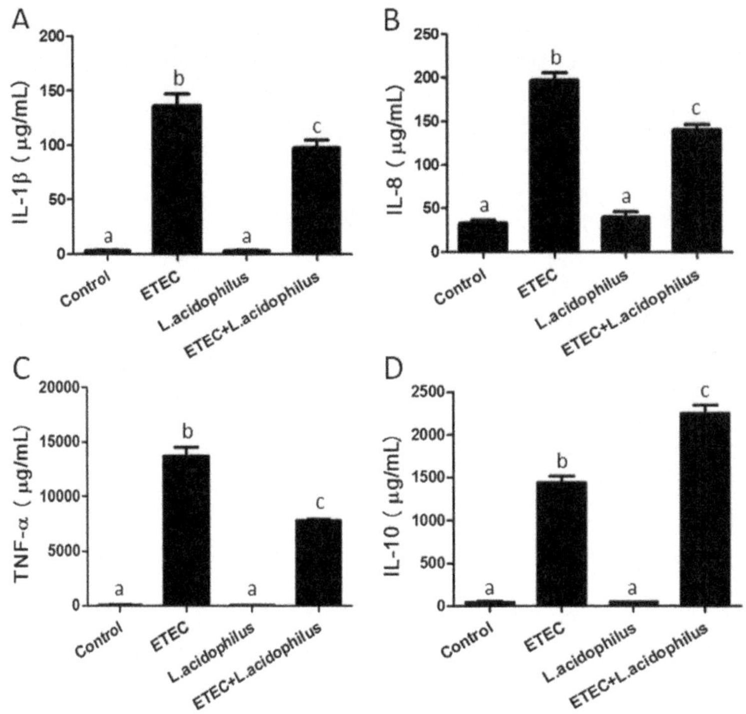


**Fig. 1** *L. acidophilus* regulates ETEC-induced serum cytokine production in piglet. Pigs were challenged with or without ETEC ($1 \times 10^9$ CFU/kg BW), the levels of IL-1β (**A**), IL-8 (**B**), TNF-α (**C**) and IL-10 (**D**) proteins in serums were evaluated 3 h after the inflammatory challenge. Values with different letters are significantly different ($P < 0.05$). ($n = 3$)

### *L. acidophilus* down regulates ETEC-induced Toll-like receptors mRNA expression in spleen and MLNs

TLRs are key components that trigger the innate immune response against pathogens. Mature lymphocytes and macrophages expressing TLRs were settled in the surrounding lymphoid organs such as spleen and mesenteric lymph nodes (MLNs). TLR2 and TLR4 play important roles in the inflammatory responses to different pathogenic bacteria [12, 17]. This prompted us to investigate whether the TLR2 and TLR4 are involved in inflammatory regulatory function of *L. acidophilus* in ETEC-challenged piglets. We examined the mRNA levels of both TLR2 and TLR4 in spleen and MLNs using real-time RT-PCR. There was no difference between *L. acidophilus* group and the control. As shown in Fig. 2a, the levels of TLR2 increased 1.5-fold and 1.0-fold in spleen and MLNs respectively, and the levels of TLR4 increased 2.5-fold both in spleen and MLNs (Fig. 2b), in ETEC group compared with control group. However, we found down-regulation 44 % and 39 % of the level of TLR2 in spleen and MLNs respectively (Fig. 2a), and 50 and 47 % of the level of TLR4 in spleen and MLNs respectively (Fig. 2b) in ETEC + *L. acidophilus* group compared with ETEC group. The data indicated that *L. acidophilus* down-regulated the ETEC-induced Toll-like receptors expression in mRNA levels.

### *L. acidophilus* inhibited NF-κB and p38 MAPK activation in ETEC-challenged piglet in spleen

Activation of TLRs activates NF-κB and MAPK signaling, which are the immune-related transcriptional factors that stimulate synthesis of cytokines and chemokines [2, 8, 9]. It has been reported that ETEC or LPS is able to induce NF-κB and MAPK activation in porcine intestinal epithelial cell line (PIE cells) [9]. We therefore evaluated the effect of *L. acidophilus* on NF-κB and MAPK activation during ETEC oral challenge. Levels of p65, p-p65, p38, and p-p38 proteins in spleen were measured by western blotting (Fig. 3a and b). As shown in Fig. 3c, the phosphorylation levels of p65 and p38 in ETEC group were significantly increased compared with control group. *L. acidophilus* group did not modify the level of p-p65 and p-p38 compared with control group. However, ETEC + *L. acidophilus* group abrogated the increase of p-p65 and p-p38 level compared with ETEC group. These results indicated that *L. acidophilus* can counteract the ETEC-induced NF-κB and MAPK activation.

### *L. acidophilus* upregulated negative regulators of TLR signaling pathway in ETEC-infected piglet in spleen

We also studied the role of negative regulators which mediates the TLR signaling pathway on relieved inflammation by

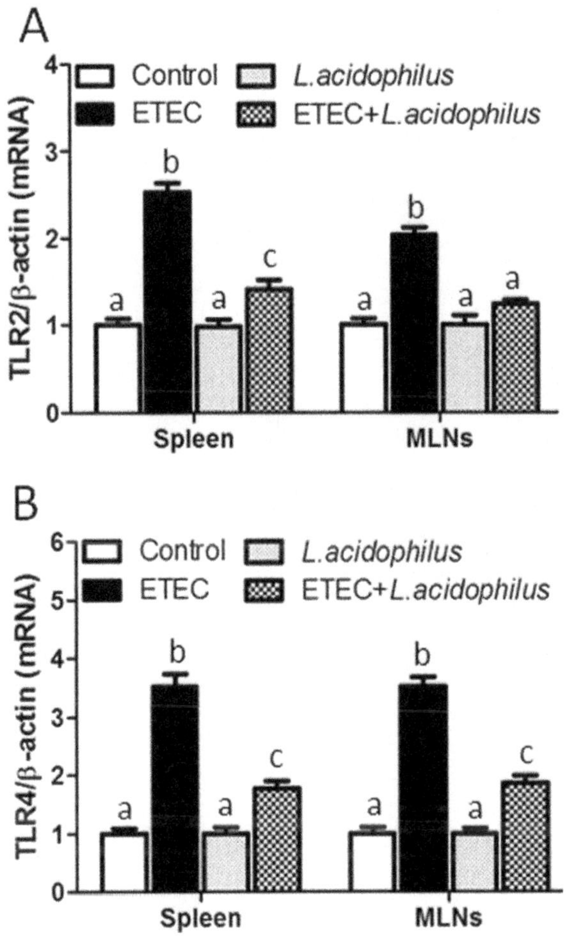


**Fig. 2** *L. acidophilus* impaired ETEC-trigged TLR2 and TLR4 expression in spleen and MLNs. Pigs were challenged with or without ETEC ($1 \times 10^9$ CFU/kg BW). Three hours post inflammatory challenge, spleen and MLNs samples were collected to examine the levels of porcine TLR2 (**A**) and TLR4 (**B**) by real-time RT-PCR. Expression of TLR2 and TLR4 was normalized to β-actin in the same sample, and presented as fold change relative to mock challenged pigs fed with basal diet (Control group). Expression of TLR2 or TLR4 in control group was normalized to 1.0. Values with different letters are significantly different ($P < 0.05$). ($n = 3$)

*L. acidophilus* in ETEC infection. The expression levels of Toll interacting protein (Tollip), interleukin-1 receptor-associated kinase M (IRAK-M), A20 and B-cell lymphoma 3-encodedprotein (Bcl-3) in spleen were determined using real-time RT-PCR. We found that ETEC group exhibited obviously lower expression of Tollip (Fig. 4a), IRAK-M (Fig. 4b), A20 (Fig. 4c) and Bcl3 (Fig. 4d) compared with the control group, but higher levels of Tollip (Fig. 4a), A20 (Fig. 4c) and Bcl-3 (Fig. 4d) were shown in *L. acidophilus* group, compared with the control group. We also found that "ETEC + *L. acidophilus*" group showed obviously higher expression of Tollip (Fig. 4a), IRAK-M (Fig. 4b), A20 (Fig. 4c) and Bcl3 (Fig. 4d) compared with *L. acidophilus* group. As our expected, there was extreme difference between "ETEC + *L. acidophilus*" group and ETEC group. The present data

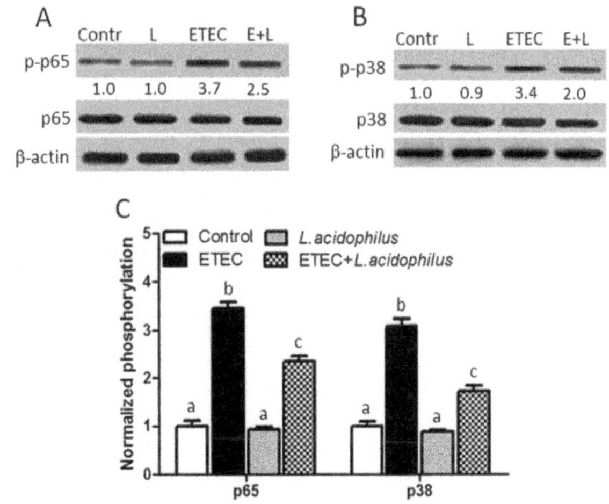


**Fig. 3** *L. acidophilus* inhibited NF-κB and p38 MAPK activation in spleen of ETEC-challenged piglet. Pigs were challenged with or without ETEC ($1 \times 10^9$ CFU/kg BW). Three hours post inflammatory challenge, spleen samples were collected to examine the expression of p65 (**A**), p-p65 (**A**), p38 (**B**) and p-p38 (**B**) by western blotting. Densitometric analysis was performed and normalized to β-actin (**C**). Phosphorylation of p65 or p38 in control group was normalized to 1.0. Values with different letters are significantly different ($P < 0.05$). ($n = 3$)

indicated that infection with ETEC strongly downregulated the levels of mRNA expression of the TLR antagonists, and this effect was impaired by *L. acidophilus*.

## Discussion

It is demonstrated that Lactic acid bacteria (LAB) are beneficial in the treatment of a variety of mucosal disorders including inflammatory damages even in the absence of infection [9, 18–20], especially during the period of weaning [12]. Studies that evaluated the effects of probiotic strains on immune responses demonstrate that probiotic-mediated protection against pathogen-induced inflammation, in part, from maintaining the balance of proinflammatory and anti-inflammatory cytokines production in immune cells [12, 21–24]. *L. acidophilus*, one of the popular probiotics, has been confirmed its role of inflammatory regulation in human MKN45 cells [22], intestinal epithelial cells [23] and in piglets [4]. Based on the increasingly supporting evidence of the beneficial effects of probiotics, it is important to explore the molecular mechanism of LAB-regulated inflammatory reaction. In this study, we used piglets orally challenged by ETEC K88 as experimental models to investigate the function of *L. acidophilus* in inflammatory response thus to elucidate the mechanisms involved in anti-inflammatory activities.

Many cytokines are of importance to influence the intestinal function [25, 26]. For examples, IL-8 has been associated with pathogen-induced alterations of tight junctions [25, 26], as well as TNF-α has remarkable

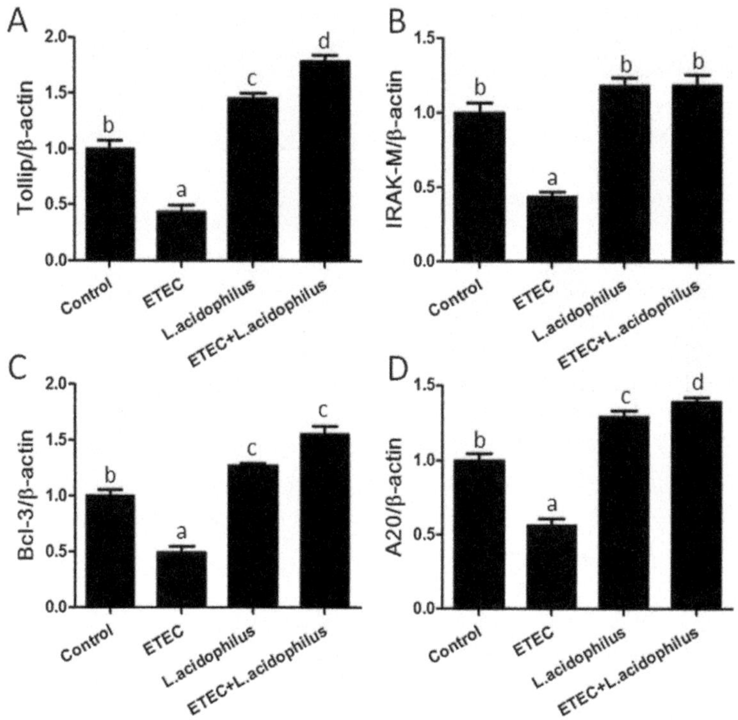


**Fig. 4** *L. acidophilus* upregulated negative regulators of TLRs signaling pathway in spleen of ETEC-infected piglet. Pigs were challenged with or without ETEC ($1 \times 10^9$ CFU/kg BW). Three hours post inflammatory challenge, spleen samples were collected to examine the levels of Tollip (**A**), IRAK-M (**B**), Bcl-3 (**C**) and A20 (**D**) by real-time RT-PCR. Expression of Tollip, IRAK, A20 and Bcl-3 was normalized to β-actin in the same sample, and presented as fold change relative to ETEC-challenged pigs fed with basal diet (ETEC group). Expression of Tollip, IRAK, A20 or Bcl-3 in ETEC group was normalized to 1.0. Values with different letters are significantly different ($P < 0.05$). ($n = 3$)

functional duality that is strongly engaged in both tissue regeneration and destruction [13]. Several studies in vitro have shown that *L. acidophilus* is able to modulate the production of proinflammatory cytokines, including IL-8, TNF-α and IL-6 in intestinal epithelial cells [8, 12, 21, 23, 27]. Here, we observed significant down-regulated proinflammatory cytokines by *L. acidophilus* including IL-1β, IL-8 and TNF-α in piglets challenged with ETEC and a remarkable rise of anti-inflammatory cytokine IL-10 production. This result was consistent with our previous findings that *L. acidophilus* alleviated inflammatory response to LPS in piglets [4]. Referring to the weakening effects on pathogenic bacteria-induced inflammation of lactic acid bacteria, some studies in vitro have shown that *L. plantarum* and *L. rhamnosus* suppress *E.coli*-induced proinflammatory cytokines of IL-1α, IL-8 and TNF-α expression [13, 28], and *L. acidophilus* can decrease Salmonella-induced TNF-α and IL-8 in human intestinal Caco-2 cells [29]. Our findings indicate that the *L. acidophilus* can also counteract the inflammation induced by a pathogenic bacterium, ETEC, in weaned piglets.

TLR2 and TLR4 are well known to initiate the inflammatory response to pathogenic bacteria [12, 17]. Intestinal damage caused by inflammation can be mediated by the inflammatory response triggered by the interaction between pathogens and TLRs. Inappropriate TLR signaling can lead to impaired tolerance to PAMPs which results in sever intestinal epithelium injury [9, 30]. TLR2 and TLR4 are believed to favor signaling the production of inflammatory cytokines [24]. Thus, it prompted us to investigate whether TLR2 and TLR4 are involved in inflammatory response to ETEC and can be regulated by *L. acidophilus*. Our data indicated that ETEC infection stimulated the expression of TLR2 and TLR4 in both spleen and MLNs in ETEC-challenged piglets, which were significantly down-regulated by the presence of *L. acidophilus* in the diet. Similar results were found in other LAB. *L. plantarum* N14 (LP14) regulates the production of proinflammatory cytokines from PIEs in response to ETEC challenge via TLR2 and TLR4 [21]. *L. delbrueckii* subsp. delbrueckii TUA4408L (Ld) suppress inflammatory cytokines production through TLR2 in PIE cells, and its acidic extracellular polysaccharide (APS) plays the role of immunomodulatory action by TLR4 [28]. *L. jensenii* TL2937 weakened the levels of proinflammatory cytokines and chemokines triggered by ETEC or LPS challenge by decreasing TLR4 expression [9]. Our findings provide evidence that both TLR2 and TLR4 signaling play important role of the inflammatory regulation of *L. acidophilus* during ETEC infection.

TLRs signal the activation of NF-κB and MAPK via multiple downstream intracellular factors [2, 24]. Activated NF-κB and MAPK stimulate synthesis of anti- and proinflammatory cytokines, including IL-10, IL-1β, TNF-a, IL-6 and IL-8, etc. [3, 8, 22–24, 31]. Thus, it will be interesting to explore whether NF-κB and MAPK are involved in *L. acidophilus*-mediated regulation on inflammatory response to ETEC. As our expected, *L. acidophilus* decreased the phosphorylation levels of both NF-κB and p38 MAPK in spleen from ETEC-infected piglets. Probiotic *L. acidophilus* is reported to decrease Salmonella-induced NF-κB activation in human intestinal Caco-2 cells [30]. Experiments in demonstrated that *L. plantarum* N14 (LP14) strongly activated NF-κB via RP105 and TLR2 in ETEC-infected PIE cells [21]. Ld strain and its APS can inhibit ETEC triggered mitogen-activated protein kinase (MAPK) and nuclear factor-κB (NF-κB) activation in PIE cells [28]. In addition, *L. jensenii* TL2937 weakened the levels of proinflammatory cytokines and chemokines induced by ETEC or LPS treatment via down-regulating NF-κB and MAPK activation [9]. Therefore, it is likely that the *L. acidophilus* alleviates the ETEC-induced inflammatory response through weakening the activation of these transcriptional factors of host cells.

The immune system needs to constantly communicate to maintain homeostasis between activation and inhibition of TLRs signals to avoid inflammatory immunological imbalance [32]. Negative regulatory mechanisms are important to weaken TLR signaling and keep the intact signaling pathway. In this regard, a ever increasing number of inhibitory factors of TLR signaling, including Tollip, IRAK-M, A20, Bcl-3 and peroxisome proliferator-activatedreceptor-γ (PPARγ) are found and characterized, which guard against chronic inflammatory and potentially detrimental TLR responses to microbe-associated molecular patterns (MAMPs) from intestinal commensal bacteria [33]. Acturally, commensal gut bacteria can regulate TLR negative regulators in intestinal epithelial cells (IECs) in turn. An anti-inflammatory mechanism of commensal *B. thetaiotaomicron* has been founded. *B. thetaiotaomicron* weakens proinflammatory cytokine expression level in IECs via contributing to nuclear translocation of the NF-κB subunit RelA, by a PPARγ-dependent pathway [34]. Another example is that *L. jensenii* TL2937 down-regulates LPS-triggering proinflammatory cytokines and chemokines production in PIE cells via enhancing the expression of A20, Bcl-3 and MKP-1, which retard the TLR4-dependent NF-κB and MAPK activation [9]. Moreover, continuously up-regulation of A20, Tollip, and SIGIRR plays a major part in the anti-inflammatory activity of *L. casei* MEP221114 in poly(I:C)-stimulated PIE cells [35]. The increased level of TLRs pathway negative regulators can also be found in other LAB which can alleviate inflammatory damages, such as *L. plantarum* [21], Ld, APS and NPS [28]. Consistently, the results here showed that *L. acidophilus* obviously up-regulated mRNA expression level of negative regulator Tollip and Bcl-3 in spleen post ETEC infection. Thus, Tollip and Bcl-3 are of importance in the anti-inflammatory effect of *L. acidophilus*. Interestingly, we found that A20, Tollip, Bcl-3 and IRAK-M were all up-regulated without the challenge by ETEC. The findings indicated that the uptake of *L. acidophilus* from diets resulted in a higher level of negative regulators of TLRs signaling pathways, thus to establish an anti-inflammatory state against inflammation damage.

## Conclusions

In summary, *L. acidophilus* suppress the activation of the different steps of NF-κB and MAPK signaling in piglets, by inhibiting ETEC-induced TLR2 and TLR4 expression, impaired the phosphorylation of NF-κB and MAPK leading to reduce the production of proinflammatory cytokines, as well as up-regulating negative regulators and anti-inflammatory cytokines. Taken together with the results presented here, we speculate that *L. acidophilus* alleviates inflammatory response to ETEC via inhibition of the NF-κB and mitogen-activated protein kinase signaling pathways in piglets. This underlying mechanism provides a theoretical basis for the clinical application of lactic acid bacteria.

#### Abbreviations

ETEC: Enterotoxigenic *Escherichia coli*; IL: Interleukin; *L. acidophilus*: *Lactobacillus acidophilus*; MAPK: Mitogen-activated protein kinase; MLNs: Mesenteric lymph nodes; NF-κB: Nuclear factor kappa B; TLR: Toll-like receptor

#### Acknowledgements

None.

#### Funding

We acknowledge financial support from the National Natural Science Foundation of China (No. 31402087), and the President Funding of Tianjin Academy of Agricultural Sciences (No. 13004).

#### Authors' contributions

HL, WW and JQ designed research. HL, LZ, LC, QZ and JQ carried out the experiments. HL, LC, QZ and JQ analyzed data. HL and JQ wrote the manuscript, JQ revised manuscript. JQ had primary responsibility for final content. All authors read and approved the final manuscript.

#### Competing interests

The authors declare that they have no competing interests.

**References**


1. Suda Y, Villena J, Takahashi Y, Hosoya S, Tomosada Y, Tsukida K, Shimazu T, Aso H, Tohno M, et al. Immunobiotic *Lactobacillus jensenii* as immune-health promoting factor to improve growth performance and productivity in post-weaning pigs. BMC Immunol. 2014;15:24.
2. Tomosada Y, Villena J, Murata K, Chiba E, Shimazu T, Aso H, Iwabuchi N, Xiao JZ, Saito T, Kitazawa H. Immunoregulatory effect of *bifidobacteria* strains in porcine intestinal epithelial cells through modulation of ubiquitin-editing enzyme A20 expression. PLoS One. 2013;8:e59259.
3. Finamore A, Roselli M, Imbinto A, Seeboth J, Oswald IP, Mengheri E. *Lactobacillus amylovorus* inhibits the TLR4 inflammatory signaling triggered by enterotoxigenic *Escherichia coli* via modulation of the negative regulators and involvement of TLR2 in intestinal Caco-2 cells and pig explants. PLoS One. 2014;9:e94891.
4. Qiao J, Li H, Wang Z, Wang W. Effects of *Lactobacillus acidophilus* dietary supplementation on the performance, intestinal barrier function, rectal microflora and serum immune function in weaned piglets challenged with *Escherichia coli* lipopolysaccharide. Antonie Van Leeuwenhoek. 2015;107:883–91.
5. Forte C, Moscati L, Acuti G, Mugnai C, Franciosini MP, Costarelli S, Cobellis G, Trabalza-Marinucci M. Effects of dietary *Lactobacillus acidophilus* and *Bacillus subtilis* on laying performance, egg quality, blood biochemistry and immune response of organic laying hens. J Anim Physiol Anim Nutr (Berl). 2016;100:977-87.
6. Alonso L, Fontecha J, Cuesta P. Combined effect of *Lactobacillus acidophilus* and beta-cyclodextrin on serum cholesterol in pigs. Br J Nutr. 2016;115:1–5.
7. Qin Y, Li H, Qiao J. The TLR2/MyD88/NF-kappaB signaling pathway regulates IL-8 production in porcine alveolar macrophages infected with porcine circovirus 2. J Gen Virol. 2016;97:445-52.
8. Jiang Y, Lu X, Man C, Han L, Shan Y, Qu X, Liu Y, Yang S, Xue Y, Zhang Y. *Lactobacillus acidophilus* induces cytokine and chemokine production via NF-kappaB and p38 mitogen-activated protein kinase signaling pathways in intestinal epithelial cells. Clin Vaccine Immunol. 2012;19:603–8.
9. Shimazu T, Villena J, Tohno M, Fujie H, Hosoya S, Shimosato T, Aso H, Suda Y, Kawai Y, et al. Immunobiotic *Lactobacillus jensenii* elicits anti-inflammatory activity in porcine intestinal epithelial cells by modulating negative regulators of the Toll-like receptor signaling pathway. Infect Immun. 2012;80:276–88.
10. McLamb BL, Gibson AJ, Overman EL, Stahl C, Moeser AJ. Early weaning stress in pigs impairs innate mucosal immune responses to enterotoxigenic *E. coli* challenge and exacerbates intestinal injury and clinical disease. PLoS One. 2013;8:e59838.
11. Long KZ, Rosado JL, Santos JI, Haas M, Al MA, DuPont HL, Nanthakumar NN, Estrada-Garcia T. Associations between mucosal innate and adaptive immune responses and resolution of diarrheal pathogen infections. Infect Immun. 2010;78:1221–8.
12. Zhang W, Zhu YH, Yang JC, Yang GY, Zhou D, Wang JF. A selected *Lactobacillus rhamnosus* strain promotes EGFR-independent akt activation in an enterotoxigenic *Escherichia coli* K88-infected IPEC-J2 cell model. PLoS One. 2015;10:e125717.
13. Li XQ, Zhu YH, Zhang HF, Yue Y, Cai ZX, Lu QP, Zhang L, Weng XG, Zhang FJ, et al. Risks associated with high-dose *Lactobacillus rhamnosus* in an *Escherichia coli* model of piglet diarrhoea: intestinal microbiota and immune imbalances. PLoS One. 2012;7:e40666.
14. Yu Q, Yuan L, Deng J, Yang Q. *Lactobacillus* protects the integrity of intestinal epithelial barrier damaged by pathogenic bacteria. Front Cell Infect Microbiol. 2015;5:26.
15. Thorgersen EB, Hellerud BC, Nielsen EW, Barratt-Due A, Fure H, Lindstad JK, Pharo A, Fosse E, Tonnessen TI, et al. CD14 inhibition efficiently attenuates early inflammatory and hemostatic responses in *Escherichia coli* sepsis in pigs. FASEB J. 2010;24:712–22.
16. Owusu-Asiedu A, Nyachoti CM, Marquardt RR. Response of early-weaned pigs to an enterotoxigenic *Escherichia coli* (K88) challenge when fed diets containing spray-dried porcine plasma or pea protein isolate plus egg yolk antibody, zinc oxide, fumaric acid, or antibiotic. J Anim Sci. 2003;81:1790–8.
17. Medzhitov R, Janeway CJ. The Toll receptor family and microbial recognition. Trends Microbiol. 2000;8:452–6.
18. Alvarez S, Villena J, Tohno M, Salva S, Kitazawa H. Modulation of innate immunity by lactic acid bacteria: impact on host response to infections. Curr Res Immunol. 2009;3:87–126.
19. Hörmannsperger G, Haller D. Molecular crosstalk of probiotic bacteria with the intestinal immune system: clinical relevance in the context of inflammatory bowel disease. Int J Med Microbiol. 2010;300:63–73.
20. D'Mello C, Ronaghan N, Zaheer R, Dicay M, Le T, MacNaughton WK, Surrette MG, Swain MG. Probiotics improve inflammation-associated sickness behavior by altering communication between the peripheral immune system and the brain. J Neurosci. 2015;35:10821–30.
21. Murofushi Y, Villena J, Morie K, Kanmani P, Tohno M, Shimazu T, Aso H, Suda Y, Hashiguchi K, et al. The toll-like receptor family protein RP105/MD1 complex is involved in the immunoregulatory effect of exopolysaccharides from *Lactobacillus plantarum* N14. Mol Immunol. 2015;64:63–75.
22. Yang Y, Chuang C, Yang H, Lu C, Sheu B. Lactobacillus acidophilus ameliorates H. pylori-induced gastric inflammation by inactivating the Smad7 and NFκB pathways. BMC Microbiol. 2012;12:38.
23. Borthakur A, Bhattacharyya S, Kumar A, Anbazhagan AN, Tobacman JK, Dudeja PK. *Lactobacillus acidophilus* alleviates platelet-activating factor-induced inflammatory responses in human intestinal epithelial cells. PLoS One. 2013;8:e75664.
24. Meshkibaf S, Fritz J, Gottschalk M, Kim SO. Preferential production of G-CSF by a protein-like Lactobacillus rhamnosus GR-1 secretory factor through activating TLR2-dependent signaling events without activation of JNKs. BMC Microbiol. 2015;15:238.
25. Nusrat A, Turner JR, Madara JL. Molecular physiology and pathophysiology of tight junctions. IV. Regulation of tight junctions by extracellular stimuli: nutrients, cytokines, and immune cells. Am J Physiol Gastrointest Liver Physiol. 2000;279:G851–7.
26. Roselli M, Finamore A, Britti MS, Konstantinov SR, Smidt H, de Vos WM, Mengheri E. The novel porcine *Lactobacillus sobrius* strain protects intestinal cells from enterotoxigenic *Escherichia coli* K88 infection and prevents membrane barrier damage. J Nutr. 2007;137:2709–16.
27. Wachi S, Kanmani P, Tomosada Y, Kobayashi H, Yuri T, Egusa S, Shimazu T, Suda Y, Aso H, et al. *Lactobacillus delbrueckii* TUA4408L and its extracellular polysaccharides attenuate enterotoxigenic *Escherichia coli*-induced inflammatory response in porcine intestinal epitheliocytes via Toll-like receptor-2 and 4. Mol Nutr Food Res. 2014;58:2080–93.
28. Chytilova M, Mudronova D, Nemcova R, Gancarcikova S, Buleca V, Koscova J, Tkacikova L. Anti-inflammatory and immunoregulatory effects of flax-seed oil and *Lactobacillus plantarum* - Biocenol LP96 in gnotobiotic pigs challenged with enterotoxigenic *Escherichia coli*. Res Vet Sci. 2013;95:103–9.
29. Huang IF, Lin IC, Liu PF, Cheng MF, Liu YC, Hsieh YD, Chen JJ, Chen CL, Chang HW, Shu CW. *Lactobacillus acidophilus* attenuates *Salmonella*-induced intestinal inflammation via TGF-beta signaling. BMC Microbiol. 2015;15:203.
30. Abreu MT, Fukata M, Arditi M. TLR signaling in the gut in health and disease. J Immunol. 2005;174:4453–60.
31. Wang L, Liu C, Chen M, Ya T, Huang W, Gao P, Zhang H. A novel *Lactobacillus plantarum* strain P-8 activates beneficial immune response of broiler chickens. Int Immunopharmacol. 2015;29:901–7.
32. Cerf-Bensussan N, Gaboriau-Routhiau V. The immune system and the gut microbiota: friends or foes? Nat Rev Immunol. 2010;10:735–44.
33. Shibolet O, Podolsky DK. TLRs in the Gut. IV. Negative regulation of Toll-like receptors and intestinal homeostasis: addition by subtraction. Am J Physiol Gastrointest Liver Physiol. 2007;292:G1469–73.
34. Kelly D, Campbell JI, King TP, Grant G, Jansson EA, Coutts AG, Pettersson S, Conway S. Commensal anaerobic gut bacteria attenuate inflammation by regulating nuclear-cytoplasmic shuttling of PPAR-γ and RelA. Nat Immunol. 2004;5:104–12.
35. Hosoya S, Villena J, Chiba E, Shimazu T, Suda Y, Aso H, Saito T, Kitazawa H. Advanced application of porcine intestinal epithelial cells for the selection of immunobiotics modulating toll-like receptor 3-mediated inflammation. J Microbiol Immun Infect. 2013;46:474–81.

2

# Decision criteria for MALDI-TOF MS-based identification of filamentous fungi using commercial and in-house reference databases

Anne-Cécile Normand[1*], Carole Cassagne[1], Magali Gautier[1], Pierre Becker[2], Stéphane Ranque[1], Marijke Hendrickx[2] and Renaud Piarroux[1]

**Abstract**

**Background:** Several Matrix-Assisted Laser Desorption/Ionization Time-of-Flight mass spectrometry protocols, which differ in identification criteria, have been developed for mold and dermatophyte identification. Currently, the most widely used approach is Bruker technology, although no consensus concerning the log(score) threshold has been established. Furthermore, it remains unknown how far increasing the number of spots to compare results might improve identification performance.

In this study, we used in-house and Bruker reference databases as well as a panel of 422 isolates belonging to 126 species to test various thresholds. Ten distinct identification algorithms requiring one to four spots were tested.

**Results:** Our findings indicate that optimal results were obtained by applying a decisional algorithm in which only the highest score of four spots was taken into account with a 1.7 log(score) threshold. Testing the entire panel enabled identification of 87.41% (in-house database) and 35.15% (Bruker database) of isolates, with a positive predictive value (PPV) of 1 at the genus level for both databases as well as 0.89 PPV (in-house database) and 0.72 PPV (Bruker database) at the species level. Applying the same rules to the isolates for which the species were represented by at least three strains in the database enabled identification of 92.1% (in-house database) and 46.6% (Bruker database) of isolates, with 1 PPV at the genus level for both databases as well as 0.95 PPV (in-house database) and 0.93 PPV (Bruker database) at the species level.

**Conclusions:** Depositing four spots per extract and lowering the threshold to 1.7, a threshold which is notably lower than that recommended for bacterial identification, decreased the number of unidentified specimens without altering the reliability of the accepted results. Nevertheless, regardless of the criteria used for mold and dermatophyte identification, commercial databases require optimization.

**Keywords:** Mass Spectrometry, Matrix-Assisted Laser Desorption/Ionization, Filamentous Fungi, Interpretation criteria, log(score) threshold

* Correspondence: anne-cecile.normand@ap-hm.fr
[1]Laboratoire de Parasitologie-Mycologie, CHU Timone, Université de la Méditerranée, Marseille, France
Full list of author information is available at the end of the article

## Background

Until a few years ago, medical laboratories performed mold identification using phenotypic characterization of colonies and occasional sequencing of informative DNA targets. Currently, the MALDI-TOF MS (Matrix-Assisted Laser Desorption/Ionization Time-of-Flight Mass Spectrometry) approach appears to be a promising method to identify molds and dermatophytes [1–8].

Many teams around the world have demonstrated that filamentous fungi could be identified by using MALDI-TOF MS. They first focused attention on a few specific mold and dermatophyte genera [1, 9–16]. Only after 2010 did research teams begin optimizing protocols to establish a MALDI-TOF MS-based identification approach for routine diagnosis of filamentous fungi and dermatophytes [2–7, 17–21]. Currently, four commercial systems are available for routine diagnosis, and an increasing number of publications have tested their performance in identifying molds and dermatophytes. Some teams have assessed the Saramis [12, 18, 22, 23], Vitek MS [24–28] or Andromas [29] systems; however, most publications refer to the Bruker technique [1–8, 16, 17, 19–21, 26, 30, 31]. According to the Bruker recommendations, the level of similarity between an unknown tested specimen and a reference sample is indicated by a log(score), which is referred to herein as "score". A score > 2.3 indicates "highly probable species identification", a score > 2 and < 2.299 indicates "secure genus identification, probable species identification", a score > 1.7 and < 1.999 indicates "probable genus identification", and a score < 1.7 indicates "unreliable identification".

However, these score thresholds have been designed for bacteria identification and are not necessarily appropriate for fungi (especially molds and dermatophytes) while using the manufacturer's reference database [4, 7, 17, 32, 33]. The majority of publications dealing with MALDI-TOF MS routine identification of filamentous fungi have used a single score threshold, usually ranging from 1.7 to 2.0, depending on the study [3, 4, 6, 8, 17]. However, a study by Shultness et al. [8] has compared the 1.7 and 2.0 score thresholds for MALDI-TOF MS identification after a liquid culture growth step in which no assessment was performed to justify these thresholds. Additionally, while the manufacturer typically recommends a single spot for bacterial identification, some authors have advocated the use of several spots of the same sample to identify molds and dermatophytes [2, 21]. According to these authors, the concordance between identification results can be tested using replicates. Theoretically, this may improve the reliability of identification results, although no study has thoroughly evaluated the optimal number of spots and log(score) threshold for mold identification. Altogether, the use of various thresholds and validation criteria for mold and dermatophyte identification complicate the interpretation of results.

The present study assessed 10 distinct identification algorithms combining several thresholds for one to four spots per sample. Because the Bruker reference database is not comprehensive enough to enable proper identification of many mold species, [4, 8] identification algorithms were assessed using an extensive in-house database (5044 references belonging to 619 species and 165 genera) and the Bruker commercial database (V3.2.1.1 : 604 references belonging to 237 species and 67 genera). Our objective was to optimize the fungal identification workflow in laboratories equipped with a Bruker MALDI-TOF MS identification platform.

## Methods

### Reference databases

The in-house reference database was constructed using a Microflex LT system (Bruker Daltonics, Bremen, Germany) in coordination with the BCCM/IHEM culture collection and the Mycology Laboratory of the AP-HM with the protocol described by Normand et al. [6] (several strains per species, four subcultures per strain and 10 spots per subculture, for a total of four references per strain). Details of the MS in-house database are listed in Additional file 1: Table S1.

### Assessment panel

The assessment panel consisted of filamentous fungus strains that were identified via both DNA sequencing and MALDI-TOF MS. The strains were either selected at the Mycology Laboratory of Marseille ($n = 224$) or the BCCM/IHEM in Brussels ($n = 198$). Of the 224 strains from Marseille, 177 were included because the identification at the species level had to be either clarified or confirmed at the time of the study, while the remaining 47 were randomly selected among frequently identified species in the clinical laboratory. The 198 Belgian strains either originated from the BCCM/IHEM collection or were clinical strains from a previous study [34]. The 422-strain panel included 126 different species belonging to 38 genera, of which 280 strains were represented three or more times in the MALDI-TOF MS reference database (191 strains for the commercial Bruker database), 130 strains were represented once or twice (76 strains for the commercial Bruker database), and 12 strains were not represented in the reference database (155 strains for the commercial Bruker database) (Tables 1 and 2). All samples were first included in the study panel, even those corresponding to species absent from the databases to ensure that the tested algorithms correctly excluded absent species. Next, the identification process was tested again, this time only including the isolates corresponding to species represented by at least three distinct strains in the

**Table 1** Panel description obtained after DNA sequencing of the strains and representation of the entire set of species in the in-house reference spectrum database

| Species not represented in the in-house reference database | TOTAL ($n = 12$) | Collection strains BCCM/IHEM | Clinical strains (Marseille Hospital) | Species represented once or twice in the in-house reference database | TOTAL ($n = 130$) | Collection strains BCCM/IHEM | Clinical strains (Marseille Hospital) | Species represented three or more times in the in-house reference database | TOTAL ($n = 280$) | Collection strains BCCM/IHEM | Clinical strains (Marseille Hospital) |
|---|---|---|---|---|---|---|---|---|---|---|---|
| *Cladosporium bruhnei* | 1 | 1 | 0 | *Acremonium breve* | 1 | 0 | 1 | *Alternaria alternata* | 2 | 1 | 1 |
| *Cladosporium halotolerans* | 2 | 2 | 0 | *Acremonium strictum* | 6 | 6 | 0 | *Aspergillus alabamensis* | 4 | 0 | 4 |
| *Colletotrichum lineola* | 1 | 1 | 0 | *Alternaria infectoria* | 2 | 1 | 1 | *Aspergillus caesiellus* | 1 | 1 | 0 |
| *Trichoderma orientale* | 1 | 1 | 0 | *Arthrinium arundinis* | 1 | 0 | 1 | *Aspergillus calidoustus* | 10 | 2 | 8 |
| *Mucor hiemalis* | 1 | 1 | 0 | *Aspergillus dimorphicus* | 1 | 0 | 1 | *Aspergillus carbonarius* | 1 | 0 | 1 |
| *Penicillium atramentosum* | 1 | 1 | 0 | *Aspergillus fischeri* | 2 | 0 | 2 | *Aspergillus carneus* | 1 | 0 | 1 |
| *Penicillium concentricum* | 1 | 1 | 0 | *Aspergillus flavipes* | 3 | 2 | 1 | *Aspergillus chevalieri* | 4 | 2 | 2 |
| *Phaeosphaeria avenaria* | 1 | 1 | 0 | *Aspergillus floccosus* | 3 | 0 | 3 | *Aspergillus clavatus* | 2 | 2 | 0 |
| *Pleospora papaveracea* | 1 | 1 | 0 | *Aspergillus fumigatiaffinis* | 2 | 0 | 2 | *Aspergillus creber* | 1 | 1 | 0 |
| *Trichoderma reesei* | 2 | 2 | 0 | *Aspergillus iizukae* | 1 | 0 | 1 | *Aspergillus flavus* | 15 | 5 | 10 |
| | | | | *Aspergillus quadrilineatus* | 1 | 0 | 1 | *Aspergillus fumigatus* | 20 | 4 | 16 |
| | | | | *Aspergillus ruber* | 2 | 2 | 0 | *Aspergillus hiratsukae* | 1 | 0 | 1 |
| | | | | *Aspergillus sclerotiorum* | 3 | 1 | 2 | *Aspergillus hollandicus* | 5 | 4 | 1 |
| | | | | *Aspergillus subolivaceus* | 1 | 0 | 1 | *Aspergillus insuetus* | 1 | 0 | 1 |
| | | | | *Aspergillus unguis* | 4 | 4 | 0 | *Aspergillus nidulans* | 8 | 4 | 4 |
| | | | | *Aureobasidium pullulans* | 3 | 2 | 1 | *Aspergillus niger* | 4 | 2 | 2 |
| | | | | *Cladosporium cladosporioides* | 1 | 1 | 0 | *Aspergillus nomius* | 1 | 0 | 1 |
| | | | | *Cladosporium pseudocladosporioides* | 2 | 2 | 0 | *Aspergillus parasiticus* | 2 | 1 | 1 |
| | | | | *Cladosporium sphaerospermum* | 1 | 1 | 0 | *Aspergillus sydowii* | 6 | 2 | 4 |
| | | | | *Cochliobolus hawaiiensis* | 2 | 0 | 2 | *Aspergillus tamarii* | 2 | 2 | 0 |
| | | | | *Curvularia lunata* | 2 | 2 | 0 | *Aspergillus terreus* | 10 | 3 | 7 |
| | | | | *Fusarium dimerum* | 2 | 1 | 1 | *Aspergillus tubingensis* | 29 | 1 | 28 |
| | | | | *Geosmithia argillacea* | 1 | 0 | 1 | *Aspergillus versicolor* | 3 | 2 | 1 |

**Table 1** Panel description obtained after DNA sequencing of the strains and representation of the entire set of species in the in-house reference spectrum database *(Continued)*

| | | | | | | | |
|---|---|---|---|---|---|---|---|
| *Geosmithia pallida* | 1 | 1 | 0 | *Aspergillus westerdijkiae* | 1 | 0 | 1 |
| *Hexagonia hydnoides* | 1 | 0 | 1 | *Beauveria bassiana* | 3 | 2 | 1 |
| *Lecytophophora sp.* | 1 | 1 | 0 | *Epicoccum nigrum* | 2 | 2 | 0 |
| *Macrophomina phaseolina* | 1 | 0 | 1 | *Exophiala dermatitidis* | 2 | 2 | 0 |
| *Mucor circinelloides* | 5 | 5 | 0 | *Fomitopsis pinicola* | 2 | 0 | 2 |
| *Penicillium asturianum* | 1 | 0 | 1 | *Fusarium equiseti* | 1 | 0 | 1 |
| *Penicillium brasilianum* | 3 | 1 | 2 | *Fusarium oxysporum* | 21 | 15 | 6 |
| *Penicillium cecidicola* | 2 | 1 | 1 | *Fusarium proliferatum* | 10 | 1 | 9 |
| *Penicillium citrinum* | 5 | 4 | 1 | *Fusarium sacchari* | 1 | 0 | 1 |
| *Penicillium corylophilum* | 4 | 2 | 2 | *Fusarium solani* | 7 | 6 | 1 |
| *Penicillium crateriforme* | 2 | 1 | 1 | *Fusarium verticillioides* | 1 | 1 | 0 |
| *Penicillium diversum* | 1 | 1 | 0 | *Galactomyces geotrichum* | 2 | 2 | 0 |
| *Penicillium expansum* | 2 | 2 | 0 | *Lichtheimia corymbifera* | 2 | 2 | 0 |
| *Penicillium funiculosum* | 2 | 0 | 2 | *Microsporum audouinii* | 1 | 0 | 1 |
| *Penicillium glabrum* | 9 | 3 | 6 | *Microsporum canis* | 2 | 0 | 2 |
| *Penicillium helicum* | 1 | 0 | 1 | *Purpureocillium lilacinus* | 3 | 2 | 1 |
| *Talaromyces marneffei* | 1 | 0 | 1 | *Paecilomyces variotii* | 3 | 2 | 1 |
| *Penicillium nalgiovense* | 1 | 0 | 1 | *Penicillium chrysogenum* | 27 | 13 | 14 |
| *Penicillium oxalicum* | 1 | 1 | 0 | *Penicillium crustosum* | 3 | 2 | 1 |
| *Penicillium pinophilum* | 3 | 0 | 3 | *Penicillium griseofulvum* | 1 | 1 | 0 |
| *Penicillium polonicum* | 6 | 6 | 0 | *Rhizomucor pusillus* | 2 | 2 | 0 |
| *Penicillium purpurogenum* | 1 | 0 | 1 | *Rhizopus oryzae* | 6 | 5 | 1 |
| *Penicillium raistrickii* | 1 | 1 | 0 | *Scedosporium apiospermum* | 11 | 2 | 9 |
| *Penicillium rolfsii* | 1 | 0 | 1 | *Scedosporium prolificans* | 3 | 2 | 1 |
| *Penicillium rugulosum* | 1 | 1 | 0 | *Schizophyllum commune* | 4 | 1 | 3 |
| *Penicillium sizovae* | 2 | 0 | 2 | *Scopulariopsis brevicaulis* | 15 | 13 | 2 |
| *Penicillium spinulosum* | 2 | 0 | 2 | *Trichoderma longibrachiatum* | 2 | 2 | 0 |
| *Penicillium terrigenum* | 1 | 0 | 1 | *Trichophyton interdigitale* | 1 | 0 | 1 |
| *Penicillium variabile* | 2 | 0 | 2 | *Trichophyton rubrum* | 8 | 0 | 8 |
| | 2 | 0 | 2 | | | | |

**Table 1** Panel description obtained after DNA sequencing of the strains and representation of the entire set of species in the in-house reference spectrum database *(Continued)*

| | | | | |
|---|---|---|---|---|
| | *Phanerochaete chrysosporium* | | | |
| | *Pithomyces chartarum* | 3 | 3 | 0 |
| | *Pseudallescheria boydii* | 1 | 0 | 1 |
| | *Pseudallescheria ellipsoidea* | 1 | 1 | 0 |
| | *Rhizomucor variabilis* | 2 | 0 | 2 |
| | *Rhizopus microsporus* | 1 | 1 | 0 |
| | *Scedosporium aurantiacum* | 6 | 6 | 0 |
| | *Tyromyces fissilis* | 3 | 0 | 3 |

**Table 2** Representation of the entire set of species included in the Bruker reference spectrum database

| Species not represented in the Bruker reference database | Total ($n = 155$) | List of the 26 species represented once or twice in the Bruker reference database | Total ($n = 76$) | List of the 25 species represented three or more times in the Bruker reference database | Total ($n = 191$) |
|---|---|---|---|---|---|
| *Acremonium breve* | 1 | *Acremonium strictum* | 6 | *Alternaria alternata* | 2 |
| *Alternaria infectoria* | 2 | *Aspergillus clavatus* | 2 | *Aspergillus flavus* | 15 |
| *Arthrinium arundinis* | 1 | *Aspergillus hollandicus* | 5 | *Aspergillus fumigatus* | 20 |
| *Aspergillus alabamensis* | 4 | *Aspergillus sclerotiorum* | 3 | *Aspergillus nidulans* | 8 |
| *Aspergillus caesiellus* | 1 | *Aspergillus sydowii* | 6 | *Aspergillus niger* | 4 |
| *Aspergillus calidoustus* | 10 | *Aspergillus tamarii* | 2 | *Aspergillus parasiticus* | 2 |
| *Aspergillus carbonarius* | 1 | *Aspergillus unguis* | 4 | *Aspergillus terreus* | 10 |
| *Aspergillus carneus* | 1 | *Beauveria bassiana* | 3 | *Aspergillus versicolor* | 3 |
| *Aspergillus chevalieri* | 4 | *Cladosporium cladosporioides* | 1 | *Aureobasidium pullulans* | 3 |
| *Aspergillus creber* | 1 | *Curvularia lunata* | 2 | *Exophiala dermatitidis* | 2 |
| *Aspergillus dimorphicus* | 1 | *Epicoccum nigrum* | 2 | *Fusarium oxysporum* | 21 |
| *Aspergillus fischeri* | 2 | *Fusarium dimerum* | 2 | *Fusarium proliferatum* | 10 |
| *Aspergillus flavipes* | 3 | *Fusarium equiseti* | 1 | *Fusarium solani* | 7 |
| *Aspergillus floccosus* | 3 | *Fusarium verticillioides* | 1 | *Lichtheimia corymbifera* | 2 |
| *Aspergillus fumigatiaffinis* | 2 | *Galactomyces geotrichum* | 2 | *Microsporum canis* | 2 |
| *Aspergillus hiratsukae* | 1 | *Mucor circinelloides* | 5 | *Purpureocillium lilacinus* | 3 |
| *Aspergillus iizukae* | 1 | *Penicillium citrinum* | 5 | *Paecilomyces variotii* | 3 |
| *Aspergillus insuetus* | 1 | *Penicillium corylophilum* | 4 | *Penicillium chrysogenum* | 27 |
| *Aspergillus nomius* | 1 | *Penicillium funiculosum* | 2 | *Penicillium expansum* | 2 |
| *Aspergillus quadrilineatus* | 1 | *Penicillium purpurogenum* | 1 | *Penicillium glabrum* | 9 |
| *Aspergillus ruber* | 2 | *Penicillium rugulosum* | 1 | *Scedosporium apiospermum* | 11 |
| *Aspergillus subolivaceus* | 1 | *Rhizopus microsporus* | 1 | *Schizophyllum commune* | 4 |
| *Aspergillus tubingensis* | 29 | *Rhizopus oryzae* | 6 | *Scopulariopsis brevicaulis* | 15 |
| *Aspergillus westerdijkiae* | 1 | *Scedosporium aurantiacum* | 6 | *Trichophyton interdigitale* | 1 |
| *Cladosporium bruhnei* | 1 | *Scedosporium prolificans* | 3 | *Trichophyton rubrum* | 8 |
| *Cladosporium halotolerans* | 2 | *Trichoderma longibrachiatum* | 2 | | |
| *Cladosporium pseudocladosporioides* | 2 | | | | |
| *Cladosporium sphaerospermum* | 1 | | | | |
| *Cochliobolus hawaiiensis* | 2 | | | | |
| *Colletotrichum lineola* | 1 | | | | |
| *Fomitopsis pinicola* | 2 | | | | |
| *Fusarium sacchari* | 1 | | | | |
| *Geosmithia argillacea* | 1 | | | | |
| *Geosmithia pallida* | 1 | | | | |
| *Hexagonia hydnoides* | 1 | | | | |
| *Trichoderma orientale* | 1 | | | | |
| *Lecythophora sp.* | 1 | | | | |
| *Macrophomina phaseolina* | 1 | | | | |
| *Microsporum audouinii* | 1 | | | | |
| *Mucor hiemalis* | 1 | | | | |
| *Penicillium asturianum* | 1 | | | | |
| *Penicillium atramentosum* | 1 | | | | |

**Table 2** Representation of the entire set of species included in the Bruker reference spectrum database *(Continued)*

| | |
|---|---|
| *Penicillium brasilianum* | 3 |
| *Penicillium cecidicola* | 2 |
| *Penicillium concentricum* | 1 |
| *Penicillium crateriforme* | 2 |
| *Penicillium crustosum* | 3 |
| *Penicillium diversum* | 1 |
| *Penicillium griseofulvum* | 1 |
| *Penicillium helicum* | 1 |
| *Talaromyces marnefei* | 1 |
| *Penicillium nalgiovense* | 1 |
| *Penicillium oxalicum* | 1 |
| *Penicillium pinophilum* | 3 |
| *Penicillium polonicum* | 6 |
| *Penicillium raistrickii* | 1 |
| *Penicillium rolfsii* | 1 |
| *Penicillium sizovae* | 2 |
| *Penicillium spinulosum* | 2 |
| *Penicillium terrigenum* | 1 |
| *Penicillium variabile* | 2 |
| *Phaeosphaeria avenaria* | 1 |
| *Phanerochaete chrysosporium* | 2 |
| *Pithomyces chartarum* | 3 |
| *Pleospora papaveracea* | 1 |
| *Scedosporium boydii* | 1 |
| *Pseudallescheria ellipsoidea* | 1 |
| *Rhizomucor pusillus* | 2 |
| *Rhizomucor variabilis* | 2 |
| *Trichoderma reesei* | 2 |
| *Tyromyces fissilis* | 3 |

database to comply with the recommendations we have issued in a previous study [6].

### Identification of the strains from the assessment panel

#### *MALDI-TOF MS-based identification*

Liquid culture technology used in the Bruker protocol being not available in the two centers that participated to the study, culturing was performed on solid agar. After culturing the molds on solid Sabouraud Chloramphenicol Gentamicin agar plates (pH 5.6; glucose 40 g/L) (OXOID, Dardilly, France) for at least 48 h at 30 °C and the dermatophytes on solid Sabouraud Actidione (Bio Rad) at 27 °C, the strains were treated as described by Cassagne et al.[2], and as advised by the manufacturer : the colonies were gently scrapped using a scalpel blade, and the fungal material (approximately 2–3 mm in diameter) was suspended in a microtube containing 900 μL of anhydrous ethyl alcohol (Carlo Erba SDS, Val de Reuil, France) and 300 μL of sterile water (Water HPLC, Prolabo, BDH, Fontenay-sous-Bois, France). After a 10-min centrifugation step at 13,000 rpm, the pellet was resuspended in 10 μL of formic acid (Sigma-Aldrich, Lyon, France). After 5 min of incubation, 10 μL of acetonitrile (Prolabo BDH) was added. The suspension was then centrifuged at 13,000 rpm for 2 min, and four spots of 1 μL of protein extract per isolate were deposited. Last, the samples were covered with 1 μL of HCCA (α-cyano-4-hydroxycinnamic acid) matrix (Sigma-Aldrich, Lyon, France). MS acquisition was performed either in Marseille or Brussels using a Microflex LT system. All raw spectra were then centralized in Marseille and compared to the two reference databases. For all 1688 tested spectra (four spots for each strain) and each database, three possible scores were collected: the highest score corresponding to a true MS identification result at the species level if available, the highest score corresponding to a true identification at the genus level but false at the species level if available, and in all cases, the highest score false at the genus level.

### *DNA sequence-based identification*

All strains belonging to the assessment panel were identified via DNA sequencing, which is the current gold standard in fungal identification. Nucleotide sequence analysis was performed either in Marseille ($n = 224$) or Brussels ($n = 198$). The rRNA ITS2 region was sequenced for each strain (using primer sequences ITS3 – GCA TCG ATG AAG AAC GCA GC and ITS4c – TCC TCC GCT TAT TGA TAT GC). An additional locus was analyzed to identify particular taxa: the partial beta-tubulin gene (primer sequences: Bt2A – GGT AAC CAA ATC GGT GCT GCT TTC and Bt2B – ACC CTC AGT GTA GTG ACC CTT GGC) was sequenced for *Aspergillus, Penicillium* and *Scedosporium* species, while the elongation factor (primer sequences: EF1 – ATG GGT AAG GAR GAC AAG AC and EF2 – GGA RGT ACC AGT SAT CAT GTT) was sequenced for *Fusarium* species. The hybridization temperatures applied for these primers were 54 °C for the ITS and beta-tubulin genes and 58 °C for the elongation factor gene. The DNA sequence-based identification criteria were as follows: a sequence longer than 350 base pairs and at least 99% homology with the NCBI and CBS nucleotide databases [35, 36].

### Identification algorithms

The following 10 distinct algorithms were used to identify the 422 isolates included in the panel.

A: Only one spot (i.e., the first of the four spots) is taken into account.

B: The first two spots are taken into account.
- $B_1$: Only the identification corresponding to the higher of the two scores is taken into account, plotted and categorized.
- $B_2$: To be accepted at the species level, both identifications corresponding to the two spots must be identical, and the same rule applies for the identification at the genus level.

C: The first three spots are taken into account.
- $C_1$: Only the identification corresponding to the highest of the three scores is taken into account, plotted and categorized.
- $C_2$: To be categorized as a concordant identification, the identification corresponding to the highest score must be identical to at least one of the other two identifications (at either the species or genus level).
- $C_3$: To be categorized as a concordant identification, the three identifications corresponding to the three spots must be identical (at either the species or genus level).

D: All four spots are taken into account.
- $D_1$: Only the identification corresponding to the highest of the four scores is taken into account, plotted and categorized.
- $D_2$: To be categorized as a concordant identification, the identification corresponding to the highest score must be identical to at least one of the other three identifications (at either the species or genus level).
- $D_3$: To be categorized as a concordant identification, the identification corresponding to the highest score must be identical to at least two

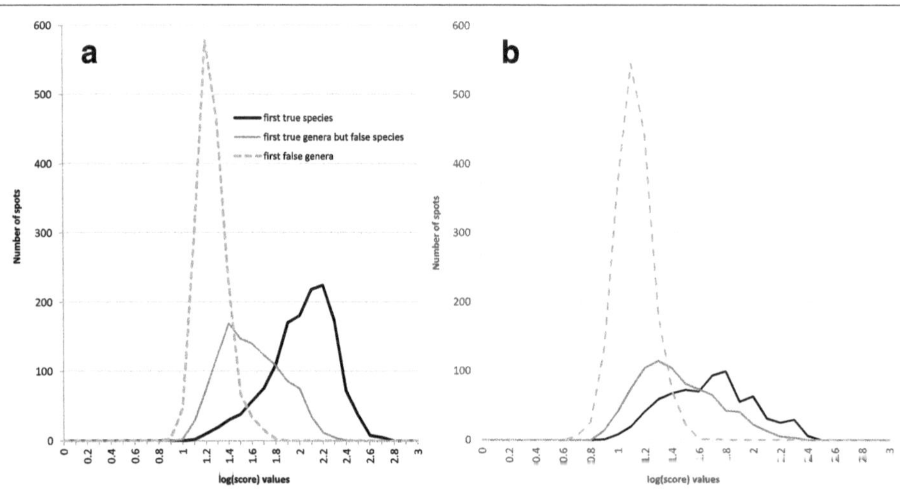


**Fig. 1** Highest log(score) value distribution of the 1688 tested spectra. **a** considering the first accurate species identification (*dark line*), the first accurate genus identification (*gray line*), and the first false genus identification (*dotted line*) using the in-house reference database. **b** considering the first accurate species identification (*dark line*), the first accurate genus identification (*gray line*), and the first false genus identification (*dotted line*) using the Bruker reference database

of the other three identifications (at either the species or genus level).

- $D_4$: To be categorized as a concordant identification, the four identifications corresponding to the four spots must be identical (at either the species or genus level).

Based on the individual spectrum results, 10 identification algorithms (or combinations of parameters) were tested per strain. For each spectrum, only the first identification (either true or false at the species and genus level) and the corresponding scores were collected. Each of the 10 algorithms was then tested using the identification thresholds 1.5, 1.6, 1.7, 1.8, 1.9, 2.0, 2.1, 2.2 and 2.3 for the entire panel consisting of 422 strains. The same process was applied for a reduced panel of 280 isolates that were represented at least three times in the in-house reference database and for another reduced panel of 191 isolates that were represented at least three times in the commercial database by Bruker.

**Statistical analyses**

An initial identification assessment was carried out by categorizing the first identification for each spot into three classes: (1) concordant at the species level with the DNA-based identification, (2) concordant at the genus level but not at the species level with the DNA-based identification, or (3) discordant at the genus level with the DNA-based identification. We then determined the best rules for accepting identification at the genus and species levels. For each threshold, algorithm, panel and database, we calculated the percentage of submitted strains that fulfilled the criteria for identification as well as

**Table 3** Comparison of the 10 different algorithms tested on the 422 strains included in the assessment panel against the Bruker reference database

| Threshold | | A | B1 | B2 | C1 | C2 | C3 | D1 | D2 | D3 | D4 |
|---|---|---|---|---|---|---|---|---|---|---|---|
| LS 1.5 | %accepted | 43.71 | 50.36 | 38.95 | 47.74 | 47.03 | 31.83 | 51.31 | 51.31 | 44.66 | 30.40 |
| | species PPV | 0.46 | 0.47 | 0.43 | 0.51 | 0.51 | 0.46 | 0.50 | 0.50 | 0.48 | 0.42 |
| | genus PPV | 0.65 | 0.67 | 0.57 | 0.70 | 0.70 | 0.61 | 0.69 | 0.69 | 0.66 | 0.54 |
| LS 1.6 | %accepted | 35.15 | 40.62 | 29.22 | 40.38 | 39.67 | 25.18 | 44.42 | 44.42 | 38.00 | 24.23 |
| | species PPV | 0.57 | 0.58 | 0.57 | 0.61 | 0.60 | 0.58 | 0.57 | 0.57 | 0.56 | 0.53 |
| | genus PPV | 0.80 | 0.83 | 0.76 | 0.83 | 0.83 | 0.77 | 0.79 | 0.79 | 0.78 | 0.68 |
| LS 1.7 | %accepted | 28.27 | 33.73 | 22.33 | 33.49 | 32.78 | 19.48 | **35.15** | 35.15 | 29.45 | 16.39 |
| | species PPV | 0.71 | 0.70 | 0.74 | 0.73 | 0.72 | 0.76 | **0.72** | 0.72 | 0.73 | 0.78 |
| | genus PPV | 1.00 | 1.00 | 1.00 | 1.00 | 1.00 | 1.00 | **1.00** | 1.00 | 1.00 | 1.00 |
| LS 1.8 | %accepted | 17.58 | 23.75 | 19.48 | 24.47 | 23.99 | 17.58 | 25.89 | 25.89 | 22.33 | 15.20 |
| | species PPV | 0.77 | 0.74 | 0.76 | 0.73 | 0.72 | 0.74 | 0.74 | 0.74 | 0.74 | 0.78 |
| | genus PPV | 1.00 | 1.00 | 1.00 | 1.00 | 1.00 | 1.00 | 1.00 | 1.00 | 1.00 | 1.00 |
| LS 1.9 | %accepted | 11.88 | 16.63 | 14.25 | 16.39 | 15.91 | 12.59 | 17.58 | 17.58 | 15.68 | 12.11 |
| | species PPV | 0.78 | 0.77 | 0.77 | 0.77 | 0.76 | 0.75 | 0.77 | 0.77 | 0.74 | 0.78 |
| | genus PPV | 1.00 | 1.00 | 1.00 | 1.00 | 1.00 | 1.00 | 1.00 | 1.00 | 1.00 | 1.00 |
| LS 2.0 | %accepted | 7.84 | 10.21 | 9.74 | 10.45 | 10.21 | 9.03 | 10.45 | 10.45 | 9.98 | 8.79 |
| | species PPV | 0.79 | 0.79 | 0.78 | 0.82 | 0.81 | 0.79 | 0.80 | 0.80 | 0.79 | 0.81 |
| | genus PPV | 1.00 | 1.00 | 1.00 | 1.00 | 1.00 | 1.00 | 1.00 | 1.00 | 1.00 | 1.00 |
| LS 2.1 | %accepted | 4.51 | 5.46 | 4.99 | 5.94 | 5.94 | 5.46 | 5.70 | 5.70 | 5.70 | 5.23 |
| | species PPV | 0.79 | 0.78 | 0.76 | 0.80 | 0.80 | 0.78 | 0.75 | 0.75 | 0.75 | 0.77 |
| | genus PPV | 1.00 | 1.00 | 1.00 | 1.00 | 1.00 | 1.00 | 1.00 | 1.00 | 1.00 | 1.00 |
| LS 2.2 | %accepted | 3.33 | 4.04 | 3.56 | 3.56 | 3.56 | 3.09 | 3.80 | 3.80 | 3.80 | 3.56 |
| | species PPV | 0.86 | 0.88 | 0.87 | 0.87 | 0.87 | 0.85 | 0.81 | 0.81 | 0.81 | 0.80 |
| | genus PPV | 1.00 | 1.00 | 1.00 | 1.00 | 1.00 | 1.00 | 1.00 | 1.00 | 1.00 | 1.00 |
| LS 2.3 | %accepted | 2.14 | 2.38 | 1.90 | 1.19 | 1.19 | 0.95 | 1.19 | 1.19 | 1.19 | 0.95 |
| | species PPV | 0.78 | 0.80 | 0.75 | 1.00 | 1.00 | 1.00 | 1.00 | 1.00 | 1.00 | 1.00 |
| | genus PPV | 1.00 | 1.00 | 1.00 | 1.00 | 1.00 | 1.00 | 1.00 | 1.00 | 1.00 | 1.00 |

% accepted: percentage of submitted strains that fulfilled the criteria for identification at either the genus or species level, depending on the applied threshold; species PPV: positive predictive value at the species level; genus PPV: positive predictive value at the genus level

the positive predictive value (PPV) of the identification result at the genus and species levels.

## Results

### MALDI-TOF MS-based identification

Using our in-house database to identify the 422 strains included in the panel, a clear separation was observed between the highest scores corresponding to true identifications at the species level and the highest scores corresponding to a false genus identification. Conversely, it was markedly more difficult to define a satisfying cut-off that differentiated between the highest scores corresponding to an accurate species identification and the highest scores corresponding to an accurate genus but false species identification (Fig. 1a). Using the commercial database led to a similar observation, although the separation between the three curves was less pronounced (Fig. 1b).

### Algorithm performance and identification score threshold

Analyses of the identification performances for the 422-strain panel with the Bruker database and the in-house database. The percentage of identified strains and PPVs at the species and genus levels obtained using the 10 distinct identification algorithms were assessed on the 422-strain panel using the Bruker database (Table 3) as well as the in-house reference database (Table 4).

Notable differences in successful identification rates were highlighted between the 10 algorithms (up to 22% difference between the best identification rate and the worst one for one defined threshold). In particular, with both references databases, the algorithm that selected

**Table 4** Comparison of the 10 different algorithms tested on the 422 strains included in the assessment panel against the in-house reference database

| Threshold | | A | B1 | B2 | C1 | C2 | C3 | D1 | D2 | D3 | D4 |
|---|---|---|---|---|---|---|---|---|---|---|---|
| LS 1.5 | %accepted | 90.97 | 93.35 | 81.00 | 92.64 | 91.69 | 73.63 | 94.54 | 94.30 | 88.36 | 72.45 |
| | species PPV | 0.80 | 0.81 | 0.82 | 0.83 | 0.83 | 0.85 | 0.82 | 0.82 | 0.82 | 0.83 |
| | genus PPV | 0.90 | 0.92 | 0.90 | 0.93 | 0.93 | 0.93 | 0.92 | 0.92 | 0.92 | 0.91 |
| LS 1.6 | %accepted | 87.41 | 90.02 | 77.67 | 89.55 | 88.60 | 70.78 | 90.26 | 90.02 | 84.32 | 68.65 |
| | species PPV | 0.84 | 0.84 | 0.86 | 0.85 | 0.86 | 0.89 | 0.86 | 0.86 | 0.86 | 0.88 |
| | genus PPV | 0.94 | 0.95 | 0.94 | 0.96 | 0.96 | 0.96 | 0.97 | 0.97 | 0.97 | 0.96 |
| LS 1.7 | %accepted | 81.95 | 85.27 | 73.16 | 85.99 | 85.27 | 68.17 | **87.41** | 87.17 | 81.47 | 65.80 |
| | species PPV | 0.89 | 0.89 | 0.91 | 0.89 | 0.89 | 0.92 | **0.89** | 0.89 | 0.89 | 0.92 |
| | genus PPV | 1.00 | 1.00 | 1.00 | 1.00 | 1.00 | 1.00 | **1.00** | 1.00 | 1.00 | 1.00 |
| LS 1.8 | %accepted | 76.96 | 81.00 | 71.97 | 79.33 | 78.86 | 65.56 | 81.71 | 81.47 | 77.20 | 64.61 |
| | species PPV | 0.89 | 0.89 | 0.91 | 0.90 | 0.91 | 0.92 | 0.90 | 0.90 | 0.90 | 0.92 |
| | genus PPV | 1.00 | 1.00 | 1.00 | 1.00 | 1.00 | 1.00 | 1.00 | 1.00 | 1.00 | 1.00 |
| LS 1.9 | %accepted | 68.88 | 74.11 | 67.46 | 70.78 | 70.31 | 61.28 | 74.11 | 73.87 | 70.55 | 60.57 |
| | species PPV | 0.90 | 0.89 | 0.90 | 0.90 | 0.91 | 0.92 | 0.89 | 0.89 | 0.89 | 0.91 |
| | genus PPV | 1.00 | 1.00 | 1.00 | 1.00 | 1.00 | 1.00 | 1.00 | 1.00 | 1.00 | 1.00 |
| LS 2.0 | %accepted | 56.77 | 63.18 | 57.72 | 63.18 | 62.71 | 55.58 | 63.66 | 63.42 | 61.28 | 55.11 |
| | species PPV | 0.92 | 0.91 | 0.92 | 0.91 | 0.92 | 0.92 | 0.91 | 0.91 | 0.91 | 0.91 |
| | genus PPV | 1.00 | 1.00 | 1.00 | 1.00 | 1.00 | 1.00 | 1.00 | 1.00 | 1.00 | 1.00 |
| LS 2.1 | %accepted | 43.47 | 50.36 | 46.08 | 49.64 | 49.41 | 43.94 | 50.59 | 50.36 | 48.93 | 44.18 |
| | species PPV | 0.94 | 0.92 | 0.93 | 0.92 | 0.93 | 0.93 | 0.92 | 0.92 | 0.92 | 0.92 |
| | genus PPV | 1.00 | 1.00 | 1.00 | 1.00 | 1.00 | 1.00 | 1.00 | 1.00 | 1.00 | 1.00 |
| LS 2.2 | %accepted | 29.69 | 35.15 | 32.30 | 33.02 | 32.78 | 29.93 | 34.44 | 34.20 | 33.97 | 31.12 |
| | species PPV | 0.95 | 0.93 | 0.93 | 0.92 | 0.93 | 0.94 | 0.92 | 0.92 | 0.92 | 0.92 |
| | genus PPV | 1.00 | 1.00 | 1.00 | 1.00 | 1.00 | 1.00 | 1.00 | 1.00 | 1.00 | 1.00 |
| LS 2.3 | %accepted | 14.25 | 18.29 | 17.34 | 16.86 | 16.86 | 16.15 | 18.05 | 18.05 | 18.05 | 17.34 |
| | species PPV | 0.95 | 0.95 | 0.95 | 0.93 | 0.93 | 0.93 | 0.92 | 0.92 | 0.92 | 0.92 |
| | genus PPV | 1.00 | 1.00 | 1.00 | 1.00 | 1.00 | 1.00 | 1.00 | 1.00 | 1.00 | 1.00 |

% accepted: percentage of submitted strains that fulfilled the criteria for identification at either the genus or species level, depending on the applied threshold; species PPV: positive predictive value at the species level; genus PPV: positive predictive value at the genus level

the best results among four spots (D1) increased the accepted identification rate (i.e., percentage of assessed strains that fulfilled the criteria for identification at either the genus or species level) (A < B1 < C1 < D1) without notably changing the identification PPV. By contrast, the inter-spot concordance criteria did not increase the identification PPV at the species level and reduced the accepted result rate (D4 < D3 < D2 ≈ D1). The main difference between the results obtained with the two databases was the percentage of accepted identifications, which was markedly lower with the Bruker database than with the in-house reference database (up to 59.8% difference between the two databases). Additionally, while the PPV for genus level identification was 100% for both databases using the 1.7 threshold, regardless of the other decision criteria applied, the PPV for species level identification was better with our in-house database than with the Bruker database (from 0.13 units for the D4 algorithm to 0.18 units for the A1 and B1 algorithms). Using the in-house reference database, the D1 combination with a threshold at 1.7 yielded 87.41% of accepted identifications, with a species identification PPV of 0.89.

We noted an increase in the species PPV with threshold values higher than 1.7, but the loss in percentage of identification is important. Using the identification score threshold of 2.00, as recommended by the manufacturer, led to a 23.75% decrease in accepted identifications and only a 2% increase in the PPV. The "highly probable species identification" criterion by Bruker (i.e., a score above the threshold of 2.3) yielded a species PPV of 0.92 but only 18.05% of accepted identifications (76 strains) using our in-house database. The identification results were even less acceptable when using the commercial database,

**Table 5** List of the 39 identification failures obtained using the Bruker database and the D1 algorithm with a score above the 1.7 threshold

| Identification via molecular biology | Nb of occurrences | Identification D1 | Score D1 | Number of references in the Bruker reference database |
|---|---|---|---|---|
| *Aspergillus tubingensis* | 10 | *Aspergillus niger* | 2.329/2.161/2.102/1.928/2.238/1.955/ 2.333/1.983/1.987/1.829 | 0 |
| *Fusarium proliferatum* | 2 | *Fusarium verticillioides* | 1.845/1.718 | 5 |
| *Curvularia hawaiensis* | 2 | *Curvularia pallescens* | 1.879/2.079 | 0 |
| *Penicillium brasilianum* | 2 | *Penicillium discolor* | 1.92/1.887 | 0 |
| *Arthrinium arundinis* | 1 | *Arthrinium phaeospermum* | 1.883 | 0 |
| *Aspergillus quadrilineatus* | 1 | *Aspergillus nidulans* | 2.065 | 0 |
| *Aspergillus calidoustus* | 4 | *Aspergillus ustus* | 1.775/2.076/1.771/1.731 | 0 |
| *Fusarium sacchari* | 1 | *Fusarium proliferatum* | 1.742 | 0 |
| *Aspergillus subolivaceus* | 1 | *Aspergillus flavus* | 1.985 | 0 |
| *Scopulariopsis brevicaulis* | 2 | *Scopulariopsis brumptii* | 1.781/1.925 | 9 |
| *Fusarium verticillioides* | 1 | *Fusarium proliferatum* | 1.953 | 2 |
| *Curvularia lunata* | 1 | *Curvularia pallescens* | 2.039 | 1 |
| *Aspergillus creber* | 1 | *Aspergillus versicolor* | 2.103 | 0 |
| *Aspergillus sydowii* | 2 | *Aspergillus versicolor* | 1.884/1.897 | 1 |
| *Aspergillus chevalieri* | 2 | *Aspergillus hollandicus* | 1.84/1.766 | 0 |
| *Fusarium equiseti* | 1 | *Fusarium incarnatum* | 1.701 | 1 |
| *Trichophyton rubrum* | 1 | *Trichophyton mentagrophytes* | 1.701 | 6 |
| *Aspergillus alabamensis* | 1 | *Aspergillus terreus* | 1.739 | 0 |
| *Aspergillus fischeri* | 1 | *Aspergillus fumigatus* | 1.898 | 0 |
| *Trichoderma orientale* | 1 | *Trichoderma longibrachiatum* | 1.802 | 0 |
| *Trichoderma reesei* | 1 | *Trichoderma koningii* | 1.865 | 0 |

with only 44 strains (10.45%) reaching the 2.0 threshold and 5 strains (1.19%) reaching the 2.3 threshold.

***Description of the misidentifications obtained with both databases***

For 32 strains, the identification best score using the D1 algorithm was higher than the 1.7 threshold, although the species level result was incorrect using the in-house database. For all of these strains, proper identification at the species level was impeded by the close similarity of the MALDI-TOF MS spectra of some species belonging to the same species complex (as it is the case with *Fusarium solani* and its closely related species *Fusarium petrolophilum*). The details of these strains are presented in Table 5. The same finding was observed for 39 misidentifications obtained using the Bruker database (Table 6).

***Analyses of the identification performances for the restricted panels with the Bruker database and the in-house database***

The robustness of the identification algorithms was tested by performing the same analysis on a restricted panel including only the strains that were represented by at least three distinct strains in the reference spectrum libraries (the restricted panel used with the Bruker database ($n = 191$) is different than the one used with the in-house database ($n = 280$) due to the differences in databases composition). As expected, the results were better than those obtained with the 422-strain panel (Tables 7 and 8), with a higher percentage of results reaching the thresholds (from

**Table 6** List of the 32 identification failures obtained using the in-house database and the D1 algorithm with a score above the 1.7 threshold

| Identification via molecular biology | Nb of occurrences | Identification D1 | Score D1 | Number of references in the in-house reference database |
|---|---|---|---|---|
| *Acremonium strictum* | 3 | *Acremonium sclerotigenum* | 1.952/2.042/2.126 | 1 |
| *Alternaria alternata* | 1 | *Alternaria soliaegyptiaca* | 2.211 | 1 |
| *Aspergillus sclerotiorum* | 1 | *Aspergillus persii* | 1.828 | 2 |
| *Cladosporium halotolerans* | 1 | *Cladosporium sphaerospermum* | 1.975 | 0 |
| *Cladosporium pseudocladosporioides* | 1 | *Cladosporium cladosporioides* | 1.969 | 1 |
| *Aspergillus nidulans* | 1 | *Aspergillus echinulatus* | 2.203 | 5 |
| *Aspergillus quadrilineatus* | 1 | *Aspergillus nidulans* | 2.339 | 2 |
| *Aspergillus hollandicus* | 1 | *Aspergillus heterocaryoticus* | 2.496 | 1 |
| *Aspergillus rubrobrunneus* | 1 | *Aspergillus hollandicus* | 2.325 | 1 |
| *Fusarium solani* | 2 | *Fusarium petroliphilum* | 1.981/2.081 | 7 |
| *Fusarium solani* | 1 | *Acremonium falciforme* | 2.036 | 7 |
| *Trichoderma orientale* | 1 | *Trichoderma longibrachiatum* | 2.026 | 0 |
| *Mucor irregularis* | 1 | *Rhizomucor variabilis* | 2.235 | 0 |
| *Penicillium asturianum* | 1 | *Penicillium oxalicum* | 1.953 | 1 |
| *Penicillium cecidicola* | 1 | *Penicillium funiculosum* | 2.067 | 1 |
| *Penicillium chrysogenum* | 1 | *Penicillium nalgiovense* | 2.328 | 6 |
| *Penicillium glabrum* | 1 | *Penicillium purpurescens* | 2.249 | 4 |
| *Penicillium nalgiovense* | 1 | *Penicillium chrysogenum* | 2.17 | 3 |
| *Talaromyces purpurogenus* | 1 | *Talaromyces amestolkiae* | 2.39 | 1 |
| *Penicillium rubrum* | 1 | *Talaromyces amestolkiae* | 2.069 | 2 |
| *Pleospora papaveracea* | 1 | *Ulocladium oudemansii* | 2.115 | 0 |
| *Pseudallescheria boydii* | 1 | *Scedosporium apiospermum* | 1.721 | 3 |
| *Scedosporium apiospermum* | 2 | *Scedosporium boydii* | 1.746/2.249 | 5 |
| *Scopulariopsis acremonium* | 1 | *Scopulariopsis brevicaulis* | 2.178 | 1 |
| *Talaromyces ramulosus* | 1 | *Penicillium cecidicola* | 2.375 | 0 |
| *Talaromyces stollii* | 1 | *Penicillium funiculosum* | 2.13 | 2 |
| *Trichoderma reesei* | 1 | *Trichoderma citrinoviride* | 1.727 | 0 |
| *Trichophyton rubrum* | 1 | *Trichophyton violaceum* | 1.95 | 12 |

**Table 7** Performance comparison of the 10 different algorithms tested on the sub-panel of 191 strains that are represented by three or more strains in the Bruker reference library

| Threshold | | A1 | B1 | B2 | C1 | C2 | C3 | D1 | D2 | D3 | D4 |
|---|---|---|---|---|---|---|---|---|---|---|---|
| LS 1.5 | %accepted | 56.02 | 61.26 | 48.17 | 58.12 | 57.07 | 38.74 | 63.35 | 63.35 | 53.93 | 35.08% |
| | species PPV | 0.63 | 0.66 | 0.59 | 0.71 | 0.71 | 0.64 | 0.69 | 0.69 | 0.66 | 0.58 |
| | genus PPV | 0.69 | 0.72 | 0.64 | 0.75 | 0.74 | 0.68 | 0.74 | 0.74 | 0.72 | 0.60 |
| LS 1.6 | %accepted | 46.07 | 49.74 | 36.65 | 49.74 | 48.69 | 31.41 | 56.54 | 56.54 | 47.64 | 28.80 |
| | species PPV | 0.76 | 0.81 | 0.77 | 0.83 | 0.83 | 0.78 | 0.77 | 0.77 | 0.75 | 0.71 |
| | genus PPV | 0.84 | 0.88 | 0.84 | 0.87 | 0.87 | 0.83 | 0.82 | 0.82 | 0.81 | 0.73 |
| LS 1.7 | %accepted | 38.74 | 43.98 | 30.89 | 43.46 | 42.41 | 26.18 | **46.60** | 46.60 | 38.74 | 20.94 |
| | species PPV | 0.91 | 0.92 | 0.92 | 0.95 | 0.95 | 0.94 | **0.93** | 0.93 | 0.92 | 0.98 |
| | genus PPV | 1.00 | 1.00 | 1.00 | 1.00 | 1.00 | 1.00 | **1.00** | 1.00 | 1.00 | 1.00 |
| LS 1.8 | %accepted | 24.61 | 31.94 | 27.23 | 31.94 | 31.41 | 23.04 | 33.51 | 33.51 | 28.80 | 18.32 |
| | species PPV | 0.98 | 0.95 | 0.94 | 0.97 | 0.97 | 0.95 | 0.98 | 0.98 | 0.98 | 1.00 |
| | genus PPV | 1.00 | 1.00 | 1.00 | 1.00 | 1.00 | 1.00 | 1.00 | 1.00 | 1.00 | 1.00 |
| LS 1.9 | %accepted | 17.28 | 23.04 | 19.90 | 21.47 | 20.94 | 15.71 | 23.56 | 23.56 | 20.42 | 15.71 |
| | species PPV | 0.97 | 0.95 | 0.95 | 1.00 | 1.00 | 1.00 | 1.00 | 1.00 | 1.00 | 1.00 |
| | genus PPV | 1.00 | 1.00 | 1.00 | 1.00 | 1.00 | 1.00 | 1.00 | 1.00 | 1.00 | 1.00 |
| LS 2.0 | %accepted | 12.04 | 15.71 | 14.66 | 14.14 | 14.14 | 11.52 | 13.61 | 13.61 | 13.09 | 11.52 |
| | species PPV | 1.00 | 0.97 | 0.96 | 1.00 | 1.00 | 1.00 | 1.00 | 1.00 | 1.00 | 1.00 |
| | genus PPV | 1.00 | 1.00 | 1.00 | 1.00 | 1.00 | 1.00 | 1.00 | 1.00 | 1.00 | 1.00 |
| LS 2.1 | %accepted | 6.81 | 7.85 | 6.81 | 8.38 | 8.38 | 7.33 | 7.33 | 7.33 | 7.33 | 6.81 |
| | species PPV | 1.00 | 1.00 | 1.00 | 1.00 | 1.00 | 1.00 | 1.00 | 1.00 | 1.00 | 1.00 |
| | genus PPV | 1.00 | 1.00 | 1.00 | 1.00 | 1.00 | 1.00 | 1.00 | 1.00 | 1.00 | 1.00 |
| LS 2.2 | %accepted | 5.24 | 6.81 | 5.76 | 5.76 | 5.76 | 4.71 | 5.24 | 5.24 | 5.24 | 4.71 |
| | species PPV | 1.00 | 1.00 | 1.00 | 1.00 | 1.00 | 1.00 | 1.00 | 1.00 | 1.00 | 1.00 |
| | genus PPV | 1.00 | 1.00 | 1.00 | 1.00 | 1.00 | 1.00 | 1.00 | 1.00 | 1.00 | 1.00 |
| LS 2.3 | %accepted | 3.14 | 3.66 | 2.62 | 2.09 | 2.09 | 1.57 | 2.09 | 2.09 | 2.09 | 1.57 |
| | species PPV | 1.00 | 1.00 | 1.00 | 1.00 | 1.00 | 1.00 | 1.00 | 1.00 | 1.00 | 1.00 |
| | genus PPV | 1.00 | 1.00 | 1.00 | 1.00 | 1.00 | 1.00 | 1.00 | 1.00 | 1.00 | 1.00 |

% accepted: percentage of submitted strains that fulfilled the criteria for identification at either the genus or species level depending on the applied threshold; species PPV: positive predictive value at the species level; genus PPV: positive predictive value at the genus level

3.30% increase in percentage of accepted identification with B1 to 24.94% increase with D4 when selecting the 1.7 threshold) and a reduced number of misidentifications. For example, using the in-house database, the species PPV for the D1 algorithm with a 1.7 score threshold on the comprehensive 422-strain panel increased from 0.89 to 0.97 when testing a sub-panel including only the 280 strains for which species were represented by at least three distinct strains. Similar trends were observed when using the commercial database (increase in the species PPV from 0.72 to 0.93).

Twenty-two species (180 isolates) of this study are represented by at least three references in both databases. We performed a per species comparison of the efficiency of the two databases for those 22 species, using the D1 algorithm. Results are compiled in Table 9. For the total of the 180 isolates, the percentage of accepted identifications is lower with the Bruker database than with the in-house database (47.22% vs 91.67%). The comparison of the species PPV obtained with both databases highlighted weaknesses for four species (*A. alternata, F. solani, P. chrysogenum* and *S. apiospermum*) with the in-house database (due to misidentifications with closely related species), and for five species (*A. parasiticus, F. proliferatum, S. brevicaulis, T. interdigitale* and *T. rubrum*) with the Bruker database. The species PPV was equal to 1.00 and identical between the two databases for 13 species of this group.

## Discussion

As shown in this study and several others, MALDI-TOF MS has recently been optimized to identify filamentous fungi at the species level, provided that an appropriate database is available [2, 4, 6–8]. The assessment panel used in this study comprised isolates belonging to 126

**Table 8** Comparison of the 10 different algorithms tested on a sub-panel of 280 strains that are represented by three or more strains in the in-house reference library

| Threshold | | A | B1 | B2 | C1 | C2 | C3 | D1 | D2 | D3 | D4 |
|---|---|---|---|---|---|---|---|---|---|---|---|
| LS 1.5 | %accepted | 93.57 | 96.07 | 95.36 | 97.14 | 97.14 | 95.36 | 98.21 | 98.21 | 97.14 | 95.71 |
| | species PPV | 0.93 | 0.94 | 0.92 | 0.94 | 0.94 | 0.91 | 0.95 | 0.93 | 0.92 | 0.90 |
| | genus PPV | 1.00 | 1.00 | 1.00 | 1.00 | 1.00 | 1.00 | 1.00 | 1.00 | 1.00 | 1.00 |
| LS 1.6 | %accepted | 90.00 | 92.50 | 92.50 | 93.93 | 93.93 | 93.21 | 95.36 | 95.36 | 94.64 | 93.93 |
| | species PPV | 0.94 | 0.94 | 0.92 | 0.95 | 0.95 | 0.91 | 0.96 | 0.95 | 0.93 | 0.91 |
| | genus PPV | 1.00 | 1.00 | 1.00 | 1.00 | 1.00 | 1.00 | 1.00 | 1.00 | 1.00 | 1.00 |
| LS 1.7 | %accepted | 86.07 | 88.57 | 88.57 | 91.07 | 91.07 | 90.36 | **92.14** | 92.14 | 91.43 | 90.71 |
| | species PPV | 0.95 | 0.95 | 0.92 | 0.96 | 0.95 | 0.92 | **0.97** | 0.95 | 0.94 | 0.91 |
| | genus PPV | 1.00 | 1.00 | 1.00 | 1.00 | 1.00 | 1.00 | **1.00** | 1.00 | 1.00 | 1.00 |
| LS 1.8 | %accepted | 81.07 | 82.86 | 82.86 | 83.93 | 83.93 | 83.21 | 86.07 | 86.07 | 85.71 | 85.36 |
| | species PPV | 0.96 | 0.96 | 0.93 | 0.97 | 0.97 | 0.93 | 0.97 | 0.96 | 0.95 | 0.92 |
| | genus PPV | 1.00 | 1.00 | 1.00 | 1.00 | 1.00 | 1.00 | 1.00 | 1.00 | 1.00 | 1.00 |
| LS 1.9 | %accepted | 69.64 | 75.36 | 75.36 | 76.79 | 76.79 | 76.07 | 79.29 | 79.29 | 78.93 | 78.57 |
| | species PPV | 0.97 | 0.96 | 0.95 | 0.97 | 0.97 | 0.93 | 0.97 | 0.96 | 0.95 | 0.93 |
| | genus PPV | 1.00 | 0.99 | 1.00 | 1.00 | 1.00 | 1.00 | 1.00 | 1.00 | 1.00 | 1.00 |
| LS 2.0 | %accepted | 60.00 | 65.00 | 65.00 | 68.57 | 68.57 | 67.86 | 70.36 | 70.36 | 70.00 | 69.64 |
| | species PPV | 0.98 | 0.97 | 0.95 | 0.97 | 0.97 | 0.94 | 0.97 | 0.96 | 0.95 | 0.94 |
| | genus PPV | 1.00 | 0.99 | 1.00 | 1.00 | 1.00 | 1.00 | 1.00 | 1.00 | 1.00 | 1.00 |
| LS 2.1 | %accepted | 46.79 | 54.64 | 54.64 | 57.14 | 57.14 | 56.79 | 59.64 | 59.64 | 59.29 | 59.29 |
| | species PPV | 0.98 | 0.97 | 0.95 | 0.98 | 0.98 | 0.94 | 0.98 | 0.97 | 0.96 | 0.94 |
| | genus PPV | 1.00 | 1.00 | 1.00 | 1.00 | 1.00 | 1.00 | 1.00 | 1.00 | 1.00 | 1.00 |
| LS 2.2 | %accepted | 30.36 | 36.79 | 36.79 | 40.00 | 40.00 | 39.64 | 42.50 | 42.50 | 42.14 | 42.14 |
| | species PPV | 0.99 | 0.97 | 0.95 | 0.97 | 0.97 | 0.95 | 0.98 | 0.97 | 0.97 | 0.96 |
| | genus PPV | 1.00 | 1.00 | 1.00 | 1.00 | 1.00 | 1.00 | 1.00 | 1.00 | 1.00 | 1.00 |
| LS 2.3 | %accepted | 15.71 | 19.29 | 19.29 | 21.43 | 21.43 | 21.43 | 22.50 | 22.50 | 22.50 | 22.50 |
| | species PPV | 0.97 | 0.98 | 0.96 | 0.97 | 0.97 | 0.95 | 0.97 | 0.95 | 0.95 | 0.95 |
| | genus PPV | 1.00 | 1.00 | 1.00 | 1.00 | 1.00 | 1.00 | 1.00 | 1.00 | 1.00 | 1.00 |

% accepted: percentage of submitted strains that fulfilled the criteria for identification at either the genus or species level depending on the applied threshold; species PPV: positive predictive value at the species level; genus PPV: positive predictive value at the genus level

different species, including many difficult-to-identify isolates. Consequently, the identification rate was low compared with previously published studies, especially studies utilizing the commercial database. Using such a challenging panel, it was possible to define a threshold discriminating correct identifications at the species level from false identifications at the genus level, although identifying a threshold that differentiated between a correct and false identification at the species level among closely related species was not feasible. Our findings indicate that the best identification results (PPV at the species level and percentage of accepted identifications) were obtained when applying a decisional algorithm in which only the highest score of four spots was taken into account, and the identification was accepted if the score was at least 1.7. This threshold is notably lower than that recommended for bacteria identification, and has already been advised by several authors for yeast identification [32, 33]. Furthermore, we also observed a marked decrease in percentage of identification between a threshold of 1.7 and 2.0. When the threshold was set too high, we noted a 20% loss in identification efficiency and only a small benefit of 2% in the differentiation of certain close species. Even if the Bruker commercial database yielded a markedly lower percentage of accepted identifications (probably due to the fact that the database was built using liquid culture media, that is does not frequently involve more than two or three references per species, and that there are errors of reference labeling), the PPV remained correct for the spectra reaching the 1.7 threshold. This result corroborates the findings of several authors who have suggested that lowering the identification score thresholds would improve MALDI-TOF MS-based fungal identification efficiency

**Table 9** Comparison of the identification efficiency of the two databases per species, using the D1 algorithm of identification with a 1.7 LS threshold, for 180 strains that are represented by three or more strains in both reference databases

| | | In-House database | | Bruker database | |
|---|---|---|---|---|---|
| | nb of isolates | % accepted | species PPV | % accepted | species PPV |
| *Absidia corymbifera* | 2 | 100.00 | 1.00 | 50.00 | 1.00 |
| *Alternaria alternata* | 2 | 100.00 | 0.50 | 100.00 | 1.00 |
| *Aspergillus flavus* | 15 | 100.00 | 1.00 | 73.33 | 1.00 |
| *Aspergillus fumigatus* | 20 | 95.00 | 1.00 | 45.00 | 1.00 |
| *Aspergillus nidulans* | 8 | 100.00 | 1.00 | 75.00 | 1.00 |
| *Aspergillus niger* | 4 | 100.00 | 1.00 | 50.00 | 1.00 |
| *Aspergillus parasiticus* | 2 | 100.00 | 1.00 | 0.00 | Not Relevant |
| *Aspergillus terreus* | 10 | 100.00 | 1.00 | 30.00 | 1.00 |
| *Aspergillus versicolor* | 3 | 66.67 | 1.00 | 33.33 | 1.00 |
| *Exophiala dermatitidis* | 2 | 50.00 | 1.00 | 50.00 | 1.00 |
| *Fusarium oxysporum* | 21 | 80.95 | 1.00 | 19.05 | 1.00 |
| *Fusarium proliferatum* | 10 | 100.00 | 1.00 | 60.00 | 0.33 |
| *Fusarium solani* | 7 | 57.14 | 0.25 | 28.57 | 1.00 |
| *Microsporum canis* | 2 | 100.00 | 1.00 | 50.00 | 1.00 |
| *Purpureocillium lilacinus* | 3 | 66.67 | 1.00 | 66.67 | 1.00 |
| *Paecilomyces variotii* | 3 | 66.67 | 1.00 | 66.67 | 1.00 |
| *Penicillium chrysogenum* | 27 | 100.00 | 0.96 | 33.33 | 1.00 |
| *Scedosporium apiospermum* | 11 | 100.00 | 0.82 | 81.82 | 1.00 |
| *Schizophyllum commune* | 4 | 100.00 | 1.00 | 100.00 | 1.00 |
| *Scopulariopsis brevicaulis* | 15 | 86.67 | 1.00 | 46.67 | 0.71 |
| *Trichophyton interdigitale* | 1 | 100.00 | 1.00 | 0.00 | Not Relevant |
| *Trichophyton rubrum* | 8 | 87.50 | 0.86 | 37.50 | 0.67 |
| Total Isolates | 180 | 91.67 | 0.95 | 47.22 | 0.92 |

Nb of isolates : number of isolates from the panel of strain that belong to each species. % accepted: percentage of submitted strains that fulfilled the criteria for identification at either the genus or species level; species PPV: positive predictive value at the species level

[7, 17, 32, 33]. However, we strongly advise against a threshold below 1.7, which could yield an increase in false identifications, in particular at the genus level, as shown in Tables 3 and 4.

The major problem encountered with the data set obtained from the assessment panel was the resemblance between the MALDI-TOF MS spectra of some species. For 32 strains, the similarity in MALDI-TOF MS spectra yielded an incorrect identification at the species level, although these identification errors systematically corresponded to taxonomically close species often belonging to the same species complex (except for *Pleospora papaveracea*, which was identified as *Ulocladium oudemansi*). Among these 32 strains (27 species), only a few species are known to be pathogenic to humans, of which none are major human pathogens. Regarding the pathogenic species that were misidentified in our study, the treatment would not vary from that prescribed based on molecular identification, and the differentiation of these species is not clinically relevant. The remaining species are often considered to be a result of contamination, and their antifungal susceptibility is unknown. Interestingly, raising the identification score threshold to 2.3 did not entirely resolve this identification issue of taxonomically close species, as six strains were incorrectly characterized despite a score above 2.3, even using our extensive in-house database. Thus, in our study, one could question the relevance of capacity and necessity to distinguish between these closely related species as none of the misidentifications implied a change in treatment.

Our results also show that the key step in efficient identification was to increase the number of replicate spots per strain. Testing four spots of each filamentous fungus yielded better identification performance, with an increase of 5% to 7% of accepted strain identifications. Such spot replications did not markedly increase the workload, as it took only a few seconds to perform the additional spots. However, the replicates may increase the per-strain acquisition time in the laboratory workflow. The four-spot method has been applied in the routine workflow at the Marseille Mycology Laboratory for the past five years, which demonstrates that this procedure is easily applicable in hospital laboratories to identify filamentous fungi isolated from clinical samples and from hospital environment samples. Multiplying the number of deposits per strains imply a small increase in the cost per strain as only one reagent (HCCA matrix) has to be used more extensively. The decisional algorithm implying concordance of the identification results of two to four spots did not enhance identification performance, regardless of the threshold selected. Furthermore, the D2, D3 and D4 algorithms tended to lower the percentage of validated identifications by disregarding identification results that would have been correct using the D1 decisional algorithm.

When focusing on the species that were represented by at least three distinct strains in the reference spectrum library, identification efficacy markedly increased. This result stresses the importance of a comprehensive reference spectrum library for the identification of filamentous fungi, including dermatophytes, as previously reported by our group [6], and as previously demonstrated for yeasts [32]. The multiplication of strains per species as well as subcultures per strain allowed us to cover the variations seen within the environment for filamentous fungi.

Our in-house reference database, which yielded markedly better results compared with the commercial Bruker database, was constructed mainly with collection strains and with some clinical strains isolated in our laboratory and obtained during routine analyses. These clinical strains were later included in the BCCM/IHEM collection in Brussels. Nevertheless, even our enhanced database still requires optimization to identify rare species. Indeed, the poor representation of rare species in the database may explain the identification percentage of 87.4% obtained with this challenging panel of 422 strains (92.1% with the 280 strains represented by more than three strains in the reference database), compared with the extremely good identification results of common species (98.1%) exposed by Gautier et al. [21].

## Conclusion

The growing number of publications on MALDI-TOF MS-based fungal identification indicates that this technology is widely used and applicable to the routine laboratory workflow. MALDI-TOF MS-based identification of fungi yields exceedingly more accurate results compared with morphology-based analyses [21]. Furthermore, this technology is less expensive and easier than the current DNA based-identification gold standard, with a turnaround time allowing achieving the analysis more rapidly. Currently, the major limitation of this approach is the lack of a comprehensive and efficient reference spectrum library. We acknowledge that in-house reference databases that are non-open-source, such as ours, are of little benefit to the global scientific community. Therefore, we plan to deploy an accessible version of our database along with an identification algorithm that can be queried online to assist scientific teams to expand the identification of fungal species. Experiments implying a multicenter approach are ongoing to finalize this project.

**Acknowledgements**

We especially thank Sandy Moore for proofreading the English version of the manuscript.

**Funding**

None.

**Authors' contributions**

ACN and PB collected the data. ACN performed the analyses. ACN and RP interpreted the data. ACN, RP, PB, SR and MH were major contributors in writing the manuscript. All authors read and approved the final manuscript.

**Competing interests**

The authors declare that they have no competing interests.

**Author details**

[1]Laboratoire de Parasitologie-Mycologie, CHU Timone, Université de la Méditerranée, Marseille, France. [2]Service of Mycology and Aerobiology, BCCM/IHEM Fungal Collection, Scientific Institute of Public Health, Brussels, Belgium.

**References**

1. Marinach-Patrice C, Lethuillier A, Marly A, Brossas J-Y, Gené J, Symoens F, Datry A, Guarro J, Mazier D, Hennequin C. Use of mass spectrometry to identify clinical Fusarium isolates. Clin Microbiol Infect Off Publ Eur Soc Clin Microbiol Infect Dis. 2009;15:634–42.
2. Cassagne C, Ranque S, Normand A-C, Fourquet P, Thiebault S, Planard C, Hendrickx M, Piarroux R. Mould routine identification in the clinical laboratory by matrix-assisted laser desorption ionization time-of-flight mass spectrometry. PLoS One. 2011;6:e28425.
3. De Carolis E, Posteraro B, Lass-Flörl C, Vella A, Florio AR, Torelli R, Girmenia C, Colozza C, Tortorano AM, Sanguinetti M, Fadda G. Species identification of Aspergillus, Fusarium and Mucorales with direct surface analysis by matrix assisted laser desorption ionization time-of-flight mass spectrometry. Clin Microbiol Infect Off Publ Eur Soc Clin Microbiol Infect Dis. 2012;18:475–84.
4. Lau AF, Drake SK, Calhoun LB, Henderson CM, Zelazny AM. Development of a clinically comprehensive database and a simple procedure for identification of molds from solid media by matrix-assisted laser desorption ionization-time of flight mass spectrometry. J Clin Microbiol. 2013;51:828–34.

5. L'Ollivier C, Cassagne C, Normand A-C, Bouchara J-P, Contet-Audonneau N, Hendrickx M, Fourquet P, Coulibaly O, Piarroux R, Ranque S. A MALDI-TOF MS procedure for clinical dermatophyte species identification in the routine laboratory. Med Mycol. 2013;51:713–20.
6. Normand A-C, Cassagne C, Ranque S, L'ollivier C, Fourquet P, Roesems S, Hendrickx M, Piarroux R. Assessment of various parameters to improve MALDI-TOF MS reference spectra libraries constructed for the routine identification of filamentous fungi. BMC Microbiol. 2013;13:76.
7. Ranque S, Normand A-C, Cassagne C, Murat J-B, Bourgeois N, Dalle F, Gari-Toussaint M, Fourquet P, Hendrickx M, Piarroux R. MALDI-TOF mass spectrometry identification of filamentous fungi in the clinical laboratory. Mycoses. 2014;57:135–40.
8. Schulthess B, Ledermann R, Mouttet F, Zbinden A, Bloemberg GV, Böttger EC, Hombach M. Use of the Bruker MALDI Biotyper for identification of molds in the clinical mycology laboratory. J Clin Microbiol. 2014;52:2797–803.
9. Li TY, Liu BH, Chen YC. Characterization of Aspergillus spores by matrix-assisted laser desorption/ionization time-of-flight mass spectrometry. Rapid Commun Mass Spectrom RCM. 2000;14:2393–400.
10. Welham KJ, Domin MA, Johnson K, Jones L, Ashton DS. Characterization of fungal spores by laser desorption/ionization time-of-flight mass spectrometry. Rapid Commun Mass Spectrom RCM. 2000;14:307–10.
11. Chen H-Y, Chen Y-C. Characterization of intact Penicillium spores by matrix-assisted laser desorption/ionization mass spectrometry. Rapid Commun Mass Spectrom RCM. 2005;19:3564–8.
12. Dong H, Kemptner J, Marchetti-Deschmann M, Kubicek CP, Allmaier G. Development of a MALDI two-layer volume sample preparation technique for analysis of colored conidia spores of Fusarium by MALDI linear TOF mass spectrometry. Anal Bioanal Chem. 2009;395:1373–83.
13. Kemptner J, Marchetti-Deschmann M, Mach R, Druzhinina IS, Kubicek CP, Allmaier G. Evaluation of matrix-assisted laser desorption/ionization (MALDI) preparation techniques for surface characterization of intact Fusarium spores by MALDI linear time-of-flight mass spectrometry. Rapid Commun Mass Spectrom RCM. 2009;23:877–84.
14. Tao J, Zhang G, Jiang Z, Cheng Y, Feng J, Chen Z. Detection of pathogenic Verticillium spp. using matrix-assisted laser desorption/ionization time-of-flight mass spectrometry. Rapid Commun Mass Spectrom RCM. 2009;23: 3647–54.
15. DE Respinis S, Vogel G, Benagli C, Tonolla M, Petrini O, Samuels G. MALDI-TOF MS of Trichoderma: A model system for the identification of microfungi. Mycol Prog. 2010;9:79–100.
16. Coulibaly O, Marinach-Patrice C, Cassagne C, Piarroux R, Mazier D, Ranque S. Pseudallescheria/Scedosporium complex species identification by Matrix-Assisted Laser Desorption Ionization Time-Of-Flight Mass Spectrometry. Med Mycol Off Publ Int Soc Hum Anim Mycol. 2011;49:621–6.
17. Del Chierico F, Masotti A, Onori M, Fiscarelli E, Mancinelli L, Ricciotti G, Alghisi F, Dimiziani L, Manetti C, Urbani A, Muraca M, Putignani L. MALDI-TOF MS proteomic phenotyping of filamentous and other fungi from clinical origin. J Proteomics. 2012;75(11):3314–30.
18. Nenoff P, Erhard M, Simon JC, Muylowa GK, Herrmann J, Rataj W, Gräser Y. MALDI-TOF mass spectrometry - a rapid method for the identification of dermatophyte species. Med Mycol. 2013;51:17–24.
19. Packeu A, Hendrickx M, Beguin H, Martiny D, Vandenberg O, Detandt M. Identification of the Trichophyton mentagrophytes complex species using MALDI-TOF mass spectrometry. Med Mycol. 2013;51:580–5.
20. Packeu A, De Bel A, l'Ollivier C, Ranque S, Detandt M, Hendrickx M. Fast and accurate identification of dermatophytes by matrix-assisted laser desorption ionization-time of flight mass spectrometry: validation in the clinical laboratory. J Clin Microbiol. 2014;52:3440–3.
21. Gautier M, Ranque S, Normand A-C, Becker P, Packeu A, Cassagne C, L'Ollivier C, Hendrickx M, Piarroux R. Matrix-assisted laser desorption ionization time of flight mass spectrometry: revolutionizing clinical laboratory diagnosis of mould infections. Clin Microbiol Infect Off Publ Eur Soc Clin Microbiol Infect Dis. 2014;20:1366–71.
22. Erhard M, Hipler U-C, Burmester A, Brakhage AA, Wöstemeyer J. Identification of dermatophyte species causing onychomycosis and tinea pedis by MALDI-TOF mass spectrometry. Exp Dermatol. 2008;17:356–61.
23. de Respinis S, Tonolla M, Pranghofer S, Petrini L, Petrini O, Bosshard PP. Identification of dermatophytes by matrix-assisted laser desorption/ ionization time-of-flight mass spectrometry. Med Mycol. 2013;51:514–21.
24. Verwer PEB, van Leeuwen WB, Girard V, Monnin V, van Belkum A, Staab JF, Verbrugh HA, Bakker-Woudenberg Ia JM, van de Sande WWJ. Discrimination of Aspergillus lentulus from Aspergillus fumigatus by Raman spectroscopy and MALDI-TOF MS. Eur J Clin Microbiol Infect Dis Off Publ Eur Soc Clin Microbiol. 2014;33:245–51.
25. De Respinis S, Monnin V, Girard V, Welker M, Arsac M, Cellière B, Durand G, Bosshard PP, Farina C, Passera M, Van Belkum A, Petrini O, Tonolla M. Matrix-assisted laser desorption ionization-time of flight (MALDI-TOF) mass spectrometry using the Vitek MS system for rapid and accurate identification of dermatophytes on solid cultures. J Clin Microbiol. 2014;52:4286–92.
26. Lévesque S, Dufresne PJ, Soualhine H, Domingo M-C, Bekal S, Lefebvre B, Tremblay C. A Side by Side Comparison of Bruker Biotyper and VITEK MS: Utility of MALDI-TOF MS Technology for Microorganism Identification in a Public Health Reference Laboratory. PLoS One. 2015;10:e0144878.
27. Iriart X, Lavergne R-A, Fillaux J, Valentin A, Magnaval J-F, Berry A, Cassaing S. Routine identification of medical fungi by the new Vitek MS matrix-assisted laser desorption ionization-time of flight system with a new time-effective strategy. J Clin Microbiol. 2012;50:2107–10.
28. McMullen AR, Wallace MA, Pincus DH, Wilkey K, Burnham CA. Evaluation of the Vitek MS Matrix-Assisted Laser Desorption Ionization-Time of Flight Mass Spectrometry System for Identification of Clinically Relevant Filamentous Fungi. J Clin Microbiol. 2016;54:2068–73.
29. Alshawa K, Beretti J-L, Lacroix C, Feuilhade M, Dauphin B, Quesne G, Hassouni N, Nassif X, Bougnoux M-E. Successful identification of clinical dermatophyte and Neoscytalidium species by matrix-assisted laser desorption ionization-time of flight mass spectrometry. J Clin Microbiol. 2012;50:2277–81.
30. Alanio A, Beretti J-L, Dauphin B, Mellado E, Quesne G, Lacroix C, Amara A, Berche P, Nassif X, Bougnoux M-E. Matrix-assisted laser desorption ionization time-of-flight mass spectrometry for fast and accurate identification of clinically relevant Aspergillus species. Clin Microbiol Infect Off Publ Eur Soc Clin Microbiol Infect Dis. 2011;17:750–5.
31. Theel ES, Hall L, Mandrekar J, Wengenack NL. Dermatophyte identification using matrix-assisted laser desorption ionization-time of flight mass spectrometry. J Clin Microbiol. 2011;49:4067–71.
32. Vlek A, Kolecka A, Khayhan K, Theelen B, Groenewald M, Boel E, Multicenter Study Group, Boekhout T. Interlaboratory comparison of sample preparation methods, database expansions, and cutoff values for identification of yeasts by matrix-assisted laser desorption ionization-time of flight mass spectrometry using a yeast test panel. J Clin Microbiol. 2014;52:3023–9.
33. Van Herendael BH, Bruynseels P, Bensaid M, Boekhout T, De Baere T, Surmont I, Mertens AH. Validation of a modified algorithm for the identification of yeast isolates using matrix-assisted laser desorption/ ionisation time-of-flight mass spectrometry (MALDI-TOF MS). Eur J Clin Microbiol Infect Dis Off Publ Eur Soc Clin Microbiol. 2012;31:841–8.
34. Becker PT, de Bel A, Martiny D, Ranque S, Piarroux R, Cassagne C, Detandt M, Hendrickx M. Identification of filamentous fungi isolates by MALDI-TOF mass spectrometry: clinical evaluation of an extended reference spectra library. Med Mycol. 2014;52:826–34.
35. Irinyi L, Serena C, Garcia-Hermoso D, Arabatzis M, Desnos-Ollivier M, Vu D, Cardinali G, Arthur I, Normand A-C, Giraldo A, da Cunha KC, Sandoval-Denis M, Hendrickx M, Nishikaku AS, de Azevedo Melo AS, Merseguel KB, Khan A, Parente Rocha JA, Sampaio P, da Silva Briones MR, e Ferreira RC, de Medeiros Muniz M, Castañón-Olivares LR, Estrada-Barcenas D, Cassagne C, Mary C, Duan SY, Kong F, Sun AY, Zeng X, et al. International Society of Human and Animal Mycology (ISHAM)-ITS reference DNA barcoding database—the quality controlled standard tool for routine identification of human and animal pathogenic fungi. Med Mycol. 2015;53:313–37.
36. Balajee SA, Borman AM, Brandt ME, Cano J, Cuenca-Estrella M, Dannaoui E, Guarro J, Haase G, Kibbler CC, Meyer W, O'Donnell K, Petti CA, Rodriguez-Tudela JL, Sutton D, Velegraki A, Wickes BL. Sequence-Based Identification of Aspergillus, Fusarium, and Mucorales Species in the Clinical Mycology Laboratory: Where Are We and Where Should We Go from Here? J Clin Microbiol. 2009;47:877–84.

# 3

# OxyR-regulated catalase CatB promotes the virulence in rice via detoxifying hydrogen peroxide in *Xanthomonas oryzae* pv. *oryzae*

Chao Yu[1†], Nu Wang[1†], Maosen Wu[1], Fang Tian[1], Huamin Chen[1], Fenghuan Yang[1], Xiaochen Yuan[2], Ching-Hong Yang[2] and Chenyang He[1*]

**Abstract**

**Background:** To facilitate infection, *Xanthomonas oryzae* pv. *oryzae* (*Xoo*), the bacterial blight pathogen of rice, needs to degrade hydrogen peroxide ($H_2O_2$) generated by the host defense response via a mechanism that is mediated by the transcriptional regulator OxyR. The catalase (CAT) gene *catB* has previously been shown to belong to the OxyR regulon in *Xoo*. However, its expression patterns and function in $H_2O_2$ detoxification and bacterial pathogenicity on rice remain to be elucidated.

**Results:** The *catB* gene encodes a putative catalase and is highly conserved in the sequenced strains of *Xanthomonas* spp. β-galactosidase analysis and electrophoretic mobility shift assays (EMSA) showed that OxyR positively regulated the transcription of *catB* by directly binding to its promoter region. The quantitative real-time PCR (qRT-PCR) assays revealed that the expression levels of *catB* and *oxyR* were significantly induced by $H_2O_2$. Deletion of *catB* or *oxyR* drastically impaired bacterial viability in the presence of extracellular $H_2O_2$ and reduced CAT activity, demonstrating that CatB and OxyR contribute to $H_2O_2$ detoxification in *Xoo*. In addition, Δ*catB* and Δ*oxyR* displayed shorter bacterial blight lesions and reduced bacterial growth in rice compared to the wild-type stain, indicating that CatB and OxyR play essential roles in the virulence of *Xoo*.

**Conclusions:** Transcription of *catB* is enhanced by OxyR in response to exogenous $H_2O_2$. CatB functions as an active catalase that is required for the full virulence of *Xoo* in rice.

**Keywords:** *Xanthomonas oryzae* pv. *oryzae*, Catalase, Hydrogen peroxide, Virulence

## Background

Plant innate immune responses to bacterial infection include an oxidative burst through elevating the levels of reactive oxygen species (ROS), which are toxic to bacterial cells and cause damages to proteins, nucleic acids, and cell membranes [1]. Hydrogen peroxide ($H_2O_2$), an important ROS, accumulates and diffuses widely between neighboring xylem elements [2]. Accordingly, pathogenic bacteria need to overcome the stress caused the by $H_2O_2$ and gradually establish their infection in plants [3].

One of the mechanisms of bacterial resistance to is $H_2O_2$ is through antioxidant enzymes including catalases [4]. Bacteria maintain basal oxidative stress resistance, but also possess a highly inducible oxidative stress response that is largely controlled by redox-sensing transcription factors which act as redox-operated genetic switches to activate genes involved in the oxidative stress response [5]. As one of the redox-sensing transcription factors, OxyR has been characterized in a number of pathogenic bacteria [6, 7]. OxyR is a DNA-binding transcription factor that is activated under oxidizing conditions by the formation of a disulfide bond between two cysteine residues [8]. When activated, OxyR regulates the expression of genes involved in detoxification by binding to their promoter regions, thus triggering cellular responses to $H_2O_2$ [5, 9].

* Correspondence: hechenyang@caas.cn
†Equal contributors
[1]State Key Laboratory for Biology of Plant Diseases and Insect Pests, Institute of Plant Protection, Chinese Academy of Agricultural Sciences, Beijing 100193, China
Full list of author information is available at the end of the article

Bacterial catalases (CATs) are central components of detoxification pathways, which prevent formation of highly reactive hydroxyl radical by catalyzing the conversion $H_2O_2$ to water and oxygen [10]. Based on their enzymatic properties, catalases fall into three classes including monofunctional heme-containing catalases, bifunctional heme-containing catalase-peroxidases, and nonheme or Mn-containing catalases [11]. Multiple catalase isozymes encoded by different genes have been identified in many bacterial species. Interestingly, these genes have different expression patterns in growth phases and in response to oxidative stress, suggesting that they may have different roles in physiological processes and bacteria-host interactions [12]. Moreover, the transcriptional levels and activities of bacterial catalases were largely induced by $H_2O_2$ [13]. For example, *Xanthomonas campestris* pv. *campestris* and *X. axonopodis* pv. *citri* displayed increased levels of catalases KatE and KatG and $H_2O_2$ resistance after $H_2O_2$ treatment [10, 14, 15]. In addition, several studies have reported that catalase activity was induced during the bacterial infection [3, 16], however, the regulatory mechanism of $H_2O_2$ detoxification and its relation to bacterial virulence remains to be demonstrated.

Bacterial leaf blight caused by *X. oryzae* pv. *oryzae* (*Xoo*) is one of the most devastating diseases of rice, which causes annual yield losses of 10–50 % in many rice growing countries [17–19]. Understanding the molecular mechanisms *Xoo* virulence is pivotal to develop effective disease control strategies [20]. *Xoo* produces several virulence-related factors including exopolysaccharide, extracellular enzymes, toxins, adhesins, and the type III secretion system and its effectors during infection [21, 22]. Furthermore, to facilitate its virulence, *Xoo* might have evolved a mechanism for suppression and evasion of basal defense response of rice, such as employing catalases to detoxify $H_2O_2$ elicited by host innate immunity [23–25]. Detoxification of endogenous $H_2O_2$ that is generated through normal metabolic processes, such as aerobic respiration, is mediated by OxyR via regulating the *ahpC* and *ahpF* genes encoding alkyl hydroperoxide reductase [26, 27]. Deletion of *ahpC* significantly affected $H_2O_2$ accumulation during the rice-*Xoo* interaction [28]. *catB* (*PXO_02830*), *katE* (*PXO_02109*) and *srpA* (*PXO_02864*), three putative catalase genes were revealed in the genome of *Xoo* wildtype strain PXO99$^A$ by *in silico* analysis [29]. The transcription of these genes were strongly induced during the bacterial interaction with rice suspension-cultured cells [30]. In addition, in-frame deletion of the *katE* gene significantly attenuated bacterial pathogenesis in rice but not $H_2O_2$ resistance [31]. The regulatory mechanisms and involvement in $H_2O_2$ detoxification and pathogenesis of CatB remain unknown.

In this study, we characterized the regulatory mechanism and function of CatB in $H_2O_2$ resistance and its contribution to virulence in rice. Promoter activities and qRT-PCR assays demonstrated that *catB* gene was transcriptionally responsive to $H_2O_2$ and positively regulated by OxyR. Gene deletion and complementation analysis revealed that CatB greatly contributed to both $H_2O_2$ detoxification and full virulence. This study demonstrates that CatB is one of the key virulence factors to facilitate pathogenesis of *Xoo* in rice via $H_2O_2$ detoxification.

## Results

### Identification, deletion and complementation of the *catB* gene

It has been reported that *catB* is one of three putative catalases-encoding genes (*catB* (*PXO_02830*), *katE* (*PXO_02109*), and *srpA* (*PXO_02864*)) that are responsible for $H_2O_2$ degradation in the genome of *Xoo* wildtype strain PXO99$^A$ [29]. The open reading frame (ORF) of *catB* is 1,524 bp in length and located in the genome at nucleotide position 4970855–4972378 (Fig. 1a). CatB contains one catalase domain (residues 28–408) with a predicted isoelectric point of 8.66 and molecular weight of 56.4 kD. CatB is also well conserved in other sequenced strains of *Xoo* (MAFF311018, KACC10331, and PXO86) (Additional file 1: Table S1). Additionally, CatB shared over 90 % amino acid identity with those in other important plant-pathogenic *Xanthomonas* species, including *X. campestris* pv. *campestris*, *X. campestris* pv. *vesicatoria*, *X. axonopodis*, *X. perforans*, *X. fragariae*, *X. arboricola*, *X. hortorum*, and *X. fuscans* (Additional file 1: Table S1). Sequence alignment analysis with experimentally validated active CATs, such as KatE from *X. axonopodis* pv. *citri* and KatA from *Bacillus subtilis* revealed that many critical residues in the catalase domain of CatB were probably conserved (Fig. 1b). These observations indicate that CatB might function as an active CAT.

To investigate the potential biological function of CatB in *Xoo*, a *catB* gene deletion mutant (Δ*catB*) and complementary strain Δ*catB*(pBBR-*catB*) were constructed as described in the Methods. DNA sequencing analysis showed that the corresponding region of *catB* was replaced by the $Gm^r$ gene (855 bp in length) in the Δ*catB* mutant. The growth for PXO99$^A$, Δ*catB* and Δ*catB*(pBBR-*catB*) were detected in M210 liquid medium. No significant differences were observed in growth rates and bacterial populations between the three strains (Additional file 2: Figure S1). In addition, our previous studies have shown that there was no difference between Δ*oxyR* and wild type in growth in vitro [26]. These results suggest that deletion of *catB* or *oxyR* does not affect the viability of *Xoo* under normal growth conditions.

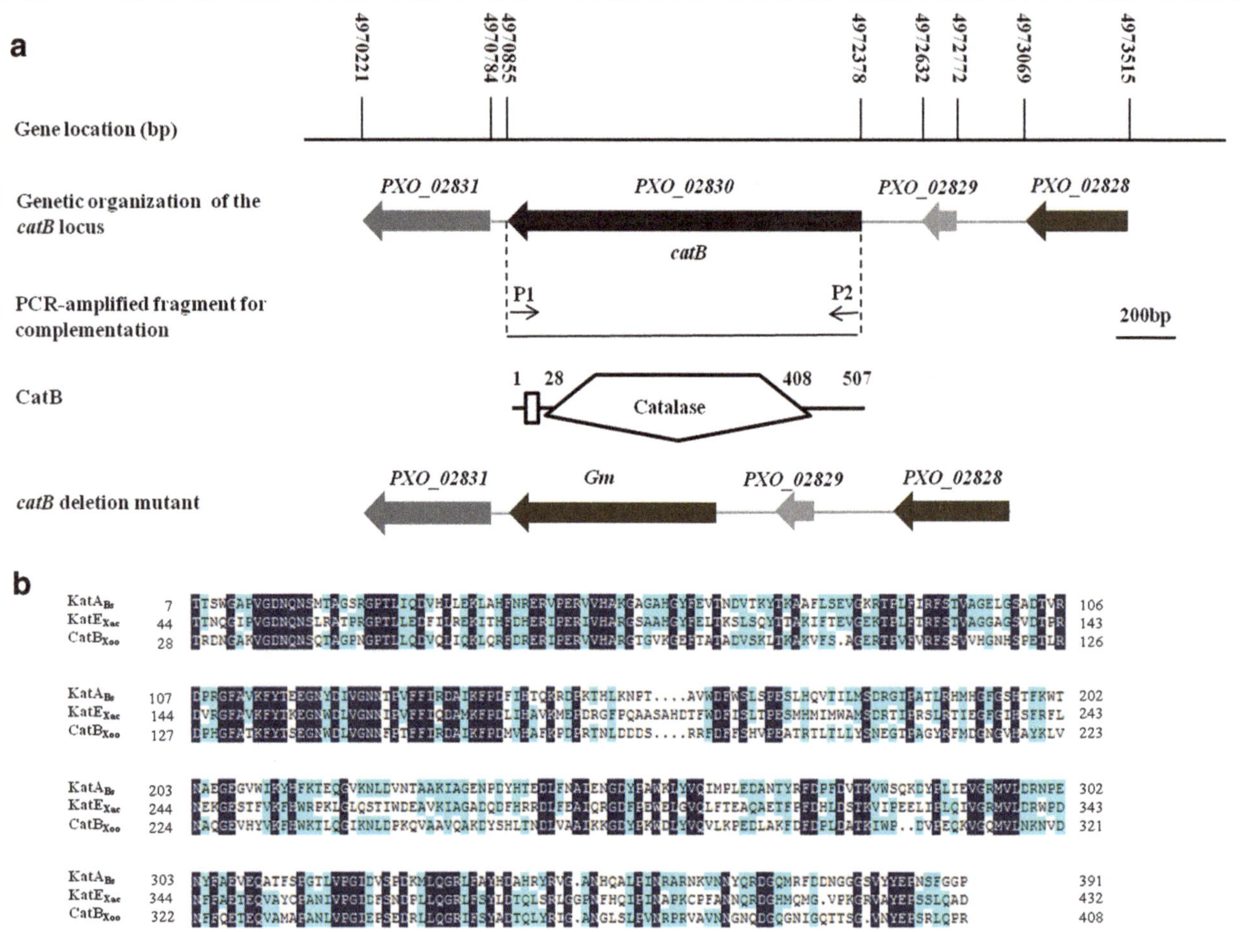


**Fig. 1** Bioinformatics analysis of *catB* gene. **a** Schematic diagram of the *catB* gene in the genome of PXO99$^A$. The open arrows indicate length, location and orientation of the ORFs. The middle element shows *catB* was amplified by PCR primers P1 and P2 and cloned into the plasmid pBBR1MCS-4 for complementation of Δ*catB*. The lower element shows domain structure analyses of the putative CatB protein. The lowest element shows *catB* was exchanged with *Gm$^r$* gene in Δ*catB*. **b** Sequence alignment of CatB (PXO_02830) in *Xoo* PXO99$^A$ was performed by using DNAMAN software. KatA and KatE were two catalases which have been experimentally validated in *Bacillus subtilis* (*Bs*) and *Xanthomonas axonopodis* pv. *citri* (*Xac*), respectively. The amino acid residues highlighted with black means the homology level is 100 %, and with blue means the homology level ≥ 50 %

**catB is transcriptionally regulated by OxyR and responsive to exogenous $H_2O_2$**

OxyR has been shown to function as a transcriptional regulator mediating $H_2O_2$ detoxification in *Xoo*. Our previous studies showed that the *catB* transcripts were significantly reduced in the *oxyR* gene deletion mutant (Δ*oxyR*) [26, 30], implying expression of *catB* is regulated by OxyR. To examine whether *catB* is a direct target of OxyR, we first expressed the recombinant OxyR protein in *E. coli* BL21 strain and obtained purified protein (Additional file 3: Figure S2). We then examined the promoter activity of *catB* by the measuring β-galactosidase activities of *catBp-lacZ* fusion in various strains. Our results showed that the β-galactosidase activity was 17-fold higher in PXO99$^A$ than in Δ*oxyR* (Fig. 2a). Then, electrophoretic mobility shift assay (EMSA) was performed to detect the binding between OxyR protein and the *catB* promoter. The results indicated that the OxyR protein bound directly to the *catB* promoter region (Fig. 3). In contrast, the negative control BSA did not bind to the *catB* promoter region. Addition of unlabeled *catB* promoter DNA fragments as a competitive probes resulted in reduced binding of OxyR with the labeled DNA fragment, while mixture with the 16S rDNA as a non-specific probe did not affected the binding (Fig. 3). These data demonstrated that OxyR regulated the transcription of *catB* gene by directly binding to its promoter region.

To investigate whether their expressions respond to $H_2O_2$ stress, the transcripts of both *catB* and *oxyR* in wild-type PXO99$^A$ in the presence or absence of exogenous $H_2O_2$ were assayed by qRT-PCR. The transcription levels of *catB* and *oxyR* were significantly elevated when exogenous $H_2O_2$ was applied (Fig. 2b). In addition, the activity of

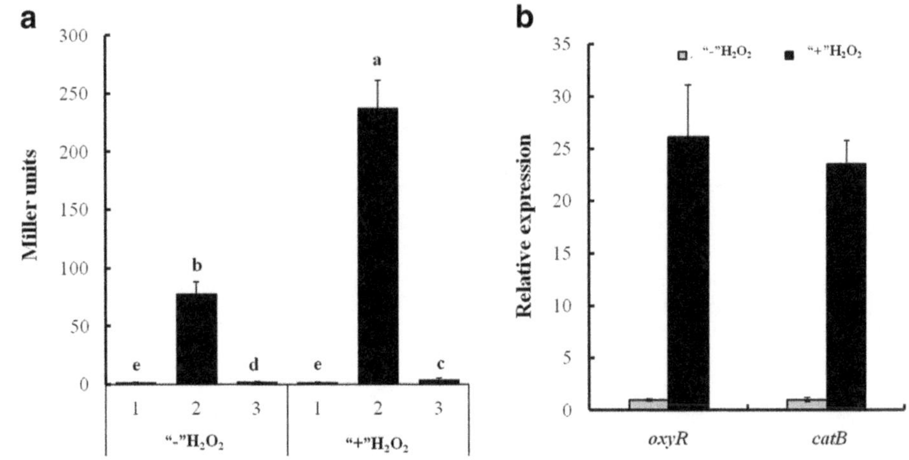


**Fig. 2** Analysis of *catB* and *oxyR* transcripts in *Xoo* strains. **a** Assays for promoter activities of *catB* in PXO99[A] and Δ*oxyR* in the presence ("+") or absence ("-") of $H_2O_2$. Overnight cultures of wildtype and Δ*oxyR* containing a pH-*catBp-lacZ* transcriptional reporter were inoculated 1:100 into fresh M210 liquid medium and shaken at 28 °C until cells reached at $OD_{600}$ of 1.0, and treated with 3 mM $H_2O_2$ for 0.5 h. The *catB* promoter activity was analyzed by measuring β-galactosidase levels. 1, WT (pH-*lacZ*); 2, WT (pH-*catBp-lacZ*); 3, Δ*oxyR* (pH-*catBp-lacZ*). pH-*lacZ* was an empty plasmid used as the control. **b** Assays for *catB* and *oxyR* transcripts in PXO99[A] treated with $H_2O_2$. Wildtype cells cultured in M210 liquid medium were exposed to $H_2O_2$ at 3 mM for 0.5 h, $H_2O_2$-untreated cells were used as the control (CK), and the total RNA was extracted with TRIzol reagent. The expression levels of *catB* and *oxyR* were detected by quantitative RT-PCR and normalized to *gyrB*. Bars represent standard errors of the means from three independent cultivations, and different letters above the bars denote statistically significant differences ($P < 0.05$, Student's *t* test)

*catB* promoter in wildtype PXO99[A] and Δ*oxyR* were measured under exogenous $H_2O_2$ conditions. As expected, the activity of *catB* promoter was dramatically induced in wildtype PXO99[A] but only slightly increased in Δ*oxyR* in the presence of exogenous $H_2O_2$ (Fig. 2a). Therefore, these results demonstrated that OxyR sensed the presence of $H_2O_2$ and then activated the transcription of *catB*.

| | | | | | |
|---|---|---|---|---|---|
| **Probe** | + | + | + | + | + |
| **OxyR** | - | + | + | + | - |
| **Cold probe** | - | - | + | - | - |
| **Negative probe** | - | - | - | + | - |
| **BSA** | - | - | - | - | + |

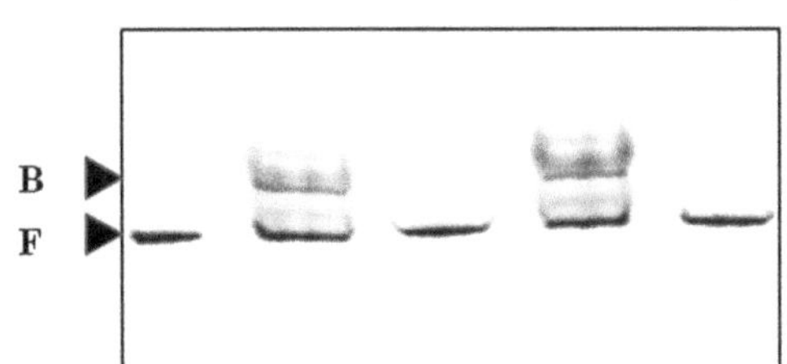


**Fig. 3** EMSA for OxyR's binding to *catB* promoter region. Purified OxyR at 5 nM was incubated with 2 nM probe (FAM-labeled *catB* promoter DNA region (length-312/+78)) at 25 °C for 30 min, and the products were run a native 4 % (W/V) polyacrylamide gel in 0.5 × TBE buffer for about 1.5 h at 100 V. Cold probe (unlabeled *catB* promoter DNA region) at 20 nM was used as specific DNA competitor and negative probe (unlabeled coding region of 16S rRNA gene) at 20 nM is used as nonspecific DNA competitor. Bovine Serum Albumin (BSA) at 5 nM was used as the non-specific protein competitor. The addition of OxyR, probes and BSA was indicated by the '+' sign and the omission was indicated by the '-' sign. B: binding probe, F: free probe

### *CatB* and *OxyR* enhance bacterial viability and CAT activity under $H_2O_2$ stress

To determine the role of *catB* and *oxyR* in $H_2O_2$ resistance, the halo assays of wildtype, Δ*catB*, Δ*oxyR* and their complementary strains Δ*catB* (pBBR-*catB*) and Δ*oxyR* (pBBR-*oxyR*) were performed with the presence of 0.25, 0.5 and 1 M of $H_2O_2$, respectively. The sensitivity of bacteria to $H_2O_2$ was indicated by the zone of inhibition. As shown in Fig. 4a and b, the diameters of inhibitory zone for Δ*catB* and Δ*oxyR* were significantly bigger than that of the wildtype at each concentration of $H_2O_2$, while no differences were observed between wildtype and the complementary strains. To further test the $H_2O_2$ sensitivity, the growth ability of PXO99[A], Δ*catB* and Δ*oxyR* and their complementary strains Δ*catB* (pBBR-*catB*) and Δ*oxyR* (pBBR-*oxyR*) in M210 with the presence of 0, 0.25, 0.5 and 1 mM of $H_2O_2$, respectively, were detected. In the absence of $H_2O_2$, there was no significant difference in growth rate and bacterial population at 12 and 24 h among these strains (Fig. 4c). Compared with the wild type, Δ*catB* and Δ*oxyR* showed more sensitivities at the concentration of 0.25 and 0.5 mM of $H_2O_2$ in M210, and their complementary strains were restored to the wild-type levels (Fig. 4c). Moreover, the growth rates of these strains were significantly inhibited and the bacterial numbers were dramatically decreased in the presence of 1 mM of $H_2O_2$ (Fig. 4c). This demonstrates the

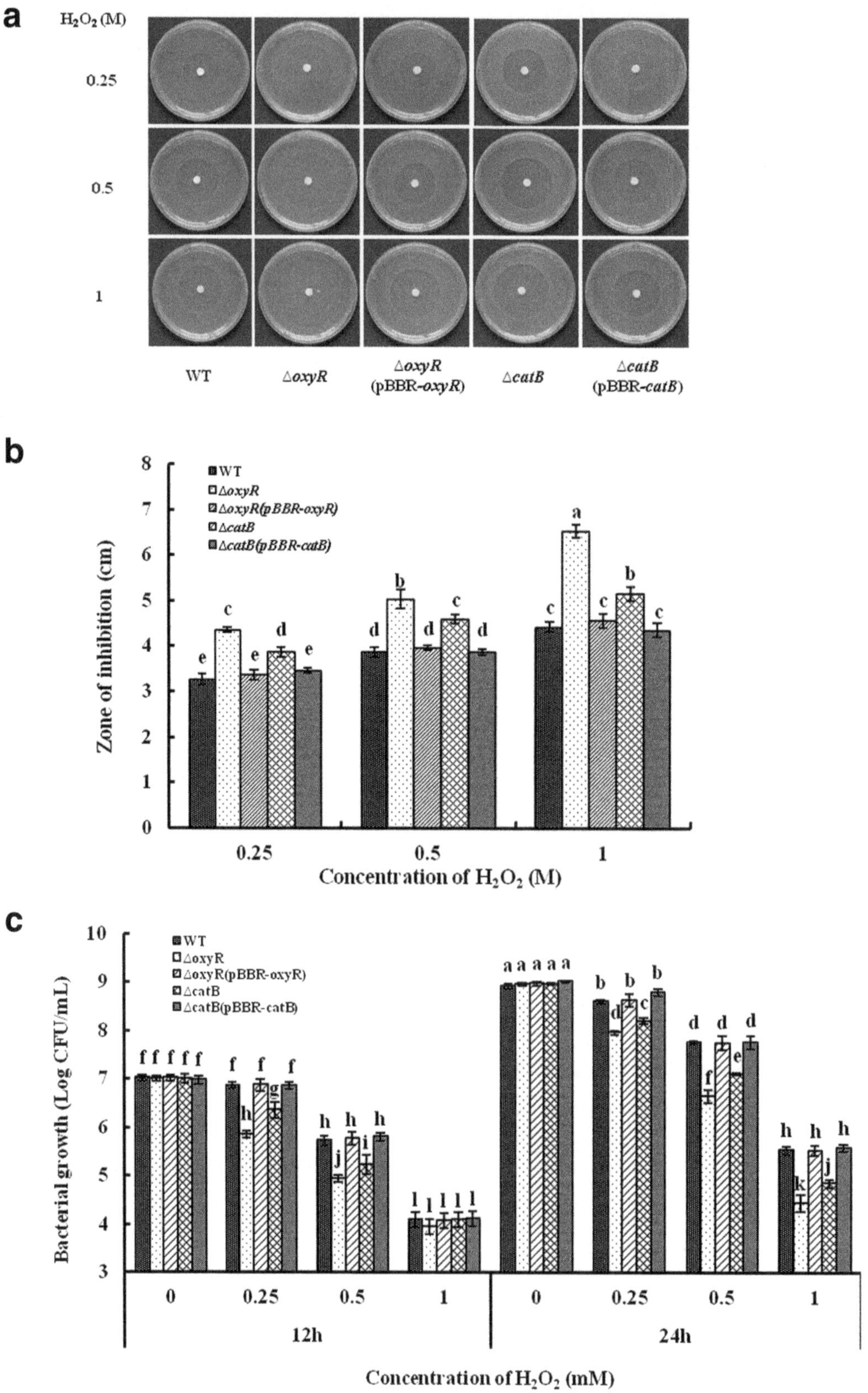


**Fig. 4** (See legend on next page.)

(See figure on previous page.)
**Fig. 4** Assays for $H_2O_2$ resistance of *Xanthomonas oryzae* pv. *oryzae* strains. **a** Disk diffusion assays. Wildtype, Δ*catB*, Δ*oxyR*, Δ*catB*(pBBR-*catB*) and Δ*oxyR*(pBBR-*oxyR*) strains at $OD_{600}$ of 1.0 were mixed with PSA medium at 1 : 100 v/v and disks saturated with different concentrations (0.25, 0.5 and 1 M) of $H_2O_2$ were placed on the central of plates. These plates were incubated at 28 °C for 72 h and the $H_2O_2$ inhibition zones were observed. **b** Diameters of the $H_2O_2$ inhibition zones. **c** $H_2O_2$ sensitivity assays. Wildtype, Δ*catB*, Δ*oxyR*, Δ*catB*(pBBR-*catB*) and Δ*oxyR*(pBBR-*oxyR*) strains at $OD_{600}$ of 1.0 were mixed with fresh M210 liquid medium at 1: 1000 v/v, and the $H_2O_2$ were added to the final concentration at 0, 0.25, 0.5 and 1 mM, respectively. The mixtures were incubated at 28 °C with 200 rpm, and the bacterial population were detected at 12 and 24 h. These experiments repeat three times, independently. Error bars represent standard derivations, and different letters above the bars denote statistically significant differences ($P < 0.05$, Student's *t* test)

essential role of CatB and OxyR in protecting bacterial viability under $H_2O_2$ stress in *Xoo*.

To further unveil the function of *catB* and *oxyR* in $H_2O_2$ degradation, we comparatively measured the CAT activities of PXO99[A], Δ*catB*, Δ*oxyR*, Δ*catB* (pBBR-*catB*) and Δ*oxyR* (pBBR-*oxyR*). The catalase activities of Δ*catB* and Δ*oxyR* were significantly decreased compared to that of PXO99[A], but were restored in their complementary strains Δ*catB* (pBBR-*catB*) and Δ*oxyR* (pBBR-*oxyR*) (Fig. 5a). In addition, the $H_2O_2$ accumulations of these strains were measured. Compared with wildtype, the $H_2O_2$ concentrations were significantly increased in Δ*catB* and Δ*oxyR*, but were restored in their complementary strains (Fig. 5b). Our results suggested that CatB and OxyR contributed to CAT activity for $H_2O_2$ detoxification in *Xoo*.

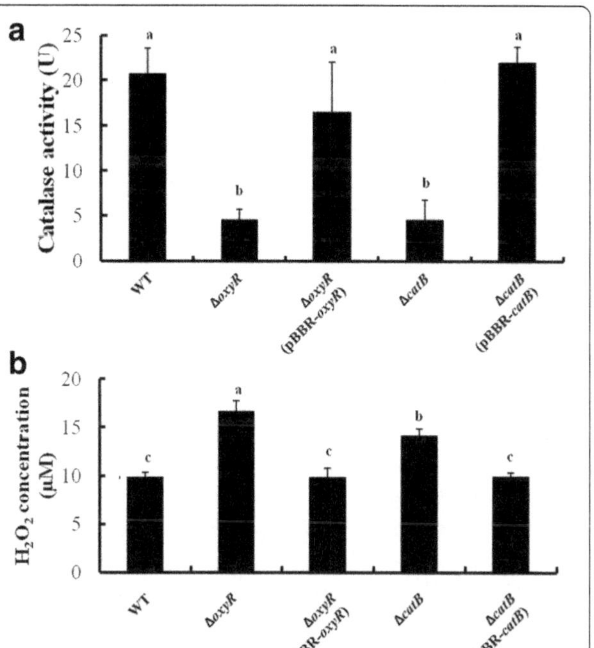


**Fig. 5** Assays for catalase activity and $H_2O_2$ accumulation of *Xanthomonas oryzae* pv. *oryzae* strains. **a** Catalase activity assays. The cells of wildtype, Δ*catB*, Δ*oxyR*, Δ*catB*(pBBR-*catB*) and Δ*oxyR*(pBBR-*oxyR*) were disrupted by sonication, and the cell extracts were separated by centrifugation at 12,000 g. Catalase activities associated with cell extracts were assessed by spectrophotometric assay. One unit (U) is defined as the amount of activity required to decompose 1 μmol of $H_2O_2$ in one minute. **b** $H_2O_2$ accumulation assays. The $H_2O_2$ concentrations of *Xoo* strains were measured as described in Methods, and the account of $H_2O_2$ was accounted by the standard curve. Error bars represent standard derivations from three replicates, and different letters above the bars denote statistically significant differences ($P < 0.05$, Student's *t* test)

### ΔcatB and ΔoxyR elicit early $H_2O_2$ production in rice

In response to pathogenic infection, host plants produce ROS to defend themselves [32]. To evaluate their abilities in the induction of ROS, $H_2O_2$ production were detected in rice leaves at 12 and 24 h post-inoculation of wildtype, Δ*catB*, Δ*oxyR*, Δ*catB*(pBBR-*catB*), and Δ*oxyR*(pBBR-*oxyR*) strains by using 3, 3′-diaminobenzidine (DAB) staining. Similar dark spots formed by DAB in the presence of $H_2O_2$ were observed in all bacterium-inoculated areas at 12 h, whereas no $H_2O_2$ accumulation was found in the $H_2O$ control (Additional file 4: Figure S3). The same results were observed for all treatments at 24 h post-inoculation. Thus, these findings indicate that wildtype, Δ*catB* and Δ*oxyR* induce $H_2O_2$ accumulation in rice at early stage of infection.

### ΔcatB and ΔoxyR showed attenuated virulence and bacterial growth in rice

To determine the functions of CatB and OxyR in virulence, pathogenicity tests for wildtype, Δ*catB*, Δ*oxyR*, Δ*catB* (pBBR-*catB*) and Δ*oxyR* (pBBR-*oxyR*) on susceptible rice plants were performed. The bacterial cells were inoculated onto the tip of rice leaves by leaf-clipping. The bacterial blight symptoms of rice were scored 14 days after bacterial inoculation. Compared with PXO99[A], Δ*catB* and Δ*oxyR* displayed reduced disease severity with shorter lesion lengths (Fig. 6a and b), and decreased bacterial growth in rice leave tissues, while Δ*catB*(pBBR-*catB*) and Δ*oxyR*(pBBR-*oxyR*) showed disease phenotypes at near-wildtype levels (Fig. 6c). These observations demonstrated that the CatB and OxyR were required for the full virulence and *in planta* growth of *Xoo* in rice.

## Discussion

Pathogenic bacteria successfully survive in the environment and infect plant tissues in part by depending on

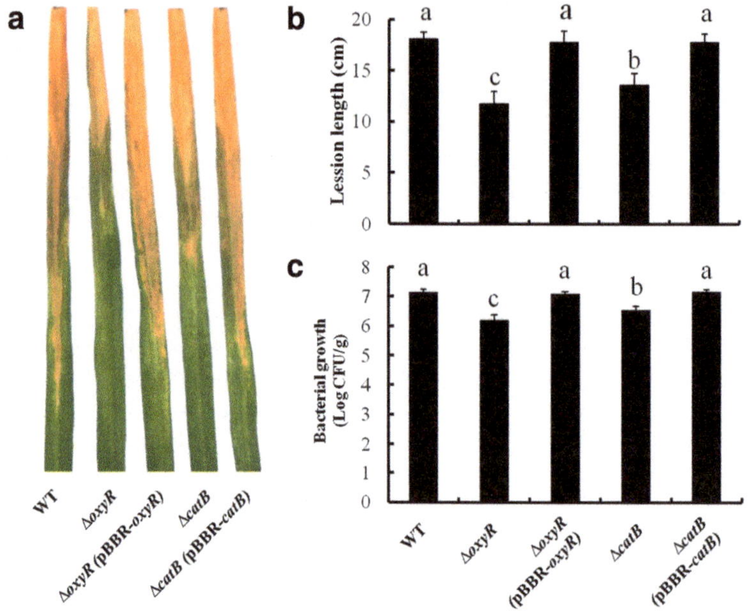


**Fig. 6** Virulence test of *Xanthomonas oryzae* pv. *oryzae* strains in rice. **a** Wildtype, Δ*catB*, Δ*oxyR*, Δ*catB*(pBBR-*catB*) and Δ*oxyR*(pBBR-*oxyR*) strains were inoculated on the rice leaves (6 weeks old) by using the leaf-clipping method. The disease symptoms were observed at 14 days post-inoculation. **b** The lesion lengths were recorded from 10 inoculated leaves for every strain. **c** Bacterial numbers in the top 20 cm of each lesion leaf were scored. Data represent the mean and standard deviations of three independent experiments, and different letters above the bars denote statistically significant differences ($P < 0.05$, Student's *t* test)

their abilities to counteract oxidative stresses including $H_2O_2$, which can penetrate through bacterial membranes to affect a variety of cellular processes [1]. Since *Xoo* is constantly exposed to $H_2O_2$, its catalases are very likely to be critical to the $H_2O_2$ detoxification process, whether it is produced endogenously through normal aerobic respiration, or the oxidative burst of rice plant cells during plant-pathogen interactions. In the current study, we revealed that CatB, working with OxyR, played a crucial role in $H_2O_2$ resistance and bacterial virulence in *Xoo* by using bioinformatics and genetic analysis approaches. To our immediate knowledge, CatB is the first catalase functionally characterized to facilitate pathogenesis via $H_2O_2$ detoxification in *Xoo*.

Bacterial catalases have been reported to be involved in the $H_2O_2$-degradation pathway and increased tolerance to oxidative stress [13]. *Xoo* depends on a diverse repertoire of antioxidative enzymes to detoxify $H_2O_2$ for its in vitro growth and survival under different $H_2O_2$ stress conditions [26]. Transcription of three catalase genes *catB*, *katE* and *srpA* were strongly induced by exogenous $H_2O_2$ and during the bacterial interaction with rice suspension-cultured cells in *Xoo* [30]. Gene deletion of *katE* remarkably reduced bacterial growth in vitro and diseases leaf lesion of rice, but did not attenuate CAT activity and bacterial resistance to exogenous $H_2O_2$ [31]. The physiological role of SrpA in *Xoo* needs to be studied in the future. In this study, we identified a catalase gene encoded CatB in *Xoo*, which shared high homology with two validated catalases KatE and KatA from *X. axonopodis* pv. *citri* and *Bacillus subtilis*, respectively (Fig. 1). Deletion of *catB* dramatically attenuated exogenous $H_2O_2$ resistance and CAT activity (Figs. 4 and 5), but did not affect the bacterial growth in vitro (Additional file 2: Figure S1). These results suggested that CatB was one of the key players participating in the $H_2O_2$ degradation pathway in *Xoo*. Furthermore, the homologues of CatB were widely found in other important plant-pathogenic *Xanthomonas* species and pathovars (Additional file 1: Table S1), suggesting that there might be a conserved and functional $H_2O_2$ resistance mechanism by CatB and its homologs in *Xanthomonas* species.

OxyR was identified as the primary $H_2O_2$ sensor responsible for $H_2O_2$ resistance in gram-negative bacteria [8, 33]. Previous studies showed that the cysteine residue (C199) of OxyR in *E. coli* was directly oxidized by $H_2O_2$ to a sulfenic acid, and generated a disulfide bond [34]. This oxidation process activated OxyR as a transcription factor. In *Pseudomonas aeruginosa*, the expression of OxyR was dramatically increased by $H_2O_2$, and mutation of three cysteine sites (C199S, C208S and C296S) for OxyR displayed hypersensitivity to $H_2O_2$ [35]. In this study, the transcription level of *oxyR* was significantly induced by exogenouse $H_2O_2$ (Fig. 2b), implying that OxyR might have similar function to sense $H_2O_2$, thereby activating the downstream target gene expression in *Xoo*.

As a transcriptional activator in response to $H_2O_2$, OxyR has the ability to directly regulate the transcription of target genes through binding to the upstream DNA region of their promoter [36]. The binding sites of four OxyR regulated genes, *katA*, *dps*, *ftn* and *cydA*, were studied in *Corynebacterium glutamicum* [37]. A 50-bp region protected by the OxyR protein in the promoter of each gene was identified using DNase I footprinting analyses. However, no significant sequence similarity was found among the four OxyR-binding sites [37], suggesting the binding sites may vary in different promoters. In this study, we demonstrated that OxyR directly bound to the DNA fragment between −312 and + 78 bp with respect to the translation start site of *catB* in *Xoo* using EMSA (Fig. 3). In addition, the promoter activity of *catB* was significantly reduced in *oxyR* deletion mutant (Fig. 2a). These results suggested that OxyR positively regulated the transcription of *catB* by binding to its promoter region. The OxyR binding sites in the *catB* promoter region will be further studied in the future.

OxyR plays a role during the peroxide stress response by regulating the transcription of catalase genes has been reported in various bacteria, but the regulatory mechanisms are different [5, 16]. For example, OxyR functioned as a positive regulator to activate the expression of catalase genes, and the *oxyR* deletion mutant was hypersensitive to hydrogen peroxide in *Escherichia coli* and *Salmonella typhimurium* [38]. In contrast, OxyR in *Corynebacterium diphtheria* repressed catalase production, and the *oxyR* deletion mutant displayed increased tolerance to $H_2O_2$ [39]. In this study, the promoter activity of *catB* in wildtype *Xoo* strain was significantly enhanced by $H_2O_2$, but dramatically decreased in Δ*oxyR* (Fig. 2a). In addition, deletion of *oxyR* or *catB* significantly reduced the tolerance to $H_2O_2$ and catalase activity (Figs. 4 and 5). These results showed that OxyR acts as a positive regulator to mediate $H_2O_2$ detoxification via controlling *catB* gene transcription in *Xoo*.

Our earlier studies have revealed that OxyR mediated endogenous $H_2O_2$-degradation by regulating the alkyl hydroperoxide reductase genes *ahpC/ahpF* in *Xoo* [26, 27]. Moreover, the transcriptional levels of catalase gene, *katE* and *srpA*, were down-regulated in Δ*oxyR* [26]. These results showed that several genes related with $H_2O_2$ detoxification were regulated by OxyR in *Xoo*. In this study, we showed that Δ*oxyR*, in comparison with Δ*catB*, was more hypersensitive to $H_2O_2$ (Fig. 4), and contained higher level of endogenous $H_2O_2$ (Fig. 5b), suggesting that there might be other genes besides *catB* that are regulated by OxyR in response to $H_2O_2$ in *Xoo*. Interestingly, no significant difference in catalase activity in Δ*catB* and Δ*oxyR* was found (Fig. 5a), implying that the CatB might be one of the major catalases in *Xoo*.

As an important host innate immune response, $H_2O_2$ is generated at the attempted invasion site in plant cells during interactions with potential pathogens and increased host disease resistance [2]. Previoulsy, we have shown that $H_2O_2$ accumulation was induced in rice by *Xoo* infection. Deletion of an alkyl hydroperoxide reductase gene *ahpC* in *Xoo* significantly decreased the endogenous $H_2O_2$ accumulation. However, the $H_2O_2$ scavenging activity was increased with a unknown compensatory mechanism, which led to a lower level of $H_2O_2$ accumulation in the *ahpC* mutant than the wild type during their interactions with rice host plants [28]. These results indicated that AhpC plays the role in the detoxification of endogenous $H_2O_2$ in *Xoo*. In this study, compared with wildtype, Δ*catB* and Δ*oxyR* showed more hypersensitivity to exogenous $H_2O_2$ (Fig. 4) and higher concentration of endogenous $H_2O_2$ (Fig. 5a), indicating that CatB and OxyR in *Xoo* might be responsible for detoxification of both endogenous and exogenous $H_2O_2$. In addition, $H_2O_2$ accumulations were also observed in rice plants infected by *Xoo* strains (Additional file 4: Figure S3), suggesting that the ability of *Xoo* strains to induce and/or degrade $H_2O_2$ is a key determinant of outcome in the *Xoo*-rice interaction. To further study the differences in $H_2O_2$ accumulation induced by wildtype, Δ*catB* and Δ*oxyR*, quantitative analysis of $H_2O_2$ production in rice and the ability of $H_2O_2$ detoxification in *Xoo* strains would be required.

During evolution, catalase activity has become inducible to help pathogen colonize its host and cause disease by detoxifying $H_2O_2$, implicating that there is a close relationship between bacterial ability to survive $H_2O_2$ stress and its virulence [3, 16]. In *X. campestris*, a catalase KatG is required for virulence in a host plant by providing protection against low levels of $H_2O_2$ [14]. Here, we showed the disease severity and bacterial population were dramatically reduced in rice leaves inoculated with Δ*catB* and Δ*oxyR* (Fig. 6), suggesting both Δ*catB* and Δ*oxyR* were significantly attenuated in detoxification of $H_2O_2$ in rice leaves, thereby resulting in the reduced pathogenicity. Based on these observations, combined with our earlier result that *catB* and *oxyR* were transcriptionally induced by $H_2O_2$ produced during interaction with rice suspension-cultured cells and also in real time of infection of rice [26, 30], we propose that both CatB and OxyR are required for full virulence and *in planta* growth of *Xoo* in rice by detoxification of $H_2O_2$. Moreover, shorter lesion lengths and fewer bacterial numbers in rice caused by Δ*oxyR* were observed than that by Δ*catB* (Fig. 6), indicating that there are other virulence factors regulated by OxyR in *Xoo*-rice interactions. This result is consistent with the previous studies that the OxyR regulon comprised of multiple genes involved in $H_2O_2$

detoxification, heme biosynthesis, reductant supply, thiol-disulfide isomerization, Fe-S center repair, iron binding and so on [5]. Accordingly, our identification of CatB regulated by OxyR in *Xoo* highlights the requirement of a functional $H_2O_2$-detoxification for bacterial pathogenesis in rice.

## Conclusions

The *Xoo catB* gene is transcriptionally up-regulated by OxyR in response to $H_2O_2$ either exogenously-applied or generated in rice upon bacterial infection. CatB functions as an active catalase to detoxify $H_2O_2$ and is also required for the full virulence. Thus, OxyR-regulated catalase CatB promotes the bacterial pathogenesis in rice through $H_2O_2$ detoxification.

## Methods

### Bacterial strains and growth conditions

The bacterial strains and plasmids used in this study are listed in Table 1. *E. coli* strains were grown in Luria-Bertani medium at 37 °C. *Xoo* wildtype strain PXO99$^A$ and derived mutants were cultured at 28 °C on peptone sucrose agar (PSA) [40] medium or M210 [41] liquid medium with appropriate antibiotics. The antibiotics used were ampicillin (Ap), kanamycin (Km), spectinomycin (Sp) and gentamycin (Gm) at concentrations of 100, 50, 50, and 50 μg mL$^{-1}$, respectively.

### Bioinformatics analysis of CatB

The domain organization of CatB was analyzed using online software available at the SMART Website (http://smart.embl-heidelberg.de/). The amino acid sequences of active CATs, which represent the conserved catalase-domain were obtained from the National Center for Biotechnology Information (NCBI) website. BLASTP was using for searching the homology in *Xanthomonas* species. Relevant sequence alignment was performed using the DNAMAN software (Lynnon Biosoft, San Ramon, USA).

**Table 1** Bacterial strains and plasmids used in this study

| Strain or plasmid | Relevant characteristics[a] | Source or Reference |
|---|---|---|
| *Escherichia coli* | | |
| DH5α | supE44 ΔlacU169(Φ80lacZΔM15) hsdR17 recA1 endA1 gyrA96 thi-1 relA1 | Hanahan 1983 [47] |
| BL21 | For protein expression | Novagen |
| *Xanthomonas oryzae* pv. *oryzae* | | |
| PXO99$^A$ | Wildtype strain, Philippine race 6 | Lab collection |
| Δ*catB* | *catB* gene deletion mutant derived from PXO99$^A$, Gm$^r$ | This study |
| Δ*catB*(pBBR-*catB*) | Complementary bacterium strain of Δ*catB*, Ap$^r$ | This study |
| Δ*oxyR* | *oxyR* gene deletion mutant derived from PXO99$^A$, Gm$^r$ | Our lab |
| Δ*oxyR*(pBBR-*oxyR*) | Complementary bacterium strain of Δ*oxyR*, Ap$^r$ | Our lab |
| Plasmid | | |
| pMD18-T | Cloning vector, Ap$^r$ | TaKaRa, Tokyo |
| pET-28a | Expression vector to generate a N-terminal His$_6$ tag, Km$^r$ | Haigene |
| pK18mobsacB | Suicidal vector carrying *sacB* gene for mutagenesis, Gm$^r$ | Schafer et al., 1994 [44] |
| pBBR1MCS-4 | Broad-host range expression vector, Ap$^r$ | Kovach et al., 1995 [48] |
| pHM1 | Broad-host range expression vector, Sp$^r$ | Hopkins et al., 1992 [49] |
| pHT304BZ | Promoterless *lacZ* vector, Ap$^r$ | Lereclus et al., 1996 [50] |
| pMDcatB | pMD18-T derivative carrying the full length of *catB*, Ap$^r$ | This study |
| pMDcatBr | pMD18-T derivative carrying the right fragment of *catB*, Ap$^r$ | This study |
| pMDcatBl | pMD18-T derivative carrying the left fragment of *catB*, Ap$^r$ | This study |
| pMDoxyR | pMD18-T derivative carrying the full length of *oxyR*, Ap$^r$ | This study |
| pKcatB | pK18mobsacB derivative carrying the full length of *catB*, Gm$^r$ | This study |
| pEToxyR | pET-28a derivative carrying the full length of *oxyR*, Km$^r$ | This study |
| pBBR-*catB* | pBBR1MCS-4 derivative carrying the full length of *catB*, Ap$^r$ | This study |
| pHTpB | pHT304BZ derivative carrying the promoter region of *catB*, Ap$^r$ | This study |
| pH-*lacZ* | pHM1 derivative carrying the promoterless *lacZ*, Sp$^r$ | This study |
| pH-*catBp-lacZ* | pHM1 derivative carrying the promoter region of *catB* and promoterless *lacZ*, Sp$^r$ | This study |

[a]Ap$^r$,Km$^r$,Sp$^r$,and Gm $^r$indicate resistant to ampicillin, kanamycin, spectinomycin and gentamicin, respectively

### Expression and purification of OxyR

The full length (942 bp) of *oxyR* (gene ID: PXO_04591) was PCR-amplified using the primer pairs oxyRF/oxyRR (Additional file 5: Table S2). The PCR fragment was gel purified and cloned to the middle vector pMD18-T (Takara, Tokyo, Japan), resulting in construct pMDoxyR, which was verified by DNA sequencing (Beijing Genomics Institute, Beijing, China). The coding region of *oxyR* was digested from pMDoxyR using BamHI and HindIII, and then cloned into expression victor pET28a, resulting in pEToxyR. The recombinant plasmid was transformed into *E. coli* BL21 strains for protein expression. The OxyR purification was performed as previously described [42]. OxyR was induced by addition of isopropyl-thio-galactopyranoside at a final concentration of 0.1 mM and the bacterial culture was then incubated at 20 °C for 6 h. Bacterial cells were chilled at 4 °C and collected by centrifugation. The supernatant containing the soluble protein was collected and mixed with pre-equilibrated Ni2_resin (GE Healthcare, Piscataway, NJ, USA) for 2 h at 4 °C, then placed into a column and extensively washed with buffer containing 1 × PBS and 20 mM imidazole. OxyR was subsequently eluted with buffer containing 100 mM imidazole. The purified OxyR protein was obtained through the gradient dialysis of 1xPBS buffer. The purified OxyR protein was analyzed by sodium dodecyl sulfate polyacrylamide gel electrophoresis.

### Electrophoretic mobility shift assay (EMSA)

The *catB* promoter DNA region (–312 to + 78, the nucleotide site upstream or downstream of translation start (+1)) was amplified by PCR using 5′ ends FAM labeled primers catBpF/catBpR (Additional file 5: Table S2). DNA binding was performed in a 10 μL reaction volume containing EMSA/Gel-Shift Binding Buffer (Beyotime, Shanghai, China), 2 nM labeled DNA fragment and 5 nM His-OxyR protein. Three controls were included in each EMSA experiment: (I) cold probe as specific DNA competitor (unlabeled catB promoter DNA region, 20 nM), (II) negative probe as nonspecific DNA competitor (unlabeled coding region of 16S rRNA gene, 20 nM), and (III) Bovine Serum Albumin (BSA, 5 nM) as nonspecific protein competitor. After incubation at 25 °C for 30 min, the products were loaded onto a native 4 % (W/V) polyacrylamide gel and electrophoresed in 0.5 × TBE buffer for about 1.5 h at 100 V. The fluorescence of samples was detected by Typhoon FLA-5100 (Fuji film, Tokyo, Japan) at 488 nm.

### Construction of *catB* promoter fusion and assay for β-galactosidase

A promoter DNA region (–312 to + 78) of *catB* was obtained by PCR with the PrimeSTAR® Max DNA Polymerase (Takara, Tokyo, Japan) and the primers catBpF and catBpR (Additional file 5: Table S2), while the PXO99$^A$ genome DNA as a template. The PCR fragment was cloned directionally into the HindIII and BamHI sites of plasmid pHT304BZ that harbors an ampicillin resistance gene (*Ap*$^r$) and a promoterless *lacZ* reporter gene. The resulting clone pHTpB was verified by DNA sequencing (Beijing Genomics Institute, Beijing, China). Next, pHTpB was treated with HindIII and KpnI, and the fragment containing *catB* promoter region and the promoterless *lacZ* reporter gene was purified with TIANgel Midi Purification Kit (Tiangen, Beijing, China), and then cloned into pHM1, resulting in plasmid pH-*catBp-lacZ*. Meanwhile, the fragment of promoterless *lacZ* was obtained from pHT304BZ using BamHI and KpnI and cloned into pHM1, resulting plasmid pH-*lacZ*. These recombinant plasmids were introduced into *Xoo* strians. The plasmid pH-*lacZ* was used as a negative control. *Xoo* strains transformed with the recombinant plasmid were grown in M210 at 28 °C till an optical density ($OD_{600}$) of 1.0 and exposed to 3 mM $H_2O_2$ or sterilized deionized and distilled water ($ddH_2O$) for 0.5 h, then harvested by centrifugation at 7,000 g for 5 min. The β-galactosidase activity in the cellular extracts was measured using the β-Galactosidase Enzyme Assay System (Promega, Wisconsin, USA). All assays were performed with three biological replicates and three repeats.

### RNA isolation and quantitative real-time PCR (qRT-PCR) analysis

The transcriptional levels of *catB* and *oxyR* at $H_2O_2$ treatment were detected as described previously with some modifications [42]. Bacterial cells were grown in M210 at 28 °C till an $OD_{600}$ of 0.8 and exposed to 3 mM $H_2O_2$ for 0.5 h, then harvested for analysis of gene expression. Total RNA was extracted with TRIzol reagent (Invitrogen, Carlsbad, CA, USA) and treated with DNase. First-stand cDNA was synthesized from total RNA using the Superscript III reverse transcriptase (Invitrogen, Carlsbad, CA, USA). RT-qPCR was performed using SYBR Green detection reagents (Quanta Biosciences, Carlsbad, CA, USA) in Applied Biosystem's 7500 Sequence Detection System (Applied Biosystems, Foster City, CA, USA) with the primers (catBqF/catBqR, oxyRqF/oxyRqR), and *gyrB* was used as a reference gene (Additional file 5: Table S2). The relative expression ratio was calculated using $2^{-\Delta\Delta Ct}$ method [43]. All experiments were performed in three biological replicates and triplicate PCR.

### Cloning, deletion and complementation of *catB*

The full length (1,524 bp) of *catB* including ribosome binding site (gene ID: PXO_02830) was amplified by polymerase chain reaction (PCR) using the primer pairs

P1/P2 (Additional file 5: Table S2). A right fragment (489 bp) and a left one (582 bp) were amplified by PCR using the primer pairs catBrF/catBrR and catBlF/catBlR (Additional file 5: Table S2), respectively. The PCR fragments were gel purified and cloned to the middle vector pMD18-T (Takara, Tokyo, Japan), resulting in constructs pMDcatB, pMDcatBr, and pMDcatBl, which were verified by DNA sequencing (Beijing Genomics Institute, Beijing, China).

The gene deletion mutant Δ*catB* derived from $PXO99^A$ was constructed by the homologous recombination as described previously by using the suicide vector pK18mob-SacB [44]. The vector pMDcatBl with the left fragment and the vector pMDcatBr with the right fragment were digested with corresponding restriction enzymes and ligated to pK18-mobsacB. A gentamicin resistance gene ($Gm^r$) at 855 bp was then inserted into the intermediate region between left and right fragment carried by pK18mobsacB, resulting in plasmids pKcatB, and then introduced into $PXO99^A$ by electroporation. The deletion mutants were screened on PSA plates containing gentamicin and 10 % sucrose. For the complementation experiment, the vector pMDcatB with the *catB* gene including ribosome binding site was digested by enzymes and cloned into pBBR1MCS-4, generating pBBR-*catB*, and then transferred into Δ*catB* by electroporation and screened on PSA plates containing ampicillin.

### Growth curve assay

The bacterial growth assay was performed as previously described [45]. In brief, *Xoo* wildtype, *catB* deletion mutant and complementary strain were grown in M210 liquid medium overnight at 28 °C, then these strains were diluted in M210 medium to a final cell density ($OD_{600}$ = 0.01). The diluted cells were cultured at 28 °C with 200 rpm, and bacterial population was measured after every 6 h. For bacterial population assay, the bacterial cells were spread onto PSA plats after optional diluted, and cultured at 28 °C for 3 days, the bacterial colonies then were counted. These experiments were repeated three times, independently.

### $H_2O_2$ resistance assay

Bacterial strains were cultured in M210 liquid medium using a shaker (200 rpm, 28 °C) until an $OD_{600}$ = 1. For $H_2O_2$ disc diffusion assays, 200 μL of each culture was taken and mixed with 20 mL soft PSA medium containing 0.5 % sodium carboxymethyl cellulose (Sigma-Aldrich, Louis, MQ, USA), then poured out the mixture quickly to PSA medium containing 1 % carboxymethyl cellulose, and ensured the mixture cover the whole PSA plate fully and smoothly. After 5 min, when the mixture became dry, put a sterilized dry Whatman 3MM filter discs of 8 mm diameter on the central of PSA plate. 10 μL of $H_2O_2$ at different concentrations (1, 0.5 and 0.25 M) was spotted onto the discs, respectively. The treated PSA plates were cultured in incubator at 28 °C, and the diameters of $H_2O_2$ inhibition zone were measured after 72 h. For $H_2O_2$ sensitivity assays, 1 mL bacterial cultures were taken and mixed with 100 mL M210 liquid medium, and $H_2O_2$ was then added to the cell suspensions to different concentrations (0, 0.25, 0.5 and 1 mM). The mixtures were cultured at 28 °C on 200 rpm, and the bacterial population was measured at 12 and 24 h, respectively. These experiments were repeated three times with three replicates.

### Catalase activity assay

The analysis of catalase activity was performed as described previously [14, 31]. The bacterial culture conditions were the same as described above. The bacterial concentration of each strain is $OD_{600}$ = 1. The bacterial cells were chilled at 4 °C, collected by concentration at 6,000 g, and then re-suspended in $ddH_2O$. Sonication was followed until the bacterial liquid visible clearly. The cell extracts were separated by centrifuge with rate 12,000 g for 30 min, and the most upper layer liquid which contained the protein was transferred to a new tube. 100 μL protein was taken and mixed with 1 mL $ddH_2O$, and the optical density of this mixture was measured at 240 nm both before and after adding $H_2O_2$ to the final concentration at 10 mM. The catalase activity was calculated by an extinction coefficient of 43.6 $M^{-1}$ $cm^{-1}$ at 240 nm. One unit of catalase activity was defined as the amount of activity required to decompose 1 μmol of $H_2O_2$ per minute under the assay conditions. The experiments were repeated three times with three replicates.

### $H_2O_2$ detection

The $H_2O_2$ accumulation in *Xoo* strains were detected as previously described. Briefly, *Xoo* strains were grown in M210 liquid medium using a shaker (200 rpm, 28 °C) until an $OD_{600}$ of 1, and harvested by centrifugation at 7,000 g for 5 min. The bacterial cells were re-suspended with 50 mM of potassium phosphate (pH 7.8), and the supernatants were collected by centrifugation again. To measure $H_2O_2$, 0.45 mL of supernatant was mixed with 0.25 mL of 200 μM Amplex red (Sigma-Aldrich, Louis, MQ, USA) and 0.25 mL of 0.02 mg/mL horseradish peroxidase (Sigma-Aldrich, Louis, MQ, USA). The amount of $OD_{610}$ was then measured and converted to $H_2O_2$ concentration using a curve obtained from standard samples. The experiment was repeated three times.

### Histochemical detection of $H_2O_2$ in rice leaves

The cultured bacteria re-suspended in sterilized $ddH_2O$ at an $OD_{600}$ of 1.0 were prepared as described above, and cells were infiltrated into rice cultivar (*Oryza sativa* L. *subsp. japonica*) leaves grown for two weeks using a needleless syringe. The $H_2O_2$ was detected by DAB staining as previously reported [46]. Briefly, leaf sections (3–5 mm) at 12 and 24 h post-inoculation were cut and placed in water with 0.01 % Triton-X-100 and DAB at 1 mg $mL^{-1}$, then this solution was infiltrated with low vacuum pressure for 10 min and the leaves were incubated for 8 h at room temperature. Finally, leaves were boiled with 95 % ethanol for 10 min and then rinsed with water, and presence of $H_2O_2$ was visualized as reddish brown colored dark spots by a light microscope (Leica, Heidelberg, Germany).

### Pathogenicity test

The bacterial cells were grown in M210 medium for 24 h at 28 °C as described above, and re-suspended in sterilized $ddH_2O$ at an $OD_{600}$ of 1.0. The susceptible rice cultivar (*Oryza sativa* L. *subsp. japonica*) plants grown for 6 weeks were used for bacterial inoculation with the leaf-clipping method [42]. At least ten leaves were inoculated for each strain, and the experiment was repeated three times. The disease lesion length was recorded 14 d after bacterial inoculation. The top 20 cm of each lesion leaf was cut down and weighted, and then grinded by the sterilized pestle. The extract was gradually diluted by $ddH_2O$, and then poured out to PSA medium and cultured in incubator with 28 °C for 3 d after it dry. The bacterial numbers in each plate were accounted, and the bacterial amount in the lesion leaf was calculated.

### Data analysis

All analysis were conducted using SPSS 14.0 (SPSS Inc., Chicago, IL, USA). The hypothesis test of percentages (*t*-test, $P = 0.05$) was used to determine significant differences in the assays for bacterial $H_2O_2$ resistance, catalase activity, endogenous $H_2O_2$ concentration, pathogenicity, gene expression and in vivo growth.

## Additional files

**Additional file 1: Table S1.** The homologs of *catB* in plant pathogenic Xanthomonas species.

**Additional file 2: Figure S1.** In vitro growth of *Xanthomonas oryzae* pv. *oryzae* wildtype, Δ*catB* and Δ*catB(pBBR-catB)* strains. The bacterial strains were cultured in M210 liquid medium at 28 °C for 200 rpm, and bacterial population were determined by $OD_{600}$ density measurement at the time points indicated. Data represent the mean and standard deviations of three independent experiments.

**Additional file 3: Figure S2.** Coomassie blue staining of recombinant OxyR protein purified from *E. coli* BL21 strains. About 10 μg of protein was loaded in the lane.

**Additional file 4: Figure S3.** Histochemical detection of $H_2O_2$ in situ by DAB staining in rice leaves. Rice plants were grown for two weeks and inoculated with $ddH_2O$ control (A), wildtype (B), Δ*catB* (C), Δ*catB*(pBBR-*catB*) (D), Δ*oxyR* (E), and Δ*oxyR*(pBBR-*oxyR*) (F) by using a needleless syringe. Dark spots represent presence of $H_2O_2$. The experiment repeats three times, independently.

**Additional file 5: Table S2.** The information of primers in this study.

### Abbreviations

Ap: Ampicillin; *Bs*: *Bacillus subtilis*; BSA: Bovine Serum Albumin; CAT: Catalase; CFU: Colony forming units; DAB: 3, 3′-diaminobenzidine; EMSA: Electrophoretic mobility shift assay; Gm: Gentamycin; $H_2O_2$: Hydrogen peroxide; Km: Kanamycin; NCBI: National Center for Biotechnology Information; OD: Optical density; PCR: Polymerase chain reaction; PSA: Peptone sucrose agar; qRT-RCR: Quantitative real-time PCR; ROS: Reactive oxygen species; Sp: Spectinomycin; U: Unit; *Xac*: *Xanthomonas axonopodis* pv. *citri*; *Xoo*: *Xanthomonas oryzae* pv. *oryzae*

### Acknowledgements

Not applicable.

### Funding

This work is supported by the grants from the National Basic Research Program (2011CB100700) and the National High-Technology Research Program (2012AA101504).

### Authors' contributions

CY, NW, MSW and CYH designed the experiments; CY and NW performed the experiments; CY, NW, MSW, FT, HMC, FHY and CYH analyzed the data; MSW, FT, HMC and FHY contributed reagents/materials/analysis tools; CY, NW, FT, XCY, CHY and CYH wrote the manuscript; All authors read and approved the final manuscript.

### Competing interests

The authors declare that they have no competing interests.

### Author details

[1]State Key Laboratory for Biology of Plant Diseases and Insect Pests, Institute of Plant Protection, Chinese Academy of Agricultural Sciences, Beijing 100193, China. [2]Department of Biological Sciences, University of Wisconsin-Milwaukee, Milwaukee, WI 53211, USA.

### References

1. Cabiscol E, Tamarit J, Ros J. Oxidative stress in bacteria and protein damage by reactive oxygen species. Int Microbiol. 2000;3(1):3–8.
2. Ros Barcelo A. Xylem parenchyma cells deliver the $H_2O_2$ necessary for lignification in differentiating xylem vessels. Planta. 2005;220(5):747–56.
3. Miller RA, Britigan BE. Role of oxidants in microbial pathophysiology. Clin Microbiol Rev. 1997;10(1):1–18.
4. Storz G, Tartaglia LA, Farr SB, Ames BN. Bacterial defenses against oxidative stress. Trends Genet. 1990;6(11):363–8.
5. Dubbs JM, Mongkolsuk S. Peroxide-sensing transcriptional regulators in bacteria. J Bacteriol. 2012;194(20):5495–503.
6. Kim SO, Merchant K, Nudelman R, Beyer Jr WF, Keng T, Deangelo J, Hausladen A, Stamler JS. OxyR: a molecular code for redox-related signaling. Cell. 2002;109(3):383–96.
7. Burbank L, Roper MC. OxyR and SoxR modulate the inducible oxidative stress response and are implicated during different stages of infection for the bacterial phytopathogen *Pantoea stewartii* subsp. *stewartii*. Mol Plant Microbe Interact. 2014;27(5):479–90.
8. Zheng M, Aslund F, Storz G. Activation of the OxyR transcription factor by reversible disulfide bond formation. Science. 1998;279(5357):1718–21.
9. Hahn JS, Oh SY, Roe JH. Role of OxyR as a peroxide-sensing positive regulator in *Streptomyces coelicolor* A3(2). J Bacteriol. 2002;184(19):5214–22.

10. Tondo ML, Petrocelli S, Ottado J, Orellano EG. The monofunctional catalase KatE of *Xanthomonas axonopodis* pv. *citri* is required for full virulence in citrus plants. PLoS One. 2010;5(5):e10803.
11. Chelikani P, Fita I, Loewen PC. Diversity of structures and properties among catalases. Cell Mol Life Sci. 2004;61(2):192–208.
12. Switala J, Triggs-Raine BL, Loewen PC. Homology among bacterial catalase genes. Can J Microbiol. 1990;36(10):728–31.
13. Loprasert S, Vattanaviboon P, Praituan W, Chamnongpol S, Mongkolsuk S. Regulation of the oxidative stress protective enzymes, catalase and superoxide dismutase in *Xanthomonas*-a review. Gene. 1996;179(1):33–7.
14. Jittawuttipoka T, Buranajitpakorn S, Vattanaviboon P, Mongkolsuk S. The catalase-peroxidase KatG is required for virulence of *Xanthomonas campestris* pv. *campestris* in a host plant by providing protection against low levels of $H_2O_2$. J Bacteriol. 2009;191(23):7372–7.
15. Tondo ML, Delprato ML, Kraiselburd I, Fernandez Zenoff MV, Farias ME, Orellano EG. KatG, the bifunctional catalase of *Xanthomonas citri* subsp. *citri*, responds to hydrogen peroxide and contributes to epiphytic survival on citrus leaves. PLoS One. 2016;11(3):e0151657.
16. Eason MM, Fan X. The role and regulation of catalase in respiratory tract opportunistic bacterial pathogens. Microb Pathog. 2014;74(1):50–58.
17. Mew TH. Current status and future prospects of research on bacterial blight of rice. Annu Rev Phytopathol. 1987;25:359–82.
18. Mansfield J, Genin S, Magori S, Citovsky V, Sriariyanum M, Ronald P, Dow M, Verdier V, Beer SV, Machado MA, et al. Top 10 plant pathogenic bacteria in molecular plant pathology. Mol Plant Pathol. 2012;13(6):614–29.
19. Liu W, Liu J, Triplett L, Leach JE, Wang GL. Novel insights into rice innate immunity against bacterial and fungal pathogens. Annu Rev Phytopathol. 2014;52:213–41.
20. Nino-Liu DO, Ronald PC, Bogdanove AJ. *Xanthomonas oryzae* pathovars: model pathogens of a model crop. Mol Plant Pathol. 2006;7(5):303–24.
21. Das A, Rangaraj N, Sonti RV. Multiple adhesin-like functions of *Xanthomonas oryzae* pv. *oryzae* are involved in promoting leaf attachment, entry, and virulence on rice. Mol Plant Microbe Interact. 2009;22(1):73–85.
22. White FF, Yang B. Host and pathogen factors controlling the rice-*Xanthomonas oryzae* interaction. Plant Physiol. 2009;150(4):1677–86.
23. Mongkolsuk S, Loprasert S, Vattanaviboon P, Chanvanichayachai C, Chamnongpol S, Supsamran N. Heterologous growth phase- and temperature-dependent expression and $H_2O_2$ toxicity protection of a superoxide-inducible monofunctional catalase gene from *Xanthomonas oryzae* pv. *oryzae*. J Bacteriol. 1996;178(12):3578–84.
24. Mongkolsuk S, Whangsuk W, Vattanaviboon P, Loprasert S, Fuangthong M. A Xanthomonas alkyl hydroperoxide reductase subunit C (ahpC) mutant showed an altered peroxide stress response and complex regulation of the compensatory response of peroxide detoxification enzymes. J Bacteriol. 2000;182(23):6845–9.
25. Grewal RK, Gupta S, Das S. *Xanthomonas oryzae* pv *oryzae* triggers immediate transcriptomic modulations in rice. BMC Genomics. 2012;13:49.
26. Zhang J, Gao J, Wu M, Chen H, He C. Roles of OxyR, a transcriptional regulator of *Xanthomonas oryzae* pv. *oryzae* in regulation of detoxification of hydrogen peroxide. Acta Microbiol Sin. 2009;49(7):874–9.
27. Wang Y, Wu M, Tian F, Chen H, He C. Transcriptional regulation analysis and prokaryotic expression of *ahpC* gene putatively-encoding alkyl hydroperoxide reductase in *Xanthomonas oryzae* pv. *oryzae*. Biotechnol Bull. 2012;12:114–9.
28. Li X, Qiao J, Yang L, Li X, Qiao S, Pang X, Tian F, Chen H, He C. Mutation of alkyl hydroperoxide reductase gene *ahpC* of *Xanthomonas oryzae* pv. *oryzae* affects hydrogen peroxide accumulation during the rice-pathogen interaction. Res Microbiol. 2014;165(8):605–11.
29. Salzberg SL, Sommer DD, Schatz MC, Phillippy AM, Rabinowicz PD, Tsuge S, Furutani A, Ochiai H, Delcher AL, Kelley D, et al. Genome sequence and rapid evolution of the rice pathogen *Xanthomonas oryzae* pv. *oryzae* PXO99$^A$. BMC Genomics. 2008;9:204.
30. Zhou JB, Wu MS, Hu J, He CY. Induction of bacterial catalase-related gene expression by $H_2O_2$ produced during interaction of rice suspension-cultured cells with *Xanthomonas oryzae* pv. *oryzae* or applied exogenously. Acta Phytopathologica Sinica. 2009;39(4):399–404.
31. Wang N, Wu M, Tian F, Chen H, He C. Effects of gene deletion of *katE* encoding putative catalase on resistance to hydrogen peroxide and pathogenicity of *Xanthomonas oryzae* pv. *oryzae*. Acta Phytopathologica Sinica. 2013;43(6):615–21.
32. Zhang J, Zhou JM. Plant immunity triggered by microbial molecular signatures. Mol Plant. 2010;3(5):783–93.
33. Jo I, Chung IY, Bae HW, Kim JS, Song S, Cho YH, Ha NC. Structural details of the OxyR peroxide-sensing mechanism. Proc Natl Acad Sci U S A. 2015; 112(20):6443–8.
34. Imlay JA. Transcription factors that defend bacteria against reactive oxygen species. Annu Rev Microbiol. 2015;15(69):93–108.
35. Bae HW, Cho YH. Mutational analysis of *Pseudomonas aeruginosa* OxyR to define the regions required for peroxide resistance and acute virulence. Res Microbiol. 2012;163(1):55–63.
36. Ni B, Zhang Y, Huang X, Yang R, Zhou D. Transcriptional regulation mechanism of ter operon by OxyR in *Yersinia pestis*. Curr Microbiol. 2014; 69(1):42–6.
37. Teramoto H, Inui M, Yukawa H. OxyR acts as a transcriptional repressor of hydrogen peroxide-inducible antioxidant genes in *Corynebacterium glutamicum* R. FEBS J. 2013;280(14):3298–312.
38. Christman MF, Storz G, Ames BN. OxyR, a positive regulator of hydrogen peroxide-inducible genes in *Escherichia coli* and Salmonella typhimurium, is homologous to a family of bacterial regulatory proteins. Proc Natl Acad Sci U S A. 1989;86(10):3484–8.
39. Kim JS, Holmes RK. Characterization of OxyR as a negative transcriptional regulator that represses catalase production in *Corynebacterium diphtheriae*. PLoS One. 2012;7(3):e31709.
40. Tsuchiya K, Mew TW, Wakimoto S. Bacteriological and pathological characteristics of wild type and induced mutants of *Xanthomonas campestris pv oryzae*. Phytopathology. 1982;72:43–6.
41. Fan S, Tian F, Li J, Hutchins W, Chen H, Yang F, Yuan X, Cui Z, Yang CH, He C. Identification of phenolic compounds that suppress the virulence of Xanthomonas oryzae on rice via the type III secretion system. Mol Plant Pathol 2016. DOI: 10.1111/mpp.12415
42. Yang F, Tian F, Sun L, Chen H, Wu M, Yang CH, He C. A novel two-component system PdeK/PdeR regulates c-di-GMP turnover and virulence of *Xanthomonas oryzae* pv. *oryzae*. Mol Plant Microbe Interact. 2012;25(10):1361–9.
43. Livak KJ, Schmittgen TD. Analysis of relative gene expression data using real-time quantitative PCR and the 2(T)(–Delta Delta C) method. Methods. 2001;25(4):402–8.
44. Schafer A, Tauch A, Jager W, Kalinowski J, Thierbach G, Puhler A. Small mobilizable multi-purpose cloning vectors derived from the *Escherichia coli* plasmids pK18 and pK19: selection of defined deletions in the chromosome of *Corynebacterium glutamicum*. Gene. 1994;145(1):69–73.
45. Tian F, Yu C, Li H, Wu X, Li B, Chen H, Wu M, He C. Alternative sigma factor RpoN2 is required for flagellar motility and full virulence of *Xanthomonas oryzae* pv. *oryzae*. Microbiol Res. 2015;170:177–83.
46. Thordalchristensen H, Zhang ZG, Wei YD, Collinge DB. Subcellular localization of $H_2O_2$ in plants. $H_2O_2$ accumulation in papillae and hypersensitive response during the barley-powdery mildew interaction. Plant J. 1997;11(6):1187–94.
47. Hanahan D. Studies on transformation of Escherichia coli with plasmids. J Mol Biol. 1983;166(4):557–80.
48. Kovach ME, Elzer PH, Hill DS, Robertson GT, Farris MA, Roop 2nd RM, Peterson KM. Four new derivatives of the broad-host-range cloning vector pBBR1MCS, carrying different antibiotic-resistance cassettes. Gene. 1995; 166(1):175–6.
49. Hopkins CM, White FF, Choi SH, Guo A, Leach JE. Identification of a family of avirulence genes from *Xanthomonas oryzae* pv. *oryzae*. Mol Plant Microbe Interact. 1992;5(6):451–9.
50. Lereclus D, Agaisse H, Gominet M, Salamitou S, Sanchis V. Identification of a *Bacillus thuringiensis* gene that positively regulates transcription of the phosphatidylinositol-specific phospholipase C gene at the onset of the stationary phase. J Bacteriol. 1996;178(10):2749–56.

# 4

# Bacterial microbiota of Kazakhstan cheese revealed by single molecule real time (SMRT) sequencing and its comparison with Belgian, Kalmykian and Italian artisanal cheeses

Jing Li[†], Yi Zheng[†], Haiyan Xu, Xiaoxia Xi, Qiangchuan Hou, Shuzhen Feng, Laga Wuri, Yanfei Bian, Zhongjie Yu, Lai-Yu Kwok, Zhihong Sun and Tiansong Sun[*]

**Abstract**

**Background:** In Kazakhstan, traditional artisanal cheeses have a long history and are widely consumed. The unique characteristics of local artisanal cheeses are almost completely preserved. However, their microbial communities have rarely been reported. The current study firstly generated the Single Molecule, Real-Time (SMRT) sequencing bacterial diversity profiles of 6 traditional artisanal cheese samples of Kazakhstan origin, followed by comparatively analyzed the microbiota composition between the current dataset and those from cheeses originated from Belgium, Russian Republic of Kalmykia (Kalmykia) and Italy.

**Results:** Across the Kazakhstan cheese samples, a total of 238 bacterial species belonging to 14 phyla and 140 genera were identified. *Lactococcus lactis* (28.93%), *Lactobacillus helveticus* (26.43%), *Streptococcus thermophilus* (12.18%) and *Lactobacillus delbrueckii* (12.15%) were the dominant bacterial species for these samples. To further evaluate the cheese bacterial diversity of Kazakhstan cheeses in comparison with those from other geographic origins, 16S rRNA datasets of 36 artisanal cheeses from Belgium, Russian Republic of Kalmykia (Kalmykia) and Italy were retrieved from public databases. The cheese bacterial microbiota communities were largely different across sample origins. By principal coordinate analysis (PCoA) and multivariate analysis of variance (MANOVA), the structure of the Kazakhstan artisanal cheese samples was found to be different from those of the other geographic origins. Furthermore, the redundancy analysis (RDA) identified 16 bacterial OTUs as the key variables responsible for such microbiota structural difference.

**Conclusion:** Our results together suggest that the diversity of bacterial communities in different groups is stratified by geographic region. This study does not only provide novel information on the bacterial microbiota of traditional artisanal cheese of Kazakhstan at species level, but also interesting insights into the bacterial diversity of artisanal cheeses of various geographical origins.

**Keywords:** Single molecule, Real-time (SMRT), Cheeses, Bacterial diversity, Kazakhstan

* Correspondence: sts9940@sina.com
[†]Equal contributors
Key Laboratory of Dairy Biotechnology and Engineering, Inner Mongolia Agricultural University, Hohhot 010018, People's Republic of China

## Background

Kazakhstan is a multi-ethnic state and livestock-based country [1] that locates in the European and Asian continents; and the country has a long history since the Neolithic Age [2]. Artisanal foods are popular with the local area and the processing technology of such traditional food is almost completely preserved. Cheeses, as one of the oldest fermented foods, have a long history from approximately the early Bronze Age [3, 4]. Moreover, the classical Kazakhstan diet comprises a large proportion of local dairy products [5], so the cheese processing technology is traditional. They are often produced by individual households or small regional factories; and they are made in a small scale with naturally occurring bacteria [6]. The traditional Kazakhstan cheeses are produced through various steps: firstly, fresh raw cow milk is collected, followed by pasteurization. The natural whey is then added to the pasteurized milk to enhance coagulation under natural conditions. Then, the whey is drained through heating and extrusion. The curds without ripening are shaped into different forms by using different molds [7]. The starter cultures for making Kazakhstan artisanal cheeses are mainly natural whey. Thus the natural microbial communities of the resultant cheeses are very rich and complex. To sum up, the artisanal cheese of Kazakhstan has its unique characteristics.

The natural microbial communities play an important role in these artisanal cheeses; and they contribute significantly to the cheese quality and properties like flavor, texture and appearance [8]. Meanwhile, it is known that cheeses and cheese-derived microorganisms can directly and/or indirectly influence the host microbiota, systemic immune responses and overall health [9–12]. Therefore, understanding the intrinsic composition of bacterial community is of high scientific interest. High-throughput sequencing has become a common technique that provides accurate depiction of the microbial communities present in traditional artisanal cheeses. Based on this approach, Quigley et al. for the first time detected the presence of several genera in artisanal cheeses, including *Arthrobacter, Brachybacterium, Faecalibacterium, Prevotella,* and *Helcococcus,* which confirmed that the level of cheese maturation had an influence on the *Lactobacillus* population [13]. The bacterial communities of some specific cheese types have been studied, such as mozzarella cheese [14], Croatian cheese [15], Herve cheese [16], and Mexican Poro cheese [17]. However, the characteristics of the microbial communities in Kazakhstan cheese have rarely been reported.

The third generation PacBio single molecule, real-time (SMRT) sequencing technology, a new and advanced high-throughput sequencing tool, generates long reads and allows high taxonomic resolution to the genus and even species level when coupled to full length 16S rRNA gene sequencing [18]. It has been successfully applied in the evaluation of milk bacterial contamination [19]. Thus, the aim of this study was to provide high resolution bacterial microbiota profiles of Kazakhstan artisanal cheese samples using the PacBio SMRT platform; a second aim of our study was to comparatively analyze the bacterial microbiota of cheese from different regions, including Kazakhstan, Belgium, Russian Republic of Kalmykia (Kalmykia) and Italy. To do so, 36 16S rRNA gene datasets were retrieved from public databases. Our study provides novel information on the microbial communities of Kazakhstan traditional cheese products.

## Methods

### Sample information

A total of six traditional artisanal cheeses were collected from two different artisanal factories of Kazakhstan (K1-K4 and K5-K6 collected respectively from Alma-Ata and Jambyl provinces, Additional file 1: Figure S1). The manufacturing process of these cheeses was similar, as described previously. Samples were collected aseptically and were stored in vacuum bags as soon as they were sampled. They were kept cold while being transported to the laboratory. The nutritional information of these cheeses is provided by the factories (Additional file 2: Table S1).

Apart from the 6 cheese samples collected from Kazakhstan. Datasets of 16S rRNA gene fragments of 36 artisanal cheese samples [14, 16] from Belgium, Russian Republic of Kalmykia (Kalmykia) and Italy were extracted from public databases for comparative analysis. The information of the samples is provided in Table 1.

### DNA extraction and PCR amplification

Total sample DNA was extracted by a combination of described methods [20–22] with slight modifications. Cheese samples were collected from different parts of the same cheese. Then, cheese samples were mixed together and crushed to uniform powders. Two grams of cheese powders were suspended in 2 mL TE buffer (10 mM Tris · Cl, 1 mM EDTA, pH8.0). The mixture was centrifuged at 8,000 × g for 5 min. Then the pellets were washed with 500 μL TE buffer at 8,000 × g for 5 min in clean 1.5 mL microcentrifuge tubes. Washed cell pellets were resuspended in 500 μL TE buffer. Suspension was frozen for 2 min by liquid nitrogen, before being incubated at 65 °Cfor 3 min; the above steps were repeated 3 times. Proteinase K solution (15 μL, 20 mg /mL in TE, Amrescolnc., USA) and 60 μL SDS solution (10%) were added, mixed and incubated at 37 °C, 300 rpm/min overnight. The mixture was incubated with 100 μL NaCl (5 M) and 80 μL CTAB solutions (10% cetyltrimethylammonium bromide, 0.7 M NaCl) at 65 °C for 30 min and extracted with 1 vol. phenol/chloroform/ isoamylalcohol. After centrifugation, the supernatant was

**Table 1** Sample information

| Sample | Sampling location | Country | Milk source | Starter | Sequenced 16S rRNA region | Sequencing technology | Cheese production process | Accession number |
|---|---|---|---|---|---|---|---|---|
| K1- K4<br>K5-K6 | Alma-Ata Province<br>Jambyl Province | Kazakhstan | Cow milk | Natural whey culture | Full length | SMRT technology, Pacific Biosciences | The fresh raw cow milk is pasteurized. The natural whey is added to the pasteurized milk and allows coagulating. Then the whey is drained through heating and extrusion. The curds (without ripening) are shaped into different forms by different molds [7]. | BioProject ID: PRJNA347428 |
| B1- B22 | Herve | Belgium | Cow milk | Not given | V1-V3 regions | GS Junior platform, 454 Life Sciences | Milk is supplemented with rennet. After coagulation, milk curds are cut in to small pieces like the size of a hazelnut, before being poured into square-shaped molds. The molds are turned over every few hours for 2 days. During the ripening period, cheese is washed 2 or 3 times per week with salt water or pure salts to enhance smear formation [16]. | BioProject ID PRJNA238292 [16] |
| R1-R6 | Yashkul | Russian Republic of Kalmykia | Cow milk | Natural whey culture | V1-V3 regions | FLX454 Titanium System, 454 Life Sciences | Fresh raw cow milk is supplemented with natural whey cultures as starters. After curding, the whey is drained through heating and extrusion. The curds without ripening are shaped by different mold | MG-RAST ID No. 4682839.3 - 4682844.3 |
| I1-I6 | Campania region | Italy | Buffalo milk | Natural whey culture | V1-V3 regions | GS Junior platform, 454 Life Sciences | Raw milk is heated at 37 °C. Then, natural whey cultures are added. After a curd-ripening phase (4.0–4.5 h at 35–37 °C), the curd is drained. The drained curd is stretched in hot water (90–95 °C) before being hand-molded to get the typical round shape [66]. | Sequence Read Archive project SRP014821 [14] |

re-extracted with chloroform/ isoamylalcohol (1:1 ratio). After another centrifugation, DNA was obtained by the addition of 500 μL isopropanol and 100 μL NaAC (0.5 M) at −20 °C for 20 min. The pellets were washed in 500 μl ethanol (70%) and dried. Then the DNA was dissolved in 50–150 μL TE at 37 °C for 1 h and added with 100 μL RNAase solution (100 μg/mL in TE, Sigma-Aldrich, USA). The quality of DNA was checked in 0.8% agarose gel electrophoresis (Liuyi Biotechnology, China) and spectrophotometer (Thermo Scientific, USA). All extracted DNA samples were stored at −20 °C for further analysis.

The forward 27 F (5'-GAGAGTTTGATCCTGGCT CAG-3') and the reverse 1541R (5'- AAGGAGGTG ATCCAGCCGCA-3') primers [23] were used to amplify the 16S rRNA gene fragments with PCRBIO *Taq* DNA polymerase (PCR Biosystems Ltd., UK). The volume of the final reaction mixture was 50 μL containing 10 μL 5× PCRBIO reaction buffer, 2 μL forward primer (10 μM), 2 μL reverse primer (10 μM), 1 μL template DNA, 1 μL PCRBIO *Taq* DNA polymerase (5 units/μL) and 34 μL double distilled $H_2O$. Conditions of PCR amplification was an initial denaturing step for 5 min at 95 °C, followed by 30 cycles of a denaturing step at 95 °C for 30 s, primer annealing at 58 °C for 45 s and elongation at 72 °C for 1 min. The program was completed with a final elongation step at 72 °C for 7 min. The PCR products were purified according to the protocol of the Pacific Biosciences (http://www.pacb.com/wp-content/uploads/2015/09/Procedure-Checklist-2-kb-Template-Preparation-and-Sequencing.pdf). The quality of PCR products was checked by using an Agilent DNA 1000 Kit and an Agilent 2100 Bioanalyser (Agilent Technologies, USA) according to the manufacturer's instructions.

### Single molecule real-time sequencing of Kazakhstan cheese 16S rRNA

The amplicons of the 16S rDNA regions purified were used to construct the DNA libraries using the Pacific Biosciences Template Prep Kit 2.0. Amplicons of the16S rDNA regions described previously were sequenced using P6/C4 chemistry on a PacBio RS II instrument (Pacific Biosciences, USA). The quality control for PCR amplifications and sequence preprocessing was performed by the methods described previously by Mosher et al., 2013 [24]. Raw data were processed by the protocol RS_ReadsOfinsert.1. Raw reads were first filtered according to the following criteria restrictively: (i) minimum full passes was up to 5; (ii) minimum predicted accuracy was 90; (iii) 1400 was the minimum read length of inserts, and (iv) 1800 was the maximum read length.

### Bioinformatics processing and statistical analyses

According to the barcode, the extracted high-quality sequences obtained by SMRT technology were sorted. Then the barcodes and primer were removed from the extracted high-quality sequences - dataset A. Datasets of 16S rRNA gene fragments of 36 artisanal cheese samples from Belgium, Kalmykia and Italy were also first filtered, with the selection for sequence read length of over 425 bp to build the dataset B. Sequences from dataset A and dataset B were then analyzed by using QIIME 1.7 (Quantitative Insights into Microbial Ecology) software. The sequences were aligned using PyNAST [25] and UCLUST [26] under 100% clustering of sequence identity to obtain representative sequences. The resultant sequences were clustered to obtain operational taxonomic units (OTU) under the threshold of 98.65% identity [27] by the method of the UCLUST algorithm. ChimeraSlayer [28] was used to remove the chimeric sequences in the representative set of OTUs. The taxonomy of each bacterial OTU representative sequence was assigned with the Ribosomal Database Project (RDP) database (version 11.4) [29] and Greengenes (version 13_8) [30] using classifier with an 80% confidence threshold. A representative chimera-checked OTU set was applied to construct a *denovo* taxonomic tree in FastTree for downstream analysis [31]. The Shannon-Wiener index, Simpson's diversity, Chao1 and rarefaction estimators were employed to measure the sequencing diversity. The weighted and unweighted principal coordinate analyses (PCoA) based on UniFrac metrics [32] were performed to assess the microbiota structure of different samples.

The diversity of bacterial communities of the 40 artisanal cheeses across 4 countries was analyzed by analysis of variance (ANOVA) based on UniFrac distances as calculated by the SAS software version 9.2 (SAS Institute Inc., Cary, NC) [33]. Multivariate ANOVA (MANOVA) and clustering of microbiota of cheeses based on Mahalanobis distances [34] were conducted by the Matlab R2011b software (the MathWorks, Natick, MA, USA). Differences in the bacterial populations between samples were evaluated with Kruskal-Wallis tests [35]. Redundancy analysis (RDA) was applied to identify the key microbial groups that contributed to the structural difference with the software Canoco for Windows 4.5 (Microcomputer Power, NY, USA). The graph presentations were generated by the R package version 3.1.2 (https://www.r-project.org/) and the Origin software version 8.5 (OriginLab Corporation, Hampton, MA).

## Results

### Richness and diversity analysis of bacterial community composition of Kazakhstan cheese

In this study, the bacterial diversity profiles of six traditional artisanal cheeses were obtained by SMRT sequencing technology. A total of 30884 of bacterial 16S rRNA raw reads were obtained. An average of 5147.33 sequences (SD = 2647.59) were generated for each sample. The number of reads and assigned OTUs of the six samples are shown in Additional file 2: Table S2. After alignment and sequence identity clustering, a total of 8875 unique OTUs were obtained. The average representative OTUs for each individual sample was 1479.17.

Additionally, the results of the Shannon index, Simpson index, Chao1 index, and the number of observed species are shown in Additional file 2: Table S2. The diversity indexes, including Chao 1 index (ranged from 893.90–8569.60), the Shannon index (4.72–8.75), Simpson index (0.84–0.99) and the number of observed species (315.89–1214.52), show that the bacterial communities varied apparently among the six cheese samples. The bacterial diversity of sample K6 was lowest compared to other samples. The Shannon diversity and the rarefaction curves indicate that the overall bacterial diversity was well captured and represented (Additional file 3: Figure S2).

### Bacterial composition in the Kazakhstan cheese samples

Each bacterial OTU representative sequence was taxonomically assigned with the Greengenes (version 13.8) and the Ribosomal Database Project (RDP) II database (version 11.4). A total of 14 phyla were identified in the sampled cheeses (Additional file 2: Table S3). The major bacterial phyla (with mean relative abundance of >0.1%) were Firmicutes (92.47%), Proteobacteria (6.95%), Actinobacteria (0.24%), and Cyanobacteria (0.17%).

At genus level, 140 bacterial genera were identified (Fig. 1a). The major genera (average relative abundance of >1%) detected in the samples included *Lactobacillus* (42.12%), *Lactococcus* (31.07%), *Streptococcus* (16.99%), *Ochrobactrum* (2.25%) and *Burkholderia* (1.98%). Large variations in bacterial composition existed among samples. *Lactobacillus* was the predominant bacterial genus of the samples K1-K4, and K6. The prevalent bacterial genus of the sample K5 was *Lactococcus*. The proportions of *Lactobacillus* and *Lactococcus* in Kazakhstan cheeses ranged from 1.53–64.51% and 2.08–78.17%, respectively.

From the six cheese samples, a total of 238 different species were detected. *Lactococcus lactis* (28.93%), *Lactobacillus helveticus* (26.43%), *Streptococcus thermophilus* (12.18%) and *Lactobacillus delbrueckii* (12.15%) were the dominant species. *Ochrobactrum lupine* (2.23%) and *Acinetobacter baumannii* (1.09%) were detected across cheese samples with varying abundances (Fig. 1b). *Lactobacillus helveticus* was dominating in samples K3, K4 and K6, contributing to 24.98%, 37.11% and 64.05% of the bacterial population, respectively. *Lactococcus lactis* (43.18%) was the predominant species of sample K1. The relative abundance of *L. delbrueckii* in sample K6 was highest among all samples. The relative abundances of *Ochrobactrum lupine* and *Acinetobacter baumannii* of the six samples ranged from 0.06–12.75% and from 0.05–5.85%, respectively. For sample K6, the relative abundances of *Ochrobactrum lupine* and *Acinetobacter baumannii* represented 12.75% and 5.85% of the total bacterial sequences.

### Comparison of bacterial profiles of artisanal cheese in different locations

Datasets of 16S rRNA gene fragments of three kinds of cheeses, respectively from Belgium, Russian Republic of Kalmykia (Kalmykia) and Italy were used for the comparative analysis together with that of Kazakhstan cheese generated from the current study.

The average UniFrac distances of the four groups were assessed with ANOVA to evaluate the differences among the samples (Fig. 2). The unweighted UniFrac distance of the Kazakhstan cheeses was significantly different from other groups ($p < 0.01$), suggesting that the Kazakhstan cheese had some unique features.

Then, PCoA based on the weighted (principal coordinate 1 and 2 accounted for 26.69 and 19.99% of the total

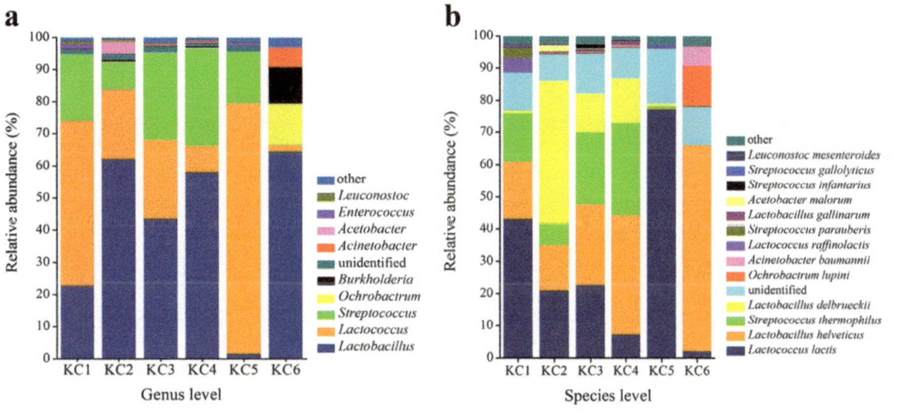


**Fig. 1** Relative abundances and bacterial diversity at genus (**a**) and species (**b**) levels

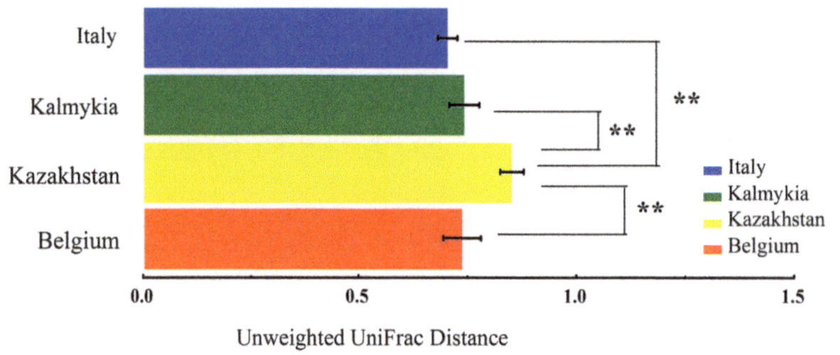


**Fig. 2** Comparison of the unweighted UniFrac distance of cheese bacterial communities from four countries "**" represents $p < 0.01$

variance, respectively) (Fig. 3a) and unweighted (accounted for 15.41 and 6.60% of the total variance, respectively) (Fig. 3b) UniFrac metrics revealed the existence of bacterial structural difference. On the unweighted PCoA score plot, clear clustering pattern based on the cheese origin was observed, while only a mild overlapping occurred on the weighted PCoA score plot, suggesting that the geographic location and origin of the cheese may be related to the distinct bacterial microbiotia composition. These results were further supported by MANOVA ($p < 0.001$) (Fig. 3c, d).

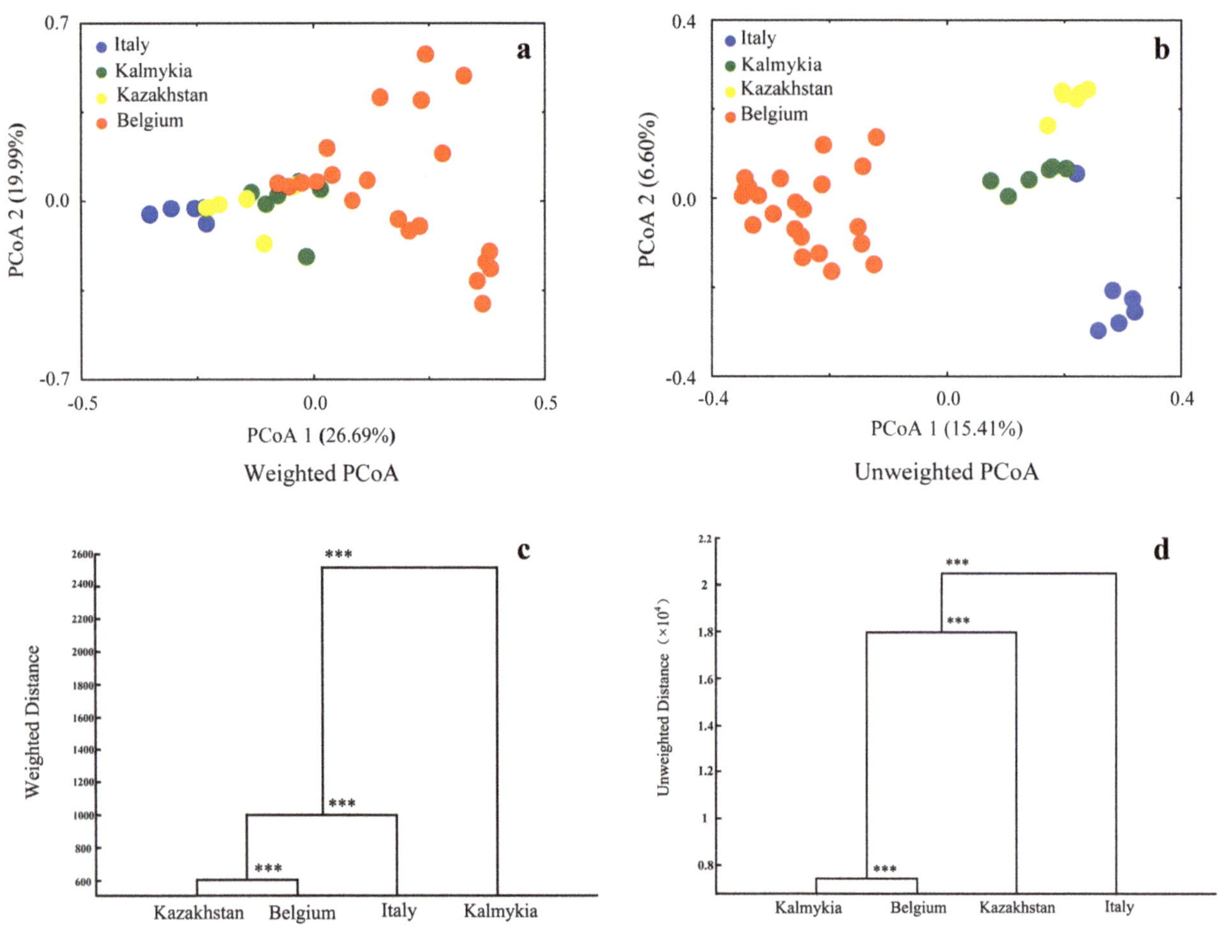


**Fig. 3** Comparison of the bacterial community structure of samples from the four countries Principal coordinate analysis based on the weighted (**a**) and unweighted (**b**) UniFrac distances (*blue = Italy, green = Kalmykia, yellow = Kazakhstan*, and *red = Belgium*). The cluster analysis was performed based on the weighted (**c**) and unweighted (**d**) Mahalanobis distances calculated by MANOVA of cheese bacteria communities, "***" represents $p < 0.001$

### Comparative analysis of the bacterial composition of the cheese samples

The proportions of the three most prevalent bacterial genera in the cheese samples from Kazakhstan, Belgium, Kalmykia and Italy were different, as shown in Table 2. In addition, the difference in bacterial composition of the 40 cheese samples was further evaluated by Kruskal-Wallis test $p < 0.05$ (Additional file 2: Table S4). The relative abundances of 43 genera were significantly different across groups. A heat map showing the distribution of these differential genera is presented in Fig. 4.

Redundancy analysis (RDA) was performed to further identify the relationship between the sample geographical origins and the key responding OTUs that contributed to the microbial community structural difference (Fig. 5). The significant difference in the bacterial composition among the 40 samples was confirmed by a Monte Carlo permutation procedure (MCPP) ($p = 0.002$). Sixteen key genera were identified, of which at least 16% of the variability in their values was explained by the canonical axis. Eleven responsive genera were key genera in samples from Kazakhstan, Italy and Kalmykia. The other five genera (*Staphylococcus*, *Marinilactibacillus*, *Psychrobacter*, *Pseudoalteromonas* and *Brevibacterium*) were located at the left side of the ordination plot, suggesting that they were enriched in the Belgium cheese (Additional file 2: Table S5).

## Discussion

Our study used the SMRT sequencing approach to describe the Kazakhstan cheese bacterial diversity and revealed the unique characteristics of the microbiota structure of the artisanal cheeses in depth. The high-throughput sequencing of 16S rRNA gene has become a vital tool for identifying members of microbial communities present in traditionally fermented dairy products [36–38]. Studies of natural bacterial communities utilizing pyrosequencing have been limited to the identification of community composition to the genus level. This study further supports that the third generation PacBio SMRT is advantageous over the more traditional second generation next sequencing technology in describing microbiota communities in dairy products like cheese because of its capacity in generating longer reads. Using the current approach, bacterial profiles at a higher taxonomic resolution to the genus and even species level can be obtained without much difficulty.

Our results show that Firmicutes and Proteobacteria are the major phyla present in traditional artisanal cheeses. This result is in line with previous studies [13, 17, 39]. As the predominant species in the artisanal cheeses, *Lactococcus lactis* is commonly used as a starter during the manufacturing process because it has rapid acidification and curd production capacities [40, 41]. *Lactococcus lactis* also significantly contributes to the organoleptic properties and microbial quality of cheese by facilitating the acidification (the formation of lactic acid) [42, 43], aroma production (the degradation of casein), nisin synthesis, and enhancement of flavor and texture characteristics. *Lactobacillus helveticus* is another prevalent species found in our samples; this species is also a common starter that is used for cheese fermentation [44–46]. *Lactobacillus helveticus* is well-known for its proteolytic activity, which is of importance in reducing the cheese bitterness [47]. Furthermore, the proteolytic activity encoded by the proteolytic system of some *L. helveticus* strains is considered to be important in releasing antihypertensive peptides [48].

*Streptococcus thermophilus* and *Lactobacillus bulgaricus* are complimentary to each other and are essential for optimizing the milk fermentation environment during yoghurt and cheese making [49]. During the fermentation process, *S. thermophilus* supports the growth of *L. bulgaricus* by producing acid and utilizes the dissolved oxygen to produce $CO_2$, and creates the anaerobic conditions for the *L. bulgaricus* growth in milk. On the other hand, *L. bulgaricus* possesses more proteolytic enzymes to release amino acids, and then stimulates the growth of *S. thermophilus* [50, 51].

*Ochrobactrum lupine* was initially isolated from soil [52, 53]. Not many previous reports have identified members of this genus in cheese, but interestingly they were identified in the Kazakhstan cheese samples (K2, K4 and K6). *Acinetobacter baumannii* is a potential pathogen that may cause nosocomial infections [54]; and it is resistant to many clinically used antibiotics [55]. The presence of 16S rDNA amplicons representing microbes of environmental or clinical concerns might have been directly related to the cheese production environment and hygienic condition. Alternatively, such contaminants may be present in the air, soil and water, which could enter the

**Table 2** The major bacterial genera in the cheese 16S rRNA datasets

| Samples | Bacterial genera | Relative abundances (%) |
|---|---|---|
| Kazakhstan cheese | *Lactobacillus* | 42.12 |
| | *Lactococcus* | 31.07 |
| | *Streptococcus* | 16.98 |
| Belgium cheese | *Lactococcus* | 44.32 |
| | *Psychrobacter* | 14.72 |
| | *Corynebacterium* | 9.80 |
| Kalmykia cheese | *Lactococcus* | 42.49 |
| | *Streptococcus* | 28.79 |
| | *Citrobacter* | 8.88 |
| Italy cheese | *Streptococcus* | 53.97 |
| | *Lactobacillus* | 39.2 |
| | *Lactococcus* | 2.88 |

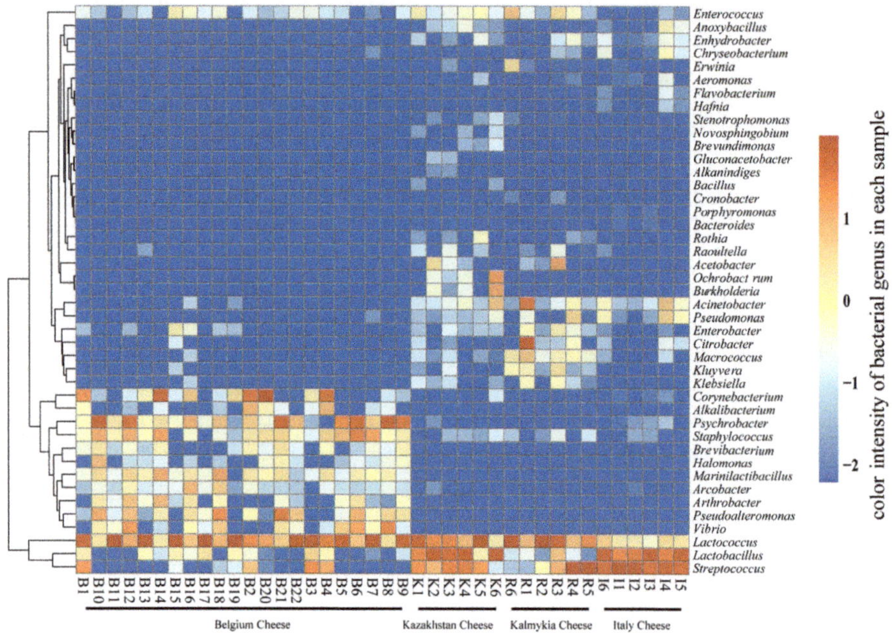


**Fig. 4** Heatmap depicting distribution of significantly different genera with $p < 0.05$ across groups Bacterial relative abundances are illustrated by the color scale

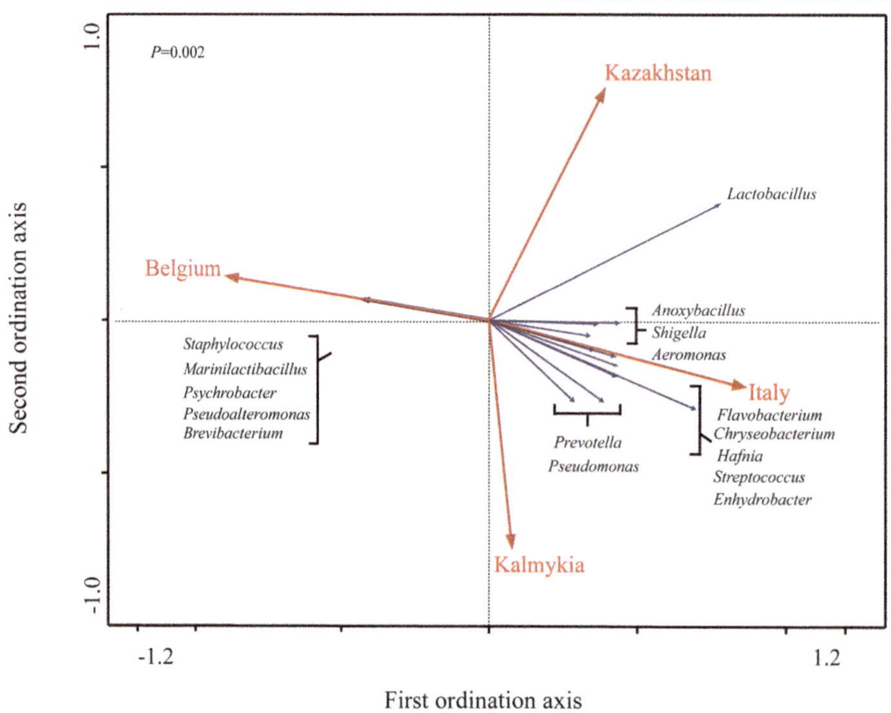


**Fig. 5** Biplot of redundancy analysis (RDA) of the cheese bacterial communities from four countries *Red* arrows represent the constrained explanatory variables, Kazahstan, Italy, Kalmykia and Belgium. Grey lines represent the response variables with the first ordination axis explaining for at least 35% and 16 genera of the variability of the bacterial microbiota. The *p*-value generated from the Monte Carlo permutation test is shown in the plot

raw materials or along the food processing procedure. It has been suggested that parallel negative controls should be included in this kind of microbiota profiling experiments, as the contamination might also source from the molecular biology grade water, PCR reagents, and DNA extraction kits that are used in the experiment [56]. For example, the *Ochrobactrum* genus has been reported to be one of those contaminants originated from laboratory reagents. Since our study did not include any parallel negative control; further study will be needed to confirm its source.

In order to further understand the characteristics of Kazakhstan cheese, we extracted 16S rRNA gene nucleotide sequences of traditionally artisanal cheese of three other regions, including Belgium, Italy and Kalmykia, from public gene databases and performed comparative analysis with the Kazakhstan cheeses collected in this study. Zhong et al. [57] comparatively analyzed the naturally fermented milk microbiota profiles based on different 16S rRNA gene regions. However, such analysis was limited by the low number of public accessible datasets of cheese 16S rRNA data, particularly full length gene sequence [58]. Based on multivariate statistical analysis, the artisanal cheese of Kazakhstan has its own characteristics. There were apparent differences in the microbiota communities between cheeses of different groups, as revealed by PCoA and MANOVA, suggesting that the geographic location may play a role in shaping the sample microbiota. Similar geographical-based variations were reported in traditional fermented vegetables [59] and home-made fermented milks [39, 44, 60]. By comparing with the cheese production methods, the four kinds of cheese varied greatly in their ripening stage (Table 1). The heatmap shows apparent differences between the microbiota composition of the Belgium cheese from the others. The type of milk might also have contributed to the differences in the microbiota composition after fermentation due to both the intrinsic nutrient components and the pre-existing natural starter cultures. Buffalo milk has a higher fat and protein content than cow milk [61], and thus water-soluble and non-protein-nitrogen. The analyses of the microbiota of raw milk, nature whey and water buffalo mozzarella cheese by high-throughput sequencing indicated that nature whey had an important effect on the final microbial composition of cheese [14, 62]. Thus, apart from the geographic location, other factors including origin and type of cheese, cheese-making technology, processing and environmental conditions may together shape the ultimate bacterial community profile [63–65]. After all, our study has provided novel and interesting information regarding the bacterial microbiota of local Kazakhstan cheeses.

## Conclusion

To sum up, the current study analyzed the bacterial communities of six traditional Kazakhstan cheese samples by SMRT. A total of 140 genera from 14 phyla were identified from the samples, with the predominant species of *Lactococcus lactis* (28.93%), *Lactobacillus helveticus* (26.43%), *Streptococcus thermophilus* (12.18%) and *Lactobacillus delbrueckii* (12.15%). By comparatively analyzing the 16S rRNA gene sequences generated from this study and those retrieved from public databases, unique bacterial community signatures could be identified in the Kazakhstan cheeses, which were largely different from those of Belgium, Kalmykia and Italy. Thus, the cheese origin does seem to play a role in shaping the sample microbiota composition.

## Additional files

**Additional file 1: Figure S1.** The appearance of the six Kazakhstan traditional artisanal cheeses.

**Additional file 2: Table S1.** The nutritional information of Kazakhstan cheeses (the content of 100 g sample). **Table S2.** Number of sequences and OTUs, microbial diversity and richness indexes of Kazakhstan cheese samples. **Table S3.** Relative abundances of bacterial phyla in the six Kazakhstan cheese samples. **Table S4.** Average relative abundances of the 43 main bacterial genera across the four sample groups. **Table S5.** Relative abundances of 16 key bacterial genera across the four sample groups.

**Additional file 3: Figure S2.** Rarefaction (left) and Shannon diversity (right) curves of samples. Each line represents the dataset of one sample.

**Acknowledgements**

The authors are indebted to the members of the Key Laboratory of Dairy Biotechnology and Engineering, Inner Mongolia Agricultural University, Hohhot, for their technical supports.

**Funding**

This research was supported by the International S&T Cooperation Program of China, ISTCP, 2014DFR31150 and the China Agriculture Research System (Grant No. CARS-37).

**Authors' contributions**

JL, YZ, ZS and TS designed the experiments. JL, HX, XX, YB, SF and WL performed the experiments. JL, YZ, QH, LK and ZY analyzed the data. JL, LK and TS drafted the manuscript. All authors read and approved the final manuscript.

**Competing interests**

The authors declare that they have no competing interest.

**References**


1. JR L, et al. Agricultural Status in Kazakhstan Republic and Korean Strategies for Agriculture Technical Cooperation with Kazakhstan. J Korean Soc Int Agric. 2014;26(4):12.
2. Miller AV, et al. Dental health, diet, and social transformations in the Bronze Age: Comparative analysis of pastoral populations in northern Kazakhstan. Quat Int. 2014;348:130–46.
3. Salque M, et al. Earliest evidence for cheese making in the sixth millennium BC in northern Europe. Nature. 2013;493(7433):522–5.
4. Yang Y, et al. Proteomics evidence for kefir dairy in Early Bronze Age China. J Archaeol Sci. 2014;45:178–86.
5. Baubekova A, et al. Biodiversity study of the yeast in fresh and fermented camel and mare's milk by denaturing gradient gel electrophoresis. J Camel Pract Res. 2015;22(1):91–5.
6. Akhmetsadykova S, et al. Microflora identification of fresh and fermented camel milk from Kazakhstan. Emirates J Food Agric. 2014;26(4):327.
7. Dagarbak N. Kazakhstan natural green food——Tarag and cheese, vol. 7. China: Macao economic digest; 2015. p. 148–9.
8. Martínez-Cuesta MC, Requena T, Peláez C. Permeabilization and lysis induced by bacteriocins and its effect on aldehyde formation by Lactococcus lactis. Biotechnol Lett. 2006;28(19):1573–80.
9. Plé C, et al., Maintaining gut ecosystems for health: Are transitory food bugs stowaways or part of the crew? Int J Food Microbiol. 2015.
10. Montel M-C, et al. Traditional cheeses: rich and diverse microbiota with associated benefits. Int J Food Microbiol. 2014;177:136–54.
11. Borresen EC, et al. Fermented foods: patented approaches and formulations for nutritional supplementation and health promotion. Recent Pat Food Nutr Agric. 2012;4(2):134–40.
12. Ceapa C, et al. Influence of fermented milk products, prebiotics and probiotics on microbiota composition and health. Best Pract Res Clin Gastroenterol. 2013;27(1):139–55.
13. Quigley L, et al. High-throughput sequencing for detection of subpopulations of bacteria not previously associated with artisanal cheeses. Appl Environ Microbiol. 2012;78(16):5717–23.
14. Ercolini D, et al. "Remake" by high-throughput sequencing of the microbiota involved in the production of water buffalo mozzarella cheese. Appl Environ Microbiol. 2012;78(22):8142–5.
15. Fuka MM, et al. Dynamics of bacterial communities during the ripening process of different Croatian cheese types derived from raw ewe's milk cheeses. 2013.
16. Delcenserie V, et al. Microbiota characterization of a Belgian protected designation of origin cheese, Herve cheese, using metagenomic analysis. J Dairy Sci. 2014;97(10):6046–56.
17. Aldrete-Tapia A, et al. High-throughput sequencing of microbial communities in Poro cheese, an artisanal Mexican cheese. Food Microbiol. 2014;44:136–41.
18. Mosher JJ, et al. Improved performance of the PacBio SMRT technology for 16S rDNA sequencing. J Microbiol Meth. 2014;104:59–60.
19. Hou Q, et al. Evaluation of bacterial contamination in raw milk, ultra-high temperature milk and infant formula using single molecule, real-time sequencing technology. J Dairy Sci. 2015;98(12):8464–72.
20. Ausubel F M. Current Protocols in Molecular Biology. Vol. 1[M]// Current protocols in molecular biology /. John Wiley & Sons, 1994.
21. Yu J, et al. Rapid identification of lactic acid bacteria isolated from home-made fermented milk in Tibet. J Gen Appl Microbiol. 2009;55(3):181–90.
22. Zhou J, Bruns MA, Tiedje JM. DNA recovery from soils of diverse composition. Appl Environ Microbiol. 1996;62(2):316–22.
23. Devereux R, Willis SG. Amplification of ribosomal RNA sequences. Netherlands: Springer; 1995. p. 277–87.
24. Mosher JJ, et al. Efficacy of a 3rd generation high-throughput sequencing platform for analyses of 16S rRNA genes from environmental samples. J Microbiol Meth. 2013;95(2):175–81.
25. Caporaso JG, et al. PyNAST: a flexible tool for aligning sequences to a template alignment. Bioinformatics. 2010;26(2):266–7.
26. Edgar RC. Search and clustering orders of magnitude faster than BLAST. Bioinformatics. 2010;26(19):2460–1.
27. Kim M, et al. Towards a taxonomic coherence between average nucleotide identity and 16S rRNA gene sequence similarity for species demarcation of prokaryotes. Int J Syst Evol Microbiol. 2014;64(Pt 2):346–51.
28. Amir A, et al. High-resolution microbial community reconstruction by integrating short reads from multiple 16S rRNA regions. Nucleic Acids Res. 2013;41(22):e205.
29. Cole JR, et al. Ribosomal Database Project: data and tools for high throughput rRNA analysis. Nucleic Acids Res. 2014;42(D1):D633–42.
30. DeSantis TZ, et al. Greengenes, a chimera-checked 16S rRNA gene database and workbench compatible with ARB. Appl Environ Microbiol. 2006;72(7):5069–72.
31. Price MN, Dehal PS, Arkin AP. FastTree: computing large minimum evolution trees with profiles instead of a distance matrix. Mol Biol Evol. 2009;26(7):1641–50.
32. Bounaadja L, et al. Real-time PCR for identification of Brucella spp.: a comparative study of IS711, bcsp31 and per target genes. Vet Microbiol. 2009;137(1):156–64.
33. Zhang J, et al. Mongolians core gut microbiota and its correlation with seasonal dietary changes. Sci Rep. 2014;4(2969):48–9.
34. Mahalanobis PC. On test and measures of group divergence, Part I: Theoretical formulae. 1930.
35. Breslow N. A Generalized Kruskal-Wallis Test for Comparing K Samples Subject to Unequal Patterns of Censorship. Biometrika. 1970;57(3):579–94.
36. O'Sullivan DJ, et al. Temporal and spatial differences in microbial composition during the manufacture of a Continental-type cheese. Appl Environ Microbiol. 2015;81(7):2525–33.
37. Erkus O, et al. Multifactorial diversity sustains microbial community stability. ISME J. 2013;7(11):2126–36.
38. Liu W, Xi X, Sudu Q, et al. High-throughput sequencing reveals microbial community diversity of Tibetan naturally fermented yak milk[J]. Annals of Microbiology. 2015;65(3):1741–51.
39. Liu W, et al. High-throughput sequencing for the detection of the bacterial and fungal diversity in Mongolian naturally fermented cow's milk in Russia. BMC Microbiol. 2015;15(1):45.
40. Cavanagh D, Fitzgerald GF, McAuliffe O. From field to fermentation: The origins of Lactococcus lactis and its domestication to the dairy environment. Food Microbiol. 2015;47:45–61.
41. del Rio B, et al. Lactose-mediated carbon catabolite repression of putrescine production in dairy Lactococcus lactis is strain dependent. Food Microbiol. 2015;48:163–70.
42. Settanni L, Corsetti A. Application of bacteriocins in vegetable food biopreservation. Int J Food Microbiol. 2008;121(2):123–38.
43. Cachon R, et al. Characterisation of lactic starters based on acidification and reduction activities. Le Lait. 2002;82(3):281–8.
44. Watanabe K, et al. Diversity of lactic acid bacteria and yeasts in Airag and Tarag, traditional fermented milk products of Mongolia. World J Microbiol Biotechnol. 2008;24(8):1313–25.
45. Sun Z, et al. Identification and characterization of the dominant lactobacilli isolated from koumiss in China. J Gen Appl Microbiol. 2010;56(3):257–65.
46. Sedláček I, Nováková D, Švec P. Ribotyping and biotyping of Lactobacillus helveticus from the koumiss. Eur Food Res Technol. 2010;230(5):753–8.
47. Griffiths MW, Tellez AM. Lactobacillus helveticus: the proteolytic system. Front Microbiol. 2013;4.
48. Chen Y, et al. Complete genome sequence of Lactobacillus helveticus H9, a probiotic strain originated from kurut. J Biotechnol. 2015;194:37–8.
49. Yang S-Y, et al. Production of γ-aminobutyric acid by Streptococcus salivarius subsp. thermophilus Y2 under submerged fermentation. Amino Acids. 2008;34(3):473–8.
50. Aghababaie M, Khanahmadi M, Beheshti M. Developing a kinetic model for co-culture of yogurt starter bacteria growth in pH controlled batch fermentation. J Food Eng. 2015;166:72–9.
51. Delbès C, Ali-Mandjee L, Montel M-C. Monitoring bacterial communities in raw milk and cheese by culture-dependent and-independent 16S rRNA gene-based analyses. Appl Environ Microbiol. 2007;73(6):1882–91.
52. Mercier A, et al. Decrease of the level of extractable polychlorinated biphenyls in soil microcosms: Influence of granular activated carbon and inoculation by natural microbial consortia. Int Biodeterior Biodegrad. 2015;105:127–36.
53. Ren X, Li H, Chen S. Cloning of the chlorothalonil-degrading gene cluster and evidence of its horizontal transfer. Curr Microbiol. 2011;62(3):1068–73.
54. Mortensen BL, Skaar EP. Host–microbe interactions that shape the pathogenesis of Acinetobacter baumannii infection. Cell Microbiol. 2012;14(9):1336–44.
55. Sengstock D, et al. Multidrug-resistant Acinetobacter baumannii: an emerging pathogen among older adults in community hospitals and nursing homes. Clin Infect Dis. 2010;50(12):1611–6.

56. Salter SJ, et al. Reagent and laboratory contamination can critically impact sequence-based microbiome analyses. Bmc Biology. 2014;12:12.
57. Zhong Z, Hou Q, Kwok L, et al. Bacterial microbiota compositions of naturally fermented milk are shaped by both geographic origin and sample type.[J]. J Dairy Sci. 2016;99(10):7832–7841.
58. Hahn A, et al. Different next generation sequencing platforms produce different microbial profiles and diversity in cystic fibrosis sputum. J Microbiol Meth. 2016;130:95–9.
59. Nam YD, Park SL, Lim SI. Microbial Composition of the Korean Traditional Food "kochujang" Analyzed by a Massive Sequencing Technique. J Food Sci. 2012;77(77):M250–6.
60 Sun Z, et al. Investigation of bacterial and fungal diversity in tarag using high-throughput sequencing. J Dairy Sci. 2014;97(10):6085–96.
61 Kapadiya DB, et al. Comparison of Surti goat milk with cow and buffalo milk for gross composition, nitrogen distribution, and selected minerals content. Vet World. 2016;9(7):710–6.
62 Pisano MB, et al. Metabolomics and microbiological profile of Italian mozzarella cheese produced with buffalo and cow milk. Food Chem. 2016;192:618–24.
63 Van Hoorde K, et al. Influence of pasteurization, brining conditions and production environment on the microbiota of artisan Gouda-type cheeses. Food Microbiol. 2010;27(3):425–33.
64 Tormo H, Lekhal DAH and Roques C. Phenotypic and genotypic characterisation of lactic acid bacteria isolated from raw goat milk and effect of farming practices on the dominant species of lactic acid bacteria. Int J Food Microbiol. 2015.
65 Koirala R, et al. Isolation and molecular characterization of lactobacilli from traditional fermented Dahi produced at different altitudes in Nepal. Dairy Sci Technol. 2014;94(4):397–408.
66 Ercolini D, et al. PCR-DGGE fingerprints of microbial succession during a manufacture of traditional water buffalo mozzarella cheese. J Appl Microbiol. 2004;96(2):263–70.

# 5

# ChMob2 binds to ChCbk1 and promotes virulence and conidiation of the fungal pathogen *Colletotrichum higginsianum*

Johannes Schmidpeter, Marlis Dahl, Jörg Hofmann and Christian Koch*

**Abstract**

**Background:** Mob family proteins are conserved between animals, plants and fungi and are essential for the activation of NDR kinases that control crucial cellular processes like cytokinesis, proliferation and morphology.

**Results:** We identified a hypomorphic allele of *ChMOB2* in a random insertional mutant (*vir-88*) of the hemibiotrophic ascomycete fungus *Colletotrichum higginsianum*. The mutant is impaired in conidiation, host penetration and virulence on *Arabidopsis thaliana*. ChMob2 binds to and co-localizes with the NDR/LATS kinase homolog ChCbk1. Mutants in the two potential ChCbk1 downstream targets *ChSSD1* and *ChACE2* show defects in pathogenicity. The genome of *C. higginsianum* encodes two more Mob proteins. While we could not detect any effect on pathogenicity in *ΔChmob3* mutants, ChMob1 is involved in conidiation, septae formation and virulence.

**Conclusion:** This study shows that ChMob2 binds to the conserved NDR/LATS Kinase ChCbk1 and plays an important role in pathogenicity of *Colletotrichum higginsianum* on *Arabidopsis thaliana*.

**Keywords:** Phytopathogenic ascomycete fungus, ATMT, *Colletotrichum higginsianum*, Conidiation, NDR Kinase Cbk1, Mob1, Mob2, Mob3, Ace2, Ssd1, Cts1

## Background

Infection of host plants by filamentous, appressoria forming fungi like the ascomycete *Colletotrichum higginsianum* depends on directed, polarized growth and morphological switches of infection structures [1–5]. This fungus employs a hemibiotrophic infection strategy that includes two phases [6]. After an initial biotrophic phase with bulbous primary hyphae, *Colletotrichum higginsianum* switches to a necrotrophic growth phase forming characteristic thin secondary hyphae. During penetration by the appressorium, which itself is a morphologically highly differentiated cell [7], the tip of the penetration peg grows directed from the penetration pore towards the host epidermis [8, 9]. The infection vesicle and the primary hyphae grow out of this structure into the first host cell, which still has an intact plasma membrane and remains alive. After the switch to necrotrophy, secondary hyphae develop. These hyphae invade neighboring cells and grow strongly polarized at the hyphal tip similar to saprophytic hyphae of other filamentous fungi [10]. For such processes, the establishment and maintenance of polarity and cell wall morphogenesis are critical. In fungi, their regulation is influenced by pathways which include a central kinase and a Mob-family protein that acts as co-activator [11]. These pathways are called RAM (Regulation of Ace2p activity and morphogenesis) and MEN (Mitotic exit network) in *S. cerevisiae* [12, 13] or MOR (Morphogenesis-related) and SIN (Septum initiation network) in *S. pombe* [14, 15]. In *S. cerevisiae*, polarized growth, cell wall morphogenesis and other processes like cell separation, daughter cell-specific gene expression and cell cycle progression are influenced by the RAM pathway [12]. Yeast Cbk1 is the terminal kinase in this pathway and belongs to the nuclear Dbf2-related (NDR)/large tumor suppressor (LATS) kinase subfamily, which is conserved from yeast to man [11]. Mob2 binds to Cbk1 and is essential for Cbk1 kinase activity and its proper localization in yeast [16]. Until now, only some Cbk1 targets have been identified. E.g. in yeast, Cbk1 binds and phosphorylates Sec2, which is involved in polarized vesicle exocytosis [17], and Ssd1, a mRNA-binding protein that associates with

* Correspondence: christian.koch@fau.de
Department of Biology, Division of Biochemistry, Friedrich-Alexander University Erlangen-Nuremberg, Staudtstrasse 5, 91058 Erlangen, Germany

many transcripts including chitinase *CTS1* mRNA [18]. Aside from polarized cell morphology, a well-established function of the RAM network in yeast is the control of activity and localization of the zinc finger transcription factor Ace2 during late mitosis [16] that regulates cell wall and cell separation genes like the chitinase *CTS1* and the glucanase *SCW11* specifically in the new daughter cell [19]. Nuclear localization of Ace2 is also coordinated with the MEN signaling pathway [20], which uses the NDR/LATS kinase Dbf2 and its associated kinase activator Mob1 [13].

The founding member of the NDR family, Cot1, is similar to yeast Cbk1 and was identified in the filamentous fungus *Neurospora crassa* as a temperature sensitive mutant allele (colonial temperature-sensitive 1) that showed impaired hyphal tip elongation [21]. Like its yeast homologs, Cot1 requires binding of a Mob co-activator protein for its activity. This function appears to be mediated by two Mob2 proteins in *N. crassa*, called Mob2a and Mob2b [22]. Filamentous fungi such as *N. crassa* and its close relative *S. macrospora* encode a third type of Mob-family protein, called Mob3 [22, 23]. Mob3 is more similar to the mammalian striatin-binding protein phocein, does not seem to be involved in NDR signalling but was found to be important for the development of protoperithecia and for hyphal fusion [22–24].

Previously, we described the identification of 75 *C. higginsianum Agrobacterium tumefaciens*-mediated transformation (ATMT) mutants, which are impaired in virulence on *Arabidopsis thaliana* [25]. Here, we show that for one of these mutants (*vir-88*), a hypomorphic allele of *ChMOB2* is responsible for the virulence phenotype and we analyzed the function of the protein encoded by this locus in *C. higginsianum*. We found that ChMob2 is required for both conidiation and formation of functional appressoria and that it binds to the NDR/LATS kinase ChCbk1. We searched for potential targets of the Mob2/Cbk1 complex and observed that mutants of two candidate genes have phenotypes similar to *vir-88*. Furthermore, we analyzed the roles of *ChMOB3* and *ChMOB1* in *C. higginsianum*.

## Results

### *vir-88* has defects in appressoria shape and production of conidia

*vir-88* was isolated from a pool of T-DNA insertion mutants with defects in virulence on *Arabidopsis thaliana* [25]. In contrast to *Arabidopsis* infected with wild-type *C. higginsianum* where all leaves showed strong necrotic lesions after 4 days, infection with *vir-88* resulted only in small and much fewer lesions (Fig. 1a). Staining infected leaves with trypan blue revealed that mutant *vir-88* formed appressoria at levels comparable to wild-type, but they were often severely deformed (Fig. 1b, Additional file 1: Figure S1). At 3 dpi, nearly two thirds of the appressoria produced by *vir-88* showed morphological defects like elongation, decreased diameter, small lateral protrusions or were less melanized. Some cells formed a melanized germ tube or a partly melanized conidium with or without an appressorium. Examples of these morphological phenotypes are shown in Fig. 1b and Additional file 1: Figure S1A. In contrast, the wild-type produced only 11% altered appressoria with less dramatic alterations (Additional file 1: Figure S1B). The ability of morphologically abnormal appressoria to establish infectious primary hyphae was strongly reduced because only 10% of all primary hyphae emanated from abnormal appressoria with morphological defects (Additional file 1: Figure S1C). As a consequence, the amount of primary hyphae formed by *vir-88* at 3 dpi was reduced to only one third of the wild-type level (Fig. 1c, d). The fraction of primary hyphae that gave rise to secondary hyphae was reduced by 62% in *vir-88* (Fig. 1c, d). On artificial hydrophobic plastic surfaces, appressoria of *vir-88* were similarly deformed in 64% of the cases compared to 7% in the WT (Fig. 1e, Additional file 1: Figure S1D).

When inoculated on potato dextrose agar or minimal medium plates, *vir-88* produced slightly smaller colonies than the wild-type (Fig. 1f). Furthermore, *vir-88* did not show different sensitivity to several stress inducing compounds than the wild-type (Additional file 2: Figure S2). In addition to its virulence phenotype, mutant *vir-88* generated less (42%) conidia on oatmeal agar plates, which were significantly smaller than wild-type conidia (Fig. 1g, h). Taken together, the T-DNA insertion mutant *vir-88* is severely impaired in virulence due to defects during appressoria differentiation and host cell penetration.

### *vir-88* encodes a hypomorphic allele of *ChMOB2*

In order to identify the mutation responsible for the phenotype of *vir-88*, we initially analyzed a T-DNA insertion site found in the original mutant screen which suggested the presence of two linked T-DNA insertions on supercontig5277 [26] close to CH063_12012 encoding a homolog to yeast Mob2 [25, 27]. By PCR analyses using a series of primers specific to that locus, both regions flanking the insertion could be isolated and were sequenced (Additional file 3: Text S1). The T-DNA insertion site was located upstream of the ATG in gene CH063_12012 and was accompanied by a small genomic deletion of 11 bp. Two T-DNA copies were inserted in a head to head fashion (Fig. 2a). Attempts to amplify the junction between the two T-DNA fragments by PCR were not successful, presumably due to the inverted arrangement of the sequences (not shown). The presence of two T-DNAs at this genomic context is consistent with the results of previous Southern Blot analysis probed with a hygromycin resistance DNA fragment [25]. CH063_12012 encodes a predicted 333 aa protein closely

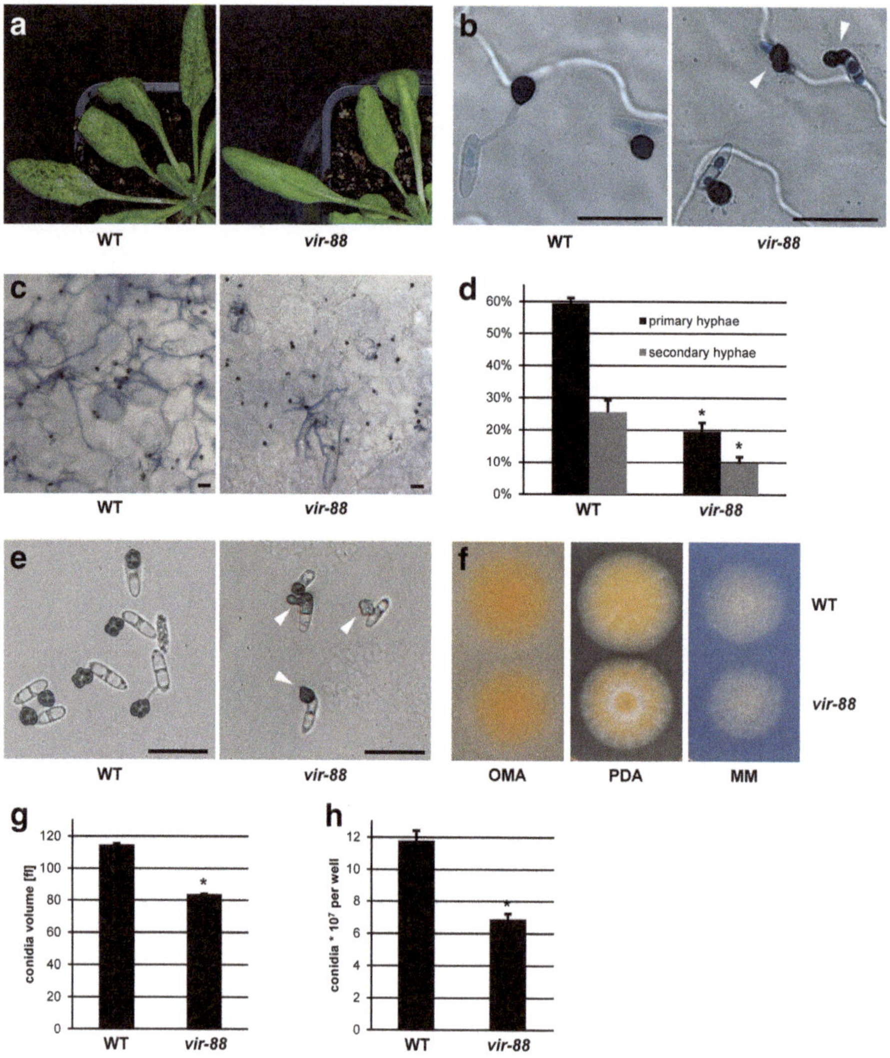


**Fig. 1** T-DNA insertion mutant *vir-88* is impaired in virulence and conidiation. *A. thaliana* Col-0 plants were infected with *C. higginsianum* WT and the *vir-88* mutant strain. **a** Macroscopic image at 4 dpi. **b** Microscopic images of germinated conidia with appressoria at 1 dpi on trypan blue stained leaves. Appressoria with unusual morphology are marked with white arrows. **c** Trypan blue stained *A. thaliana* leaves at 3 dpi. **d** Quantification of trypan blue stained primary hyphae (relative to appressoria) and secondary hyphae (relative to primary hyphae) at 3 dpi. **e** Appressoria of WT and *vir-88* on artificial hydrophobic surface. Appressoria with unusual morphology are marked with arrows. Scale bars = 20 μm. **f** Vegetative growth of WT and *vir-88* on oatmeal agar (OMA), potato dextrose agar (PDA) and Czapek-Dox minimal medium (MM). **g, h** Average conidial cell volume **g** and conidial titer **h** of WT and *vir-88* strains recovered from oatmeal agar after 7 days. The data of 3 separate plates is given. Error bars are represented as the standard error from three replicates. Significant differences based on t-tests ($p < 0.05$) are marked with asterisks

related to yeast Mob2. Because the T-DNA insertion was close to the annotated start codon and as the annotated N-terminus of CH063_12012 did not match the N-terminus of the yeast protein, it was important to verify the annotated gene model. In order to identify the correct coding region and complete transcript, we performed 5'- and 3'-RACE PCR (rapid amplification of cDNA ends) using total wild-type RNA as template (Additional file 3: Text S2). As shown in Fig. 2a, the resulting revised gene model for the *ChMOB2* gene consists of two 49 nt introns in the coding region, a 5'-UTR of 164 nt, a 3'-UTR of 216 nt and 834 nt of coding sequence. Accordingly, the T-DNA insertions of *vir-88* are located 161 bp upstream of the *ChMOB2* coding sequence. Interestingly, the putative *ChMOB2* transcription start at -164, as determined by 5'-RACE PCR, is located in the small genomic sequence which was replaced by T-DNA sequence in the genome of *vir-88*. Since the insertion is positioned upstream of *ChMOB2* and deletes the putative transcription start, we analyzed *ChMOB2* mRNA accumulation in RNA samples from *vir-88* and wild-type *C. higginsianum* by qualitative and quantitative PCR (Fig. 2b, c). In wild-type samples, *ChMOB2* RNA was detected in conidia, in appressoria formed on artificial surfaces, in mycelium and in infected

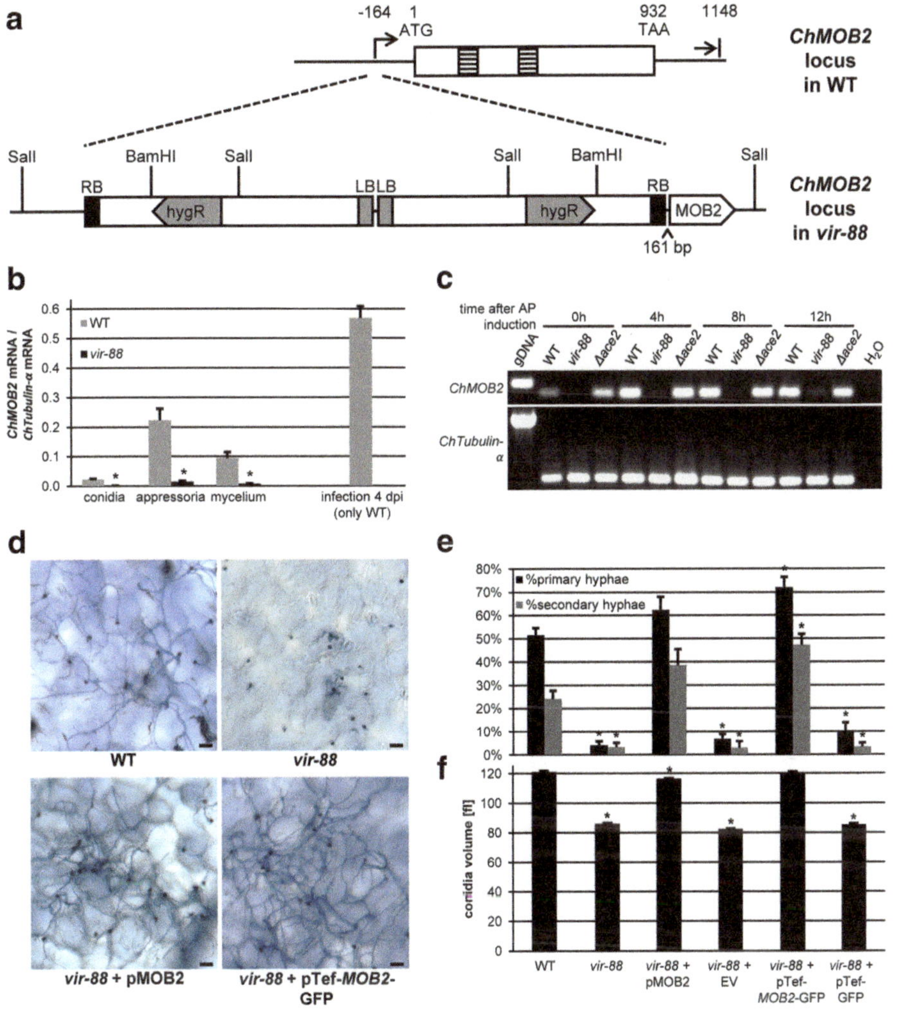


**Fig. 2** *vir-88* encodes a hypomorphic *ChMOB2* allele. **a** Schematic representation of the *ChMOB2* locus in wild-type (top) and *vir-88* (bottom). Beginning and end of the *ChMOB2* transcript are shown relative to the start codon. Introns are depicted as striped boxes. LB = left border; RB = right border. **b** Quantitative RT-PCR of *ChMOB2* mRNA abundance in conidia, appressoria, mycelium and in infected plants (4 dpi; in this case only data from WT is shown). The data were normalized against α-TUBULIN (CH063_01222) and are shown as the average of three biological replicates. **c** Semi-quantitative RT-PCR of *ChMOB2* and α-TUBULIN in RNA samples from in vitro appressoria 0, 4, 8 and 12 h after inoculation. cDNAs from WT, *vir-88* and *ΔChace2* (CY6353, see below) strains were used as template for PCR. Genomic WT DNA (gDNA) or $H_2O$ were used as controls. **d**, **e** *A. thaliana* leaves at 3 days after infection with WT, *vir-88* and *vir-88* strains complemented with either the *ChMOB2* wild-type allele (pMOB2, pCK3110) or with a pTef-*MOB2*-GFP construct (pCK4129). Typical trypan blue stained leaves are shown in **d**. Quantifications are shown in **e**. Scale bars = 20 µm. **f** Average volume of conidia from the indicated strains recovered from oatmeal agar after 7 days. The data from 3 separate plates is given. Error bars are represented as the standard error from three replicates. Significant differences based on t-tests ($p < 0.05$) are marked with asterisks

*A. thaliana* leaves. RNA levels reached 20% relative to the housekeeping gene *ChTUBULIN-α* (CH063_01222) in appressoria and 55% in RNA from *A. thaliana* leaves 4 days after infection with wild-type *C. higginsianum*. *ChMOB2* was expressed strongly during appressoria formation as early as 4 h after inoculation (Fig. 2c). In contrast, *ChMOB2* transcript levels were reduced more than ten-fold in all *vir-88* samples (Fig. 2b, c). *ChMOB2* expression in *vir-88* infected *A. thaliana* could not be analyzed because virulence of this strain was too limited.

In order to verify that the insertion upstream of *ChMOB2* and the resulting strong reduction of *ChMOB2* expression is responsible for the phenotype of the *vir-88* mutant, we complemented *vir-88* both with a plasmid (pMOB2, pCK3110) harboring the *ChMOB2* wild-type allele from -1374 to +1797 and with a translational *ChMOB2*-GFP fusion under control of the strong and constitutive elongation factor 1 alpha promoter (pTef-*MOB2*-GFP, pCK4129). After reintroduction of wild-type *ChMOB2* by ATMT with pMOB2, virulence was fully restored (Fig. 2d, e). When conidia were recovered from oatmeal plates, their cell titers (not shown) and average cell size were restored to nearly wild-type levels (Fig. 2f). Transformation with pTef1-*MOB2*-GFP fully rescued the conidia cell size phenotype (Fig. 2f), while conidiation was restored to 80% of wild-type (not shown), similar to strains complemented with the wild-type allele. In the pTef1-*MOB2*-GFP transformants, virulence appeared to be even increased relative to the wild-type (Fig. 2d, e) possibly due to overexpression of *ChMOB2*. In all cases, empty vector controls ("EV" and "pTef-GFP") did not affect the *vir-88* phenotypes. Since all observed phenotypes of *vir-88* can be complemented by reintroduction of *ChMOB2*, the phenotype of this mutant must be linked to the identified insertion. In summary, *vir-88* encodes a hypomorphic *ChMOB2* allele expressing less than 10% mRNA, which was generated by insertion of T-DNA in the 5'-UTR separating the coding region from its promoter.

### The *C. higginsianum* genome encodes three members of the Mob1/phocein protein family

Mob2 belongs to the Mob1/phocein protein family, which is widely conserved among eukaryotes and many of its members have been shown to bind and activate NDR kinases [11, 15, 16, 22]. In contrast to metazoans, which contain up to six different Mob proteins [11], fungal genomes contain between two and four Mob coding genes. The resulting proteins can be grouped into three clades (Fig. 3): The first contains proteins with high similarity to yeast Mob1, which binds the Dbf2 kinase [13], while the second clade clusters together with yeast Mob2, which binds to Cbk1 [16]. The third clade contains phocein, the founding member of the Mob1/phocein family from *Rattus norvegicus* [24], and proteins from filamentous fungi like *S. macrospora* or *N. crassa* Mob3, which have been reported to be essential for development of reproductive protoperithecia and for hyphal fusion [22, 23]. All filamentous fungi analyzed in Fig. 3 contain one protein from each clade, except *S. macrospora* and *N. crassa*, which both harbor two *MOB2*-like genes. In *N. crassa*, Mob2a and Mob2b have partly overlapping functions and were both shown to interact with the Cbk1 like kinase Cot1 [22]. As *C. higginsianum* contains one protein from each clade, we assume that each Mob protein has functions similar to the respective homologs from its clade.

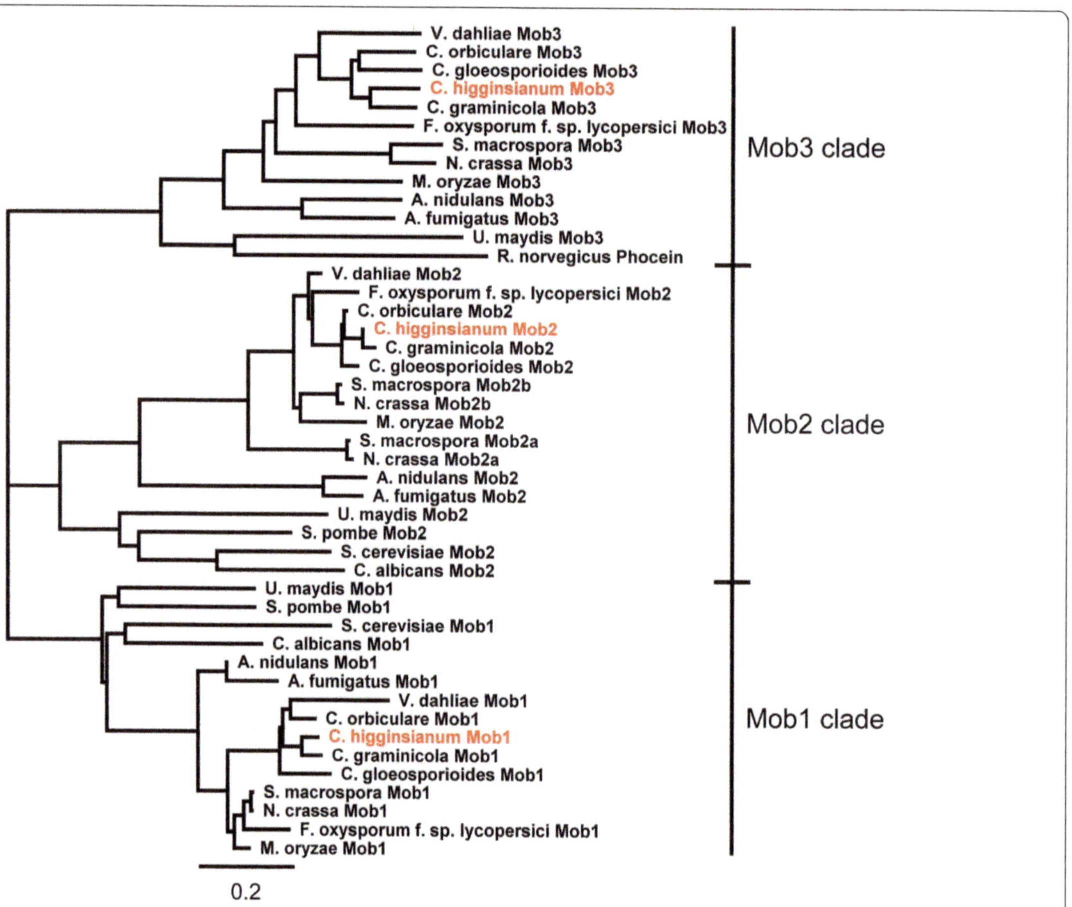


**Fig. 3** The genome of *C. higginsianum* encodes three Mob-like proteins. Unrooted phylogenetic tree of Mob proteins from *C. higginsianum* (marked in red) and other fungi. The tree was generated with the Geneious Tree builder plugin using a Blosum55 cost matrix and the Jukes-Cantor tree building method. The founding member of the Mob family (Phocein) from rat was included as a reference. Branch lengths correspond to substitutions per site. Scale bar = 0.2 substitutions per site

### *ChMOB2* and *ChCBK1* may be essential genes in *C. higginsianum*

In order to obtain *ΔChmob2* null mutants, we attempted to perform targeted gene knockout of *ChMOB2* by transformation of the non-homologous end-joining defective mutant *ΔChku80* [25] with the *ChMOB2* deletion plasmid pCK3712. This plasmid harbors the hygromycin resistance cassette (hph) flanked by *ChMOB2* upstream (-1015 to -198) and downstream sequence (+704 to +1513) on the T-DNA. In the *ΔChku80* background >90% of all transforming DNA should integrate by homologous recombination [25]. In three independent experiments, however, we recovered only transformants that harbored ectopically integrated DNA (a total of six). In contrast, transformations of the NHEJ proficient wild-type strain yielded >160 transformants. We therefore expect that *ChMOB2* may be an essential gene in *C. higginsianum*, at least in the *ΔChku80* background. Interestingly, *MOB2* can be inactivated in *S. cerevisiae* and in *N. crassa* [22, 27]. As described for *S. cerevisiae*, Mob family proteins often act as co-activators for NDR/LATS kinases. In particular, Mob2 binds to Cbk1 kinase in yeast [16]. We assumed that ChMob2 has a similar function in *C. higginsianum* and may act through activation of ChCbk1. We attempted to delete *ChCBK1* (CH063_12968) in the *ΔChku80* background. In three independent experiments this failed similar to *ChMOB2* (not shown). It is therefore likely that both *ChMOB2* and *ChCBK1* have essential functions in *C. higginsianum*. In addition, attempts to generate knock-down mutants of these genes by expression of antisense-RNAs did not lead to transformants with significantly reduced levels of the respective transcripts (Additional file 4: Figure S3).

### ChMob2 co-localizes with the potential kinase ChCbk1 in appressoria

In order to further examine the function of *C. higginsianum* Mob2, we investigated the subcellular localization of ChMob2 and its potential binding partner ChCbk1. We constructed a strain (CY6720: *ΔChku80 ΔChcbk1*:: *CBK1*-mCherry *ΔChmob2*:: *MOB2*-GFP) where the genomic copy of *ChCBK1* was replaced by a *CBK1*-mCherry fusion and where *ChMOB2* was simultaneously replaced by *MOB2*-GFP. The resulting strain had no significant defects in either virulence, conidiation or vegetative growth except for producing slightly smaller conidia (-12% compared to the parental strain *ΔChku80*, Additional file 5: Figure S4). Since both *ChMOB2* and *ChCBK1* may be essential in *C. higginsianum*, reduced ChMob2 or ChCbk1 function should have obvious effects. We therefore concluded that both fusion proteins were functional. In conidia isolated from OMA plates (Fig. 4a), both fusion proteins were more or less evenly distributed over the

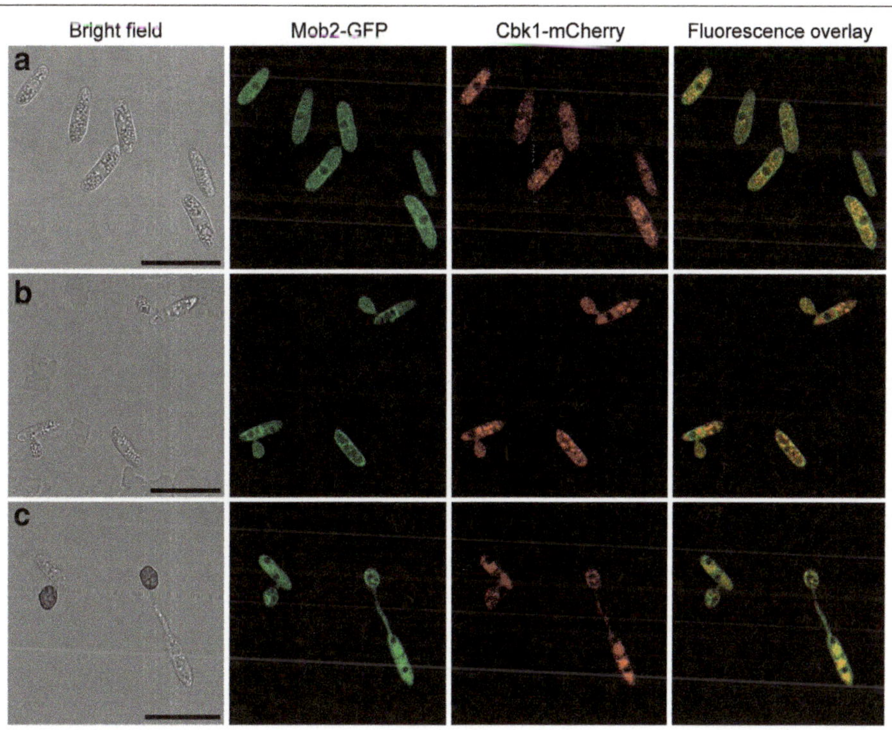


**Fig. 4** ChMob2 co-localizes with ChCbk1 in the cytoplasm. Conidia of *ΔChcbk1*:: *CBK1*-mCherry *ΔChmob2*:: *MOB2*-GFP (CY6720) were rinsed off oatmeal plates and analyzed by CLSM during in vitro appressoria formation. **a** Conidia, **b** in vitro AP at 8 hpi, **c** in vitro AP at 18 hpi. Bright field, GFP and mCherry channels are shown from left to right. The rightmost picture shows an overlay of GFP and mCherry channels. The intensities between panels are not directly comparable, because different settings were used to optimally visualize protein localization. Scale bars = 20 μm

whole cell except for the nucleus, which appeared to be partly excluded. The two proteins co-localized to some extent. Early upon in vitro appressoria formation, when appressoria were unmelanized (Fig. 4b), both Cbk1-mCherry and Mob2-GFP were completely overlapping. In the germinated conidia attached to these appressoria, however, the fusion proteins seemed to be localized to different compartments. In mature appressoria and their corresponding germinated conidia (Fig. 4c), both fusion proteins were again co-localized all over the cells except in nuclei, where both were absent. In axenic cultures, co-localization could not be addressed because the Cbk1-mCherry fusion was barely detectable (not shown). *In planta,* we could not detect either of the two fusion proteins, presumably because of the high background fluorescence in green leaves (not shown). Taken together, Mob2-GFP and Cbk1-mCherry show the same subcellular localization during appressoria formation.

### Mob2-GFP physically interacts with Cbk1-HA

ChMob2 and ChCbk1 can occupy the same subcellular compartment. Consequently, they might physically interact with each other during certain times in the cell cycle. In order to test their possible interaction, we performed co-immunoprecipitation of ChMob2 and ChCbk1. We constructed a strain (CY6681) where *ChCBK1* was fused to a 6xHA tag at its own genetic locus. In addition, this strain ectopically expressed *MOB2*-GFP under control of the Tef1a promoter. Both proteins could be detected in axenic mycelium after western blotting with anti-GFP and anti-HA antibodies (Fig 5, input). Mob2-GFP was detected at about 60 kDa (Fig. 5, input). Cbk1-HA produced several bands between 60 and 95 kDa, possibly because of different phosphorylation states of the protein or due to protein degradation (Fig. 5, Input). We used protein extracts from mycelium to investigate the interaction of ChCbk1 with ChMob2. First, Cbk1-HA was immunoprecipitated with anti-HA antibodies coupled to magnetic beads (Fig. 5, IP: α-HA). The successful pull down of Cbk1-HA is shown in the western blot decorated with anti-HA antibodies (Fig. 5 IP: α-HA, lower panel). The presence of Mob2-GFP in these immunoprecipitates was determined by decorating the western blots with anti-GFP antibodies (Fig. 5, IP: α-HA, upper panel). In extracts containing both fusion proteins (Fig. 5, IP: α-HA, lane 1), Mob2-GFP was efficiently co-immunoprecipitated with Cbk1-HA (Fig. 5, IP: α-HA, upper panel, lane 1), but not from control extracts lacking Cbk1-HA (Fig. 5, IP: α-HA, upper panel, lane 4). Similarly, when Mob2-GFP was pulled down with anti-GFP antibody coated beads, Cbk1-HA was co-precipitated (Fig. 5, IP: α-GFP, lower panel, lane 1). Some Cbk1-HA bands were missing from the anti-GFP IP, suggesting that at least some of the ChCbk1 forms might not bind to ChMob2. The successful co-precipitation of ChMob2 and ChCbk1 indicates that the two proteins can interact in axenic *C. higginsianum* cells. While technically not testable, it is likely that this interaction also occurs during infection.

Next, we tried to verify this interaction by mass spectrometry using Cbk1-HA expressing cell extracts without overexpressing Mob2-GFP. This approach could potentially also identify downstream targets of ChCbk1-phosphorylation and could show whether or not ChMob2 is the sole activator of ChCbk1. We pulled down Cbk1-

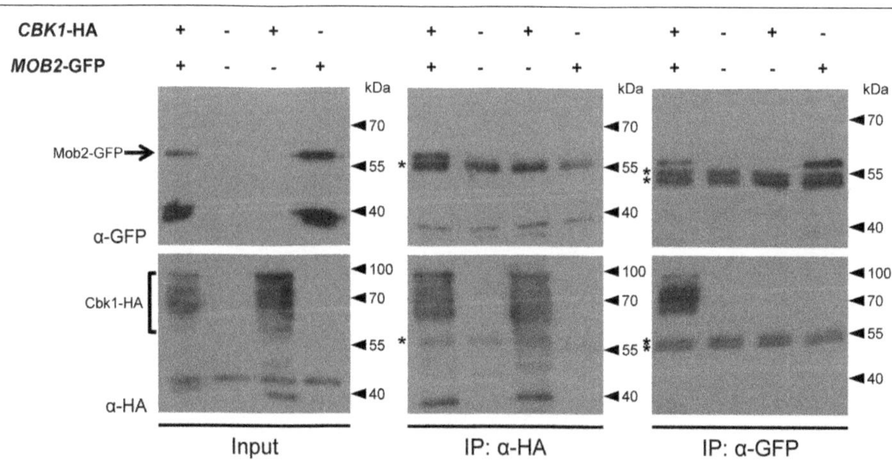


**Fig. 5** ChMob2 interacts with ChCbk1. Whole cell extracts from strains expressing different combinations of ChMob2-GFP and ChCbk1-HA were used to analyze the interaction of ChMob2 and ChCbk1 by immunoprecipitation. Strains used from left to right: *ΔChcbk1*:: *CBK1*-HA pTef1a-*MOB2*-GFP (CY6681), the parental strain *ΔChku80* (CY6021), *ΔChcbk1*:: *CBK1*-HA (CY6678) and pTef1a-*MOB2*-GFP (CY6543). Western blots of whole cell protein extracts (input), immunoprecipitates with anti-HA antibody coated beads (IP: α-HA) and immunoprecipitates with anti-GFP antibody coated beads (IP: α-GFP) were decorated either with anti-GFP (upper panels) or with anti-HA (lower panels) primary antibodies. The band corresponding to Mob2-GFP is marked with an arrow. The multiple bands corresponding to Cbk1-HA are marked with a bracket. Asterisks mark bands corresponding to primary antibody which are present in the IP and therefore are recognized by the secondary antibody

HA from extracts of axenic mycelium of CY6678 (*ΔChcbk1*:: *CBK1*-HA) and from extracts of the parental strain as control in two biological replicates using anti-HA antibody-coated magnetic beads. The eluates were analyzed by Nano LC-MS/MS and quantified label-free by integrating peak areas (Additional file 6: Table S1). ChMob2 was only identified in the *CBK1*-HA IPs but not in samples from the untagged parental strain, further verifying the interaction of these two proteins in *C. higginsianum*. The other peptides identified were probably contaminants as they consisted mainly of highly expressed enzymes from the citrate cycle, from glycolysis or proteins associated with ribosomes. The lack of potential ChCbk1 targets in the pull-downs is likely due to transient interactions of the kinase with its substrates. Alternatively, the important proteins may not be present or not targeted by ChCbk1 in samples of in vitro mycelium.

**Two potential downstream targets of the ChMob2/ChCbk1 complex are required for full pathogenicity**

Two well described Cbk1 targets in yeast are the transcription factor Ace2 and the mRNA-binding protein Ssd1 [16, 18]. We therefore tested whether or not their orthologs from *C. higginsianum* are involved in pathogenicity. We identified possible orthologs in the *C. higginsianum* genome [26]. The closest homolog to ScAce2 was CH063_01293 with 24.1% identity. After comparison with the sequence of *S. cerevisiae* Ace2 and the more closely related AfAce2 of *A. fumigatus* [28], we concluded that the automatic annotation of CH063_01293 was not complete and extended the N-terminus by 271 amino acids resulting in a revised *C. higginsianum ChACE2* gene model encoding for 815 amino acids. This extension could be verified by sequencing *ChACE2* cDNA (not shown). The revised protein has 19.5% sequence identity to ScAce2 and 31.9% to AfAce2 over the full sequence. The region containing the C2H2 zinc finger DNA binding domain showed 65.8% sequence identity to AfAce2. The potential *C. higginsianum* Ssd1 (Accession BAI59007 [29]) shows 34.5%, 77.9% and 93.2% sequence identity to the potential orthologs from *Saccharomyces cerevisiae* (Accession NP_010579), *Magnaporthe grisea* (Accession XP_003715039) and *Colletotrichum lagenarium* (Accession BAE66713), respectively. Although both ChSsd1 and ChAce2 show less than 40% identity to their respective yeast homologs, they are clearly the best match in reciprocal BLAST searches (not shown). We generated *ΔChace2* (CY6353) and *ΔChssd1* (CY6649) deletion strains in the *ΔChku80* background and examined their phenotype (Fig. 6). Consistent with a role of ChAce2 and ChSsd1 downstream of the ChMob2/ChCbk1 complex, both mutants exhibited significantly reduced virulence (Fig. 6b). Furthermore, *ΔChace2* and *ΔChssd1* strains showed decreased vegetative fitness, as their colonies on minimal medium were smaller by about 35% and 50% than wild-type, respectively (Fig. 6c). None of the mutants showed significantly altered sensitivity to osmotic stress, heat, $H_2O_2$, or cell wall stress (Additional file 2: Figure S2). Interestingly, conidial cell size and especially the conidial titer recovered from OMA plates were reduced in the *ΔChace2* mutant (Fig. 6d) similar to the phenotype observed in *vir-88*. The phenotype of *ΔChace2* could be rescued by reintroduction of wild-type *ChACE2* into the *ΔChku80* locus of this strain (Additional file 7: Figure S5), further confirming the involvement of ChAce2 in virulence and conidiation. The decrease in fitness of the *ΔChace2* and *ΔChssd1* mutants could contribute to their virulence phenotype. The conidiation phenotype of the *ΔChace2* mutant, however, is likely to be a specific phenotype, as slow growing mutants do not a priori produce less conidia (Fig. 6d). It was not possible to generate a *ΔChace2 ΔChssd1* double mutant (not shown).

In order to analyze if ChAce2 is regulated by the ChMob2/ChCbk1 complex in *C. higginsianum* as in *S. cerevisiae* [16], we analyzed the mRNA amount of potential orthologs of the ScAce2 targets *CTS1* and *SCW11* [19, 30] and of the *S. pombe* Ace2 target *MID2* [31] in samples from *vir-88* and *ΔChace2*. We expected that if ChAce2 activity were regulated by ChMob2, the transcript amount of ChAce2 targets might be different in *vir-88*, which expresses only little *ChMOB2* (Fig. 2b), and in *ΔChace2* mutants, which lacks *ChACE2*. However, using both semi-quantitative and quantitative RT-PCR we could not detect significant *ChCTS1*, *ChSCW11* or *ChMID2* transcript changes in appressoria samples in comparison to the wild-type (Additional file 8: Figure S6; see also Fig. 2c). The mRNA level of *ChACE2* was not significantly different between *vir-88* and the wild-type (Additional file 8: Figure S6). This could indicate that similar to the situation in yeast [16], ChAce2 is not regulated on the level of mRNA accumulation but by ChCbk1-dependent phosphorylation. Furthermore, *ΔChcts1* strains (CY7110 and CY7111), which lack one of the genes predicted to encode chitinase, did not show any defects in virulence, conidiation or cell separation (Additional file 8: Figure S7). A cell separation phenotype was observed in yeast *Δcts1* mutants [32]. This may indicate that Ace2 regulates different targets in *C. higginsianum* than it does in yeasts. In summary, we were not able to identify downstream targets of ChMob2/ChCbk1 in *C. higginsianum*, but deletion of two Cbk1 targets known from other systems resulted in pathogenicity and conidiation phenotypes resembling those of *vir-88*.

**The roles of ChMob3 and ChMob1 in *C. higginsianum***

While *MOB1* and *MOB2* seem to be present in all fungal genomes, *MOB3* is only found in filamentous fungi and

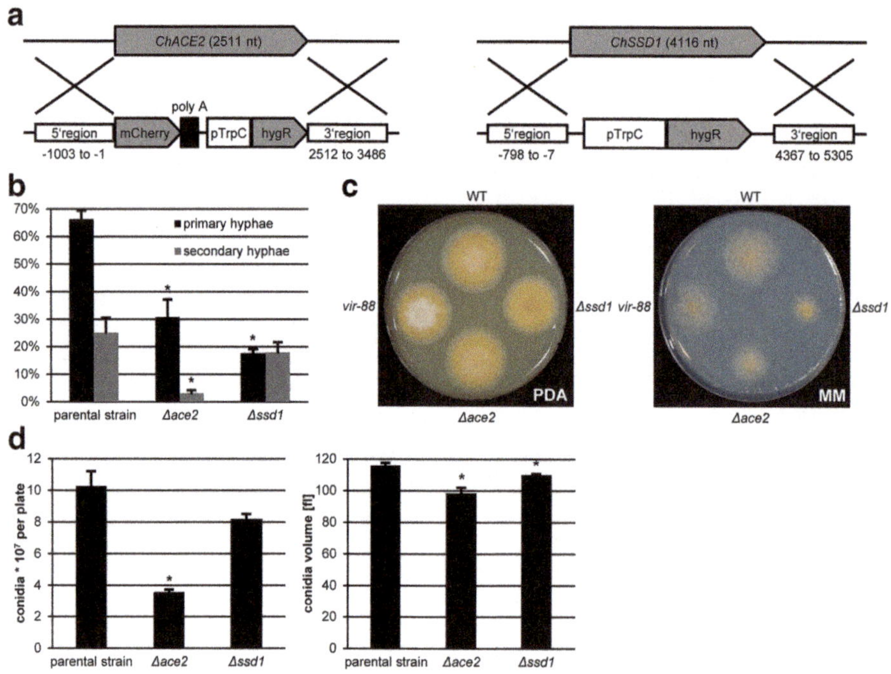


**Fig. 6** *ChACE2* and *ChSSD1* are required for full virulence of *C. higginsianum*. **a** Schematic representation of the genomic loci of *ChACE2* and *ChSSD1* (top) and the knockout cassettes of the respective deletion plasmids (bottom). The position of 5′ and 3′ flanking regions are given relative to the respective start codon. **b** Spray infection of *A. thaliana* at 3 dpi. Primary hyphae (relative to appressoria) and secondary hyphae (relative to primary hyphae) produced by the indicated strains. **c** Vegetative growth of the indicated strains on potato dextrose agar (PDA) and Czapek-Dox agar minimal medium (MM). **d** Average number of conidia (left) and cell volume (right) for the indicated strains recovered from oatmeal agar after 7 days. The data from 3 separate plates is shown. The strains used in **b**-**d** were wild-type (CY5535), *vir-88*, *ΔChace2* (CY6353), *ΔChssd1* (CY6649) and the *ΔChku80* parental strain of the last two (CY6021). Error bars are represented as the standard error from three replicates. Significant differences based on t-tests ($p < 0.05$) are marked with asterisks

higher eukaryotes [22]. The mammalian ortholog of Mob3 (Phocein) was identified in rat dendritic cells and has been shown to bind to striatin, a calmodulin binding protein [24]. In fungi, *MOB3* has been reported to be essential for sexual development and for hyphal fusion in *N. crassa* and *S. macrospora* [22, 23]. Although *C. higginsianum* has not been described to perform a sexual cycle, a potential ortholog of this gene is encoded in the *C. higginsianum* genome. We therefore investigated the potential roles of ChMob3 in *C. higginsianum* by targeted gene knockout of *ChMOB3 (CH063_02262)*. Except for a very mild reduction in colony size on both PDA and minimal medium, we did not observe significant changes either in virulence or conidiation of the resulting *ΔChmob3* mutants (Additional file 9: Figure S8). The expression pattern of *ChMOB3* did not give any further hint regarding its physiological function, as it was expressed at low, constitutive levels under all tested conditions (not shown). We cannot exclude that ChMob3 has a specific function under different, specific conditions.

In yeasts and *N. crassa*, Mob1 and Mob2 have distinct, but similar functions, namely the binding and activation of a NDR/LATS-kinase [13, 15, 16, 22]. In *S. cerevisiae*, Mob2 binds to Cbk1 in the RAM pathway and Mob1 binds to Dbf2 in the mitotic exit network. Although binding of Mob1 or Mob2 to their respective kinases is thought to be specific [15, 22], there may be crosstalk between MEN and RAM networks in *S. cerevisiae* [20, 33]. We therefore investigated if ChMob1 and ChMob2 have overlapping functions in *C. higginsianum* by genetically inactivating *ChMOB1* (CH063_02209). Transformation of *ΔChku80* with a *ChMOB1* deletion plasmid resulted in mutants with a strong growth phenotype on all tested media (Fig. 7a). In addition, these mutants showed strong sensitivity to the cell wall stress inducing and chitin binding dye congo red (Additional file 2: Figure S2), indicating a defect in chitin turnover in the cell wall [34]. Furthermore, *ΔChmob1* mutants hardly produced any conidia on oatmeal medium (not shown). The few conidia that could be recovered were smaller by about 50% (not shown) and were significantly reduced in formation of appressoria, primary hyphae and secondary hyphae after droplet inoculations (Fig. 7b). In contrast to wild-type conidia which form a septum during germination, *ΔChmob1* conidia were never septated (Additional file 9: Figure S9). Furthermore, *ΔChmob1* conidia were often connected to 1 or 2 other conidia, possibly because they did not undergo proper cell

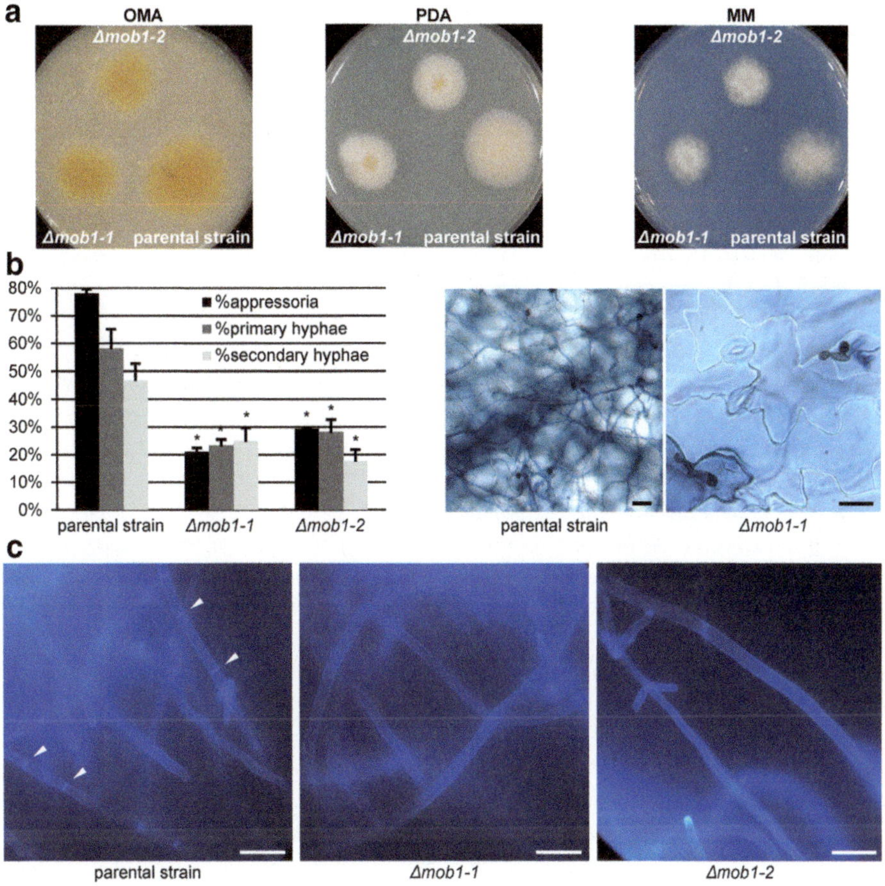


**Fig 7** *ChMOB1* is required for septum formation and cell division. Two independent *ΔChmob1* transformants (CY7242 and CY7243) and the parental strain *ΔChku80* (CY6021) were tested for vegetative growth, virulence and the ability to produce septae. **a** Vegetative growth on oatmeal agar (OMA), potato dextrose agar (PDA) and Czapek-Dox minimal medium (MM). **b** *A. thaliana* 4 days after droplet inoculation. Representative images of trypan blue stained leaves are shown on the right. The quantification of appressoria (relative to total conidia titer used for infection), primary hyphae (relative to appressoria) and secondary hyphae (relative to primary hyphae) of this infection is shown on the left. The data of 3 plants per strain with 2 leaves per plant with at least 100 counted appressoria per leaf is given. Error bars are represented as the standard error from three replicates. Significant differences based on t-tests ($p < 0.05$) are marked with asterisks. **c** Axenic mycelium stained with calcofluor white. Septae are marked with white arrows. Scale bars = 20 µm

separation during conidiation (Additional file 9: Figure S9). Hyphae from axenic mycelium of *ΔChmob1* strains also hardly produced any septae at all, in contrast to the parental strain (Fig. 7c). The strong septation phenotype of *ΔChmob1* strains is similar to mutants in *S. cerevisiae* and *S. pombe*, where Mob1 is required for completion of mitosis [27], initiation of cytokinesis and septation [35].

In summary, *C. higginsianum* contains two additional *MOB* genes. Consistent with a function of Mob3 during sexual development in other systems [22, 23], deletion of *ChMOB3* did not lead to any significant phenotype in *C. higginsianum* which is lacking a sexual cycle. In contrast, knockout of *ChMOB1* leads strong defects in appressoria formation, vegetative growth, septation, conidiation and infection.

## Discussion

Screening for genes involved in pathogenicity though aimed at identification of genes with functions exclusively required for certain aspects of the infection process, will always also identify genes with vegetative functions at many stages of their life cycle. ATPases [25, 36], intracellular transport [37], nutrition [38] and morphogenesis [39] are notable examples. Appressorial pathogens like *Colletotrichum higginsianum* undergo switches in polarity during infection of plant cells. In particular, the directed formation of a penetration peg underneath the appressorium is essential for infection. It therefore is not too surprising to identify polarity determinants like Mob2 in a screen for pathogenicity genes.

We identified a weak allele of *ChMOB2* in the *vir-88* mutant that in most cases produces appressoria with morphologic defects like elongation or secondary outgrowths. These appressoria are not able to penetrate and to initiate the infection process. Since a fraction of the *vir-88* appressoria looked normal and appeared functional, the mutant shows residual weak symptoms in some spots on the leaf, resulting in a mutant with reduced pathogenicity. In addition, *vir-88* produces less and smaller conidia. Although *vir-88* expresses less than 10% of ChMob2 mRNA relative to the wild-type, it is not a null mutant. Since the T-DNA insertion in the genome of *vir-88* is located just in front of the *ChMOB2* reading frame, it is likely that residual ChMob2 activity may be present, at least in some cells. The phenotype of a true null mutant could not be analyzed because all attempts to generate a deletion allele by homologous recombination failed. The likely explanation, therefore, is that *ChMOB2* is an essential gene at least in the *ΔChku80* background used for all knockouts. A low expression, which may vary from cell to cell, could also be the reason why not all mutant cells behave identical, resulting in incomplete penetrance of the underlying genetic defect. While this makes interpretation of the phenotype more difficult, the fortuitous isolation of a weak allele allowed us to characterize the functions of ChMob2. The defect of most *vir-88* appressoria to penetrate the host tissue correlated with morphological defects. In agreement with a possible essential function for ChMob2, its binding partner, the NDR/LATS kinase ChCbk1, could also not be knocked out. While a failure to generate knockout mutants is no evidence for an essential function, it resembles the situation in yeast and *S. pombe*, where Mob2 and Cbk1 have been found to be essential. [15, 40]. Interestingly, the lethality of RAM mutations in yeast is suppressed by loss of Ssd1 [41–43], showing that the consequences of RAM mutations may be strain specific. In most systems, *mob2* and *cbk1* mutations are indeed associated with pleiotropic phenotypes and morphological defects. In the opportunistic human pathogen *C. albicans*, all mutants in the RAM pathway including CaMob2 and CaCbk1 are viable but show hyperpolarization [44]. In the basidiomycetes *C. neoformans* and *U. maydis*, mutants of the Cbk1 ortholog Ukc1, Mob2 and all other associated proteins thought to be involved in the same pathway lead to hyperpolarized growth and decreased virulence [45, 46]. In *A. nidulans*, mutants in Cbk1 (AnCotA) and Mob2 orthologs (AnMobB) show growth defects, strongly reduced ability to form conidia, altered conidia cell size and increased number of nuclei in spores [47, 48]. In *N. crassa*, which is more closely related to *C. higginsianum*, the *cot-1 (ts)* mutant and the *mob2a mob2b* double mutant show increased rate of branching in axenic mycelium, reduced growth rate and reduced conidiation [21, 22]. Most phenotypes observed in our study for the *Chmob2* mutant *vir-88* are consistent with functions reported for Mob2 and Cbk1 in filamentous fungi, including a function in conidiation. Reduced cell size of conidia was an obvious phenotype of *vir-88*. Altered conidial cell size may be the consequence of altered cell division or septation in conidial precursor cells [49]. Altered conidial cell size was also observed in *N. crassa cot1* mutants. They were, however, 4 times larger than wild-type cells, rather than smaller as in the case of *C. higginsianum* [48]. We also observed about 50% reduction in conidiation ability in the *vir-88* mutant. Considering that *vir-88* is not a null allele of *ChMOB2*, this phenotype is reminiscent of *A. nidulans* and *N. crassa mob2* mutants which largely fail to form conidia [22, 47]. Conidiation is critically dependent on septation, which in turn is also dependent on the Mob1/DBF2-dependent MEN (Mitotic Exit Network) or SIN (Septation Initiation Network) signalling pathways in many fungi [13, 22, 50–52]. In fact, *N. crassa* and *A. nidulans Δmob1* [22, 50] show sporulation defects similar to the *C. higginsianum ΔChmob1* mutants described in this study. It is possible that the Mob1/Dbf2 and Mob2/Cbk1 complexes have overlapping functions or that they control each other. The latter may be the case in related systems. In *N. crassa*, Mob1 appears to affect Cot1 activity more than its direct interactor and co-activator Mob2b [22]. In *S. pombe*, SpSid1 (= Dbf2) controls SpOrb6 (= Cbk1) activity [53], while Mob1 and Cdc14 activity is required for proper Mob2 and Ace2 localization in yeast [16]. For comparison, a simplified model of the potential interplay between MEN and RAM of *S. cerevisiae* is shown in Additional file 10: Figure S10.

In addition to a strong conidiation phenotype, *ΔChmob1* mutants exhibited attenuated ability to form appressoria, primary and secondary hyphae upon infection and showed a severe septation defect. These phenotypes are similar to the corresponding mutants in *N. crassa* [22] and *A. nidulans* [50]. Mob1 is essential both in S. *cerevisiae* [27] and *S. pombe* [14], but cells harboring conditional alleles of *MOB1* either arrest in late mitosis or, in the case of *S. pombe*, do not septate and become multinucleate. In addition to nuclear division, *S. cerevisiae* Mob1 also seems to have a role during cytokinesis and cell separation [54]. Based on these similarities, it is likely that the function of Mob1 and its associated signalling pathway is conserved among ascomycetes.

What are the critical targets of the Mob2/Cbk1 pathway in filamentous fungi? Similar to the situation in yeast, it was found that deletion of the *N. crassa* homolog of *SSD1* (*GUL1*) can suppress defects of the *cot1 (ts)* allele [55, 56] suggesting that Ssd1 is an important component of the RAM pathway also in filamentous fungi. In *S. cerevisiae*, Ssd1 inactivation is thought to occur through Cbk1 phosphorylation [41–43]. Because Ssd1 may be a conserved Cbk1 target we analyzed the phenotype of a *ΔChssd1* mutant allele. We found that the lack of ChSsd1 leads to attenuated virulence in *C. higginsianum*. The effect on virulence was similar to *vir-88* and to *Δssd1* mutants in the plant pathogens *C. lagenarium* and *Magnaporthe grisea* [57] supporting the notion that Ssd1 is a conserved virulence associated target of the RAM pathway. In *C. lagenarium* [57], *Δssd1* mutants mostly failed to penetrate and intracellular hyphae were only detected very rarely. In *M. grisea* [57], *Δssd1* mutants were still able to produce intracellular primary hyphae on rice plants, but they were mostly restricted to dead host cells.

We also analyzed the effect of ChAce2 on pathogenicity in *C. higginsianum* because it may be a conserved target of the RAM pathway. Again, we saw a reduction of pathogenicity and, interestingly, a comparable effect on cell volume in conidia and reduced conidiation. However, we found no evidence for posttranslational regulation of ChAce2 by the ChMob2/ChCbk1 complex, as described for *S. cerevisiae* [12].

We observed that *C. higginsianum* Mob2 and Cbk1 localize to the cytoplasm and are excluded from nuclei in conidia and during in vitro appressoria formation. In baker's yeast, however, Mob2 and Cbk1 proteins can be found at sites of polarized growth like the cortex of the growing bud and the mating projection or in the daughter cell nucleus and the septum during cell separation, where they supposedly regulate cytokinesis genes through the transcription factor ScAce2 [16]. This difference in localization could indicate that Ace2 is not a conserved Mob2/Cbk1 target in filamentous fungi. In the absence of verified targets for the transcription factor ChAce2, it also remains unclear whether or not ChAce2 and yeast Ace2 have a common function. A protein homologous to ScAce2 was also analyzed in *A. fumigatus* [28], where knockout mutants had attenuated ability to produce conidia, abnormal cell wall architecture and were hypervirulent in a mouse model. This protein (AfAce2) is more closely related to ChAce2 (31.9% sequence identity) than to the yeast protein and, given their common conidiation phenotype, they may be orthologous gene products.

## Conclusion

This study showed that the Mob-family protein ChMob2 from the plant pathogen *Colletotrichum higginsianum* is involved in virulence on *Arabidopsis thaliana* and has a role in conidiation. ChMob2 forms a complex with the conserved ChCbk1 Kinase. The study further analyzed the functions of the other two Mob proteins encoded in *C. higginsianum* by targeted gene knockouts. *ChMOB1* is required for conidiation, cytokinesis and plant infection, while *ΔChmob3* mutants have no obvious phenotype in vegetative cells or during infection.

## Methods

### Strains and media

*E. coli*, *A. tumefaciens* and *C. higginsianum* strains were obtained, cultured and transformed as described [25] with the following modification: For selection of bialaphos-resistant *C. higginsianum* transformants, the transformation mixture of conidia and *Agrobacteria* was co-cultivated on Czapek Dox minimal medium plates containing 2.64 g/l $(NH_4)_2SO_4$. For selection, the plates were supplemented with 10 – 20 μg/ml bialaphos and 100 μg/ml cefotaxime. *A. thaliana* Col-0 was grown and infected with *C. higginsianum* as described [25]. In vitro appressoria were induced on petri dishes as described [25]. For analysis of conidiation, *strains* were grown for 7 days on oatmeal plates and then stored at 4 °C. Conidia were rinsed off with sterile water, washed once by centrifugation for 5 min at 3000 rpm and measured by cell counting. Three wells of an oatmeal agar 12-well petri dish were inoculated with 100 μl of conidia suspension containing 1000 conidia. After incubation for 7 to 8 days, each well was thoroughly harvested three times with 500 μl MQ. The three aliquots were pooled and measured by cell counting. This experiment was repeated at least once for every strain. All *C. higginsianum* strains used in this study are listed in Additional file 11: Text S3.

### Standard techniques

DNA manipulations, PCR reactions, Southern blotting and plasmid DNA isolations followed standard protocols as described [25, 58]. Extraction of RNA, RT-PCR and qRT-PCR were performed as described [25]. Oligonucleotide sequences and plasmid constructions are listed in Additional file 11: Text S3.

### Microscopy and histochemical staining

Histochemical samples or samples containing fluorescent reporter proteins were stained and analyzed by microscopy as described [25] with the following modifications. Confocal GFP fluorescence was detected between 498 and 547 nm and mCherry fluorescence between 570 and 637 nm. Quantification of trypan blue stained fungal infection structures after spray infection was performed with 3 infected plants per strain at 3 days post infection. Two leaves were analyzed per plant. At least 400 appressoria were counted per leaf. Septae were stained with calcofluor white (100 fold dilution with PBS buffer of a 10 mg/ml stock solution in DMSO) by addition of the working solution to the specimen and rinsing 2 to 3 times with water. Fluorescence was observed using the DAPI filter.

### Sequence analyses and accessions

*C. higginsianum* DNA sequences [59] were obtained from the *Colletotrichum* Sequencing Project, Broad Institute of Harvard and MIT [26] and from the EnsemblFungi Server

[60]. Where indicated, DNA sequences were obtained from the Max Planck Institute for Plant Breeding Research *Colletotrichum higginsianum* Database [61] and [62]. Sequence alignments and phylogenetic trees were performed with Geneious Alignment or ClustalW algorithms as implemented in the Geneious 5.5.6 software package (Biomatters Limited).

Genbank Accessions: *ChMOB1* (CCF39800), *ChMOB2* (KP261084), *ChMOB3* (CCF40036), *ChACE2* (BK009983), *ChSSD1* (AB508804). Additional accessions can be found in Additional file 11: Text S3.

### Rapid amplification of cDNA ends (RACE)

5'- and 3'- RACE PCR was performed as described in the SMART™ RACE cDNA Amplification Kit User Manual (Clontech) with the following modifications. 5'RACE PCR-ready cDNA was prepared using 1 µg of total RNA isolated from conidia, 1 µl of 5'CDS primer (12 µM, CK4502), 1 µl of SMART II A oligo (12 µM, CK3983) and 1 µl Powerscript reverse transcriptase (Clontech) in a total volume of 8 µl. 3'RACE PCR-ready cDNA was prepared identically except for 1 µl of 3'CDS primer (12 µM) instead of the 5'CDS primer and without SMART II A oligo. ChMOB2 5' RACE PCR was performed using 2.5 µl of 5'RACE PCR-ready cDNA as template and 5 µl universal primer mix (UPM, 0.4 µM CK3951 and 2 µM CK3952) together with 1 µl CK4050 (10 µM). ChMOB2 3'RACE PCR was performed using 2.5 µl 3'RACE PCR-ready cDNA and 5 µl UPM together with 1 µl CK4048 (10 µM). These reactions were set up in a total volume of 50 µl and contained 1.25 U HotStart Taq (Peqlab), 5 µl reaction buffer S and 1 µl dNTPs (10 mM each) in addition to template and primers. The reaction products were diluted 50 fold in Tricine-EDTA buffer and used as templates for the nested ChMOB2 RACE PCR with the nested universal primer mix (NUPM) together with CK4049 (5'RACE PCR) or CK3971 (3'RACE PCR). The resulting ChMOB2 PCR products of approximately 250 bp (5'RACE PCR) and 1100 bp (3'RACE PCR) were subcloned into CloneJet (Thermo Fisher Scientific) and sequenced.

### Co-immunoprecipitation of *C. higginsianum* extracts

6 x $10^5$ conidia (washed once) were inoculated in 300 ml liquid Modified Mathur's medium [63] and incubated for 48 h at 28 °C in a water bath while shaking at 160 rpm. The mycelium was filtered, washed with water, squeezed out and pestled to a fine powder. 200 mg of mycelial powder was resuspended in 1 ml IPP150 buffer (10 mM Tris–HCl pH 8.0, 150 mM NaCl, 0.1% NP40) containing Complete™ Protease Inhibitor Cocktail (Roche). After addition of 500 µl glass beads, the suspension was beaten at 4 °C for 20 min and centrifuged at 13000 rpm for 10 min. The supernatant was taken off, recentrifuged and taken off again. 50 µl magnetic Pan Mouse IgG Dynabeads® were washed three times with 1 ml PBS buffer containing 5 mg/ml BSA and subsequently incubated over-night in PBS/BSA solution containing either 2.5 µg mouse anti-GFP (Roche applied science, mixture of clones 7.1 and 13.1) or 2.5 µg mouse anti-HA antibody (12CA5). Afterwards, the beads were washed three times with PBS/BSA. 350 µl of mycelial raw extract (= input) was added to the beads and incubated for 2 h. The beads were washed 6 times with 1 ml IPP150 containing protease inhibitor. Purified proteins were eluted from the beads by incubation for 7 min at 95 °C in 50 µl 3xSDS sample buffer (300 mM Tris–HCl pH 6.8, 6% SDS, 30% glycerol, 60 µg/ml Bromophenol blue, 10% 2-Mercaptoethanol).

### Mass spectrometric analysis

Mycelium was grown in two replicates as described above for co-immunoprecipitation experiments. 400 mg of mycelial powder was harvested, resuspended in IPP150 buffer containing cOmplete™ Protease Inhibitor and PhosSTOP Phosphatase Inhibitor Cocktails (Roche) and beaten with glass beads for 30 min at 4 °C. After centrifugation at 13000 rpm for 10 min, the supernatant was taken off and purified again by centrifugation. The raw extract was incubated for 2 h with anti-HA antibody covered magnetic beads as described above. The beads were washed 6 times with IPP150 buffer containing protease and phosphatase inhibitor. Purified proteins were eluted from the beads using two times 50 µl of 0.2 M glycine-HCl buffer (pH 2.5) and then neutralized with 10 µl of 1 M Tris–HCl (pH 10.4). Tryptic digestion of the eluted samples, their processing and Nano-LC-MS/MS analysis was performed as described [64]. Raw data files were evaluated using Peaks7 (Bioinformatics Solutions Inc., Waterloo, ON, Canada) and a *Colletotrichum higginsianum* proteome database [26]. The amino acid sequence of Cbk1-HA was added manually to this database. For quantification of peak areas, two groups (*CBK1*-HA vs parental strain) containing the respective two biological replicates were used. The samples were normalized by the total ion count (TIC) and only proteins that were at least 10 times more abundant in the *CBK1*-HA samples with a significance of 10 or more were included in the analysis.

## Additional files

**Additional file 1: Figure S1.** *vir-88* produces irregularly shaped appressoria. (PPTX 3922 kb)

**Additional file 2: Figure S2.** *ΔChmob1* mutants are sensitive against cell wall stress induced by congo red. (PPTX 5212 kb)

**Additional file 3: Text S1 S2.** T-DNA flanking sequence of *vir-88* and *ChMOB2* RACE PCR Sequences. (DOCX 15 kb)

**Additional file 4: Figure S3.** Silencing constructs directed against *ChMOB2* and *ChCbk1* did not lead to significant reduction of the respective transcripts.

**Additional file 5: Figure S4.** *MOB2*-GFP and *CBK1*-mCherry in locus fusions are functional.

**Additional file 6: Table S1.** Nano LC-MS/MS identification of proteins co-purified with Cbk1-HA.

**Additional file 7: Figure S5.** The *ΔChace2* mutant can be complemented by reintroduction of *ChAce2*.

**Additional file 8: Figure S6, S7.** Potential ChAce2 targets.

**Additional file 9: Figure S8, S9.** Functions of ChMob1 and ChMob3.

**Additional file 10: Figure S10.** The RAM and MEN pathways in *Saccharomyces cerevisiae*.

**Additional file 11: Text S3.** Additional accessions, plasmid constructions, strains, plasmids and oligonucleotides.

**Acknowledgements**

Batian Föhr, Lisa Knipfer, Jasmin Friedrich and Paul Wollschläger for help during construction of various plasmids; Martin Korn and Lars Voll for helpful discussions and Peter-Louis Plaumann for critically reading the manuscript.

**Funding**

Supported by the Universitätsbund Erlangen-Nürnberg and by the Open-Access-Fonds of the FAU.

**Authors' contributions**

CK conceived the project. JS and MD performed all biological experiments. JH performed the mass spectrometric analysis. JS prepared the figures. JS and CK wrote the manuscript. All authors read and approved the final manuscript.

**Competing interests**

The authors declare that they have no competing interests.

**References**

1. Gupta YK, Dagdas YF, Martinez-Rocha AL, Kershaw MJ, Littlejohn GR, Ryder LS, Sklenar J, Menke F, Talbot NJ. Septin-Dependent Assembly of the Exocyst Is Essential for Plant Infection by Magnaporthe oryzae. Plant Cell. 2015;27:3277–89.
2. Giraldo MC, Dagdas YF, Gupta YK, Mentlak TA, Yi M, Martinez-Rocha AL, Saitoh H, Terauchi R, Talbot NJ, Valent B. Two distinct secretion systems facilitate tissue invasion by the rice blast fungus Magnaporthe oryzae. Nat Commun. 2013;4:1996.
3. Nesher I, Minz A, Kokkelink L, Tudzynski P, Sharon A. Regulation of pathogenic spore germination by CgRac1 in the fungal plant pathogen Colletotrichum gloeosporioides. Eukaryot Cell. 2011;10:1122–30.
4. Sakaguchi A, Miyaji T, Tsuji G, Kubo Y. Kelch repeat protein Clakel2p and calcium signaling control appressorium development in Colletotrichum lagenarium. Eukaryot Cell. 2008;7:102–11.
5. Fukada F, Kubo Y. Colletotrichum orbiculare Regulates Cell Cycle G1/S Progression via a Two-Component GAP and a GTPase to Establish Plant Infection. Plant Cell. 2015;27:2530–44.
6. O'Connell R, Herbert C, Sreenivasaprasad S, Khatib M, Esquerre-Tugaye MT, Dumas B. A novel Arabidopsis-Colletotrichum pathosystem for the molecular dissection of plant-fungal interactions. Mol Plant Microbe Interact. 2004;17:272–82.
7. Perfect SE, Hughes HB, O'Connell RJ, Green JR. Colletotrichum: A model genus for studies on pathology and fungal-plant interactions. Fungal Genet Biol. 1999;27:186–98.
8. Latunde-Dada AO. Colletotrichum: tales of forcible entry, stealth, transient confinement and breakout. Mol Plant Pathol. 2001;2:187–98.
9. Howard RJ, Valent B. Breaking and entering: host penetration by the fungal rice blast pathogen Magnaporthe grisea. Annu Rev Microbiol. 1996;50:491–512.
10. Steinberg G. Hyphal growth: a tale of motors, lipids, and the Spitzenkorper. Eukaryot Cell. 2007;6:351–60.
11. Hergovich A, Stegert MR, Schmitz D, Hemmings BA. NDR kinases regulate essential cell processes from yeast to humans. Nat Rev Mol Cell Biol. 2006;7:253–64.
12. Nelson B, Kurischko C, Horecka J, Mody M, Nair P, Pratt L, Zougman A, McBroom LD, Hughes TR, Boone C, Luca FC. RAM: a conserved signaling network that regulates Ace2p transcriptional activity and polarized morphogenesis. Mol Biol Cell. 2003;14:3782–803.
13. Komarnitsky SI, Chiang YC, Luca FC, Chen J, Toyn JH, Winey M, Johnston LH, Denis CL. DBF2 protein kinase binds to and acts through the cell cycle-regulated MOB1 protein. Mol Cell Biol. 1998;18:2100–7.
14. Hou MC, Salek J, McCollum D. Mob1p interacts with the Sid2p kinase and is required for cytokinesis in fission yeast. Curr Biol. 2000;10:619–22.
15. Hou MC, Wiley DJ, Verde F, McCollum D. Mob2p interacts with the protein kinase Orb6p to promote coordination of cell polarity with cell cycle progression. J Cell Sci. 2003;116:125–35.
16. Weiss EL, Kurischko C, Zhang C, Shokat K, Drubin DG, Luca FC. The Saccharomyces cerevisiae Mob2p-Cbk1p kinase complex promotes polarized growth and acts with the mitotic exit network to facilitate daughter cell-specific localization of Ace2p transcription factor. J Cell Biol. 2002;158:885–900.
17. Kurischko C, Kuravi VK, Wannissorn N, Nazarov PA, Husain M, Zhang C, Shokat KM, McCaffery JM, Luca FC. The yeast LATS/Ndr kinase Cbk1 regulates growth via Golgi-dependent glycosylation and secretion. Mol Biol Cell. 2008;19:5559–78.
18. Jansen JM, Wanless AG, Seidel CW, Weiss EL. Cbk1 regulation of the RNA-binding protein Ssd1 integrates cell fate with translational control. Curr Biol. 2009;19:2114–20.
19. Doolin MT, Johnson AL, Johnston LH, Butler G. Overlapping and distinct roles of the duplicated yeast transcription factors Ace2p and Swi5p. Mol Microbiol. 2001;40:422–32.
20. Brace J, Hsu J, Weiss EL. Mitotic exit control of the Saccharomyces cerevisiae Ndr/LATS kinase Cbk1 regulates daughter cell separation after cytokinesis. Mol Cell Biol. 2011;31:721–35.
21. Yarden O, Plamann M, Ebbole DJ, Yanofsky C. cot-1, a gene required for hyphal elongation in Neurospora crassa, encodes a protein kinase. EMBO J. 1992;11:2159–66.
22. Maerz S, Dettmann A, Ziv C, Liu Y, Valerius O, Yarden O, Seiler S. Two NDR kinase-MOB complexes function as distinct modules during septum formation and tip extension in Neurospora crassa. Mol Microbiol. 2009;74:707–23.
23. Bernhards Y, Poggeler S. The phocein homologue SmMOB3 is essential for vegetative cell fusion and sexual development in the filamentous ascomycete Sordaria macrospora. Curr Genet. 2011;57:133–49.
24. Baillat G, Moqrich A, Castets F, Baude A, Bailly Y, Benmerah A, Monneron A. Molecular cloning and characterization of phocein, a protein found from the Golgi complex to dendritic spines. Mol Biol Cell. 2001;12:663–73.
25. Korn M, Schmidpeter J, Dahl M, Muller S, Voll LM, Koch C. A Genetic Screen for Pathogenicity Genes in the Hemibiotrophic Fungus Colletotrichum higginsianum Identifies the Plasma Membrane Proton Pump Pma2 Required for Host Penetration. PLoS One. 2015;10:e0125960.
26. Colletotrichum Sequencing Project, Broad Institute of Harvard and MIT http://www.broadinstitute.org/scientific-community/science/projects/fungal-genome-initiative/colletotrichum-genome-project?sp=S7000006370958884. Accessed 2016.
27. Luca FC, Winey M. MOB1, an essential yeast gene required for completion of mitosis and maintenance of ploidy. Mol Biol Cell. 1998;9:29–46.

28. Ejzykowicz DE, Cunha MM, Rozental S, Solis NV, Gravelat FN, Sheppard DC, Filler SG. The Aspergillus fumigatus transcription factor Ace2 governs pigment production, conidiation and virulence. Mol Microbiol. 2009;72:155–69.
29. Ushimaru T, Terada H, Tsuboi K, Kogou Y, Sakaguchi A, Tsuji G, Kubo Y. Development of an efficient gene targeting system in Colletotrichum higginsianum using a non-homologous end-joining mutant and Agrobacterium tumefaciens-mediated gene transfer. Mol Genet Genomics. 2010;284:357–71.
30. Dohrmann PR, Butler G, Tamai K, Dorland S, Greene JR, Thiele DJ, Stillman DJ. Parallel pathways of gene regulation: homologous regulators SWI5 and ACE2 differentially control transcription of HO and chitinase. Genes Dev. 1992;6:93–104.
31. Petit CS, Mehta S, Roberts RH, Gould KL. Ace2p contributes to fission yeast septin ring assembly by regulating mid2+ expression. J Cell Sci. 2005;118:5731–42.
32. Kuranda MJ, Robbins PW. Chitinase is required for cell separation during growth of Saccharomyces cerevisiae. J Biol Chem. 1991;266:19758–67.
33. Mazanka E, Weiss EL. Sequential counteracting kinases restrict an asymmetric gene expression program to early G1. Mol Biol Cell. 2010;21:2809–20.
34. Ram AF, Klis FM. Identification of fungal cell wall mutants using susceptibility assays based on Calcofluor white and Congo red. Nat Protoc. 2006;1:2253–6.
35. Sparks CA, Morphew M, McCollum D. Sid2p, a spindle pole body kinase that regulates the onset of cytokinesis. J Cell Biol. 1999;146:777–90.
36. Balhadere PV, Talbot NJ. PDE1 encodes a P-type ATPase involved in appressorium-mediated plant infection by the rice blast fungus Magnaporthe grisea. Plant Cell. 2001;13:1987–2004.
37. Fujihara N, Sakaguchi A, Tanaka S, Fujii S, Tsuji G, Shiraishi T, O'Connell R, Kubo Y. Peroxisome biogenesis factor PEX13 is required for appressorium-mediated plant infection by the anthracnose fungus Colletotrichum orbiculare. Mol Plant Microbe Interact. 2010;23:436–45.
38. Huser A, Takahara H, Schmalenbach W, O'Connell R. Discovery of pathogenicity genes in the crucifer anthracnose fungus Colletotrichum higginsianum, using random insertional mutagenesis. Mol Plant Microbe Interact. 2009;22:143–56.
39. Wang L, Lin X. Morphogenesis in fungal pathogenicity: shape, size, and surface. PLoS Pathog. 2012;8:e1003027.
40. Verde F, Wiley DJ, Nurse P. Fission yeast orb6, a ser/thr protein kinase related to mammalian rho kinase and myotonic dystrophy kinase, is required for maintenance of cell polarity and coordinates cell morphogenesis with the cell cycle. Proc Natl Acad Sci U S A. 1998;95:7526–31.
41. Du LL, Novick P. Pag1p, a novel protein associated with protein kinase Cbk1p, is required for cell morphogenesis and proliferation in Saccharomyces cerevisiae. Mol Biol Cell. 2002;13:503–14.
42. Kurischko C, Weiss G, Ottey M, Luca FC. A role for the Saccharomyces cerevisiae regulation of Ace2 and polarized morphogenesis signaling network in cell integrity. Genetics. 2005;171:443–55.
43. Jorgensen P, Nelson B, Robinson MD, Chen Y, Andrews B, Tyers M, Boone C. High-resolution genetic mapping with ordered arrays of Saccharomyces cerevisiae deletion mutants. Genetics. 2002;162:1091–9.
44. Song Y, Cheon SA, Lee KE, Lee SY, Lee BK, Oh DB, Kang HA, Kim JY. Role of the RAM network in cell polarity and hyphal morphogenesis in Candida albicans. Mol Biol Cell. 2008;19:5456–77.
45. Sartorel E, Perez-Martin J. The distinct interaction between cell cycle regulation and the widely conserved morphogenesis-related (MOR) pathway in the fungus Ustilago maydis determines morphology. J Cell Sci. 2012;125:4597–608.
46. Walton FJ, Heitman J, Idnurm A. Conserved elements of the RAM signaling pathway establish cell polarity in the basidiomycete Cryptococcus neoformans in a divergent fashion from other fungi. Mol Biol Cell. 2006; 17:3768–80.
47. Shi J, Chen W, Liu Q, Chen S, Hu H, Turner G, Lu L. Depletion of the MobB and CotA complex in Aspergillus nidulans causes defects in polarity maintenance that can be suppressed by the environment stress. Fungal Genet Biol. 2008;45:1570–81.
48. Johns SA, Leeder AC, Safaie M, Turner G. Depletion of Aspergillus nidulans cotA causes a severe polarity defect which is not suppressed by the nuclear migration mutation nudA2. Mol Genet Genomics. 2006;275:593–604.
49. Park HS, Yu JH. Genetic control of asexual sporulation in filamentous fungi. Curr Opin Microbiol. 2012;15:669–77.
50. Kim JM, Lu L, Shao R, Chin J, Liu B. Isolation of mutations that bypass the requirement of the septation initiation network for septum formation and conidiation in Aspergillus nidulans. Genetics. 2006;173:685–96.
51. Krapp A, Simanis V. An overview of the fission yeast septation initiation network (SIN). Biochem Soc Trans. 2008;36:411–5.
52. Mah AS, Jang J, Deshaies RJ. Protein kinase Cdc15 activates the Dbf2-Mob1 kinase complex. Proc Natl Acad Sci U S A. 2001;98:7325–30.
53. Kanai M, Kume K, Miyahara K, Sakai K, Nakamura K, Leonhard K, Wiley DJ, Verde F, Toda T, Hirata D. Fission yeast MO25 protein is localized at SPB and septum and is essential for cell morphogenesis. EMBO J. 2005;24:3012–25.
54. Luca FC, Mody M, Kurischko C, Roof DM, Giddings TH, Winey M. Saccharomyces cerevisiae Mob1p is required for cytokinesis and mitotic exit. Mol Cell Biol. 2001;21:6972–83.
55. Seiler S, Vogt N, Ziv C, Gorovits R, Yarden O. The STE20/germinal center kinase POD6 interacts with the NDR kinase COT1 and is involved in polar tip extension in Neurospora crassa. Mol Biol Cell. 2006;17:4080–92.
56. Herold I, Yarden O. Regulation of Neurospora crassa cell wall remodeling via the cot-1 pathway is mediated by gul-1. Curr Genet. 2016. doi:10.1007/s00294-016-0625-z.
57. Tanaka S, Yamada K, Yabumoto K, Fujii S, Huser A, Tsuji G, Koga H, Dohi K, Mori M, Shiraishi T, et al. Saccharomyces cerevisiae SSD1 orthologues are essential for host infection by the ascomycete plant pathogens Colletotrichum lagenarium and Magnaporthe grisea. Mol Microbiol. 2007;64:1332–49.
58. Green MR. Molecular cloning : a laboratory manual / Michael R. Green, Joseph Sambrook. Cold Spring Harbor, N.Y: Cold Spring Harbor Laboratory Press; 2012.
59. O'Connell RJ, Thon MR, Hacquard S, Amyotte SG, Kleemann J, Torres MF, Damm U, Buiate EA, Epstein L, Alkan N, et al. Lifestyle transitions in plant pathogenic Colletotrichum fungi deciphered by genome and transcriptome analyses. Nat Genet. 2012;44:1060–5.
60. EnsemblFungi Server. http://fungi.ensembl.org/Colletotrichum_higginsianum/Info/Index. Accessed 2016.
61. Max Planck Institute for Plant Breeding Research Colletotrichum higginsianum Database. http://gbrowse.mpiz-koeln.mpg.de/cgi-bin/gbrowse/colletotrichum_higginsianum_public/. Accessed 2016.
62. Zampounis A, Pigne S, Dallery JF, Wittenberg AH, Zhou S, Schwartz DC, Thon MR, O'Connell RJ. Genome Sequence and Annotation of Colletotrichum higginsianum, a Causal Agent of Crucifer Anthracnose Disease. Genome Announc. 2016;4.
63. Mathur RS, Barnett HL, Lilly VG. Sporulation of Colletotrichum lindemuthianum in culture. Phytopathology. 1950;1:104–14.
64. Lamm CE, Scherer M, Reuter N, Amin B, Stamminger T, Sonnewald U. Human promyelocytic leukemia protein is targeted to distinct subnuclear domains in plant nuclei and colocalizes with nucleolar constituents in a SUMO-dependent manner. FEBS Open Biol. 2016;6:1141–54.

6

# Spontaneous mutations in the *flhD* operon generate motility heterogeneity in *Escherichia coli* biofilm

Shelley M. Horne, Joseph Sayler, Nicholas Scarberry, Meredith Schroeder, Ty Lynnes and Birgit M. Prüß*

**Abstract**

**Background:** Heterogeneity and niche adaptation in bacterial biofilm involve changes to the genetic makeup of the bacteria and gene expression control. We hypothesized that i) spontaneous mutations in the *flhD* operon can either increase or decrease motility and that ii) the resulting motility heterogeneity in the biofilm might lead to a long-term increase in biofilm biomass.

**Results:** We allowed the highly motile *E. coli* K-12 strain MC1000 to form seven- and fourteen-day old biofilm, from which we recovered reduced motility isolates at a substantially greater frequency (5.4 %) than from a similar experiment with planktonic bacteria (0.1 %). Biofilms formed exclusively by MC1000 degraded after 2 weeks. In contrast, biofilms initiated with a 1:1 ratio of MC1000 and its isogenic *flhD::kn* mutant remained intact at 4 weeks and the two strains remained in equilibrium for at least two weeks. These data imply that an 'optimal' biofilm may contain a mixture of motile and non-motile bacteria.

Twenty-eight of the non-motile MC1000 isolates contained an IS1 element in proximity to the translational start of FlhD or within the open reading frames for FlhD or FlhC. Two isolates had an IS2 and one isolate had an IS5 in the open reading frame for FlhD. An additional three isolates contained deletions that included the RNA polymerase binding site, five isolates contained point mutations and small deletions in the open reading frame for FlhC. The locations of all these mutations are consistent with the lack of motility and further downstream within the *flhD* operon than previously published IS elements that increased motility. We believe that the location of the mutation within the *flhD* operon determines whether the effect on motility is positive or negative.

To test the second part of our hypothesis where motility heterogeneity in a biofilm may lead to a long-term increase in biofilm biomass, we quantified biofilm biomass by MC1000, MC1000 *flhD::kn*, and mixtures of the two strains at ratios of 1:1, 10:1, and 1:10. After 3 weeks, biofilm of the mixed cultures contained up to five times more biomass than biofilm of each of the individual strains.

**Conclusion:** Mutations in the *flhD* operon can exert positive or negative effects on motility, depending on the site of the mutation. We believe that this is a mechanism to generate motility heterogeneity within *E. coli* biofilm, which may help to maintain biofilm biomass over extended periods of time.

**Keywords:** Motility heterogeneity, Biofilm, *Escherichia coli* K-12, Mutations, IS elements

* Correspondence: Birgit.Pruess@ndsu.edu
Department of Veterinary and Microbiological Sciences, North Dakota State University, Fargo, ND 58103, USA

## Background

Bacteria use diverse mechanisms to adapt to their environment. Adaptation at the level of gene expression is often short term; for example to survive at high [1] or low [2] temperature after a sudden shift. Selection for specific mutations in response to selective pressure can lead to antibiotic resistance (for a recent review, see [3]). This study investigates a combination of adaptive gene expression with selective mutation by individual bacteria within a biofilm. Biofilm is defined as bacterial colonies encased in a matrix that form on surfaces, at air-liquid interfaces, or within eukaryotic cells. Distinct environmental niches within the biofilm are characterized by the availability of oxygen or nutrients, as well as exposure to antibiotics and/or the immune system. While bacteria respond to these niches at an individual level, the differential responses of many bacteria can lead to phenotypic heterogeneity across the biofilm, in particular in cases where the selective pressure is insufficient to clearly favor one or the other phenotype.

Three examples of adaptive gene expression leading to motility heterogeneity in a biofilm come from *Bacillus subtilis* [4], *Escherichia coli* K-12 [5], and *Salmonella enterica* [6–8]. In *B. subtilis*, genes of motility, matrix production, and sporulation are expressed in a spatiotemporal order [4]. In the study by our own laboratory, expression of the *E. coli* flagellar master regulator $FlhD_4 FlhC_2$ (FlhD/FlhC) was limited to the outermost edge of the biofilm at specific times during biofilm development [5]. In *S. enterica*, motility bistability enabled the bacteria to perform a division of labor, where motile cells contribute to the infection and non-motile cells serve as a reservoir [6–8]. Many examples of selective mutation include a range of DNA transfer and recombination events, such as horizontal DNA transfer, conjugation, transformation, and intraspecies DNA transfer by mobilizable elements [9]. An intriguing combination of adaptive gene expression and selective mutation are IS elements, where the initial evolutionary event of the IS element insertion into genes or operons can lead to changes in gene expression, dysfunctional proteins, and phenotypic changes (for a review on mutations by transposons, see [10]).

The *flhD* operon that controls expression of the FlhD/FlhC transcriptional activator and hence the expression of all other flagellar genes is subject to IS element insertion. Barker and coworkers determined that insertion of an IS5 element into the *flhD* promoter of the originally non-motile version of *E. coli* MG1655 Fnr- resulted in a 2.7-fold increase in *flhD* expression and a 7-fold increase in the rate of migration through semi-solid agar [11]. Wang and Wood determined that an IS5 in the *flhD* promoter of BW25113 increased both motility and biofilm amounts [12]. In both these previous reports [11, 12], IS5 inserted into the same 4-bp target site (5'-TTAA-3') located 96–99 bp upstream of the transcriptional start of *flhD*, causing a poorly motile strain to become considerably more motile. A third group led by C. Park selected motile isolates from the non-motile version of MG1655 and found IS elements even further upstream of the transcriptional start for *flhD*. Relative to the transcription start site, two IS5 elements inserted 315 and 166 bp upstream, while one IS1 inserted 303 bp upstream [13]. The final study that identified insertion elements in the *flhD* promoter is by Fahrner and Berg, who found IS elements up to 476 bp from the transcriptional start that still had stimulating effects on expression of the *flhD* operon [14]. The authors concluded that a large upstream region is involved in the topology of *flhD* expression.

The aforementioned IS elements in the *flhD* operon of different *E. coli* K-12 backgrounds have several characteristics in common: i) the mutants were obtained from genetic background strains that were non-motile prior to the experiment and ii) IS elements inserted upstream of the transcriptional start of the operon, where they increased *flhD* expression and enhanced motility.

With this study, we used the highly motile genetic background strain MC1000. We recovered non-motile isolates from biofilm formed by this strain that carried mutations in their *flhD* operon further downstream than the previously reported isolates with increased motility. We conclude that the location of the mutation within the *flhD* operon determines the effect on motility and propose this as a mechanism to generate motility heterogeneity in *E. coli* K-12 biofilm. We hypothesize that this motility heterogeneity may help the bacteria to maintain a biofilm for a longer period of time, leading to an increased biofilm biomass relative to the parental strain after prolonged incubation.

## Results

### Motility heterogeneity in *E. coli* K-12 biofilm

Bacterial isolates were recovered from biofilms produced by the highly motile MC1000 after 7 and 14 days. From the 7-day old biofilm, we recovered 368 isolates; from the 14-day old biofilm, we recovered 1217 isolates. We tested the total of 1585 isolates for motility (Table 1, motility column). From the 7-day old biofilm, we identified 4 non-motile isolates and 9 isolates with reduced motility. From the 14-day old biofilm, we recovered 58 non-motile isolates and 14 isolates with reduced motility. To determine whether the tendency towards reduced motility is a feature of biofilm formation, we tested planktonic MC1000 bacteria. In three replicate experiments, 2070 isolates were recovered from 14-day old aerated cultures; three of these isolates were non-motile. The reduced or non-motile MC1000 biofilm isolates were designated JS17 through JS101 (Table 1, colony designations column); the planktonic MC1000 isolates

**Table 1** Summary of phenotypes for isolates that were recovered from biofilm of MC1000

| Incubation | # of col. | Motility[a] | Colony designations[b] | pXL27 compl.[c] | PCR1/PCR2[d] | | | | PCR3[e] |
|---|---|---|---|---|---|---|---|---|---|
| | | | | | Partially Motile | Non-Motile | | | |
| | | | | | | Category I | Category II | Category III | |
| 7d | 368 | 4 NM (1 %) | JS28-31 | 4 col. PM | | 4 col. with parental PCR fragments | | | ND |
| | | 9 PM (2.4 %) | JS19-JS27 | ND | 9 col. with parental PCR fragments | | | | ND |
| 14 d | 1217 | 58 NM (4.8 %) | JS18, JS42-79, JS82-JS100 | 54 col. FM, 4 col. PM | | 21 col. with parental PCR fragments | 3 col. lacking a PCR1 product, 3 col. lacking a PCR2 product | 31 col. with extended PCR fragments in both reactions | 31 col. tested: 6 col. JS44 orient, 22 col. JS90 orient, 3 col. no PCR3 fragments |
| | | 14 PM (1.1 %) | S17, JS32-JS41, JS80, JS81, JS101 | ND | 14 col. with parental PCR fragments | | | | ND |

[a]Motility was determined on motility plates: NM, the isolate was completely non-motile; PM, the isolate exhibited partial motility relative to its parent strain

[b]All 85 MC1000 biofilm isolates that exhibited reduced motility relative to MC1000 were given JS designations

[c] Non-motile isolates from MC1000 biofilm were tested for complementation with pXL27: FM, at least 7 of 8 transformants exhibited the full motility of the parent; PM, fewer than 7 of 8 transformants were fully motile or all colonies were partially motile. ND, not determined

[d]Reduced motility isolates from MC1000 biofilm were subjected to PCR1 and PCR2 to test for insertions/deletions within the *flhD* operon

[e]The 31 isolates from category III were subjected to PCR3 to identify IS1 insertions

were designated JS107 to JS109. The greater frequency of reduced or non-motile isolates isolated from the biofilm relative to the planktonic bacteria (5.4 % versus 0.1 %) suggests that the biofilm environment of MC1000 may have selected for non-motile isolates. Note that 'Additional file 1' that is provided as online supplement contains raw data from experiments that were performed with isolates JS1 through JS101 at the level of the individual isolate. Table 1 of this manuscript categorizes isolates.

To determine whether extended times of incubation could increase the percentage of non-motile bacteria in MC1000 biofilm, we tried to recover isolates from 1-month old biofilms. In duplicate experiments, the biofilm disintegrated after 2 weeks. At 4 weeks, we recovered only between $10^4$ and $10^5$ colony forming units (CFU) from a surface area of 28.3 $cm^2$ (data not shown).

We next initiated biofilm formation with a 1:1 mixture of MC1000 and its isogenic MC1000 *flhD*::*kn* mutant (designated BP19) and monitored the number of total and mutant bacteria over the course of 4 weeks. After the first day, the maximal total number of biofilm bound bacteria was $2 \times 10^9$ CFU/well on a surface area of 9.6 $cm^2$ (Fig. 1, open diamonds, dashed line). Bacterial numbers gradually decreased to about$10^8$ CFU/well. The percentage of *flhD* mutants started at approximately 50 %, which is consistent with the intended 1:1 inoculation of the two bacterial strains. During the first 2 weeks, mutant percentage fluctuated between 50 and 100 %; over the next 2 weeks, it increased steadily to 100 % (Fig. 1, closed squares, solid line). An initial sharp peak in the percentage of *flhD* mutants will be addressed in the Discussion. We conclude that MC1000 biofilm maintains heterogeneity for at least 2 weeks.

### Characterization of the non-motile MC1000 isolates

Non-motile isolates recovered from MC1000 biofilm were characterized further. To test for mutations in the *flhD* operon, we attempted to complement the lack of motility with the *flhD*-expressing plasmid pXL27 (Table 1, pXL27 complementation column). Of the 62 isolates tested, 54 complemented to full MC1000 motility, suggesting the presence of mutations within the *flhD* operon. The other 8 non-motile isolates complemented to partial MC1000 motility. The lack of full complementation suggests that the loss of motility resulted from a non-*flhD* mutation, perhaps in other flagella genes; these isolates were not further characterized.

The presence of insertions and/or deletions within the *flhD* promoter in the partially or non-motile MC1000 biofilm isolates was determined with polymerase chain reactions PCR1 and PCR2 (Fig. 2; Table 1, entire PCR1/PCR2 block of columns; [11]). All partially motile isolates, yielded PCR products that were identical in length to those of the MC1000 parent strain. As representative of this group, JS81 is included in Fig. 2b, sequence analysis of its *flhD* operon did not reveal any mutations from the IS5 to the end of the FlhC open reading frame. The non-motile isolates divided into three categories.

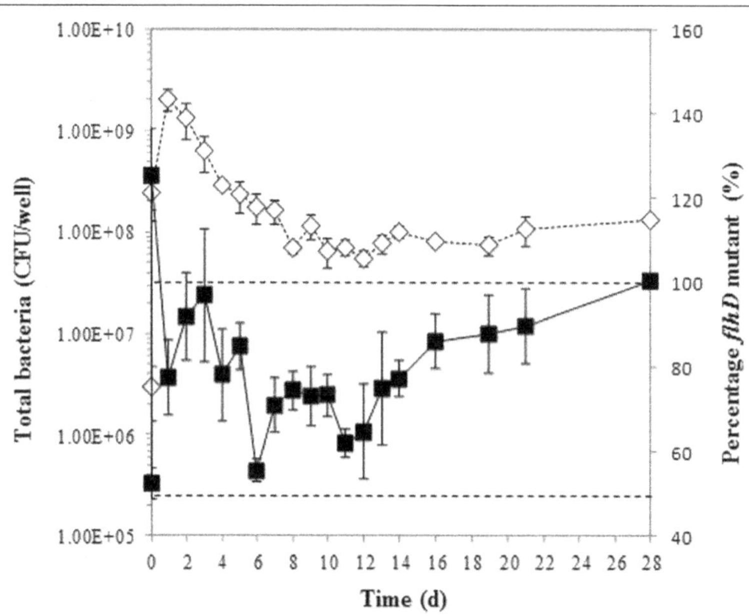


**Fig. 1** Competitive biofilm experiment. MC1000 and BP19 were grown at a starting ratio of 1:1. Biofilm bound bacteria were enumerated on LB and LB/kn plates. The total number of bacteria (*white diamonds, dashed line*) is presented as colony forming units per well (CFU/well) and was calculated from the LB plates. To determine the percentage of *flhD* mutants (black squares, solid line), bacterial counts from the kanamycin plates were divided by total bacterial numbers. Dashed lines mark 50 and 100 % *flhD* mutants

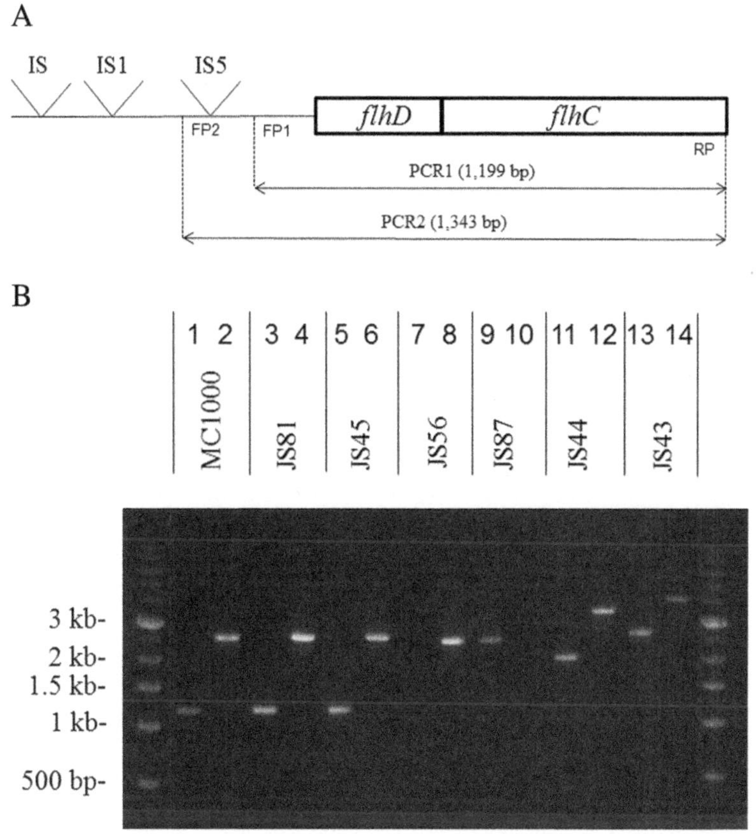


**Fig. 2** PCR1 and PCR2. **a** details the *flhD* operon and the two PCR reactions [11]. PCR1 is expected to yield a 1199 bp product, PCR2 is expected to yield a 1344 bp product in the absence of insertions or deletions. FP1stands for Forward Primer for PCR1, FP2 stands for Forward Primer for PCR2, RP stands for Reverse Primer for both PCR1 and PCR2. IS5 refers to the IS5 element at position −99 to −96 from the transcriptional start that was described by Barker *et al.* [11] and Wang and Wood [12]. IS1 refers to the IS1 elements that was described by Lee and Park [13] at −303 pb. IS refers to the IS5 described by Fahrner and Berg as the most distal (−476) IS element that had a positive effect on *flhD* expression [14]. **b** contains the 1 kb ladder (Amresco, Solon OH) and 14 lanes of PCR products: lanes 1 & 2, MC1000; lanes 3 & 4, JS81; lanes 5 & 6, JS45; lanes 7 & 8, JS56; lanes 9 & 10, JS87; lanes 11 & 12, JS44; lanes 13 and 14, JS43. Odd numbered lanes show products of PCR1, even numbers lanes those of PCR2

Category I contained 4 isolates from the 7 day biofilms and 21 isolates from the 14 day biofilms that produced PCR products that were of identical length as those produced by MC1000. As representative of the category I isolates, JS45 is included in Fig. 2b. Sequence analysis revealed that JS45 had a frame shift mutation in codon 96 of FlhC that led to a TGA stop codon in position 97. JS59 and JS68 had a point mutation in codon 67 of FlhC that led to a Trp to Ser change, JS84 had a deletion of 14 base pairs, leading to a frame shift in codon 136 of FlhC and a changed protein sequence thereafter. JS98 had a point mutation in codon 103 of FlhC that converted this codon to a TAG stop codon. Sequence data for these five category I isolates are included in Additional file 2: Figure S1.

Category II contained 6 isolates that failed to produce a PCR product in one of the two reactions. JS56 (Fig. 2b), JS62 and JS79 failed to produce a PCR1 product and produced a PCR2 product that was slightly shorter than that of MC1000. Sequence analyses of the *flhD* operons of these isolates revealed that they had deletions that started at the parental IS5 element and extended to 79 bp (JS62) and 14 bp (JS56 and JS79) downstream of the transcriptional start site of the operon (Additional file 2: Figure S2). For these three isolates, the deleted region included the binding site for the forward primer 1. JS87 is an example of an isolate that did not produce a PCR2 product (Fig. 2b). It produced a PCR1 product of ~ 2.5 kb. Sequence analysis of JS87 revealed that the *flhD* operon contained an IS5 at 139 bp into the open reading frame of FlhD, which is consistent with the length of the PCR1 product. Since we were unable to obtain a PCR2 product, it is likely that there are additional mutations in the area of the forward 2 primer. The other two isolates in this group were JS76 and JS82, both of which produced PCR1 and PCR2 products that were very similar in length to the respective products of JS87.

Category III consisted of 30 isolates that produced PCR products larger than those of MC1000 in both reactions. Of the 30 isolates in this category, 28 isolates produced PCR fragments of 2 and 3.3 kb in PCR1 and PCR2, an example of this group is JS44 (Fig. 2b). This combination of PCR products is indicative of the continued presence of the IS5 from MC1000, supplemented by an insertion of approximately 800 bp within the part of the *flhD* operon that is amplified with the forward 1 primer. The remaining two category III isolates produced PCR1 and PCR2 products that were even larger. These are JS43 (Fig. 2b) and JS70, whose PCR products were 2.6 and 4 kb, respectively.

As examples of category III isolates that produced PCR1 and PCR2 products of 2 and 3.3 kb, we sequenced the *flhD* operons of JS44, JS51, JS58, and JS90 (Additional file 2: Figure S3). JS44 contained an IS1 element in the reverse orientation 54 bp upstream of the ATG start codon of FlhD. On the upstream side, the IS1 was flanked by a 9 bp duplication from the *flhD* promoter sequence. JS51 contained an IS1 element in the forward orientation within the open reading of FlhD at 95 bp downstream of the ATG. JS58 contained an IS1 element in the reverse orientation 5 bp upstream of the ATG start codon of FlhD. This IS1 was flanked downstream by a duplication of the sequence AATAATG, which does not interrupt FlhD's ATG start codon or its open reading frame. JS90 contained an IS1 element in the forward orientation within the open reading frame for FlhC, 49 bp downstream of the ATG. Sequence analysis with the forward 2 primer revealed the continued presence of the IS5 element from MC1000 in JS44, JS58, and JS90.

To determine the presence of IS1 elements in the remaining category III isolates, we performed PCR3 (Table 1, PCR3 column, Fig. 3). Among the isolates that had produced PCR1 and PCR2 products of 2 and 3.3 kb in PCR1 and PCR2, we found four more isolates (JS61, JS75, JS78, and JS92) that had IS1 elements in the reverse orientation and at the approximate location as

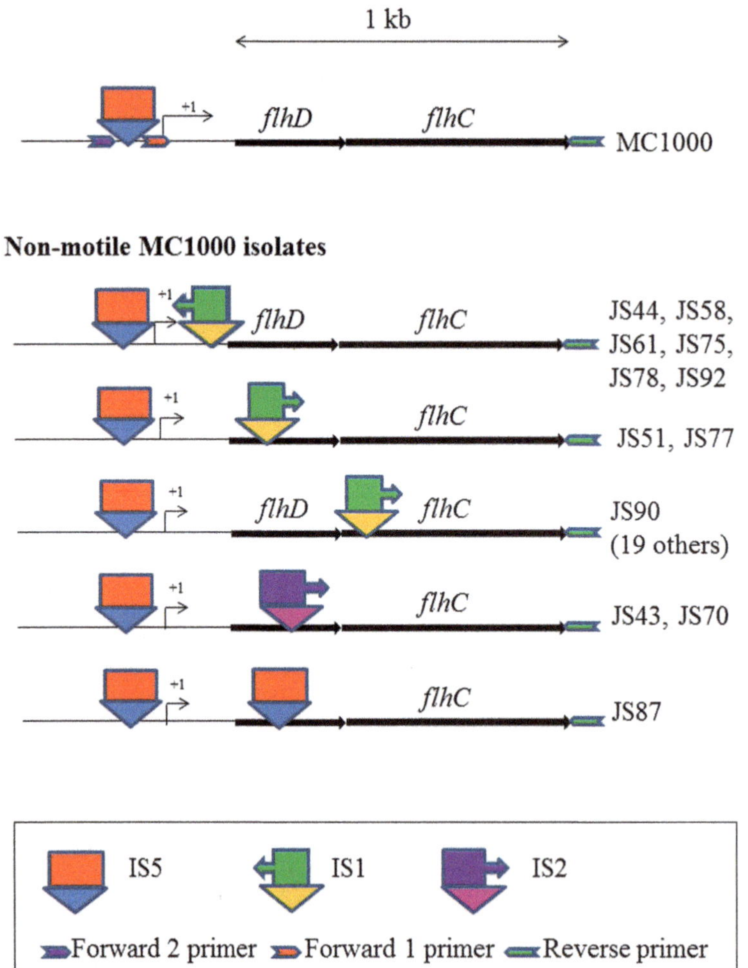


**Fig. 3** Outcome of the sequence analysis of the category III isolates and PCR3. IS elements and primers are indicated, as are open reading frames. A 1 kb size marker is provided at the top of the Figure. Size is only an approximation. In particular, primers are drawn larger than they really are, while IS elements are drawn smaller

JS44 and JS58. JS77 had the IS1 element inserted in the same orientation at approximately the same location as JS51. An additional 19 isolates had IS1 in the forward orientation and at the approximate location of JS90. These are noted in Additional file 1. Finally, we sequenced the *flhD* operons of the two category III isolates that had produced PCR1 and 2 products larger than 2 and 3.3 kb. JS43 and JS70 had an IS2 element 181 bp into the open reading frame of FlhD. The location of all IS1, IS2, and IS5 elements that were found in the *flhD* operon within this study are included in Fig. 3.

In summary, we identified IS1 elements in the *flhD* operons of 28 of the 62 non-motile isolates that were tested. Two isolates carried an IS2 in their *flhD* gene, one carried an IS5 in *flhD*. Three isolates had deletions in the promoter region of *flhD*, five had small mutations in the open reading frame for FlhC. The IS element insertions were downstream of the transcriptional start of the operon, either between the transcriptional and the translational start of *flhD* or within the open reading frames of FlhD or FlhC. The small mutations were also found in the open reading frame for FlhC. The deletions included the region from the IS5 to the transcriptional start and a portion of the untranslated mRNA transcript.

### *Motility heterogeneity leads to an increase in biofilm biomass after three weeks of incubation*

Biofilm was incubated for up to 3 weeks, composed of MC1000, BP19 (MC1000 *flhD::kn*), or mixtures of MC1000 and BP19 at ratios of 1:1, 10:1, and 1:10 (Fig. 4). Quantification of biofilm with CV after 1 week determined that the 1:1 mixture of MC1000 and BP19 (Fig. 4a, 1 week, green bar) had slightly more biofilm than the pure culture of MC1000 (Fig. 4a, 1 week, blue bar) or BP19 (Fig. 4, 1 week, orange bar) and significantly more biofilm than the remaining two mixtures (Fig. 4a, 1 week, yellow and purple bars). This increase in CV detectable biofilm biomass was not paralleled by an increase in cell viability (Fig. 4b). At this time, non-motile isolates had almost completely taken over the biofilms of the three mixed cultures, whereas MC1000 biofilm contained 15 % non-motile isolates (Fig. 4c). Intriguingly, the biofilm by BP19 contained two motile isolates (out of 400 tested) that were still kanamycin resistant. We can not explain this phenomenon and did not investigate these isolates further.

After 2 weeks, the 1:1 mixture, and the 1:10 mixture contained more biofilm than the other cultures (Fig. 4a). For the 1:1 mixture, this increase in biofilm biomass was paralleled by an increase in viable cell counts (Fig. 4b), for the 1:10 mixture, this was not the case. With the exception of MC1000, biofilms contained primarily non-motile isolates.

After 3 weeks, the three mixed cultures contained up to five times more biofilm than MC1000. For the 1:1 mixture and the 10:1 mixture, this increase was paralleled by an increase in viable cell counts, though admittedly with a large standard error. Biofilm of MC1000 and the three mixed cultures contained a mixture of motile and non-motile isolates. It is noteworthy that biofilm by the non-motile BP19 contained such a small number of viable cells that a CFU count could not be computed (missing bar in Fig. 4b) and motility could not determined (missing bar in Fig. 4c).

To determine whether the non-motile isolates of MC1000 carried mutations similar to the ones earlier described in this study, the non-motile isolates from one week old MC1000 biofilm, designated JS111 to JS125, were further investigated. Transforming the isolates with pXL27 yielded 47 % of motile transformants, indicating that the respective isolate had acquired a mutation in its *flhD* operon. All isolates were subjected to PCR1 and PCR2, the *flhD* operons of JS111and JS119 were sequenced. JS111 had the 1200 bp PCR1 product that is characteristic for MC1000 and a 1400 bp PCR2 product. In agreement with this, sequence analysis revealed the lack of the IS5 element of the MC1000 parental strain. There was a single base pair change from C to A at −63 bp from the transcriptional start. JS119 carried an IS5 at 91 bp before the ATG start codon of FlhD (Additional file 2: Figure S2). As described in Additional file 2: Figure S2 and Fig. 3, we obtained isolates that carried mutations in the *flhD* operon. The nature of the mutations may vary between the experiments.

To visualize the increased biofilm biomass of the heterogeneous biofilm, we combined all data points, regardless of the strain/mixture and the time point (Fig. 4d). For clarification, we maintained the color code from the bar plots (Figs. 4a to c), which means that yellow circles represent the 1:1 mixture of the two strains. Specifically, the yellow 5.5 fold data point at 50 % motility was computed from the yellow bars at 3 weeks in Figs. 4a and c. Throughout all data points, the biofilm that consisted of 50 % motile and 50 % non-motile isolates had by far the largest amount of biomass. In contrast, cultures that consisted of either 100 % motile isolates or 100 % non-motile isolates formed between 5 and 10 fold less biofilm.

## Discussion

This study presents several pieces of evidence to support our hypothesis that i) mutants can arise in the *flhD* operon that decrease motility and that ii) the resulting motility heterogeneity in the biofilm may help the bacteria to maintain biofilm biomass over time.

The first set of experiments led to the conclusion that the biofilm environment may favor a mixture of motile and non-motile bacteria. From biofilms formed by the highly motile MC1000, we recovered 62 non-motile isolates, but no hyper-motile ones. We conclude that

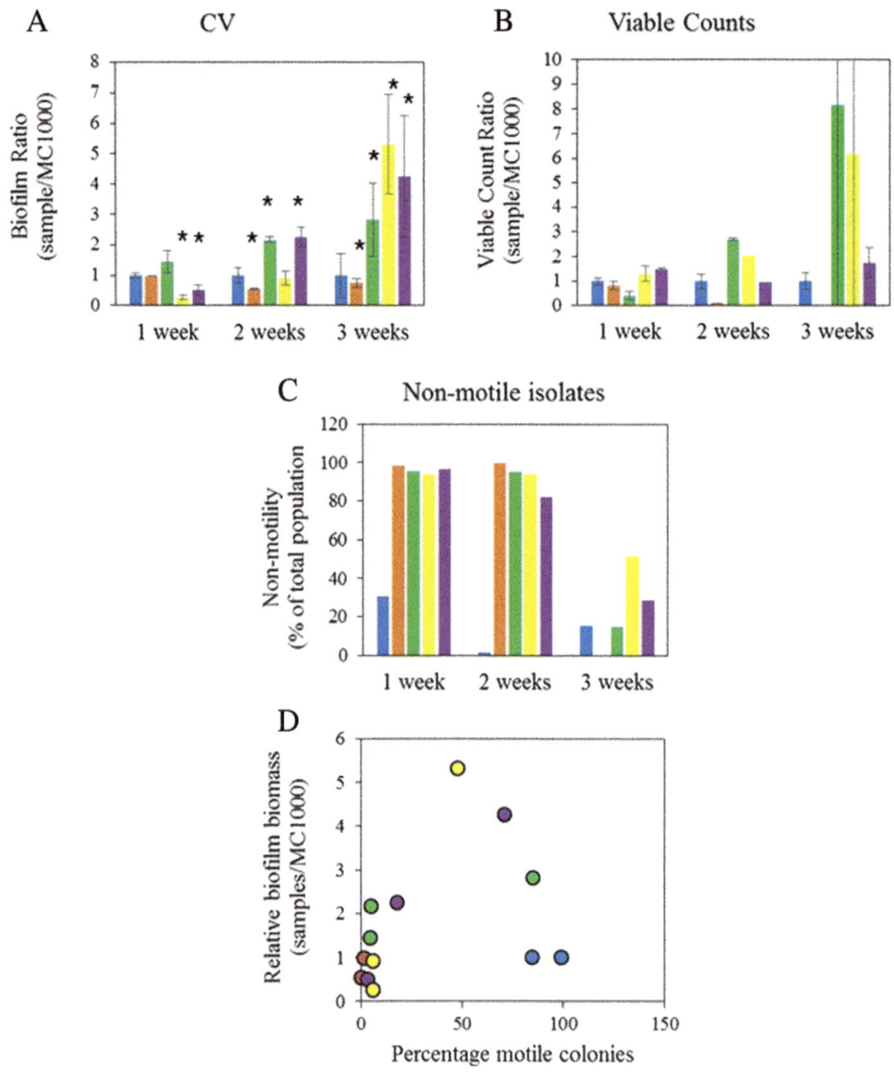


**Fig. 4** Biofilm biomass in mixed biofilms. Biofilm was grown by MC1000, BP19 and mixtures of MC1000 and BP19 in ratios of 1:10, 1:1, and 10:1. Biofilm amounts were determined weekly for 3 weeks by the CV assay **a**, viable cell counts were determined simultaneously (**b**). **c** indicates the percentage of non-motile bacteria in the biofilm, determined on motility plates. For **a** through **c**, blue bars are used for MC1000, orange bars for BP19, green bars for the 1:1 mixture, yellow for the 10:1 mixture, and purple for the 1:10 mixture. For **d**, data from **a** and **c** were combined and the relative biofilm biomass was plotted versus the percentage of motile isolates. The color code for the circles is identical to the color codes for the bars in (**a**) to (**c**)

moderate selective pressure towards the opposing motility phenotype might operate in this biofilm. If this were true, then the biofilm environment may favor a mixture of motile and non-motile bacteria. To test this hypothesis, we initiated biofilms with MC1000 or with a mixture of MC1000 and the non-motile BP19 and followed the composition of the biofilm for about one month. The biofilm formed by the former degraded, whereas the latter formed a biofilm that maintained its integrity for a month with a mixture of both strains for most of the 1 month time period of the experiment (Fig. 1).

An intriguing phenomenon was observed during this competitive experiment. At 6 h, the *flhD* mutant strain BP19 had completely taken over the biofilm to a percentage that was even on top of 100 % (Fig. 1, second data point of the black squares). While we don't know why more bacteria had grown on the kanamycin plates than on the plain LB plates, we believe that the massive increase in the *flhD* mutant population might be due to the increase in the cell division rate of the mutants relative to the parent strain that was previously observed [11, 15, 16]. In summary of these previous publications, *flhD* mutants divided to 37 times higher cell densities than isogenic parent bacteria due to a depletion of serine from the bacterial growth media that the *flhD* mutants failed to sense. In agreement with these previous publications, we observe large numbers of *flhD* mutants after 24 h of incubation in all of our experiments with this strain.

Motility heterogeneity has been observed in other environments. One such environment is the mouse intestine. In the streptomycin-treated mouse intestine, *flhD* deletion mutants derived from the originally motile *E. coli* MG1655 parent eventually took over the population, but only after prolonged incubation [17, 18]. Intriguingly, 10 to 20 % of the remaining bacteria had *envZ* missense mutations [19]. EnvZ is the histidine kinase of the osmoregulation system EnvZ/OmpR. The $envZ_{P41L}$ mutation increased the levels of phospho-OmpR [20], an inhibitor of *flhD* expression [21]. The authors concluded that the intestine likely contains niches where high *flhD* expression might be the advantage, alongside niches where low *flhD* expression might be an advantage [18]. The researchers did not attribute the entirety of this effect to motility. Instead, they proposed a 'restaurant' hypothesis [19], where the many metabolic genes that are regulated by *flhD* [22, 23] contribute to the niche adaptation. Our own studies with the aerotaxis sensor Aer, which impacts the sugar acid degradation pathway, support this hypothesis [24].

A second example of motility heterogeneity can be found in *S. enterica*, where FliZ and YdiV are responsible for the co-existing sub-populations of motile and non-motile bacteria during host colonization. FliZ is a post-translational activator of *flhD* expression [6] and constitutes a positive feed-back loop [25]. The other half of this switch is YdiV, which binds to FlhD/FlhC and inhibits its action upon downstream flagellar genes [26]. FliZ is responsible for the motile sub-population, whereas YdiV partitions cells into the non-motile sub-population [8]. On the basis of our current study, we propose a fundamentally different mechanism by which bacteria can partition their populations into motile and non-motile sub-populations, namely by mutating the *flhD* operon of individual bacteria, modulating FlhD/FlhC expression or producing dys-functional proteins, and thus generating motility heterogeneity within the biofilm.

This study investigates both, the mechanism by which heterogeneity is generated and the long-term impact of such heterogeneity on the biofilm. As mechanism, we propose that the location of the mutation within the *flhD* operon determines whether the effect on motility is positive or negative. All the previously published *E. coli* K-12 biofilm isolates had increased motility [11–14]. Barker and coworkers postulated that the positive effect on *flhD* expression of these insertions may be due to interference with some of the negative regulators (*e.g.* OmpR, RcsB [13, 21, 27–29]) that have binding sites in this region. Many of the non-motile isolates that were recovered from MC1000 in this study carried IS elements in their *flhD* operons that were located either between the transcriptional and the translational start of the operon or within the open reading frames of FlhD or FlhC. This could be attributed to different post-transcriptional causes. For JS58, the IS1 is only 5 bp upstream of the ATG start codon for FlhD. It is more than likely that this would interfere with binding of the ribosome. For JS51 and JS90, the IS1 elements are within the open reading frames of FlhD and FlhC, respectively, an arrangement that has been described for other bacterial species. Within the genus *Shigella*, *Shigella dysenteriae* contains an IS1 in the open reading frame of FlhC, whereas *Shigella flexneri* has an IS1 in the open reading frame of FlhD [30]. Inactivation of FlhC is the obvious reason for the lack of motility in these bacteria.

Other mutations that were found in non-motile isolates of MC1000 biofilm support the conclusions that were drawn from the IS elements. Deletions in the *flhD* operon of JS56, JS62, and JS79 covered the binding site for the forward 1 primer that coincides with the −35 binding site of the RNA polymerase. Likewise, the *flhD* operons of JS45, JS59, JS68, JS84, and JS98 contained point mutations and a 14 bp deletion within the open reading frame for FlhC. These mutations, too, would be expected to render the bacteria non-motile.

The fact that previous studies recovered hyper-motile isolates from *E. coli* biofilm and this study describes the recovery of non-motile isolates from *E. coli* biofilm may appear as a contradiction. However, all previous studies were performed with genetic backgrounds of *E. coli* K-12 that were poorly motile at the beginning of the experiment. In contrast, this study used a highly motile genetic background in MC1000. Together, these observations indicate that there may be a moderate selective pressure towards the respective other motility phenotype in all these experiments. We interpret this as a confirmation of our conclusion that biofilm strives towards a mixture of both motile and non-motile bacteria.

To study the long-term impact of motility heterogeneity on the biofilm, we quantified biofilm biomass from either pure cultures of MC1000, BP19, or mixtures of MC1000 and BP19 (Fig. 4). After 3 weeks, the different mixtures yielded up to 5 times more biofilm biomass than either of the two pure cultures. This was true regardless of the composition of the inoculated mixture. While the viable bacterial counts did not always parallel the results from the CV assay, there is no doubt that the majority of the tested biofilms did indeed contain live bacteria. After 3 weeks, these viable bacteria constituted mixtures of both motile and non-motile bacteria. Again, this was true regardless of the composition of the mixture at time of inoculation. Intriguing observations were made with the *flhD* mutant, BP19. We were unable to perform the experiment for the entirety of the 3 weeks because the biofilm disintegrated. In fact, the 2 weeks biofilm already contained almost no viable bacteria anymore (Fig. 4b). The question remains how the mutant was able to make a biofilm in the first place, considering that flagella are thought to be needed for the initial attachment. It is possible that the bacteria used fimbriae/pili/curli as attachment tools straight from the beginning, as was proposed by A.J. Wolfe's lab [31] for some of the acetate mutants which were able to produce a fimbriae based biofilm, albeit with a different integrity and smaller amounts of biomass.

Altogether, this experiment leads to the conclusion that motility heterogeneity helps the biofilm to maintain its integrity for a longer period of time. Assuming that a larger biofilm biomass made up primarily of viable bacteria may constitute as an advantage, one could consider this experiment first evidence towards a new hypothesis that motility heterogeneity in a biofilm may constitute a long-term advantage.

## Conclusions

This study attempted to combine investigations of two important mechanisms of adaptation that help bacteria adapt to the heterogenous niches within the biofilm: spontaneous mutations and phenotypic expression. The experiments presented support our hypothesis that mutations can arise in the *flhD* operon of biofilm bound *E. coli* that render the bacteria non-motile. We propose this is a mechanism to generate motility heterogeneity in biofilm and provided evidence that such heterogeneity leads to increases in biofilm biomass after prolonged incubation.

## Methods

### Bacterial strains

MC1000 is a highly motile (5.9 mm/h on motility plates [32]) strain, due to the presence of an IS5 element in its *flhD* promoter [11]. MC1000 forms biofilm poorly [32]. MC1000 *flhD::kn*, designated BP19, does not express *flhD*, is non-motile and resistant towards kanamycin [23, 32, 33]. All JS designated isolates were recovered as non-motile or reduced motility colonies from either biofilm or planktonic bacteria of MC1000. Bacterial strains were maintained as freezer stocks at −80 °C and streaked onto Luria Bertani (LB; 1 % tryptone, 0.5 % NaCl, 0.5 % yeast extract) agar plates prior to each experiment.

### Biofilm formation, colony recovery, and motility screening

Biofilms were grown from MC1000 in duplicate, separate wells of a 6 well plate in LB at 32 °C. After 7 and 14 days, liquid growth media was removed and biofilms were washed twice with PBS. Bacteria were suspended in PBS and homogenized by pipetting them up and down the pipette tip several times. The homogenates were serially diluted, and plated on LB agar plates. As a control experiment, we grew bacteria in bottles under shaking and serially diluted the planktonic bacteria in the growth medium. Within each experiment, we picked several hundreds of isolated colonies (referred to as isolates) from the plates and screened them for motility. This was done on semi solid agar plates made from tryptone broth (TB; 1 % tryptone, 0.5 % NaCl) and 0.3 % agar [34]. Plates were incubated for 4–8 h at 32 °C in a humid environment. The diameters of the swarm rings were compared between the parent strain and the respective derivatives. Motility phenotypes were confirmed by testing the respective isolate on motility plates two more times. Isolates that reproducibly produced a motility phenotype that was different from their parent were maintained as freezer stocks and given a JS designation. Note that colony designations take the different strains and time points under consideration, but do not distinguish between the two replicate experiments.

### Long-term biofilm experiments

The first long-term experiment followed the protocol described for the recovery of the isolates, but expanded the incubation time to 4 weeks. Since the biofilm disintegrated during this time, we then inoculated the cultures with a 1:1 starting mixture of MC1000 and its isogenic MC1000 *flhD::kn* mutant. This was done in LB on multiple 35 mm petri dishes made of polystyrene, which were incubated at 32 °C. Biofilms were grown over the course of 28 days. At each time point, individual petri dishes were removed from the incubator and processed. Bacterial growth medium was removed, biofilms were washed twice with PBS, bacteria were serially diluted, and plated onto LB agar plates and LB plates, supplemented with 50 μg/ml kanamycin. The total number of bacteria was presented as colony forming units per well (CFU/well) and was calculated from the LB plates. To determine the percentage of *flhD* mutants, bacterial counts from the kanamycin plates were divided by total bacterial numbers. The experiment was done twice in two replicates each, averages and standard deviations were calculated across the four experiments.

### Molecular characterization of the non-motile isolates from MC1000 biofilm

To test whether some of the non-motile MC1000 biofilm isolates might contain mutations within their *flhD* operons in addition to the IS5, all 85 isolates were transformed with the *flhD* and *flhC* expressing plasmid pXL27 [35] and tested for complementation of the motility phenotype. Plasmid pXL27 was moved into the respective bacterial isolates via chemical transformation, taking advantage of the penicillin resistance gene on the plasmid as a selective marker. For each non-motile colony, 8 independent transformants were tested on motility plates as described [34].

To detect insertions and deletions within the *flhD* operon, two PCR reactions (PCR1 and PCR2) were performed that were originally designed to detect insertions of IS elements within the *flhD* promoter [11]. The PCR 1 fragment (Fig. 2a) starts downstream of the published [11, 12] hot spot for IS1 and IS5 in the *flhD* promoter and ends at the 3'-end of the *flhC* open reading frame. This fragment is expected to be 1199 bp in length in the absence of insertions. The PCR2 fragment (Fig. 2a) starts upstream of the hotspot. PCR2 is expected to yield a 1343 bp fragment in the absence of IS elements. PCR1 was done with forward primer 1, PCR2 with forward primer 2. Both reactions used the same reverse primer. Note that the sequences for these and all other PCR and sequencing primers are listed in Additional file 2: Table S1.

For the MC1000 parent strain, the expectation was a PCR2 fragment of 2538 bp together with a PCR1 fragment of 1199 bp length. This is due to the IS5 between the binding sites of the forward primers 1 and 2. For MC1000 isolates that produced PCR1 and PCR2 fragments of those same lengths, this would indicate that there are not any large deletions or insertions besides the parental IS5 (category I isolates). The lack of a PCR1 or PCR2 product might point towards mutations in the area of the respective primer binding site (category II isolates). Should both PCR fragments be longer than 1343 bp and 1199 bp, this would be indicative of an insertion downstream of the forward primer 1 (category III isolates).

As one example of the partially motile isolates, we sequenced the *flhD* operon of JS81 and compared it to that of MC1000. As representatives of the category I isolates that produced PCR1 and PCR2 products of parental length, we sequenced the *flhD* operons of JS45, JS59, JS68, JS84, and JS98. As examples of the category II isolates that did not produce a PCR1 product, we sequenced the *flhD* operons of JS56, JS62 and JS79, as one example of a category II isolate that did not produce a PCR2 product, we sequenced the *flhD* operon of JS87. Among the category III isolates that produced PCR1 and PCR2 products of 2 kb and 3.3 kb respectively, the *flhD* operons from JS44, JS51, JS58, JS78, and JS90 were sequenced. Also sequenced were the *flhD* operons of JS43 and JS70, the category III isolates that produced PCR1 and PCR2 products longer than 2 and 3.3 kb. Sequence analysis was done from PCR product with the Sanger technique by Macrogen (Rockville, MD). Primers that were used to produce the PCR product, as well as primers that were used for the sequencing reactions are listed in Additional file 2: Table S2.

PCR3 was performed to test for the presence, approximate location, and orientation of IS1 elements in the remaining isolates that had produced larger PCR1 and PCR2 products than their parent MC1000 strain. We used forward primer 3A and 3B together with the reverse primer to detect the two different possible orientations of the IS1. The presence of a PCR product by forward primer 3A is indicative of IS1 in forward orientation. A PCR product yielded by forward primer 3B is indicative of IS1 in reverse orientation. The length of the respective PCR fragment is indicative of the distance from the binding site for the reverse primer.

**Determination of the effect of motility heterogeneity on biofilm biomass after prolonged incubation**

We inoculated 6 well polystyrene plates with MC1000, BP19, and mixtures of MC1000 and BP19 at ratios of 1:1, 10 MC1000/1 BP19 (designated 10:1), and 1 MC1000/10 BP19 (designated 1:10) in TB. Plates were covered with adhesive film (VWR International, Chicago IL) and incubated at 34 °C. Biofilm amounts were determined after 1, 2, and 3 weeks with the crystal violet assay (CV) [36]. Briefly, the planktonic bacteria were removed and the biofilms were washed twice with PBS, dried for 10 min, and stained for 10 min with 0.1 % crystal violet in $ddH_2O$. The CV solution was removed and the biofilms were washed twice with PBS. CV was extracted from the bacteria with a mixture of 80 % ethanol and 20 % acetone, diluted 1:10, and the $OD_{600}$ was determined with a Synergy H1 plate reader from Biotek Instruments (Winooski, VT). Relative biofilm biomass is expressed as ratios where the $OD_{600}$ of each of the cultures was divided by that of MC1000. Averages and standard deviations were calculated across the three replicates for each strain/mixture. A Student's *t*-test was used to determine the statistical significance of the difference between each mixture (or the non-motile mutant) and the MC1000 parent strain. A *p*-value below 0.05 was considered significant.

On separate 6 well plates from the above experiment, biofilms were re-suspended, serially diluted, and bacteria were isolated by spreading the dilutions onto LB agar plates. Viability counts were calculated from these plates and expressed as ratios as described for the CV ratio. For the motility test, at least 100 isolated colonies per culture and time point were spotted on motility plates. An exception from this was BP19, whose biofilm did not permit the recovery of 100 colonies after 3 weeks of incubation. Motility was expressed as percentage of non-motile isolates within the total population, calculated across all isolates for the respective strain/mixture and time point. In a separate analysis, all data points were combined, regardless of strain/mixture or time point. Each data point was plotted as relative biofilm biomass from the CV assay versus the percentage of motile isolates.

Non-motile colonies from 1 week old MC1000 biofilm were designated JS111 through JS125 and kept at −80 °C. These isolates were transformed with pXL27 and tested for complementation of the motility phenotype. PCR1 and

PCR2 were performed with all these isolates, the *flhD* operons of JS111 and JS119 were sequenced.

**Abbreviation**

CV: Crystal violet

**Acknowledgements**

The authors thank Dr. Alan J. Wolfe (Loyola University Chicago, Maywood IL) for critical reading of the manuscript.

**Funding**

The work was funded by grant 1R15AI089403 from the NIH/NIAID. BMP and SMH were funded by the North Dakota Experiment Station. BMP was supported by the USDA National Institute of Food and Agriculture, Hatch project number ND02429. The Synergy plate reader was purchased from grant 2012-67006-19659 from the USDA/NIFA.

**Authors' contributions**

SMH performed the vast majority of the PCR reactions, designed PCR3, and did the sequence analysis. JS isolated the colonies and did the motility and complementation screens. NS performed the colony analysis of the second set of non-motile MC1000 isolates. MI contributed to the colony experiments, the PCR3 reactions, and provided Fig. 2b. TL did one of the long-term experiments (Fig. 1). BMP designed the project, performed the replicate of the long-term experiment and all replicates of the biofilm biomass experiment, contributed to the analysis and interpretation of all data, and wrote the manuscript. SMH and MI were heavily involved in the revision of the manuscript. JS, NS, and TL provided input to the manuscript, as related to their respective experiments. All authors read the manuscript and approved it.

**Competing interests**

The authors declare that they have no competing interests.

**References**

1. Murata M, Fujimoto H, Nishimura K, Charoensuk K, Nagamitsu H, Raina S, Kosaka T, Oshima T, Ogasawara N, Yamada M. Molecular strategy for survival at a critical high temperature in *Escherichia coli*. PLoS One. 2011;6:e20063.
2. White-Ziegler CA, Um S, Perez NM, Berns AL, Malhowski AJ, Young S. Low temperature (23° C) increases expression of biofilm-, cold-shock- and RpoS-dependent genes in *Escherichia coli* K-12. Microbiology. 2008;154:148–66.
3. Chang HH, Cohen T, Grad YH, Hanage WP, O'Brien TF, Lipsitch M. Origin and proliferation of multiple-drug resistance in bacterial pathogens. Microbiol Mol Biol Rev. 2015;79:101–16.
4. Vlamakis H, Aguilar C, Losick R, Kolter R. Control of cell fate by the formation of an architecturally complex bacterial community. Genes Dev. 2008;22:945–53.
5. Samanta P, Clark ER, Knutson K, Horne SM, Prüß BM. OmpR and RcsB abolish temporal and spatial changes in expression of *flhD* in *Escherichia coli* biofilm. BMC Microbiol. 2013;13:182.
6. Saini S, Brown JD, Aldridge PD, Rao CV. FliZ Is a posttranslational activator of $FlhD_4C_2$-dependent flagellar gene expression. J Bacteriol. 2008;190:4979–88.
7. Stewart MK, Cookson BT. Non-genetic diversity shapes infectious capacity and host resistance. Trends Microbiol. 2012;20:461–6.
8. Stewart MK, Cookson BT. Mutually repressing repressor functions and multi-layered cellular heterogeneity regulate the bistable *Salmonella fliC* census. Mol Microbiol. 2014;94:1272–84.
9. Relman DA. Microbial genomics and infectious diseases. N Engl J Med. 2011;365:347–57.
10. Zhang Z, Saier Jr MH. Transposon-mediated adaptive and directed mutations and their potential evolutionary benefits. J Mol Microbiol Biotechnol. 2011;21:59–70.
11. Barker CS, Prüß BM, Matsumura P. Increased motility of *Escherichia coli* by insertion sequence element integration into the regulatory region of the *flhD* operon. J Bacteriol. 2004;186:7529–37.
12. Wang X, Wood TK. IS5 inserts upstream of the master motility operon *flhDC* in a quasi-Lamarckian way. ISME J. 2011;5:1517–25.
13. Lee C, Park C. Mutations upregulating the *flhDC* operon of *Escherichia coli* K-12. J Microbiol. 2013;51:140–4.
14. Fahrner KA, Berg HC. Mutations that stimulate *flhDC* expression in *Escherichia coli* K-12. J Bacteriol. 2015;197:3087–96.
15. Prüß BM, Matsumura P. A regulator of the flagellar regulon of *Escherichia coli*, *flhD*, also affects cell division. J Bacteriol. 1996;178:668–74.
16. Prüß BM, Markovic D, Matsumura P. The *Escherichia coli* flagellar transcriptional activator *flhD* regulates cell division through induction of the acid response gene *cadA*. J Bacteriol. 1997;179:3818–21.
17. Gauger EJ, Leatham MP, Mercado-Lubo R, Laux DC, Conway T, Cohen PS. The role of motility and the *flhDC* operon in *Escherichia coli* MG1655 colonization of the mouse intestine. Infect Immun. 2007;75:3315–24.
18. Leatham MP, Stevenson SJ, Gauger EJ, Krogfelt KA, Lins JJ, Haddock TL, Autieri SM, Conway T, Cohen PS. Mouse intestine selects non-motile *flhDC* mutants of *Escherichia coli* MG1655 with increased colonizing ability and better utilization of carbon sources. Infect Immun. 2005;73:8039–49.
19. Leatham-Jensen MP, Frimodt-Moller J, Adediran J, Mokszycki ME, Banner ME, Caughron JE, Krogfelt KA, Conway T, Cohen PS. The streptomycin-treated mouse intestine selects *Escherichia coli envZ* missense mutants that interact with dense and diverse intestinal microbiota. Infect Immun. 2012;80:1716–27.
20. Adediran J, Leatham-Jensen MP, Mokszycki ME, Frimodt-Moller J, Krogfelt KA, Kazmierczak K, Kenney LJ, Conway T, Cohen PS. An *Escherichia coli* Nissle 1917 missense mutant colonizes the streptomycin-treated mouse intestine better than the wild type but is not a better probiotic. Infect Immun. 2014;82:670–82.
21. Shin S, Park C. Modulation of flagellar expression in *Escherichia coli* by acetyl phosphate and the osmoregulator OmpR. J Bacteriol. 1995;177:4696–702.
22. Prüß BM, Liu X, Hendrickson W, Matsumura P. FlhD/FlhC-regulated promoters analyzed by gene array and *lacZ* gene fusions. FEMS Microbiol Lett. 2001;197:91–7.
23. Prüß BM, Campbell JW, Van Dyk TK, Zhu C, Kogan Y, Matsumura P. FlhD/FlhC is a regulator of anaerobic respiration and the Entner-Doudoroff pathway through induction of the methyl-accepting chemotaxis protein Aer. J Bacteriol. 2003;185:534–43.
24. Horne SM, Mattson KR, Prüß BM. An *Escherichia coli aer* mutant exhibits a reduced ability to colonize the streptomycin-treated mouse large intestine. Ant van Leeuwenhoek Internat J Gen Mol Microbiol. 2009;95:149–158.
25. Saini S, Koirala S, Floess E, Mears PJ, Chemla YR, Golding I, Aldridge C, Aldridge PD, Rao CV. FliZ induces a kinetic switch in flagellar gene expression. J Bacteriol. 2010;192:6477–81.
26. Wada T, Hatamoto Y, Kutsukake K. Functional and expressional analyses of the anti-$FlhD_4C_2$ factor gene *ydiV* in *Escherichia coli*. Microbiology. 2012;158:1533–42.
27. Hagiwara D, Sugiura M, Oshima T, Mori H, Aiba H, Yamashino T, Mizuno T. Genome-wide analyses revealing a signaling network of the RcsC-YojN-RcsB phosphorelay system in *Escherichia coli*. J Bacteriol. 2003;185:5735–46.
28. Lehnen D, Blumer C, Polen T, Wackwitz B, Wendisch VF, Unden G. LrhA as a new transcriptional key regulator of flagella, motility and chemotaxis genes in *Escherichia coli*. Mol Microbiol. 2002;45:521–32.
29. Ko M, Park C. H-NS dependent regulation of flagellar synthesis is mediated by a LysR family protein. J Bacteriol. 2000;182:4670–2.

30. Al Mamun AA, Tominaga A, Enomoto M. Detection and characterization of the flagellar master operon in the four *Shigella* subgroups. J Bacteriol. 1996;178:3722–6.
31. Wolfe AJ, Chang DE, Walker JD, Seitz-Partridge JE, Vidaurri MD, Lange CF, Prüß BM, Henk MC, Larkin JC, Conway T. Evidence that acetyl phosphate functions as a global signal during biofilm development. Mol Microbiol. 2003;48:977–88.
32. Prüß BM, Verma K, Samanta P, Sule P, Kumar S, Wu J, Horne SM, Christianson DA, Stafslien SJ, Wolfe AJ, Denton A. Environmental and genetic factors that contribute to *Escherichia coli* K-12 biofilm formation. Arch Microbiol. 2010;192:715–28.
33. Malakooti J. Molecular analysis of the *Escherichia coli* motor components and regulation of internal control elements of the *flaA* operon. Dissertation, University of Illinois at Chicago; 1989.
34. Wolfe AJ, Berg HC. Migration of bacteria in semisolid agar. Proc Natl Acad Sci U S A. 1989;86:6973–7.
35. Liu X, Matsumura P. The FlhD/FlhC complex, a transcriptional activator of the *Escherichia coli* flagellar class II operons. J Bacteriol. 1994;176:7345–51.
36. O'Toole GA, Pratt LA, Watnick PI, Newman DK, Weaver VB, Kolter R. Genetic approaches to study of biofilms. Methods Enzymol. 1999;310:91–109.

# 7

# Functional and comparative genome analysis of novel virulent actinophages belonging to *Streptomyces flavovirens*

A. Sharaf[1,2*], F. Mercati[3], I. Elmaghraby[4], R. M. Elbaz[5] and E. M. Marei[6]

**Abstract**

**Background:** Next Generation Sequencing (NGS) technologies provide exciting possibilities for whole genome sequencing of a plethora of organisms including bacterial strains and phages, with many possible applications in research and diagnostics. No *Streptomyces flavovirens* phages have been sequenced to date; there is therefore a lack in available information about *S. flavovirens* phage genomics. We report biological and physiochemical features and use NGS to provide the complete annotated genomes for two new strains (Sf1 and Sf3) of the virulent phage *Streptomyces flavovirens*, isolated from Egyptian soil samples.

**Results:** The *S. flavovirens* phages (Sf1 and Sf3) examined in this study show higher adsorption rates (82 and 85%, respectively) than other actinophages, indicating a strong specificity to their host, and latent periods (15 and 30 min.), followed by rise periods of 45 and 30 min. As expected for actinophages, their burst sizes were 1.95 and 2.49 virions per mL. Both phages were stable and, as reported in previous experiments, showed a significant increase in their activity after sodium chloride (NaCl) and magnesium chloride ($MgCl_2.6H_2O$) treatments, whereas after zinc chloride (ZnCl2) application both phages showed a significant decrease in infection.
The sequenced phage genomes are parts of a singleton cluster with sizes of 43,150 bp and 60,934 bp, respectively. Bioinformatics analyses and functional characterizations enabled the assignment of possible functions to 19 and 28 putative identified ORFs, which included phage structural proteins, lysis components and metabolic proteins.
Thirty phams were identified in both phages, 10 (33.3%) of them with known function, which can be used in cluster prediction. Comparative genomic analysis revealed significant homology between the two phages, showing the highest hits among Sf1, Sf3 and the closest *Streptomyces* phage (VWB phages) in a specific 13Kb region. However, the phylogenetic analysis using the Major Capsid Protein (MCP) sequences highlighted that the isolated phages belong to the BG *Streptomyces* phage group but are clearly separated, representing a novel sub-cluster.

**Conclusion:** The results of this study provide the first physiological and genomic information for *S. flavovirens* phages and will be useful for pharmaceutical industries based on *S. flavovirens* and future phage evolution studies.

**Keywords:** Bacteriophage, Biological stability, Whole genome sequence, NGS, Comparative genomics

* Correspondence: sharaf@paru.cas.cz
[1]Genetic Department, Faculty of Agriculture, Ain Shams University, Cairo 11241, Egypt
[2]Institute of Parasitology, Biology Centre, Czech Academy of Sciences, 37005 České Budějovice, Czechia
Full list of author information is available at the end of the article

## Background

Bacteriophages (phages), natural viral predators of bacteria, are engaged in a constant evolutionary arms race with their hosts [1], playing major roles in the ecological balance of microbial life and in microbial diversity.

Most double-stranded DNA (dsDNA) phages share the same gene pool [2]; however, sequence comparisons reveal a widespread horizontal exchange of sequences among genomes, mediated by both non-homologous and homologous recombination. High frequency exchange among phages occupying similar ecological niches results in a high rate of mosaic diversity in local populations [3]. Studies confirm that phage genomes are mosaics and represent a large common genetic pool due to horizontal exchange [4, 5].

The screening of microbial natural products continues to constitute an important route to the discovery of chemicals for developing new therapeutic agents and evaluating the therapeutic potential of bacterial taxa [6–8]. In this respect, actinomycetes are a group of microorganisms mostly used in biotechnology for handling bioactive compounds. [9, 10]. Moreover, bacteriophages can be used to detect antiviral compound production by actinomycetes. Finally, actinophages are isolated and investigated because they can influence antibiotic production in bacterial strains, causing problems in the pharmaceutical industry. The vast majority of actinophages were isolated from sediments, but direct isolation from soil generally yields extremely low titers [11, 12]. However, although it is difficult to grow bacteriophages from soil without enrichment, a wide range of counts has been reported [13, 14].

Recently, there has been expanding interest in bacteriophages that infect *Streptomyces* species, since the phages can support the development of cloning vectors [15]. Such vectors could open the way for genetic manipulation as an important tool for *Streptomyces* improvement. Moreover, the mechanisms of the system for phage infection and multiplication could be useful in the fermentation industry and lead to the development of phage cloning vectors [16]. To date, no studies on phages isolated from *S. flavovirens*, an important source for several pharmaceutical drugs, such as actinomycin complex, mureidomycin and pravastatin [17, 18], have been carried out.

The development of high-throughput NGS (Next Generation Sequencing) technologies [19, 20] and the possibility to sequence entire genomes or transcriptomes more efficiently and economically than with first generation sequencing strategies permitted the collection of large amounts of information and the analysis of sequences from hundreds of thousands of species. Therefore, the dawn of next generation sequencing technologies has opened up exciting possibilities for whole genome sequencing in a wide range of organisms and the bacterial viruses have not been excluded from this revolution, despite the fact that their genomes are orders of magnitude smaller in size compared with bacteria and other organisms.

The Actinophage Sequence Databases (http://phagesdb.org/) currently include 5861 genomes from putative actinophages, 120 of which infect *Streptomyces* species and sixty-five of which are sequenced, but no genomes of phages isolated form *S. flavovirens* are currently available. The NCBI genome database contains around 600 Caudovirales genomes to date but the number of complete bacteriophage genomes published is growing slowly [21].

Until now, no phages belonging to *S. flavovirens* have been sequenced and relatively little is known about *S. flavovirens* phage genomics. In the present work, we report the first whole genome sequencing study and annotation of two *S. flavovirens* virulent phages. The results will provide an important genomic resource for future investigations in the bacteriophages related to *S. flavovirens* and for phage evolution studies.

## Methods

### Source of lytic actinophages

Two isolates of *Streptomyces flavovirens* phages, named Sf1 and Sf3, were obtained from the virology lab, Agric. Microbiology Department, Faculty of Agriculture, Ain Shams University, Cairo, Egypt. Phages were isolated from soil and the morphological properties were analyzed by standard methodology and reported in Marei and Elbaz (2013) [22].

### Purification of lytic actinophages

The high titer phage suspension of each isolated phage was prepared using a liquid culture enrichment technique. The high titer phage suspension of each phage was ultracentrifuged at 30000 rpm for 90 min. at 4 °C in a Beckman L7-35 ultracentrifuge. The pellet was gently resuspended in 0.5 ml of 0.2 M phosphate buffer pH 7.2 [23].

### Adsorption rate and one-step growth experiments

The adsorption experiments were carried out with two isolated phage suspensions added to spores of their indicator host (*S. flavovirens*). Suspensions of each phage were incubated at 30 °C with gentle shaking. Samples were withdrawn at regular intervals after inoculation.

The mycelial fragments of the indicator strain were removed by centrifugation and the concentration of phage remaining in the supernatant was counted. The adsorption rates of the two phages were determined by measuring residual plaque-forming ability in membrane-filtered samples of an attachment mixture [24] and the adsorption rate constant k (mL/min) was calculated [25]. The one-step growth experiment was performed as described by Dowding (1973) [24].

**Physiochemical stability**

To evaluate the phages' stability three different chemicals (NaCl, MgCl2.6H2o and ZnCl2), were used. Five concentrations (0.1, 0.2, 0.3, 0.4 and 0.5 mM) for each salt were employed [26]. To test the effect of different treatments phage solutions for both tested strains with final concentrations of $10^7$ PFU/ml were utilized. The mixture was incubated for 10 min at room temperature (RT). The number of plaques was determined using the double layer method (plaque assay test) [27]. A control test was prepared by mixing bacterial suspension with phage without the tested chemicals.

**DNA isolation, library preparation and whole genome sequencing**

Genomic DNA was isolated from the propagated phages according to the procedure described by Kieser et al. [28]. DNA quality was assessed using a Nanodrop Bioanalyzer ND1000 (ThermoScientific). Sequencing libraries were prepared by shearing 1 µg of DNA in blunt-ended fragments by linking the Ion adapters using an Ion XpressTM Plus Fragment Library Kit (Life Technologies, Carlsbad, USA) according to the manufacturer's specifications. The sized and ligated fragments were amplified by emulsion-PCR using the Ion OneTouch 200 Template kit (Life Technologies, Carlsbad, USA). Quality and insert size distribution were assessed using an Agilent Bioanalyzer DNA 1000 chip. Libraries were sequenced on an Ion Torrent PGM semiconductor sequencer (Life Technologies, Carlsbad, USA) using the 200 bp protocol and an Ion Torrent 314 chip following the manufacturer instructions (Life Technologies, Carlsbad, USA).

**Assembly and bioinformatics analyses**

Raw reads resulting from Sf1 and Sf3 sequencing were trimmed using Trimmomatic with single end mode (no quality encoding was specified to allow the program to determine it automatically [29]) and assembled separately using the gsAssembler (Roche Applied Science, Indianapolis, IN); the Graphical User Interface (GUI) version was used with the default parameters. The collected contigs were visualized and validated using Hawkeye [30]. Resulting contigs for each phage showed approximately 60-fold sequence read coverage. The expected sequence accuracy was 95% with a statistical error of less than 1 in 10,000 bp. Sequence homologies were determined by using BLASTn against the actinophage database to assign the phages to a cluster [31].

**Open reading frame (ORF) analysis and gene prediction**

Open reading frames (ORFs) were identified and the genome sequences of each phage were annotated as described previously in Dobbins et al., 2004 by using DNA Master (J. G. Lawrence) (http://cobamide2.bio.pitt.edu) software and visual inspection [32]. For a genome-wide viewpoint an association with the annotation refinement, functional analysis and other explorations was developed using Phamerator. Protein sequence relationships and conserved domains within genes were also studied. Gene products were grouped into "Phamilies" generally referred to as "Phams", or groups of proteins with a high degree of similarity to one another. The pairwise alignment scores and significant rate were determined using BLASTp and ClustalW [33].

**Genomic comparisons between the sequenced and the close related phages**

Sequence comparisons were performed by using the BLAST algorithm available at NCBI [34] and Mauve software [35]. A comparison map among Sf1 and Sf3 *Streptomyces* phages and closely related phages (VWB and SV1) with available genomes in the National Center for Biotechnology Information (NCBI) nucleotide database (https://www.ncbi.nlm.nih.gov/) was generated by Circoletto (http://tools.bat.infspire.org/circoletto/) [34, 36]. For pictogram construction, bit-score values were used to describe the quality of the alignment at a given point. The bit-score is a normalized version of the score value returned by the BLAST searches, expressed in bits [37].

The phylogenetic tree of Major Capsid Protein (MCP) genes from two new isolated phages (Sf1 and Sf3) and 20 related *Streptomyces* phages available in the NCBI database was constructed with Geneious software version (R8) (http://www.geneious.com) [38] based on the Neighbor-Joining (NJ) algorithm.

## Results and discussion

**Adsorption rate constant and growth characteristics of isolated phages**

Adsorption of Sf1 and Sf3 was determined using *S. flavovirens* cells grown in phage medium to the early exponential phase of growth (15-h cultures). About 82 and 85% of all infective Sf1 and Sf3 particles, respectively, were adsorbed within 20 min of contact. The adsorption reached a maximum after 30 min. for both phages. The adsorption constant K was 3.66 pL/min for Sf1 and 3.80 pL/min for Sf3, determined by the Adams's formula [27]. The phages adsorption rates were higher than other actinophages [39], which was probably due to the strong specificity of the Sf1 and Sf3 phages to their host.

The production of Sf1 and Sf3 phages were determined in a one-step growth experiment at 30 °C. Results revealed that the latent periods of Sf1 and Sf3 were approximately 15 and 30 mins, respectively. After 30 and 45 mins the maximum rise period was shown and the burst sizes were 1.95 and 2.49 PFU/mL for Sf1 and Sf3, respectively (Fig. 1). The present results are in agreement with the data obtained from a study on 24

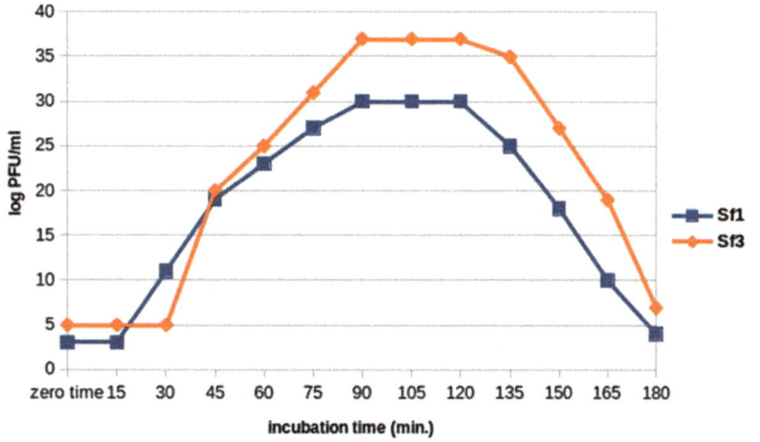


**Fig. 1** One-step growth experiment for Sf1 and Sf3 phages development on *S. flavovirens* at 30 °C

actinophages [40], underlining that under controlled cultural conditions the infection of isolated *Streptomycetes* cells by phages was varied.

### Physiochemical stability of isolated actinophages

Sodium and magnesium chloride treatments yielded a significant increase in both phages' activity for all concentrations used compared with the control, while zinc chloride application with concentrations > 0.3 mM caused a significant decrease of activity for Sf1 and Sf3 (Fig. 2). Similar results were reported in previous studies [41–43]. Absence of calcium and magnesium ions prevents adsorption and the lysis cycle, while their presence stimulates a significant increase in phage activity, probably due to the increase of adsorption and penetration rates. On the contrary, zinc and aluminum chloride showed significant loss of infectivity in both phages. This is in accordance with the experiments performed by Robert and Charles, which suggested that aluminum caused viral inactivation related to the dissociation of viral capsid proteins [44].

### Genome organization of phages Sf1 and Sf3

Genome sequencing generated 69,719 and 107,273 reads for each phage with around 60-fold coverage and 43,150 bp, and 60,934 bp assembled sequences for Sf1 and Sf3, respectively. The pair-wise alignment [45] revealed that the genomes of Sf1 and Sf3 shared an overall high level of similarity, with conserved regions of high identity (100% identity) interspersed between regions with high variability (ranging from 23.9% to 87.5%) (Fig. 3a). A similar mosaic genome structure has been observed in most other phage genomes, indicating extensive horizontal genetic exchange among phages [46–49]. No close relatives (Singleton) from modeling of both genome construction were revealed (Fig. 1).

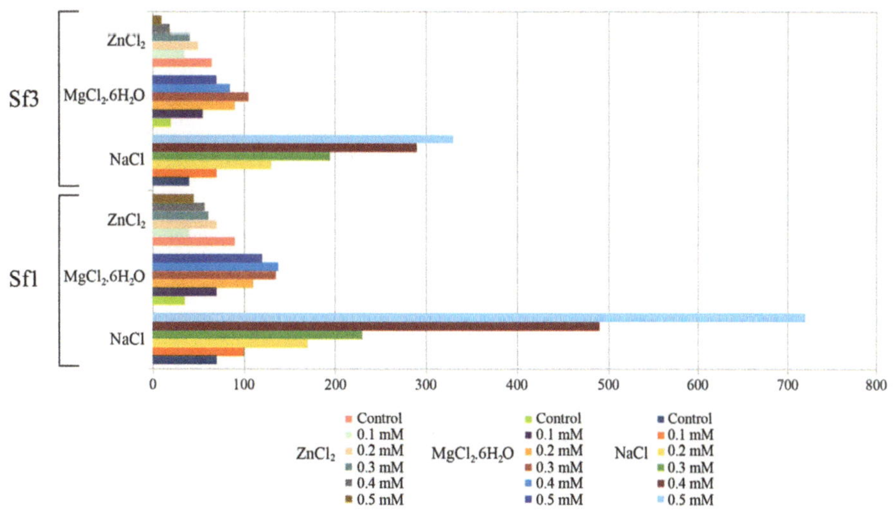


**Fig. 2** Effect of three different chemicals on the Sf1 and Sf3 infectivity

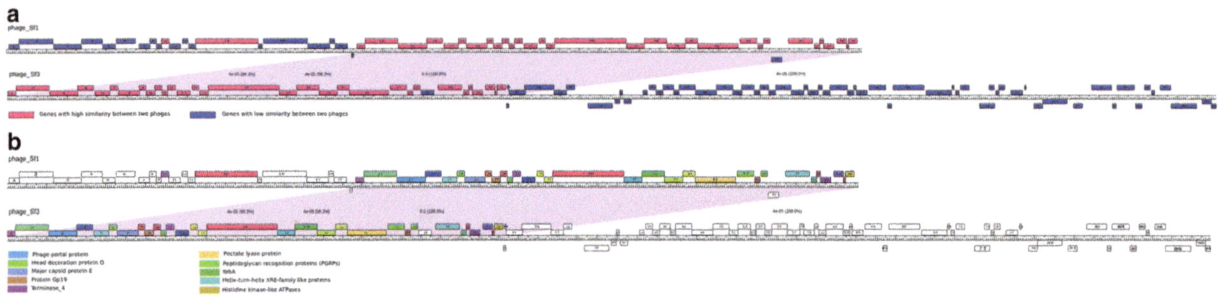

**Fig. 3** Genomic organization of Sf1 and Sf3 phages. Phages were mapped using Phamerator; the purple lines between phages underline the regions with high similarity, while the ruler corresponds to genome base pairs. The predicted genes are shown as boxes either above or below the genome (ruler), depending on whether are rightwards or leftwards transcribed, respectively. Gene numbers are shown within each box; pink boxes refereed to the genes with high similarity between two phages while the blue boxes refereed to the genes that show low similarity. **a** The phages maps showed by cluster conservation. **b** The phages maps showed by phams; genes are colored according to their function categories "phams"

Annotation of Sf1 and Sf3 genomes revealed 52 and 91 putative ORFs, respectively. According to their homology, 19 out of 52 ORFs (36.5%) from Sf1 and 28 out of 91 (30.8%) from the Sf3 genome have been assigned functions compared to known conserved domains [50] (Tables 1 and 2). Isolated genes were mainly involved in DNA replication and repair, nucleotide metabolism, lysis, phage structural proteins and other enzymes. The results obtained are in agreement with other bacteriophage studies [51–53]. Phage Sf1 showed 52 ORFs (Table 1), named gp1 - gp52, while 91 ORF were identified from Phage Sf3, from gp1 to gp91 (Table 2). The majority of members of identified families are bacteriophage proteins, while others (75%) have unknown function [54, 55].

### Phage structure and assembly genes

Several genes code for terminase subunit proteins, such as gp1 and 2 which code for terminase_4 (pfam05119) and terminase_1 (pfam03354) super-families, respectively. The gp3 and gp23 genes encode for the phage portal protein (pfam05133), an important protein involved in DNA transport during its packaging and ejection. Another relevant gene is gp6 which, together with gp27,codes for the major capsid protein (PHA00665) [56] and the major capsid protein E domain (pfam03864) [57], respectively, involved in the stabilization of the condensed form of DNA in phage heads. Some genes involved in tail development, gp14 (pfam10145) and gp17 (pfam13550), were also identified.

In Sf3we found a gene (gp3) encoding phage portal protein (pfam05133), crucial for DNA migration and building the junction between head and tail proteins [58], and others, such as gp7 and gb16, that encode for the major capsid protein E domain (pfam03864) [57] or for lyase (gp21), like pectate lyase_3 superfamily protein (pfam12708). A phage putative head morphogenesis protein (TIGR01641) of 110 amino acids found exclusively in phage-related proteins, was encoded by gp84. Putaive head morphogenesis proteins such as gp85, which encodesthe transcriptional activator RhaR (PRK13502), and gp89, involved in the phage terminase_3 (COG1783) synthesis, are activated during the beginning of double-stranded viral DNA packaging [59].

### DNA replication and metabolic genes

The gp44 gene encodes YabA (COG4467), a protein that interacts with the DnaA initiator and the DnaN sliding clamp and drives the control of DNA replication initiation [60, 61]. gp46 and gp52 encode for helix-turn-helix XRE-family like proteins (cd00093) [62] and histidine kinase-like ATPases (cd00075) [63], respectively, two important binding proteins with roles in the replication, repair, storage and modification of DNA. gp4 encodes a protein belonging to the MATE family (cd13126), which functions as a translocase for lipopolysaccharides [64], while gp5 codes for the golgin subfamily protein A5, a protein responsible for maintaining Golgi structure in intra-Golgi retrograde transport [65].

ORFs with the same biological roles were also identified in Sf3 phage. Indeed gp35 encodes for a HhH-GPD superfamily base excision DNA repair protein (pfam00730). This group includes endonuclease III, 8-oxoguanine DNA glycosylases and DNA-3-methyladenine glycosylase II [66]. Other members include different types of DNA and RNA exonucleases such as RNase T, oligoribonuclease, and RNA exonuclease (REX) [67]; Holliday junction resolvases (HJRs) (cd00529), endonucleases structurally similar to RNase H and Hsp70, which specifically resolve Holliday junction DNA intermediates during homologous recombination was encoded by gp52 [68]. Gp76 encodes for HNH nucleases (cd00085), an endonuclease signature which is found in viral, prokaryotic and eukaryotic proteins [69].

**Table 1** Overview of Sf1 phage ORFs, summary of homology searches and annotations

| ORF | Product | Strand | Begin | End | AA | Motif | Predicted functions | Homology score | E-value |
|---|---|---|---|---|---|---|---|---|---|
| ORF 1 | gp1 | + | 126 | 650 | 175 | pfam05119 | Terminase_4 superfamily | 65,72 | 1.77e-14 |
| ORF 2 | gp2 | + | 643 | 2349 | 569 | pfam03354 | Terminase_1 superfamily | 243 | 3.32e-73 |
| ORF 3 | gp3 | + | 2376 | 3785 | 470 | pfam05133 | Phage portal protein | 147 | 3.13e-39 |
| ORF 4 | gp4 | + | 3796 | 4842 | 349 | cd13126 | (MATE) proteins | 36,9 | 5.07e-03 |
| ORF 5 | gp5 | + | 4857 | 5480 | 208 | pfam09787 | Golgin subfamily A5 | 35,58 | 6.35e-03 |
| ORF 6 | gp6 | + | 5535 | 6479 | 315 | PHA00665 | major capsid protein | 42,17 | 9.33e-05 |
| ORF 7 | gp7 | + | 6661 | 7194 | 178 | - | - | - | - |
| ORF 8 | gp8 | + | 7191 | 7532 | 114 | - | - | - | - |
| ORF 9 | gp9 | + | 7532 | 7783 | 84 | PRK14573 | bifunctional D-alanyl-alanine synthetase | 34,02 | 2.66e-03 |
| ORF 10 | gp10 | + | 7783 | 8178 | 132 | - | - | - | - |
| ORF 11 | gp11 | + | 8168 | 8740 | 191 | - | - | - | - |
| ORF 12 | gp12 | + | 8843 | 9109 | 89 | - | - | - | - |
| ORF 13 | gp13 | + | 9145 | 9489 | 115 | - | - | - | - |
| ORF 14 | gp14 | + | 9493 | 12630 | 1046 | pfam10145 | Phage-related minor tail protein | 88,94 | 1.30e-19 |
| ORF 15 | gp15 | + | 12631 | 12837 | 69 | - | - | - | - |
| ORF 16 | gp16 | + | 12896 | 15148 | 751 | - | - | - | - |
| ORF 17 | gp17 | + | 15163 | 16281 | 373 | pfam13550 | Putative phage tail protein | 43,42 | 1.33e-05 |
| ORF 18 | gp18 | + | 16281 | 16538 | 86 | - | - | - | - |
| ORF 19 | gp19 | + | 16563 | 17216 | 218 | - | - | - | - |
| ORF 20 | gp20 | - | 17411 | 17518 | 36 | - | - | - | - |
| ORF 21 | gp21 | + | 17716 | 18126 | 137 | - | - | - | - |
| ORF 22 | gp22 | + | 18141 | 19799 | 553 | - | - | - | - |
| ORF 23 | gp23 | + | 19824 | 21251 | 476 | pfam05133 | Phage portal protein | 129 | 4.91e-33 |
| ORF 24 | gp24 | + | 21244 | 22044 | 267 | - | - | - | - |
| ORF 25 | gp25 | + | 22107 | 22856 | 250 | - | - | - | - |
| ORF 26 | gp26 | + | 22870 | 23265 | 132 | pfam02924 | Bacteriophage lambda head decoration protein D | 47,27 | 2.61e-08 |
| ORF 27 | gp27 | + | 23280 | 24326 | 349 | pfam03864 | Phage major capsid protein E | 62,35 | 3.13e-11 |
| ORF 28 | gp28 | + | 24323 | 24646 | 108 | - | - | - | - |
| ORF 29 | gp29 | + | 24652 | 25095 | 148 | pfam09355 | Phage protein Gp19 | 33,61 | 5.91e-03 |
| ORF 30 | gp30 | + | 25092 | 25445 | 118 | - | - | - | - |
| ORF 31 | gp31 | I | 25442 | 25726 | 95 | | | - | - |
| ORF 32 | gp32 | + | 25726 | 26127 | 134 | - | - | - | - |
| ORF 33 | gp33 | + | 26200 | 26865 | 222 | - | - | - | - |
| ORF 34 | gp34 | + | 26969 | 27292 | 108 | - | - | - | - |
| ORF 35 | gp35 | + | 27337 | 27765 | 143 | - | - | - | - |
| ORF 36 | gp36 | + | 27772 | 31344 | 1191 | cd00254 | Lytic Transglycosylase (LT) | 56,65 | 1.82e-09 |
| ORF 37 | gp37 | + | 31349 | 32239 | 297 | - | - | - | - |
| ORF 38 | gp38 | + | 32239 | 33384 | 382 | - | - | - | - |
| ORF 39 | gp39 | + | 33386 | 34312 | 309 | - | - | - | - |
| ORF 40 | gp40 | + | 34326 | 34943 | 206 | - | - | - | - |
| ORF 41 | gp41 | + | 34953 | 36977 | 675 | pfam12708 | Pectate lyase_3 superfamily protein | 73,24 | 8.54e-15 |
| ORF 42 | gp42 | + | 37059 | 37886 | 276 | cd06583 | Peptidoglycan recognition proteins (PGRPs) | 58,45 | 4.03e-11 |
| ORF 43 | gp43 | + | 37933 | 38184 | 84 | - | - | - | - |
| ORF 44 | gp44 | + | 38228 | 38560 | 111 | COG4467 | YabA | 34,76 | 9.25e-04 |

**Table 1** Overview of Sf1 phage ORFs, summary of homology searches and annotations *(Continued)*

| | | | | | | | | | |
|---|---|---|---|---|---|---|---|---|---|
| ORF 45 | gp45 | - | 38602 | 39129 | 176 | - | - | - | - |
| ORF 46 | gp46 | + | 39475 | 40689 | 405 | cd00093 | Helix-turn-helix XRE-family like proteins | 45,24 | 1.45e-06 |
| ORF 47 | gp47 | + | 40777 | 41046 | 90 | - | - | - | - |
| ORF 48 | gp48 | + | 41043 | 41237 | 65 | - | - | - | - |
| ORF 49 | gp49 | + | 41234 | 41752 | 173 | - | - | - | - |
| ORF 50 | gp50 | + | 41901 | 42383 | 161 | - | - | - | - |
| ORF 51 | gp51 | + | 42383 | 42481 | 33 | - | - | - | - |
| ORF 52 | gp52 | + | 42474 | 42923 | 150 | cd00075 | Histidine kinase-like ATPases | 36,09 | 4.50e-04 |

### Cell lysis genes

Crucial genes implicated in lysis activities, such as the cell wall degradation process in bacteria during host infection, were identified in the Sf1 genome. Indeed, gp36 encodes for the lytic transglycosylase (LT) (cd00254) that catalyzes the cleavage of the beta-1,4-glycosidic bond between N-acetylmuramic acid and N-acetyl-D-glucoseamine, similar to "goose-type" lysozymes. gp42 encodespeptidoglycan recognition proteins (PGRPs) (cd06583), namely receptors that bind and hydrolyze peptidoglycans of bacterial cell walls, and contains two conserved histidines and a cysteine, typical residues of zinc binding sites [70].

While gp21 is included in the pectate lyase superfamily (pfam12708), proteins with a beta helical structure like pectate lyase and most closely related to glycosyl hydrolase family and gp22 encodes to Peptidoglycan recognition proteins (PGRPs) (cd06583) [70], were identified in Sf3 genome.

Both phage genomes show up to bring a modular organization, with genes of related function clustered together (Fig. 3a and b). DNA sequences of the first 13 kb in Sf3 are highly similar to the last DNA sequences in Sf1 and encode for DNA packaging structural proteins (Fig. 3b).

On the basis of the amino acid sequence similarity between the gene products, the conserved pfam05133 motif and the gene locations, orf3 is predicted to encode a portal protein in both phages. No small terminase-encoding gene could be identified in either genome. The largest gene in Sf1 genome is located in orf36 (3.5 kb) encoding the lytic transglycosylase (LT), while the largest one in Sf3 genome with the same length is orf16, encoding the major capsid protein E domain. [48, 71, 72]. A possible lyase gene is positioned distinctively in both phage genomes (orf41 for Sf1 and orf21 for Sf3). Those genes located downstream in both phage genomes encode proteins involved in DNA synthesis, metabolism and repair (Fig. 3b).

### Evolutionary relationship of Sf1 and Sf3

Sf1 and Sf3 phages show 30 phams, where 29 out of 30 phams contain two members (Table 3), while three members belong to pham number 12. Ten phams (33.3%) were assigned with known functionality; the others are unknown. Therefore, some of these phams are informative and can be used in evolutionary studies. Indeed, as reported for mycobacteriophages [73], single, ubiquitous, semi-conserved genes can be utilized for cluster prediction, useful when the whole genome sequence is unavailable. The 30 identified phams, which include important genes (see below), underline a close phylogenetic relationship between the two isolated phages and provide important information that can be used in future evolutionary relationship studies by comparing the genes identified in the *Streptomyces flavovir*ens phages and homologous genes in other bacteriophages.

orf27 (Sf1) and orf7 (Sf3) as members of pham n.7 were assigned as phage major capsid protein (MCP) E domains; this important class of genes was also used as a single gene prediction system for the mycobacteriophage clusters analysis [73]. orf23 (Sf1) and orf3 (Sf3), members of pham n. 3, were classified as phage portal proteins. These proteins were used in some previous investigations as a marker of diversity indicating, in some cases, the connections between habitat properties, microbial community structure and phage community composition [74]. orf29 (Sf1) and orf9 (Sf3) are the members of pham n.9, were assigned to phage protein gp19, an important tail component. Most of the proteins forming the phage tail components as well as other needle-like assemblies (e.g. secretion systems and bacteriocins) have a common origin from a single protein module [74]. This evidence emphasizes the importance of phage protein diversification and specialization in the evolution of different and complex bacterial systems and in bacterial adaptation, developing new functions and providing a distinct selective advantage [74].

As expected, the virulent phages developed phams involved in lysogenic pathways. Indeed, orf41 (Sf1) and orf21 (Sf3), grouped in pham n.20, showed high homology to the pectate lyase superfamily protein that can modify the properties of polysaccharides. Since the pectinolytic protein family is commonly represented in prokaryotic and eukaryotic microorganisms and, in plants,

**Table 2** Overview of Sf3 phage, ORFs, summary of homology searches and annotations

| ORF no. | Product | Strand | Begin | End | AA | Motif | Predicted functions | Homology score | E-value |
|---|---|---|---|---|---|---|---|---|---|
| ORF 1 | gp1 | + | 16 | 426 | 137 | - | - | - | - |
| ORF 2 | gp2 | + | 441 | 2099 | 553 | - | - | - | - |
| ORF 3 | gp3 | + | 2124 | 3551 | 476 | pfam05133 | Phage portal protein_Gp6 | 129 | 4.91e-33 |
| ORF 4 | gp4 | + | 3544 | 4344 | 267 | - | - | - | - |
| ORF 5 | gp5 | + | 4407 | 5156 | 250 | - | - | - | - |
| ORF 6 | gp6 | + | 5176 | 5565 | 130 | pfam02924 | Bacteriophage lambda head decoration protein D | 46,5 | 5.02e-08 |
| ORF 7 | gp7 | + | 5580 | 6626 | 349 | pfam03864 | Phage major capsid protein E | 62,35 | 3.13e-11 |
| ORF 8 | gp8 | + | 6623 | 6946 | 108 | - | - | - | - |
| ORF 9 | gp9 | + | 6952 | 7395 | 148 | pfam09355 | Phage protein Gp19 | 33,61 | 5.91e-03 |
| ORF 10 | gp10 | + | 7392 | 7745 | 118 | - | - | - | - |
| ORF 11 | gp11 | + | 7748 | 8026 | 93 | - | - | - | - |
| ORF 12 | gp12 | + | 8026 | 8427 | 134 | - | - | - | - |
| ORF 13 | gp13 | + | 8500 | 9165 | 222 | - | - | - | - |
| ORF 14 | gp14 | + | 9269 | 9592 | 108 | - | - | - | - |
| ORF 15 | gp15 | + | 9637 | 10065 | 143 | - | - | - | - |
| ORF 16 | gp16 | + | 10072 | 13644 | 1191 | pfam03864 | Phage major capsid protein E | 62,35 | 3.13e-11 |
| ORF 17 | gp17 | + | 13649 | 14539 | 297 | - | - | - | - |
| ORF 18 | gp18 | + | 14539 | 15684 | 382 | - | - | - | - |
| ORF 19 | gp19 | + | 15686 | 16612 | 309 | - | - | - | - |
| ORF 20 | gp20 | + | 16626 | 17243 | 206 | - | - | - | - |
| ORF 21 | gp21 | + | 17253 | 19277 | 675 | pfam12708 | Pectate lyase superfamily protein | 73,24 | 8.54e-15 |
| ORF 22 | gp22 | + | 19359 | 20186 | 276 | cd06583 | Peptidoglycan recognition proteins (PGRPs) | 58,45 | 4.03e-11 |
| ORF 23 | gp23 | + | 20233 | 20484 | 84 | - | - | - | - |
| ORF 24 | gp24 | + | 20528 | 20860 | 111 | COG4467 | YabA | 34,76 | 9.25e-04 |
| ORF 25 | gp25 | + | 20908 | 21546 | 213 | PHA03169 | hypothetical protein; Provisional | 35,72 | 5.62e-03 |
| ORF 26 | gp26 | + | 21757 | 22989 | 411 | cd00093 | Helix-turn-helix XRE-family like proteins. | 45,24 | 1.33e-06 |
| ORF 27 | gp27 | + | 23077 | 23346 | 90 | - | - | - | - |
| ORF 28 | gp28 | + | 23343 | 23537 | 65 | - | - | - | - |
| ORF 29 | gp29 | + | 23534 | 24052 | 173 | - | - | - | - |
| ORF 30 | gp30 | + | 24201 | 24683 | 161 | - | - | - | - |
| ORF 31 | gp31 | + | 24668 | 24781 | 38 | - | - | - | - |
| ORF 32 | gp32 | + | 24774 | 25223 | 150 | cd00075 | Histidine kinase-like ATPases | 36,09 | 4.50e-04 |
| ORF 33 | gp33 | - | 25247 | 25363 | 39 | - | - | - | - |
| ORF 34 | gp34 | + | 25319 | 25381 | 21 | - | - | - | - |
| ORF 35 | gp35 | + | 25382 | 26221 | 280 | pfam00730 | HhH-GPD superfamily base excision DNA repair protein | 47,36 | 3.71e-07 |
| ORF 36 | gp36 | + | 26181 | 27680 | 500 | - | - | - | - |
| ORF 37 | gp37 | + | 27677 | 28327 | 217 | cd01672 | Thymidine monophosphate kinase (TMPK) | 112 | 1.37e-30 |
| ORF 38 | gp38 | + | 28324 | 28755 | 144 | cd04683 | the Nudix hydrolase superfamily | 153 | 3.17e-48 |
| ORF 39 | gp39 | - | 29387 | 30592 | 402 | - | - | - | - |
| ORF 40 | gp40 | - | 30712 | 30963 | 84 | - | - | - | - |
| ORF 41 | gp41 | + | 30962 | 31105 | 48 | - | - | - | - |
| ORF 42 | gp42 | - | 31162 | 31524 | 121 | cd00093 | Helix-turn-helix XRE-family like proteins. | 41 | 1.99e-06 |
| ORF 43 | gp43 | + | 32113 | 32337 | 75 | - | - | - | - |
| ORF 44 | gp44 | + | 32425 | 32778 | 118 | - | - | - | - |
| ORF 45 | gp45 | + | 32771 | 33139 | 123 | - | - | - | - |

**Table 2** Overview of Sf3 phage, ORFs, summary of homology searches and annotations *(Continued)*

| | | | | | | | | | |
|---|---|---|---|---|---|---|---|---|---|
| ORF 46 | gp46 | + | 33136 | 33678 | 181 | - | - | - | - |
| ORF 47 | gp47 | + | 33675 | 33947 | 91 | - | - | - | - |
| ORF 48 | gp48 | + | 33944 | 34774 | 277 | pfam12705 | PD-(D/E)XK nuclease superfamily | 34,99 | 5.68e-03 |
| ORF 49 | gp49 | + | 34777 | 35838 | 354 | - | - | - | - |
| ORF 50 | gp50 | + | 35835 | 36647 | 271 | cd06127 | DEDDh 3′–5′ exonuclease domain family | 111 | 4.30e-30 |
| ORF 51 | gp51 | + | 36644 | 37099 | 152 | - | - | - | - |
| ORF 52 | gp52 | + | 37096 | 37713 | 206 | cd00529 | Holliday junction resolvases (HJRs) | 38,38 | 3.22e-04 |
| ORF 53 | gp53 | + | 37710 | 37985 | 92 | - | - | - | - |
| ORF 54 | gp54 | + | 37991 | 38428 | 146 | - | - | - | - |
| ORF 55 | gp55 | + | 38425 | 38775 | 117 | - | - | - | - |
| ORF 56 | gp56 | + | 38788 | 39564 | 259 | - | - | - | - |
| ORF 57 | gp57 | + | 39567 | 40202 | 212 | - | - | - | - |
| ORF 58 | gp58 | + | 40199 | 40402 | 68 | - | - | - | - |
| ORF 59 | gp59 | + | 40399 | 40926 | 176 | - | - | - | - |
| ORF 60 | gp60 | + | 40923 | 41120 | 66 | - | - | - | - |
| ORF 61 | gp61 | + | 41153 | 41506 | 118 | - | - | - | - |
| ORF 62 | gp62 | + | 41503 | 42369 | 289 | - | - | - | - |
| ORF 63 | gp63 | + | 42366 | 42692 | 109 | - | - | - | - |
| ORF 64 | gp64 | + | 42689 | 42814 | 42 | pfam10969 | Protein of unknown function (DUF2771) | 35,51 | 1.35e-04 |
| ORF 65 | gp65 | + | 42811 | 43368 | 186 | - | - | - | - |
| ORF 66 | gp66 | + | 43466 | 44308 | 281 | - | - | - | - |
| ORF 67 | gp67 | - | 44375 | 44590 | 72 | pfam02604 | Antitoxin Phd_YefM | 30,73 | 5.22e-03 |
| ORF 68 | gp68 | + | 44674 | 46254 | 527 | - | - | - | - |
| ORF 69 | gp69 | + | 46345 | 47619 | 425 | - | - | - | - |
| ORF 70 | gp70 | + | 47651 | 47743 | 31 | - | - | - | - |
| ORF 71 | gp71 | + | 47817 | 47996 | 60 | - | - | - | - |
| ORF 72 | gp72 | + | 48073 | 48510 | 146 | cd00397 | DNA breaking-rejoining enzymes | 40,54 | 2.27e-05 |
| ORF 73 | gp73 | - | 49011 | 49772 | 254 | - | - | - | - |
| ORF 74 | gp74 | + | 49506 | 49766 | 87 | - | - | - | - |
| ORF 75 | gp75 | + | 49841 | 49996 | 52 | - | - | - | - |
| ORF 76 | gp76 | + | 49993 | 50292 | 100 | cd00085 | HNH nucleases | 38,22 | 1.45e-05 |
| ORF 77 | gp77 | + | 50587 | 51171 | 195 | COG4983 | Uncharacterized protein | 79,98 | 9.19e-18 |
| ORF 78 | gp78 | + | 51278 | 51349 | 24 | - | - | - | - |
| ORF 79 | gp79 | - | 51714 | 52163 | 150 | - | - | - | - |
| ORF 80 | gp80 | - | 52167 | 53399 | 411 | - | - | - | - |
| ORF 81 | gp81 | - | 53709 | 54161 | 151 | - | - | - | - |
| ORF 82 | gp82 | + | 54647 | 55672 | 342 | pfam06381 | Protein of unknown function (DUF1073) | 39,22 | 1.07e-03 |
| ORF 83 | gp83 | - | 55695 | 55889 | 65 | - | - | - | - |
| ORF 84 | gp84 | + | 55948 | 56844 | 299 | TIGR01641 | phage putative head morphogenesis protein | 59,7 | 1.40e-11 |
| ORF 85 | gp85 | - | 56884 | 57198 | 105 | PRK13502 | transcriptional activator RhaR | 32,72 | 8.64e-03 |
| ORF 86 | gp86 | + | 57346 | 57642 | 99 | - | - | - | - |
| ORF 87 | gp87 | + | 57698 | 57847 | 50 | - | - | - | - |
| ORF 88 | gp88 | + | 58134 | 58682 | 183 | - | - | - | - |
| ORF 89 | gp89 | - | 58679 | 59956 | 426 | COG1783 | Phage terminase_3 | 161 | 6.32e-45 |
| ORF 90 | gp90 | - | 60233 | 60691 | 153 | - | - | - | - |
| ORF 91 | gp91 | - | 60684 | 60932 | 83 | - | - | - | - |

**Table 3** Related Conserved Domains (CD) to the detected Phamilies using Phamerator

| Pham | Conserves Domains (CD) | Number of members | Mean translation length | Phage Sf1 | Phage Sf3 |
|---|---|---|---|---|---|
| 1 | - | 2 | 136 | ORF 21 | ORF 1 |
| 2 | - | 2 | 552 | ORF 22 | ORF 2 |
| 3 | Phage portal protein | 2 | 475 | ORF 23 | ORF 3 |
| 4 | - | 2 | 266 | ORF 24 | ORF 4 |
| 5 | - | 2 | 249 | ORF 25 | ORF 5 |
| 6 | Bacteriophage lambda head decoration protein D | 2 | 130 | ORF 26 | ORF 6 |
| 7 | Phage major capsid protein E | 2 | 348 | ORF 27 | ORF 7 |
| 8 | - | 2 | 107 | ORF 28 | ORF 8 |
| 9 | Phage protein Gp19 | 2 | 147 | ORF 29 | ORF 9 |
| 10 | - | 2 | 117 | ORF 30 | ORF 10 |
| 11 | - | 2 | 93 | ORF 31 | ORF 11 |
| 12 | Terminase_4 superfamily | 3 | 132,3333 | ORF 1, ORF 32 | ORF 12 |
| 13 | - | 2 | 221 | ORF 33 | ORF 13 |
| 14 | - | 2 | 107 | ORF 34 | ORF 14 |
| 15 | - | 2 | 142 | ORF 35 | ORF 15 |
| 16 | - | 2 | 296 | ORF 37 | ORF 17 |
| 17 | - | 2 | 381 | ORF 38 | ORF 18 |
| 18 | - | 2 | 308 | ORF 39 | ORF 19 |
| 19 | - | 2 | 205 | ORF 40 | ORF 20 |
| 20 | Pectate lyase superfamily protein | 2 | 674 | ORF 41 | ORF 21 |
| 21 | Peptidoglycan recognition proteins (PGRPs) | 2 | 275 | ORF 42 | ORF 22 |
| 22 | - | 2 | 83 | ORF 43 | ORF 23 |
| 23 | YabA | 2 | 110 | ORF 44 | ORF 24 |
| 24 | Helix-turn-helix XRE-family like proteins | 2 | 407 | ORF 46 | ORF 26 |
| 25 | - | 2 | 89 | ORF 47 | ORF 27 |
| 26 | - | 2 | 64 | ORF 48 | ORF 28 |
| 27 | - | 2 | 172 | ORF 49 | ORF 29 |
| 28 | - | 2 | 160 | ORF 50 | ORF 30 |
| 29 | - | 2 | 34,5 | ORF 51 | ORF 31 |
| 30 | Histidine kinase-like ATPases | 2 | 149 | ORF 52 | ORF 32 |

is involved in remodelling cell walls, it is clear that the divergence from the ancestral protein over time has allowed different micro-organisms to target a range of pectin-like substrates while the overall structure has been maintained [75]. orf42 (Sf1) and orf22 (Sf3) are members of pham n.21 and classified as peptidoglycan recognition proteins (PGRPs), an innate class of immunity molecules present in insects, mollusks, echinoderms, and vertebrates that by interacting with peptidoglycan in the cell wall, rather than permeabilizing bacterial membranes, kills bacteria. These proteins were reported, at least in one carboxy-terminal domain, as homologous in bacteriophage and bacteria [76]. orf46 (Sf1) and orf26 (Sf3) are grouped in pham n.24 and were identified as helix-turn-helix (HTH) XRE-family-like proteins, one of the early studied regulatory DNA-binding proteins involved in metabolic regulation in bacteria. This class of genes encodes components to process environmental metabolites (e.g. lactose) and to produce interacting constituents in the development of a lytic or lysogenic pathway in phages. A common ancestor for all DNA-binding domains was suggested and, through its duplication and divergence, the diversity of transcription regulators that drive bacterial and phage genes was generated. The HTH fold investigations confirmed the significance of this module in DNA–protein interactions across a wide phylogenetic spectrum including a wide variety of phages [77].

orf26 (Sf1) and orf6 (Sf3), members of pham n. 6, were classified as bacteriophage lambda head decoration protein D. Since the protein allows for the display of many copies of a foreign protein, which is advantageous for displaying weak ligands for affinity selection, a useful platform for phage polypeptide display was recently developed [78]. Interestingly, orf32 in Sf1 and orf12 in Sf3 were not assigned functions previously, although they belong to the pham n. 12 together with orf 1 (Sf1) which is classified as terminase_4.

A standard Nucleotide NCBI BLAST (blastn) search was developed using both Sf1 and Sf3 phage whole genome sequences as a query against a non-redundant nucleotide sequences database. Starting from a whole phage dataset (https://www.ncbi.nlm.nih.gov/) the available phage genomes with the best identity percentages (VWB and SV1) were chosen and a pictogram was developed (Fig. 4). Seventy-eight percent identity for both *S. flavovirens* phages compared to the complete genome of

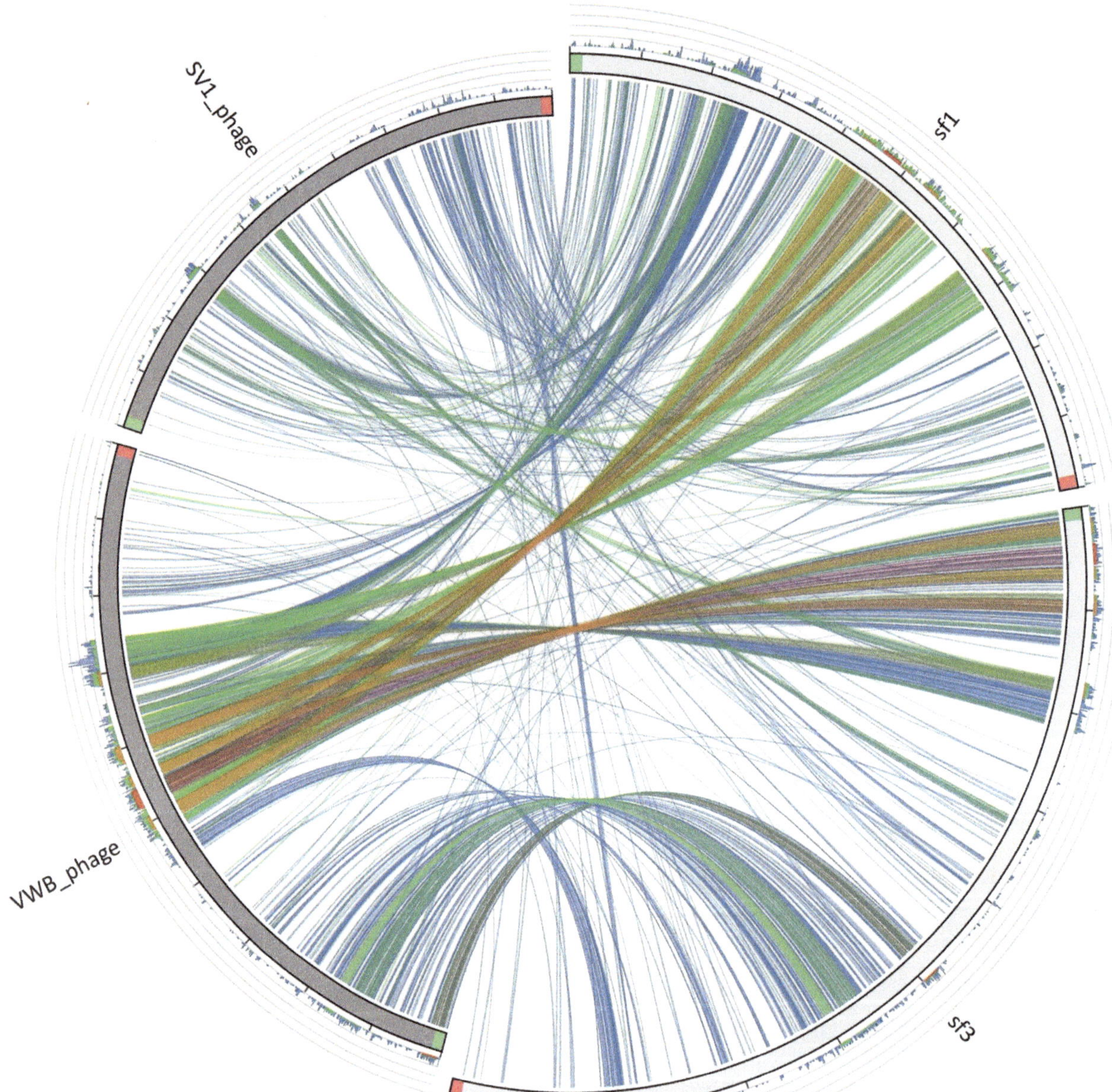


**Fig. 4** Sequence similarities among Sf1, Sf3, VWB and SV1 phages. The picture shows the results of the BLAST local alignments using Sf1 and Sf3 as a query against the VWB and SV1 phages sequences. The different colours (blue, green, orange and red) represent the overall quality of the aligned segments along the phage sequences, evaluated on the basis of the bit-score values from the worst to the best score (blue to red). The bit-score is a normalized version of the score value obtained by BLAST searches, expressed in bits. The height of the coloured bars in the histogram shows how many times each colour hits a specific fragment of the other phage sequences. A twist in a ribbon indicates that the local alignment is inverted (query and database sequence on opposite strands)

bacteriophage VWB, isolated from *S. venezuelae* strain ETH 14630 (AY320035.2), was exhibited (with 29% and 36% of coverage for Sf1 and Sf3, respectively), while 75% of identity for both studied phages with *S. venezuelae* phage SV1 (JX182371.1) was reported, but with low query coverage (11% for Sf1 and 14% for Sf3), probably due to the phylogenetic distance between the compared phages.

The alignment of both Sf1 and Sf3 genomes against the sequences of VWB phage, carried out by Mauve software, revealed that most hits occurred around a 13Kb region (Fig. 4). The approximate location of this region were (18000–31000) within the Sf1 genome, (1–13000) in the Sf3 genome and (23000–36000) in the VWB genome. On the contrary, the alignment of both *S. flavovirens* phage genomes versus the sequences of SV1 showed only a short region (~1Kb) with moderate bit score ranging from 9691–10707 and 10300–11208 in the genomes of Sf1 and Sf3, respectively, consistent with the low sequence coverage obtained.

The MCPs diversity between Sf1, Sf3 and 20 related *Streptomyces* phages, due to a combination of illegitimate and homologous recombination [79] and mutational drift, was also evaluated. The current investigation highlighted the hybrid generation between phage genera [80] or phage families [81]. Twenty-two *Streptomyces* phages were grouped in five main branches (Fig. 5). The Lannister MCP shared a close evolutionary relationship with the Izzy, Aaronocolus, and Caliburn sequences, demonstrating that phages may undergo genetic exchange by horizontal gene transfer from a large shared pool [4] and that horizontal gene transfer between phages is a component of their evolution. Numerous gene exchanges

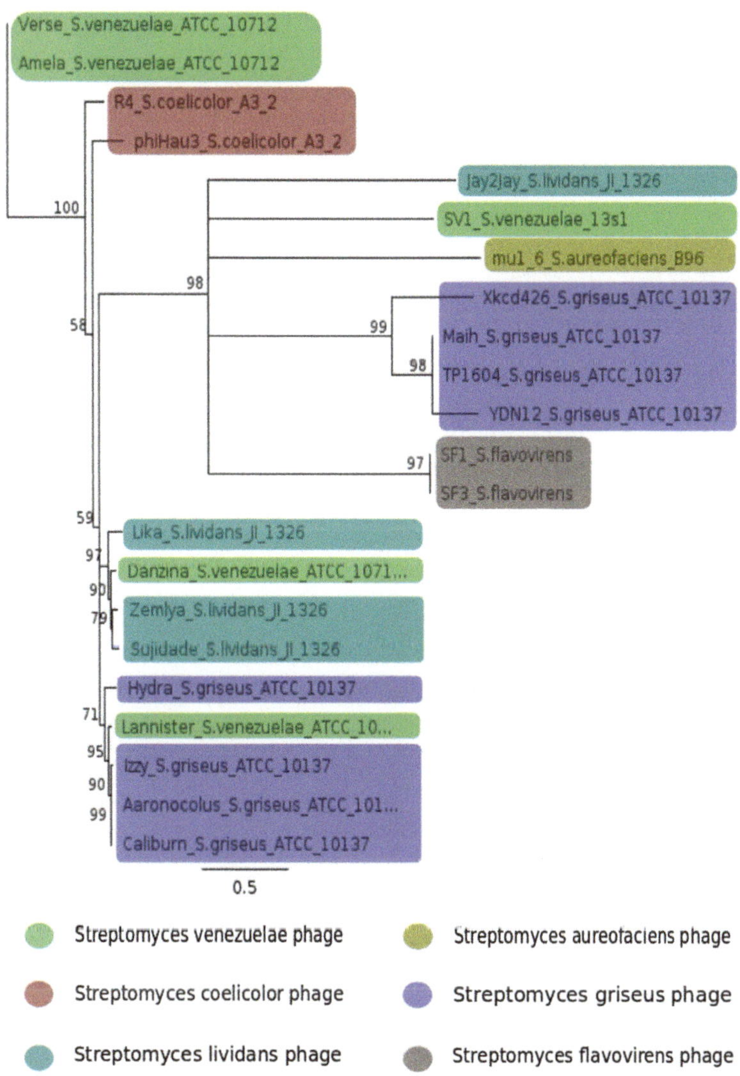


**Fig. 5** Phylogenetic analysis of studied phages and other members (20) of the Streptomyces phages group based on MCPs. Bootstrap values indicate the number of times a node was supported in 1000 resampling replications

within each major clade and core phage functions do not appear to have co-evolved with specific hosts [82].

Our phylogenetic analysis is useful for further studies, since both Sf1 and Sf3 were recovered in a clade that included phages that infect *Streptomyces* species but most of these phages (Maih, YDN12, Xkcd426 and TP1604) were members of the BG phage cluster; this clustering does not represent a phylogenetic or taxonomic grouping but rather provides a framework for reflecting their overall genome relationships and for identifying genes that have been recently exchanged and their genomic context [83, 84]. Moreover, Sf1 and Sf3 grouped in a separate branch, indicating that isolated phages belong to the BG phage cluster but represent a different sub-cluster.

## Conclusion

Recently, large advances have occurred in phage genomics; nevertheless,the full extent of phage diversity and evolutionary pathways are yet unknown. With the advent of NGS technologies a much greater volume of transcriptome and genome sequences is available and we can therefore expect an increased flow of new data in upcoming years. Current assessment suggests that more than 1031 phages exist on earth, representing more than ten million phage "species". Of these, less than 6000 have been observed using electron microscopy and fewer than 1000 genomes have been sequenced. The available sequences show that the majority of phages analyzed are tailed phages belonging to the family Siphoviridae, but less is known about the degree of their genetic diversity. The genomic characterization of phages is necessary to evaluate their important ecological impact. In spite of their ubiquity, phages have not yet been characterized for many bacterial genera. In the present study, biological, physiochemical and genome sequences of two new virulent *Streptomyces* phages are presented, representing the first genomic report of *S. flavovirens* phages which may represent a new sub-cluster of the BG *Streptomyces* phage cluster.

### Abbreviations

dsDNA: Double-stranded DNA; HJRs: Holliday junction resolvases; HTH: Helix-turn-helix; LT: Lytic transglycosylase; MCP: Major capsid protein; NCBI: National Center for Biotechnology Information; NGS: New Generation Sequencing; ORFs: Open reading frames; PGRPs: Peptidoglycan recognition proteins; Phages: Bacteriophages; REX: RNA exonuclease; TDP: Thymidine diphosphate; TMP: Thymidine monophosphate; TMPK: Thymidine monophosphate kinase; TTP: Thymidine triphosphate

### Acknowledgement

The authors would like to thank the Center for Research in Agricultural Genomics (CRAG) service laboratory, Barcelona, Spain for providing the sequencing instruments and reagents used in the study, Ezio Fontana, IBBR-CNR, Palermo, Italy for his advice and discussion about the whole genome data analysis and Heather Esson, Biology Center ASCR, Institute of Parasitology, Czech Republic for assistance with language editing.

### Funding

This work does not obtained any fund.

### Authors' contributions

All authors conceived and designed the experiments; AS carried out the experiments and performed the bioinformatics and statistical analysis; AS and FM compiled the results and drafted the manuscript. All the authors read and approved the final manuscript.

### Competing interests

The authors declare that they have no competing interests.

### Author details

[1]Genetic Department, Faculty of Agriculture, Ain Shams University, Cairo 11241, Egypt. [2]Institute of Parasitology, Biology Centre, Czech Academy of Sciences, 37005 České Budějovice, Czechia. [3]Institute of Biosciences and Bioresources (IBBR), National Research Council (CNR) of Italy, 90129 Palermo, Italy. [4]Central Lab. of Organic Agriculture, Agricultural Research Center, Giza 12619, Egypt. [5]Botany and Microbiology Department, Faculty of Science, Helwan University, Ain-Helwan, Cairo 11970, Egypt. [6]Microbiology Department, Faculty of Agriculture, Ain Shams University, Cairo 11241, Egypt.

### References

1. Klumpp J, Fouts DE, Sozhamannan S. Next generation sequencing technologies and the changing landscape of phage genomics. Bacteriophage. 2012;2(3):190–9.
2. Sandmeier H. Acquisition and rearrangement of sequence motifs in the evolution of bacteriophage tail fibres. Mol Microbiol. 1994;12(3):343–50.
3. Juhala RJ, Ford ME, Duda RL, et al. Genomic sequences of bacteriophages HK97 and HK022: pervasive genetic mosaicism in the lambdoid bacteriophages. J Mol Biol. 2000;299:27–51.
4. Hendrix RW, Smith MCM, Burns RN, et al. Evolutionary relationships among diverse bacteriophages and prophages: all the world's a phage. Proc Natl Acad Sci U S A. 1999;96:2192–7.
5. Hendrix RW. Bacteriophage genomics. Current Opinion in Microbiol. 2003;6:506–11.
6. Bradley SG, Ritzi D. Electron microscopical studies of actinophage multiplication. J Gen Virol. 1967;1(3):285–90.
7. Bull AT, Ward AC, Goodfellow M. Search and discovery strategies for biotechnology: the paradigm shift. Microbiol Mol Biol Rev. 2000;64:573–606.
8. Ipek KD. Actinophages as indicators of actinomycete taxa in marine environments. Antonie Van Leeuwenhoek. 2005;87:19–28.
9. Korn-Wendish F, Schneider J. Phage typing-a useful tool in actinomycete systematics. Gene. 1992;115:243–7.
10. Williams ST, Locci R, Beswick A, et al. Detection and identification of novel actinomycetes. Res Microbiol. 1993;144(8):653–6.
11. Ackerman HW, Berthiaume L, Jones LA. New actinophage species. InterVirology. 1985;23:121–30.
12. Othman BA, Askora AA, Awny NM, et al. Virulent Bacteriophages for Streptomyces griseoflsvus Isolated from Soil. Pak J Biotechnol. 2008;5:109–18.
13. Ashelford KE, Day MJ, Bailey MJ, et al. In situ population dynamics of bacterial viruses in a terrestrial environment. Appl Environ Microbiol. 1999;65:169–74.
14. Ashelford KE, Day MJ, Fry JC. Elevated Abundance of Bacteriophage Infecting Bacteria in Soil. Appl Environ Microbiol. 2003;69(1):285–9.
15. Chater KF, Hopwood DA, Kieser T, et al. Gene cloning in Streptomyces. Curr Top Microbiol Immunol. 1982;96:69–95.

16. Harris JE, Chater KF, Bruton CJ, et al. The restriction mapping of c gene deletions in Streptomyces bacteriophage 4C31 and their use in cloning vector development. Gene. 1983;22(2–3):167–74.
17. Oliver K. Grundwissen Pharmazeutische Biotechnologie. Vieweg + Teubner Verlag: Wiesbaden; 2002. ISBN 3-322-80124-1.
18. Satinder KB, Gurpreet SD, Carlos RS. Biotransformation of Waste Biomass into High Value Biochemicals. New York, NY: Springer New York; 2014.
19. Eisenstein M. Oxford Nanopore announcement sets sequencing sector abuzz. Nat Biotechnol. 2012;30:295–6.
20. Srivatsan A, Han Y, Peng J, *et al.* High-precision, whole-genome sequencing of laboratory strains facilitates genetic studies. PLoS Genet 2008, doi:10.1371/journal.pgen.1000139.
21. Hatfull GF. Bacteriophage genomics. Curr Opin Microbiol. 2008;11:447–53.
22. Marei EM, Elbaz RM. Isolation and Molecular Characterization of Three Virulent Actinophages Specific for Streptomyces flavovirens. J Virol Res. 2013;2:12–7.
23. Lanning S, Williams ST. Methods for the direct isolation and enumeration of actinophages in soil. J Gen Microbiol. 1982;128:2063–71.
24. Dowding JE. Characterization of a bacteriophage virulent for Streptomyces coelicolor A3(2). J Gen Microbiol. 1973;76:163–76.
25. ADAMS MH. Methods of study of bacterial viruses. Methods in Med Research. 1950;2:1–73.
26. Paunikar WN, Sanmukh SG, Ghosh TK. Effect of metal ions and chemical solvents on the adsorption of Salmonella phage on Salmonella choleraesuis subspecies indica. Int J of Pharma and Biosci. 2012;3(1):181–90.
27. Adams MH. Bacteriophages. New York: Interscience; 1959.
28. Kieser T, Bibb MJ, Buttner MJ, et al. Practical Streptomyces Genetics. Norwich: The John Innes Foundation; 2000.
29. Lohsem M, Bolger AM, Nagel A, *et al.* RobiNA: a user-friendly, integrated software solution for RNA-Seq-based transcriptomics. Nucleic Acids Res 2012, doi:10.1093/nar/gks540.
30. Schatz MC, Phillippy AM, Shneiderman B, *et al.* Hawkeye: an interactive visual analytics tool for genome assemblies. Genome Biology 2007, doi:10.1186/gb-2007-8-3-r34.
31. Hatfull GF, Jacobs-Sera D, Lawrence JG, et al. Comparative genomic analysis of sixty mycobacteriophage genomes: Genome clustering, gene acquisition and gene size. J Mol Biol. 2010;397(1):119–43.
32. Dobbins AT, George Jr M, Basham DA, et al. Complete Genomic Sequence of the Virulent Salmonella Bacteriophage SP6. J Bacteriol. 2004;186(7):1933–44.
33. Cresawn SG, Bogel M, Day N, et al. Phamerator: a bioinformatic tool for comparative bacteriophage genomics. BMC Bioinformatics. 2011;12:395.
34. Altschul SF, Gish W, Miller W, et al. Basic local alignment search tool. J Mol Biol. 1990;215(3):403–10.
35. Darling AE, Mau B, Perna NT. ProgressiveMauve: multiple genome alignment with gene gain, loss and rearrangement. PLoS One 2010, doi:10.1371/journal.pone.0011147.
36. Krzywinski M, Schein J, Birol I, et al. Circos: an information aesthetic for comparative genomics. Genome Res. 2009;19(9):1639–45.
37. Darzentas N. Circoletto: visualizing sequence similarity with Circos. Bioinformatics. 2010;26(20):2620–1.
38. Kearse M, et al. Geneious basic: an integrated and extendable desktop software platform for the organization and analysis of sequence data. Bioinformatics. 2012;28:1647–9.
39. Anné J, Van Mellaert L, Decock B, et al. Further biological and molecular characterization of actinophage VWB. J Gen Microbiol. 1990;136:1365–72.
40. Pringsulaka O, Chvanich S, Doi K, et al. Characteristics and Distribution of Actinophages Isolated from Thai Soil Samples. Sci Asia. 2004;30:223–30.
41. AlKhazindar M, Sayed ETA, Khalil MS, et al. Isolation and Characterization of Two Phages Infecting *Streptomyces scabies*. Res J Pharm, Biol Chem Sci. 2016;7(2):1351–63.
42. Binetti A, Quiberoni A, Reinheimer J. Phage adsorption to *Streptococcus thermophilus*. Influence of environmental factors and characterization of cell-receptors. Food Res Int. 2002;35:73–83.
43. Quiberoni A, Guglielmotti D, Binetti A, et al. Characterization of three *Lactobacillus delbrueckii* sub sp. *bulgaricus* phages and the physiochemical analysis of phage adsorption. J Appl Microbiol. 2004;96:340–51.
44. Robert B, Charles PJ. Characterization of the effect of aluminum metal on poliovirus. J of Indus Microbiol. 1988;3:33–8.
45. Brudno M, Do CB, Cooper GM, et al. NISC Comparative sequencing program LAGAN and multi-LAGAN: efficient tools for large-scale multiple alignment of genomic DNA. Genome Res. 2003;13(4):721–31.
46. Casjen SR, Thuman-Commike PA. Evolution of mosaically related bacteriophage genomes seen through the lens of phage P22 virion assembly. Virology. 2011;411:393–415.
47. Hendrix RW, Lawrence JG, Hatfull GF, et al. The origins and ongoing evolution of viruses. Trends Microbiol. 2000;8(11):504–8.
48. Petrovski S, Tillett D, Seviour RJ. Genome sequences and characterization of the Gordonia phages GTE5 and GRU1 and their use as potential biocontrol agents. Appl Environ Microbiol. 2012;78:42–7.
49. Rohwer F, Edwards R. The phage proteomic tree: a genome-based taxonomy of phage. J Bacteriol. 2002;184:4529–5.
50. Marchler-Bauer A, Derbyshire MK, Gonzales NR, et al. CDD: NCBI's conserved domain database. Nucleic Acids Res. 2015;43(Database issue):222–6.
51. Van Dessel W, Van Mellaert L, Liesegang H, et al. Complete genomic nucleotide sequence and analysis of the temperate bacteriophage VWB. J Anne Virology. 2005;331:325–37.
52. Grande L, Michelacci V, Tozzoli R, et al. Whole genome sequence comparison of vtx2-converting phages from Enteroaggregative Haemorrhagic Escherichia coli strains. BMC Genomics. 2014. doi:10.1186/1471-2164-15-574.
53. Leon-Velarde CG, Kropinski AM, Chen S, et al. Complete genome sequence of bacteriophage vB_YenP_AP5 which infects Yersinia enterocolitica of serotype O:3. Virol J. 2014. doi:10.1186/1743-422X-11-188.
54. Schouler C, Ehrlich SD, Chopin MC. Sequence and organization of the lactococcal prolate-headed bIL67 phage genome. Microbiology. 1994;140:3061–9.
55. Ford ME, Sarkis GJ, Belanger AE, et al. Genome structure of mycobacteriophage D29: implications for phage evolution. J Mol Biol. 1998;279(1):143–64.
56. Kropinski AM, Kovalyova IV, Billington SJ, et al. The genome of epsilon15, a serotype-converting, Group E1 Salmonella enterica-specific bacteriophage. Virology. 2007;369(2):234–44.
57. Katsura I. Structure and inherent properties of the bacteriophage lambda head shell. VI. DNA-packaging-defective mutants in the major capsid protein. J Mol Biol. 1989;205(2):397–405.
58. Lurz R, Orlova EV, Günther D, et al. Structural organisation of the head-to-tail interface of a bacterial virus. J Mol Biol. 2001;310(5):1027–37.
59. Gual A, Camacho AG, Alonso JC. Functional analysis of the terminase large subunit, G2P, of Bacillus subtilis bacteriophage SPP1. J Biol Chem. 2000;275(45):35311–9.
60. Noirot-Gros MF, Velten M, Yoshimura M, et al. Functional dissection of YabA, a negative regulator of DNA replication initiation in Bacillus subtilis. Proc Natl Acad Sci U S A. 2006;103(7):2368–73.
61. Hayashi M, Ogura Y, Harry EJ, et al. Bacillus subtilis YabA is involved in determining the timing and synchrony of replication initiation. FEMS Microbiol Lett. 2005;247(1):73–9.
62. Luscombe NM, Austin SE, Berman HM, *et al.* An overview of the structures of protein-DNA complexes. Genome Biol 2000, doi:10.1186/gb-2000-1-1-reviews001.
63. Dutta R, Inouye M. GHKL, An emergent ATPase/kinase superfamily. Trends Biochem Sci. 2000;25(1):24–8.
64. Hvorup RN, Winnen B, Chang AB, et al. The multidrug/oligosaccharidyl-lipid/polysaccharide (MOP) exporter superfamily. Eur J Biochem. 2003; 270(5):799–813.
65. Satoh A, Wang Y, Malsam J, et al. Golgin-84 is a rab1 binding partner involved in Golgi structure. Traffic. 2003;4(3):153–61.
66. Bruner SD, Norman DP, Verdine GL. Structural basis for recognition and repair of the endogenous mutagen 8-oxoguanine in DNA. Nature. 2000;403(6772):859–66.
67. Ibrahim H, Wilusz J, Wilusz CJ. RNA recognition by 3'-to-5' exonucleases: the substrate perspective. Biochim Biophys Acta. 2008;1779(4):256–65.
68. Aravind L, Makarova KS, Koonin EV. SURVEY AND SUMMARY: holliday junction resolvases and related nucleases: identification of new families, phyletic distribution and evolutionary trajectories. Nucleic Acids Res. 2000;28(18):3417–32.
69. Ku WY, Liu YW, Hsu YC, et al. The zinc ion in the HNH motif of the endonuclease domain of colicin E7 is not required for DNA binding but is essential for DNA hydrolysis. Nucleic Acids Res. 2002;30(7):1670–8.
70. Cheng X, Zhang X, Pflugrath JW, et al. The structure of bacteriophage T7 lysozyme, a zinc amidase and an inhibitor of T7 RNA polymerase. Proc Natl Acad Sci U S A. 1994;91(9):4034–8.
71. Petrovski S, Seviour RJ, Tillett D. Genome characterization of the polyvalent lytic bacteriophage GTE2 with the potential for biocontrol of Gordonia, Rhodococcus and Nocardia stabilized foams in activated sludge plants. Appl Environ Microbiol. 2011;77(12):3923–9.

72. Petrovski S, Dyson ZA, Seviour RJ, et al. Small but sufficient: the Rhodococcus phage RRH1 has the smallest known Siphoviridae genome at 14.2 kb. J Virol. 2012;86:358–63.
73. Smith KC, Castro-Nallar E, Fisher JNB, et al. Phage cluster relationships identified through single gene analysis. BMC Genomics. 2013;14:410.
74. Sullivan MB, Coleman ML, Quinlivan V, et al. Portal protein diversity and phage ecology. Environ Microbiol. 2008;10(10):2810–23.
75. Hooton SP, Timms AR, Rowsell J, et al. Salmonella Typhimurium-specific bacteriophage FSH19 and the origins of species specificity in the Vi01-like phage family. Virol J. 2011;8:498.
76. Dziarski R, Gupta D. Protein family review The peptidoglycan recognition proteins (PGRPs). Genome Biol. 2006;7:232.
77. Aravind L, Anantharaman V, Balaji S, et al. The many faces of the helix-turn-helix domain: Transcription regulation and beyond. FEMS Microbiol Rev. 2005;29:231–62.
78. Wendt JL, Feiss M. A fragile lattice: replacing bacteriophage E's head stability gene D with the shp gene of phage 21 generates the Mg2 + –dependent virus, E shp. Virology. 2004;326:41–6.
79. Clark AJ, Inwood W, Cloutier T, Dhillon TS. Nucleotide sequence of coliphage HK620 and the evolution of lambdoid phages. J Mol Biol. 2001;311:657–79.
80. Canchaya C, Fournous G, Chibani-Chennoufi S, Dillmann ML, Brussow H. Phage as agents of lateral gene transfer. Curr Opin Microbiol. 2003;6:417–24.
81. Recktenwald J, Schmidt H. The nucleotide sequence of Shiga toxin (Stx) 2e-encoding phage phiP27 is not related to other Stx phage genomes, but the modular genetic structure is conserved. Infect Immun. 2002;70:1896–908.
82. Dekel-Bird NP, Avrani S, Sabehi G, Pekarsky I, Marston MF, Kirzner S, Lindell D. Diversity and evolutionary relationships of T7-like podoviruses infecting marine cyanobacteria. Environ Microbiol. 2013;15:1476–91.
83. Lawrence JG, Hatfull GF, Hendrix RW. Imbroglios of viral taxonomy: genetic exchange and failings of phenetic approaches. J Bacteriol. 2002;184:4891–905.
84. Lima-Mendez G, Van Helden J, Toussaint A, Leplae R. Reticulate representation of evolutionary and functional relationships between phage genomes. Mol Biol Evol. 2008;25:762–77.

8

# Improving the production of acetyl-CoA-derived chemicals in *Escherichia coli* BL21(DE3) through *iclR* and *arcA* deletion

Min Liu[1], Yamei Ding[3], Hailin Chen[1,2], Zhe Zhao[1,2], Huizhou Liu[1], Mo Xian[1*] and Guang Zhao[1,4*]

**Abstract**

**Background:** Acetyl-CoA-derived chemicals are suitable for multiple applications in many industries. The bio-production of these chemicals has become imperative owing to the economic and environmental problems. However, acetate overflow is the major drawback for acetyl-CoA-derived chemicals production. Approaches for overcoming acetate overflow may be beneficial for the production of acetyl-CoA-derived chemicals.

**Results:** In this study, a transcriptional regulator *iclR* was knocked out in *E.coli* BL21(DE3) to overcome acetate overflow and improve the chemicals production. Two important acetyl-CoA-derived chemicals, phloroglucinol (PG) and 3-hydroxypropionate (3HP) were used to evaluate it. It is revealed that knockout of *iclR* significantly increased expressions of *aceBAK* operon. The cell yields and glucose utilization efficiencies were higher than those of control strains. The acetate concentrations were decreased by more than 50% and the productions of PG and 3HP were increased more than twice in *iclR* mutants. The effects of *iclR* knockout on cell physiology, cell metabolism and production of acetyl-CoA-derived chemicals were similar to those of *arcA* knockout in our previous study. However, the *arcA-iclR* double mutants couldn't gain higher productions of PG and 3HP. The mechanisms are unclear and needed to be resolved in future.

**Conclusions:** Knockout of *iclR* significantly increased gene expression of *aceBAK* operon and concomitantly activated glyoxylate pathway. This genetic modification may be a good way to overcome acetate overflow, and improve the production of a wide range of acetyl-CoA-derived chemicals.

**Keywords:** Acetyl-CoA, *Escherichia coli*, *iclR*, *arcA*, 3-hydroxypropionate, Phloroglucinol

## Background

Acetyl-CoA-derived chemicals, including polyhydroxyalkanoates, isoprenoids, polyketides, lipids and butyrate, are suitable for multiple applications in food additive, medicine, agriculture, cosmetic, and chemical industries [1–3]. The bio-production of acetyl-CoA-derived chemicals has become imperative owing to the diminishing petroleum reserves and growing environmental concerns [4]. To date, *Escherichia coli* remains one of the most widely used hosts in recombinant bioprocesses because of its low manufacturing cost and easy manipulation [5, 6]. The bio-production of acetyl-CoA-derived chemicals through engineering *Escherichia coli* has made significant progresses in recent years. However, it is worth mentioning that acetyl-CoA can be converted to acetate through the acetate kinase/phosphate acetyl transferase (AckA-Pta) pathway when cells grow rapidly on glucose abundant conditions, also known as 'acetate overflow' [7–9] (Fig. 1). Acetate excretion affects the cell density, biomass accumulation and macromolecule biosynthesis even at concentrations as low as 0.5 g/l [10, 11]. Mutations in both *ackA* and *pta* have been reported a strong reduction of acetate production. However, this is at the expense of the cell growth rate and is accompanied by an increase in the production of other by-products such as formate and lactate [12]. So, decreasing the undesirable conversion from acetyl-CoA to acetate may be beneficial for acetyl-CoA-derived chemicals production. As predicted, our

* Correspondence: xianmo1@qibebt.ac.cn; zhaoguang@qibebt.ac.cn
[1]CAS Key Laboratory of Biobased Materials, Qingdao Institute of Bioenergy and Bioprocess Technology, Chinese Academy of Sciences, Qingdao 266101, China
Full list of author information is available at the end of the article

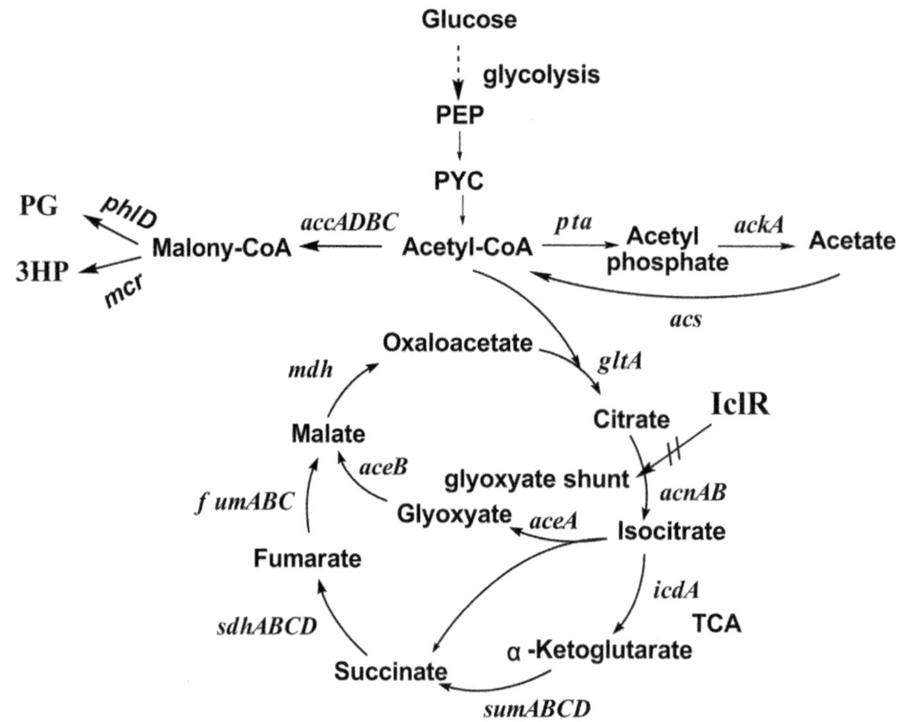


**Fig. 1** The main carbon metabolic pathways in recombinant *E.coli* strains

previous study has revealed that knockout of *arcA* in *E.coli* BL21(DE3) exhibited surprising efficacy of overcoming acetate formation, and the productions of acetyl-CoA-derived chemicals PG and 3HP were improved by 2.25-fold and 2.08-fold, respectively [13].

Acetate overflow is observed in all *Escherichia coli* strains but the extent can differ greatly between the various strains [7, 14]. *E.coli* BL21 strains showed lower acetate yields as compared to K12 strains [15, 16]. $C^{13}$-Constrained metabolic flux analysis revealed that the lower acetate yields in BL21 were caused by the higher activation of glyoxylate pathway and acetate assimilation pathway [15, 17]. IclR (isocitrate lyase regulator) is a local regulator, repressing the gene expression of *aceBAK* operon, which codes for the glyoxylate pathway enzymes malate synthase (AceB), isocitrate lyase (AceA) and isocitrate dehydrogenase kinase/phosphatase(AceK) [18–20] (Fig. 1). Decreasing the gene expression of *iclR* may activate glyoxylate pathway and decrease acetate formation. If not, lots of acetyl-CoA would be transformed to acetate through AckA-Pta pathway, which in turn starve the precursor for the acetyl-CoA-derived chemicals production and hurt the yields.

In this study, the *iclR* gene of *E.coli* BL21(DE3) was knocked out, and the resultant strain was used to produce two acetyl-CoA-derived chemicals, phloroglucinol (PG) and 3-hydroxypropionate (3HP) which are both important bulk chemicals [21, 22]. The effects of *iclR* deletion on gene expression of *aceBAK* operon, the cell physiology and metabolism were also determined and discussed. As *arcA* deletion also had positive effects on PG and 3HP production, the cell physiology and metabolism and production of acetyl-CoA-derived chemicals in *E.coli* BL21(DE3) *arcA-iclR* double mutant were also investigated in this study.

## Methods

### Strains construction

The oligonucleotide primers used in this study were listed in Table 1, and the recombinant plasmids and strains used in this study were listed in Table 2. The *iclR* mutants were constructed using P1 phage transduction from BL21(DE3) and *arcA* single mutant Q1949 as previously described [23]. The donor strain of *E.coli* BW25113 *iclR::kan* (JW3978) was purchased from the Keio collection [24]. The kanamycin genes were removed by a temperature-sensitive plasmid of pCP20, and the resultant strains were named as Q2280 and Q2138, respectively . The four subunits of *accA*, *accD* and *accBC* from *E.coli* were cloned into pACYCDuet-1, resulting in plasmid pA-*accADBC*. The *phlD* gene from *P. fluorescens* Pf5 (ATCC BAA -477) and *mar* gene from *E.coli* were cloned into pET30a, resulting in plasmid pET-*phlDmar*. [13, 22] The plasmids pA-*accADBC* and pET-*phlDmar* were transformed into *E.coli* BL21(DE3), *arcA* mutant Q1949, *iclR* mutant Q2280 and *arcA-iclR*

**Table 1** Primers used in this study

| Primer name | Sequence |
|---|---|
| Strain construction | |
| ID-*iclR*_F | TACGAAATGCCGGATCGTTG |
| ID-*iclR*_R | TCTTGTTTATCAAGAGTGTC |
| pKD13-Kan_R | GGTGAGATGACAGGAGATCC |
| RT-PCR Reaction | |
| 16S rRNA_F | TGGTCTGAGAGGATGACCAG |
| 16S rRNA_R | TGCTTCTTCTGCGGGTAACG |
| *aceB*_F | ATGACTGAACAGGCAACAAC |
| *aceB*_R | CGAATGGAAGCT GTTTCCGA |
| *aceA*_F | ATGAAAACCCGTACACAACA |
| *aceA*_R | CGAGGCTGTTGATGTAGCCT |
| *aceK*_F | CGTGGCCTGGAATTATTGAT |
| *aceK*_R | GTTAGTAATGCAGCGCAGTT |

double mutant Q2138 to generate PG-producing strains Q1944, Q1963, Q2283 and Q2176. The *mcr*-N and *mcr*-C ($N_{940}V$/ $K_{1106}W$/ $S_{1114}R$) from *Chloroflexus aurantiacus* were cloned into pETDuet-1, resulting in plasmid pMCR-N-C [25]. The plasmids pA-*accADBC* and pMCR-N-C were also transformed into *E.coli* BL21(DE3), Q1949, Q2280 and Q2138 to generate 3HP -producing strains Q2204, Q2284, Q2285 and Q2286.

**Gene expression by RT-PCR**

Total RNAs from BL21(DE3) and Q2280 were isolated using Bacteria RNA Kit (omega) according to the manufacturer's recommendations. The quantity and purity of RNAs were determined by optical density measurements at 260nm and 280nm by 1% agarose gel electrophoresis. Reverse-transcription (RT)-PCR reactions were carried out in a PCR thermocycler using TransScrip® One-step gDNA removal and cDNA synthesis kit. The reaction mix was first incubated 10min at 25°C and then incubated 30 min at 42°C for reverse transcription. Inactivated gDNA remover by heating 5 s at 85°C. Gene expression levels of *aceBAK* operon in BL21(DE3) and *iclR* mutant Q2280 were analyzed by PCR amplification with primers in Table 1. The gene transcription levels of 16S rRNA were used as inner reference for normalization.

**Table 2** Plasmids and strains used in this study

| Plasmids and strains | Description | Source |
|---|---|---|
| Plasmids | | |
| pETDuet-1 | $Amp^r$ *oripBR322 lacI^q T_7p* | Novagen |
| pACYCDuet-1 | $Cm^r$ *oriP15A lacI^q T_7p* | Novagen |
| pCP20 | $Cm^r$, $Amp^r$ pSC101 ori *CI857* | Novagen |
| pA-*accADBC* | $rep_{p15A}$ $Cm^R$ *lacI* $P_{T7}$ *accA* $P_{T7}$ *accD* $P_{T7}$ *accBC* | [22] |
| pET-*phlDmar* | $rep_{pBR322}$ $kan^R$ *lacI* $P_{T7}$ *phlD* $P_{T7}$*mar* | [22] |
| pMCR-N-C | $rep_{pBR322}$ $Amp^R$ *lacI* $P_{T7}$ *His$_6$-mcr$_{1-549}$* $P_{T7}$ *His$_6$-mcr$_{550-1219}$*($N_{940}V$/ $K_{1106}W$/ $S_{1114}R$) | [25] |
| Strains | | |
| *E. coli* DH5α | $F^-$ *supE44* Δ*lacU169* (φ80 *lacZ* Δ*M15*) *hsdR17 recA1 endA1 gyrA96 thi-1 relA1* | Invitrogen |
| *E. coli* BL21(DE3) | $F^-$ *ompT gal dcm lon hsdSB* ($rB^-$ $mB^-$) λ(DE3) | Invitrogen |
| JW3978 | *E. coli* BW25113 *iclR::kan* | Keio collection |
| Q1949 | *E. coli* BL21(DE3)Δ*arcA* | [13] |
| Q2133 | *E. coli* BL21(DE3) *iclR::kan* | This study |
| Q2280 | *E. coli* BL21(DE3)Δ*iclR* | This study |
| Q2135 | *E. coli* BL21(DE3)Δ*arcA iclR::kan* | This study |
| Q2138 | *E. coli* BL21(DE3)Δ*arcA*Δ*iclR* | This study |
| Q1944 | *E. coli* BL21(DE3)/pA-*accADBC*/pET-*phlDmar* | This study |
| Q1963 | Q1949/pA-*accADBC*/pET-*phlDmar* | This study |
| Q2283 | Q2280/pA-*accADBC*/pET-*phlDmar* | This study |
| Q2176 | Q2138/pA-*accADBC*/pET-*phlDmar* | This study |
| Q2204 | *E. coli* BL21(DE3)/pA-*accADBC*/pMCR-N-C | This study |
| Q2284 | Q1949/pA-*accADBC*/pMCR-N-C | This study |
| Q2285 | Q2280/pA-*accADBC*/pMCR-N-C | This study |
| Q2286 | Q2138/pA-*accADBC*/pMCR-N-C | This study |

**Shake-flask cultivation of the recombinant *E.coli* strains**

Shake-flask experiments were carried out in triplicate in 250 ml flask containing 50 ml fermentation medium. The strains were grown overnight at 37°C with shaking in LB broth, and then 1:50 diluted into 50 ml fermentation medium. When $OD_{600}$ of the culture reached about 0.6, IPTG was added to a final concentration of 0.1 mM/l and further incubated at 30°C. 10 mg/l Biotin and 20mM/l $NaHCO_3$ were needed for 3HP production. The fermentation medium for PG production contains 9.8 g/l $K_2HPO_4{\bullet}3H_2O$, 2.1 g/l citric acid•$H_2O$, 0.3 g/l ferric ammonium citrate, 3.0 g/l $(NH_4)_2SO_4$, 0.2 g/l $MgSO_4{\bullet}7H_2O$, 20 g/l glucose and 1000 × trace mental (3.7 g/l $(NH_4)_6Mo_7O_{24}{\bullet}4H_2O$, 2.9 g/l $ZnSO_4{\bullet}7H_2O$, 24.7 g/l $H_3BO_3$, 2.5 g/l $CuSO_4{\bullet}5H_2O$, 15.8 g/l $MnCl_2{\bullet}4H_2O$). The fermentation medium for 3HP production contains 14 g/l $K_2HPO4{\bullet}3H_2O$, 5.2 g/l $KH_2PO_4$, 1 g/l NaCl, 1 g/l $NH_4Cl$, 0.25 g/l $MgSO_4{\bullet}7H_2O$, 0.2 g/l yeast extract, 20 g/l glucose. The $OD_{600}$ of the culture and the concentrations of residual glucose were measured during the whole fermentation course.

**Analytical methods**

The cell concentration was measured by the optical density of the culture at 600 nm, and the value of density was converted to cell dry weight (CDW) based on that one unit of $OD_{600}$ was equivalent to 0.36 g/l CDW. The concentration of residual glucose was quantified by using an SBA-40D Biological Sensing Analyzer. The concentration of PG in the fermentation supernatant was quantified using the colorimetric reaction at $OD_{446}$ between cinnamaldehyde and PG. [26] The concentrations of 3HP and acetate in medium were determined by HPLC as described previously [13, 27].

## Results

**Knockout of *iclR* increased gene expression of *aceBAK* operon**

IclR is a transcriptional regulator that regulates gene expression of *aceBAK* operon [18, 20]. In order to investigate the effect of *iclR* deletion on gene expression of *aceBAK* operon, the total RNAs of wild strain BL21(DE3) and *iclR* mutant strain Q2280 were extracted to carry out the RT-PCR reaction. The gene transcription levels of 16S rRNA were used as inner reference for normalization. As shown in Fig. 2, the mRNA levels of *aceB*, *aceA* and *aceK* in *iclR* mutant Q2280 were much higher than those of the wild-type strain BL21(DE3), while the mRNA levels of 16S rRNA were very similar in these two strains, suggesting that *iclR* deletion could up-regulate the gene expression of *aceBAK* operon.

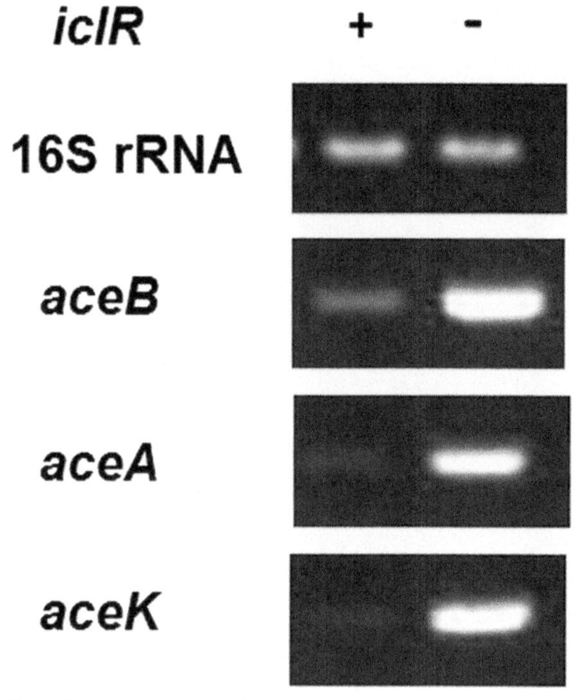


**Fig. 2** Genes transcription of *aceBAK* operon in *E.coli* BL21(DE3) and *iclR* mutant by RT-PCR. Constitutively, gene transcriptional level of 16S rRNA was used as inner reference

**Effects of *iclR* knockout on cell physiology and metabolism**

The *iclR* mutants were constructed using P1 phage transduction and the recombinant plasmids were transformed for PG and 3HP production. During the cultivation process, the $OD_{600}$ and residual glucose concentrations of the control strains (Q1944 and Q2204) and *iclR* mutants (Q2283 and Q2285) were measured to investigate the effects of *iclR* deletion on cell growth and glucose utilization. As shown in Fig. 3 and Fig. 4, the *iclR* mutants showed similar cell dry weight (CDW) as compared to their corresponding control strains. After fermentation, the cell dry weight of PG-producing strains Q1944 and Q2283 were 2.05 ± 0.06 g/l and 2.08 ± 0.09 g/l, while 3HP-producing strains Q2204 and Q2285 accumulated 2.64 ± 0.13 g/l and 2.76 ± 0.22 g/l CDW, respectively. These results drew a conclusion that deletion of *iclR* had no obvious effect on cell growth during the bio-production of PG and 3HP. Meanwhile, the glucose consumptions and cell yields were also compared. The glucose consumptions of *iclR* mutants obviously decreased since IPTG induction. Finally, the residual glucose concentrations of *iclR* mutants Q2283 and Q2285 were significantly higher than their corresponding control strains (Fig. 3). Knockout of *iclR*

decreased the glucose consumptions without impairing the cell dry weight, therefore, *iclR* mutants possessed higher cell yields. It is speculated that the genetic

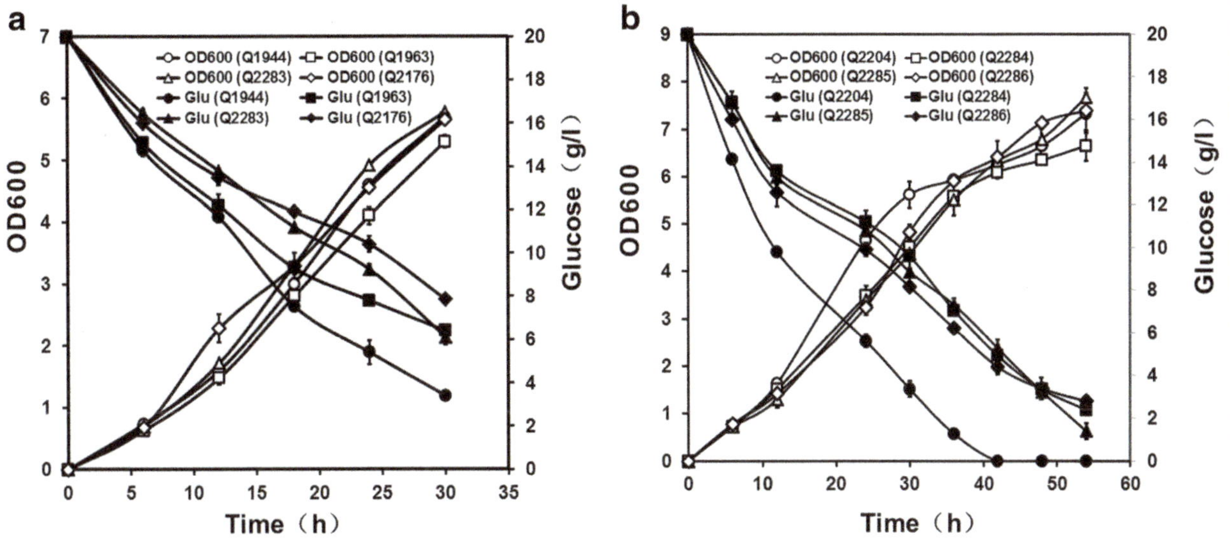


**Fig. 3** Cell growth and glucose consumption of recombinant *E.coli* strains during the process of cultivation in shaking flasks. **a** The OD600 and residual glucose concentrations of PG-producing strains Q1944, Q1963, Q2283 and Q2176 in their cultures. **b** The OD600 and residual glucose concentrations of 3HP-producing strains Q2204, Q2284, Q2285 and Q2286 in their cultures

modification of *iclR* knockout could improve the glucose utilization efficiency and reduce the manufacturing cost in industrial applications.

Acetate is the major undesirable metabolite in the production of acetyl-CoA-derived chemicals. After cultivation, the control strain Q1944 and Q2204 produced 3.14 ± 0.13 g/l and 5.86 ± 0.05 g/l acetate, while the *iclR* mutants Q2283 and Q2285 produced 1.41 ± 0.01 g/l and 2.12 ± 0.13 g/l acetate, decreased by 55% and 64%, respectively. In conclusion, *iclR* deletion showed dramatic effects on glucose utilization efficiency and acetate formation during the process of PG and 3HP production. However, the cell growth was very similar between the control strains and *iclR* mutants.

### Knockout of *iclR* improved PG and 3HP production

The effects of *iclR* deletion on PG and 3HP production were also investigated (Fig. 4). The PG production of *iclR* mutant Q2283 was 0.71 ± 0.06 g/l, showing 1.7-time higher than that of the control strain Q1944. The

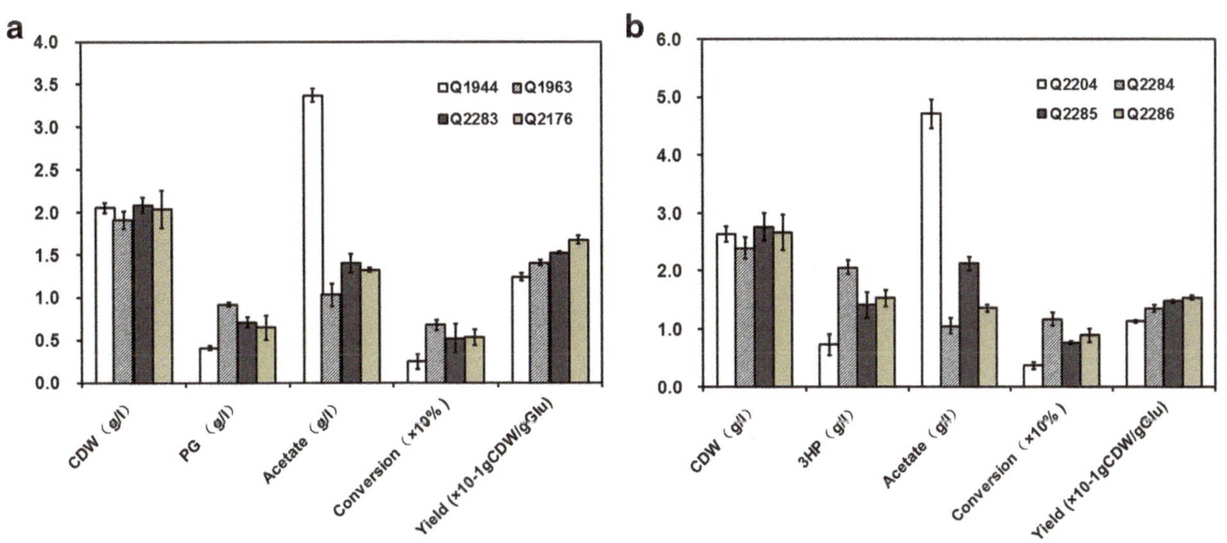


**Fig. 4** The effects of *arcA* and *iclR* knockout on the cell phycology and metabolism of recombinant *E.coli* strains during the process of cultivation in shaking flasks. **a** The cell dry weight, PG productions, acetate concentrations, glucose conversion efficiencies and cell yields of PG-producing strains. **b** The cell dry weight, 3HP productions, acetate concentrations, glucose conversion efficiencies and cell yields of 3HP-producing strains

glucose conversion efficiency to PG was increased by 2.1-fold, from 2.47% to 5.19%. In 3HP production, the *iclR* mutant Q2285 produced 1.41 ± 0.22 g/l 3HP, showing 1.9-time higher than that of the control strain Q2204. The glucose conversion efficiency to 3HP was increased by 2.0-fold, from 3.65% to 7.58%. Knockout of *iclR* significantly improved the production of acetyl-CoA-derived chemicals and the glucose conversion efficiency.

#### Effects of *arcA* and *iclR* knockouts on cell physiology and metabolism

Summarizing the above results, deletion of *arcA* or *iclR* can improve the cell yield and glucose utilization efficiency and repress the acetate formation, so we are curious whether the double knockout *of arcA* and *iclR* should further affect the cell physiology and metabolism. The detailed comparison of the cell growth, glucose consumption, cell yield and acetate production in *arcA* mutants, *iclR* mutants and *arcA-iclR* double mutants were also made. As shown in Fig. 3 and Fig. 4, the cell growth was seriously impaired through *arcA* deletion. While, it was not affected in *iclR* mutants and *arcA-iclR* double mutants. The double mutants Q2176 and Q2286 produced 2.03 ± 0.22 g/l and 2.66 ± 0.31 g/l CDW, the same as the control strains of 2.05 ± 0.06 g/l and 2.64 ± 0.12 g/l CDW. In addition, the cell yields of *arcA-iclR* double mutants were slightly higher than those of the control strains, the *arcA* single mutants and the *iclR* single mutants. It was assumed that the combined deletion of *arcA* and *iclR* was beneficial for the conversion from glucose to the biomass.

All the mutant strains showed reduced acetate productions as compared to the control strains (Fig. 4). However, the ability to overcome acetate formation of *arcA-iclR* double mutants was less than that of the *arcA* single mutants. The regulatory mechanisms remained unclear and needed to be resolved.

#### Effects of *arcA* and *iclR* knockouts on PG and 3HP production

The *arcA-iclR* double mutants showed lower productions of PG and 3HP than the *arcA* single mutants. The productions of PG and 3HP in *arcA* single mutants were increased by 2.25-fold and 2.83-fold than those in the control strains Q1944 and Q2204, while they were only increased by 1.58-fold and 2.15-fold in *arcA-iclR* double mutants. Consistent with that effects, the *arcA-iclR* double mutants also showed lower glucose conversion efficiencies than the *arcA* single mutants. The glucose conversions efficiencies to PG and 3HP in *arcA* single mutants were 6.76% and 11.7% respectively, while they were only 5.35% and 8.89% in *arcA-iclR* double mutants.

Taken together, both mutants of *arcA* and *iclR* showed positive effects on the productions and glucose conversion efficiencies to PG and 3HP. However the *arcA* single mutants had the most prominent results, and the *iclR* single mutants, the *arcA-iclR* double mutants showed the similar effects on these aspects.

### Discussion

The excretion and assimilation of acetate usually undergo a metabolic switch when cells growing on glucose. Acetate is produced by the Pta-AckA pathway in exponential growth phase, and it is converted to acetyl-CoA by the catalysis of acetyl-CoA synthetase (Acs) with the cells convert to stationary phase [28]. As the intracellular enzyme pyroposphatase removes the intermediate pyrophosphate, the reaction from acetate to acetyl-CoA is irreversible. Acetyl-CoA was utilized to generate ATP and metabolic intermediates for the cellular activities and exogenous genes expression via TCA cycle and glyoxylate shunt. So, acetate overflow is not only detrimental to cell growth and macromolecular biosynthesis, it is also a loss of carbon and therefore an economic sink [12]. To date, numerous approaches have been reported to overcome acetate excretion. If some of these approaches could also be used to improve the production of target compounds and decrease the production cost, these studies will have important significance and market value. In this study, the effects of *arcA* and/or *iclR* deletion on the production of acetyl-CoA-derived chemicals were determined and discussed. The results revealed that the *arcA* and/or *iclR* deletion significantly decreased the acetate concentrations. All the mutants showed higher cell yields and higher glucose utilization efficiencies. More importantly, the productions of PG and 3HP were increased than those of the control strains (Fig. 3 and Fig. 4). So, overcoming acetate overflow through *arcA* and/or *iclR* deletion could improve the production of acetyl-CoA-derived chemicals. Acetate overflow is observed in all *Escherichia coli* strains, however, the extent is very different between the K-12 and B strains, which are the most common *E.coli* strains for laboratorial and biotechnological applications [6]. BL21 strains showed less acetate accumulations than K-12 strains during the process of high cell-density fermentation with glucose as the sole carbon. A comparison of the genome sequences between BL21 and K12 strains showed that BL21 strains possessed rare codons in *arcA* and two mutations in the promoter region of *iclR* [29]. These variations of BL21 resulted in the reduced expressions of *arcA* and *iclR*, presumably contributing to the lower acetate accumulation. A previous research revealed that the central metabolic fluxes of BL21(DE3), especially with respect to the TCA fluxes, the glyoxylate pathway and the pentose pathway were similar to that of the K12 strains with *arcA* and *iclR* deletion [29]. Meanwhile, the acetate concentration of *E. coli* MG1655 Δ *arcA* Δ *iclR* was decreased to the value of BL21(DE3) [30]. Considering all that, the expression levels of *arcA* and *iclR* in BL21 strains were lower than K12 strains due to the variations in rare codons and promoter region, respectively. A previous study considered that the levels of *iclR* gene and its gene products in *E. coli* BL21 strains were much lower compared to in K12 strains, probably, the *iclR* levels were so low that a complete deletion of *iclR* [14]. However, our results revealed that knockout of *iclR* in BL21(DE3) could also significantly decrease acetate producitons by more than 50% and improve the PG and 3HP productions by more than twice. Thus, genetic modifications of *iclR* in K12 and BL21 strains both had obvious effect on overcoming acetate overflow, though the *iclR* levels were already very low in BL21 strains.

ArcA (anaerobic redox control) is the cytosolic response regulator of the dual-component regulator system ArcAB [31, 32]. It is reported that phosphorylated ArcA can repress genes expression of TCA cycle and glyoxylate shunt, such as *gltA*, *acnAB*, *icdA*, *sucABCD*, *sdhCDAB*, *fumA*, *mdh*, and *aceB* [33–35]. Our previous results demonstrated that the acetate concentrations of *arcA* mutants in BL21(DE3) were decreased by more than 70% and the productions of PG and 3HP were increased by more than twice. The decreased acetate concentrations and increased PG and 3HP productions of *arcA* mutants could be caused by the following reasons: the higher activity of TCA cycle and glyoxylate pathway; the more metabolic flux through pyruvate dehydrogenase (PDH) complex, which catalyzes pyruvate to acetyl-CoA [13, 32, 36]. IclR (isocitrate lyase regulator) is a local regulator, inhibiting genes expression of *aceBAK* operon in glyoxylate pathway. Knockout of *iclR* significantly increased gene expression of *aceBAK* operon according to the results of RT-PCR (Fig. 2). In glyoxylate pathway, isocitrate is directly split into succinate and glyoxylate without $CO_2$ loss, it may be used to explain the fact that *iclR* mutants showed higher cell yields than their corresponding control strains. Glyoxylate pathway is the main pathway for acetate assimilation, knockout of *iclR* can activate the glyoxylate pathway and promote the conversion from acetate to acetyl-CoA [17, 37]. In addtion, the metabolic flux from phosphoenolpyruvic acid (PEP) to acetyl-CoA is obviously increased in *iclR* mutant strains [29]. So, knockout of *iclR* can elevate the metabolic flux from acetate to acetyl-CoA and from PEP to acetyl-CoA. On the one hand, acetyl-CoA is metabolized to provide the energy and precursors for exogenous genes expression in recombinant pathways. And on the other hand, acetyl-CoA is the direct substrate for the production of acetyl-CoA-derived chemicals. Therefore, the *iclR* knockout can decrease the acetate formation and increase the production of acetyl-CoA-derived chemicals.

To summarize, the *arcA* deletion and *iclR* deletion were observed to behave similar effects on the cell physiology and metabolism, except the cell growth. The cell growth of *arcA* mutants was much lower than their corresponding control strains, however it was not impaired in *iclR* mutants. It still remains unclear what causes this phenomenon. Then, the *arcA* and *iclR* double mutants were also constructed to study the effects on the cell physiology, the metabolism and the production of acetyl-CoA-derived chemicals. Surprisingly, the double mutants had no better results for PG and 3HP productions than those of *arcA* and *iclR* single mutants. Meanwhile, the glucose conversion efficiencies and the abilities for overcoming acetate formation were also lower than those of the *arcA* single mutants. These results proved that the effects of polygenic modification on the produciton of target compounds may not be better than the single genic modification, especially by the global regulators. A global regulator usually controls several operons that belong to different groups [36]. Engineering of a global regulator could show pleiotropic phenotypes and variations in metabolic pathways and carbon fluxes. It is hard to determine the certain gene or the single pathway with a global regulator was modified. Furthermore, the results will be more unpredictable and complex when the several regulators are combined modified. In this study, why the effects of *arcA-iclR* double mutant on overcoming acetate overflow and improving the acetyl-CoA-derived chemicals production were less than the *arcA* single mutants? A previous study reported that *arcA-iclR* deletion of *E. coli* K12 MG1655 had dramatic effects on the cell physiology and metabolism. However, no significant differences of metabolic flux were observed between the wild type BL21(DE3) and the *arcA-iclR* double mutant [14]. The tiny variations of metabolic flux in BL21(DE3) *arcA-iclR* double mutant may lead to the less effects on the cell physiology and metabolism. However, the exact mechanisms would be unravelled by determining the distribution of metabolic fluxes in *arcA* single mutant, *iclR* single mutant and *arcA-iclR* double mutant in future.

## Conclusions

Acetate overflow is the major drawback for acetyl-CoA-derived chemicals production. Approaches for overcoming acetate overflow may be beneficial for the production of acetyl-CoA-derived chemicals. IclR is a transcriptional regulator that regulates gene expression of *aceBAK* operon in glyoxylate pathway. In this work, *iclR* was knocked out in *E. coli* BL21(DE3) to construct a mutant strain, and the productions of two important acetyl-CoA-derived chemicals, phloroglucinol and 3-hydroxypropionate were used to evaluate it. The results revealed that knockout of *iclR* significantly improved gene expression of *aceBAK* operon. The *iclR* mutants showed higher cell yields and higher glucose utilization efficiencies without sacrificing cell growth. The acetate concentrations were decreased by more than 50%. More importantly, the productions of PG and 3HP were 2-times higher than those of control strains. These phenomena of *iclR* knockout were similar to those of the global regulator *arcA* knockout in our previous study. However, the *arcA-iclR* double mutants could not produce higher productions of PG and 3HP. Considering all that, the genetic modification of *iclR* would be a good choice to improve the production of a wide range of acetyl-CoA-derived chemicals in industry.

### Abbreviations

3HP: 3-hydroxypropionate; Ace: Acetate; DCW: Dry cell weight; Glu: Glucose; PEP: Phosphoenolpyruvate; PG: Phloroglucinol; RT-PCR: Reverse-transcription polymerase chain reaction; SDS-PAGE: Sodium dodecyl sulfate-polyacrylamide gel electrophoresis

### Acknowledgement

The authors would like to thank the financial support from the Chinese Academy of Sciences, Shandong Province and the postdoctoral Science Foundation of China.

### Funding

This work was supported by100-Talent Project of CAS (to GZ), Key Program of CAS (ZDRW-ZS-2016-3M), Scientific and Technological Project of Tianjin (14ZCZDSY00047), the 60-class General Financial Grant from the China Postdoctoral Science Foundation (2016M600564) and Taishan Scholars Climbing Program of Shandong (No.tspd20150210).

### Authors' contributions

ML and MX participated in the design of the study. ML and GZ were responsible for the review of literature and preparation of the manuscript. YD and HL helped to analyze the statistic dates and revise the manuscript. ML constructed the recombinant strains, carried out the RT-PCR experiments. ML, HC and ZZ participated in the fermentation researches. All authors read and approved the final manuscript.

### Competing interests

The authors declare that they have no competing interests.

### Author details

[1]CAS Key Laboratory of Biobased Materials, Qingdao Institute of Bioenergy and Bioprocess Technology, Chinese Academy of Sciences, Qingdao 266101, China. [2]University of Chinese Academy of Sciences, Beijing 100049, China. [3]Institute of Oceanology, Chinese Academy of Sciences, Qingdao 266071, China. [4]Randian Technology Company Limited, Tianjin 300457, China.

### References

1. Chen Y, Daviet L, Schalk M, Siewers V, Nielsen J. Establishing a platform cell factory through engineering of *yeast* acetyl-CoA metabolism. Metab Eng. 2013;15:48–54.
2. Ajikumar PK, Xiao W-H, Tyo KE, Wang Y, Simeon F, Leonard E, Mucha O, Phon TH, Pfeifer B, Stephanopoulos G. Isoprenoid pathway optimization for Taxol precursor overproduction in *Escherichia coli*. Science. 2010;330:70–4.
3. Anderson AJ, Dawes EA. Occurrence, metabolism, metabolic role, and industrial uses of bacterial polyhydroxyalkanoates. Microbiol Rev. 1990;54:450–72.
4. Zheng Y, Li Z, Feng S, Lucas M, Wu G, Li Y, Li C, Jiang G. Biomass energy utilization in rural areas may contribute to alleviating energy crisis and global warming: A case study in a typical agro-village of Shandong. China Renew Sust Energ Rev. 2010;14:3132–9.
5. Eiteman MA, Altman E. Overcoming acetate in *Escherichia coli* recombinant protein fermentations. Trends Biotechnol. 2006;24:530–6.
6. Terpe K. Overview of bacterial expression systems for heterologous protein production: from molecular and biochemical fundamentals to commercial systems. Appl Microbiol Biotechnol. 2006;72:211–22.
7. Luli GW, Strohl WR. Comparison of growth, acetate production, and acetate inhibition of *Escherichia coli* strains in batch and fed-batch fermentations. Appl Environ Microbiol. 1990;56:1004–11.
8. Valgepea K, Adamberg K, Nahku R, Lahtvee PJ, Arike L, Vilu R. Systems biology approach reveals that overflow metabolism of acetate in *Escherichia coli* is triggered by carbon catabolite repression of acetyl-CoA synthetase. BMC Syst Biol. 2010;4:166.
9. Yu M, Shimizu K. Catabolite regulation analysis of *Escherichia coli* for acetate overflow mechanism and co-consumption of multiple sugars based on systems biology approach using computer simulation. J Biotechnol. 2013; 168:155–73.
10. Farmer WR, Liao JC. Reduction of aerobic acetate production by *Escherichia coli*. Appl Environ Microbiol. 1997;63:3205–10.
11. Nakano K, Rischke M, Sato S, Märkl H. Influence of acetic acid on the growth of *Escherichia coli* K12 during high-cell-density cultivation in a dialysis reactor. Appl Microbiol Biotechnol. 1997;48:597–601.
12. De MM, De MS, Soetaert W, Vandamme E. Minimizing acetate formation in *E. coli* fermentations. J Ind Microbiol Biot. 2007;34:689–700.
13. Liu M, Yao L, Mo X, Ding Y, Liu H, Zhao G. Deletion of arcA increased the production of acetyl-CoA-derived chemicals in recombinant *Escherichia coli*. Biotechnol Lett. 2016;38:97–101.
14. Waegeman H, Maertens J, Beauprez J, De Mey M, Soetaert W. Effect of *iclR* and *arcA* deletions on physiology and metabolic fluxes in *Escherichia coli* BL21 (DE3). Biotechnol Lett. 2012;34:329–37.
15. Noronha SB, Yeh HJC, Spande TF, Shiloach J. Investigation of the TCA cycle and the glyoxylate shunt in *Escherichia coli* BL21 and JM109 using $13^{C}$-NMR/MS. Biotechnol Bioeng. 2000;68:316–27.
16. Phue J-N, Shiloach J. Transcription levels of key metabolic genes are the cause for different glucose utilization pathways in *E. coli* B (BL21) and *E. coli* K (JM109). J Biotechnol. 2004;109:21–30.
17. Phue JN, Noronha SB, Hattacharyya R, Wolfe AJ, Shiloach J. Glucose metabolism at high density growth of *E. coli* B and *E. coli* K: differences in metabolic pathways are responsible for efficient glucose utilization in *E. coli* B as determined by microarrays and Northern blot analyses. Biotechnol Bioeng. 2005;90:805–20.
18. Yamamoto K, Ishihama A. Two different modes of transcription repression of the *Escherichia coli* acetate operon by IclR. Mole Microbiol. 2003;47:183–94.
19. Gui L, Sunnarborg A, Pan B, Laporte DC. Autoregulation of *iclR*, the gene encoding the repressor of the glyoxylate bypass operon. J Bacteriol. 1996; 78:321–4.
20. Molina-Henares AJ, Krell T, Guazzaroni ME, Segura A, Ramos JL. Members of the IclR family of bacterial transcriptional regulators function as activators and/or repressors. Fems Microbio Rev. 2006;30:157–86.
21. Rathnasingh C, Raj SM, Jo JE, Park S. Development and evaluation of efficient recombinant *Escherichia coli* strains for the production of 3-hydroxypropionic acid from glycerol. Biotechnol Bioeng. 2009;104:729–39.
22. Cao Y, Jiang X, Zhang R, Xian M. Improved phloroglucinol production by metabolically engineered *Escherichia coli*. Appl Microbiol Biotechnol. 2011; 91:1545–52.
23. Moore SD. Assembling new *Escherichia coli* strains by transduction using phage P1. In: Strain Engineering. Springer; 2011;765:155–69. doi:10.1007/978-1-61779-197-0_10.
24. Baba T, Ara T, Hasegawa M, Takai Y, Okumura Y, Baba M, Datsenko KA, Tomita M, Wanner BL, Mori H. Construction of *Escherichia coli* K-12 in-frame, single-gene knockout mutants: the Keio collection. Mole syst biol. 2006;2:8.
25. Liu C, Ding Y, Zhang R, Liu H, Xian M, Zhao G. Functional balance between enzymes in malonyl-CoA pathway for 3-hydroxypropionate biosynthesis. Meta Eng. 2016;34:104–11.

26. Zha W, Rubin-Pitel SB, Zhao H. Exploiting genetic diversity by directed evolution: molecular breeding of type III polyketide synthases improves productivity. Mole BioSys. 2008;4:246–8.
27. Rathnasingh C, Raj SM, Lee Y, Catherine C, Ashok S, Park S. Production of 3-hydroxypropionic acid via malonyl-CoA pathway using recombinant *Escherichia coli* strains. J Biotechnol. 2012;157:633–40.
28. Kumari S, Beatty C. Df, Busby S, Simel E, Hovel MG, Wolfe A: Regulation of acetyl coenzyme A synthetase in *Escherichia coli*. J Bacteriol. 2000;182:4173–9.
29. Waegeman H, Beauprez J, Moens H, Maertens J, De Mey M, Foulquie-Moreno MR, Heijnen JJ, Charlier D, Soetaert W. Effect of *iclR* and *arcA* knockouts on biomass formation and metabolic fluxes in *Escherichia coli* K12 and its implications on understanding the metabolism of *Escherichia coli* BL21 (DE3). BMC Microbiol. 2011;11:70.
30. Waegeman H, De Lausnay S, Beauprez J, Maertens J, De Mey M, Soetaert W. Increasing recombinant protein production in *Escherichia coli* K12 through metabolic engineering. N Biotechnol. 2013;30:255–61.
31. Loui C, Chang AC, Lu S. Role of the ArcAB two-component system in the resistance of *Escherichia coli* to reactive oxygen stress. BMC Microbiol. 2009;9:1–14.
32. Nizam SA, Zhu J, Ho PY, Shimizu K. Effects of *arcA* and *arcB* genes knockout on the metabolism in *Escherichia coli* under aerobic condition. Biochem Eng J. 2009;44:240–50.
33. Iuchi S, Lin EC. *arcA* (dye), a global regulatory gene in *Escherichia coli* mediating repression of enzymes in aerobic pathways. Proc Natl Acad Sci USA. 1988;85:1888–92.
34. Chao G, Park SJ. Aerobic regulation of the *sucABCD* genes of *Escherichia coli*, which encode...-ketoglutarate. J Bacteriol. 1997;179:4138–42.
35. Salmon KA, Hung SP, Steffen NR, Krupp R, Baldi P, Hatfield GW, Gunsalus RP. Global gene expression profiling in *Escherichia coli* K12: effects of oxygen availability and ArcA. J Biol Chem. 2005;280:15084–96.
36. Perrenoud A, Sauer U. Impact of global transcriptional regulation by ArcA, ArcB, Cra, Crp, Cya, Fnr, and Mlc on glucose catabolism in *Escherichia coli*. J Bacteriol. 2005;187:3171–9.
37. Yoon SH, Jeong H, Kwon S-K, Kim JF: Genomics, biological features, and biotechnological applications of *Escherichia coli* B:"Is B for better?!". In: Systems Biology and Biotechnology of *Escherichia coli*. Springer; 2009: 1–17. doi:10.1007/978-1-61779-197-0_10.

9

# Adaptation response of *Pseudomonas fragi* on refrigerated solid matrix to a moderate electric field

Wenbo Chen[1,2], Honghai Hu[1,2], Chunjiang Zhang[1,2], Feng Huang[1,2], Dequan Zhang[1] and Hong Zhang[1,2*]

**Abstract**

**Background:** Moderate electric field (MEF) technology is a promising food preservation strategy since it relies on physical properties—rather than chemical additives—to preserve solid cellular foods during storage. However, the effectiveness of long-term MEF exposure on the psychrotrophic microorganisms responsible for the food spoilage at cool temperatures remains unclear.

**Results:** The spoilage-associated psychrotroph *Pseudomonas fragi* MC16 was obtained from pork samples stored at 7 °C. Continuous MEF treatment attenuated growth and resulted in subsequent adaptation of M16 cultured on nutrient agar plates at 7 °C, compared to the control cultures, as determined by biomass analysis and plating procedures. Moreover, intracellular dehydrogenase activity and ATP levels also indicated an initial effect of MEF treatment followed by cellular recovery, and extracellular β-galactosidase activity assays indicated no obvious changes in cell membrane permeability. Furthermore, microscopic observations using scanning and transmission electron microscopy revealed that MEF induced sublethal cellular injury during early treatment stages, but no notable changes in morphology or cytology on subsequent days.

**Conclusion:** Our study provides direct evidence that psychrotrophic *P. fragi* MC16 cultured on nutrient agar plates at 7 °C are capable of adapting to MEF treatment.

**Keywords:** Adaptation response, Moderate electric field, Nutrient agar plates, Psychrotrophic microorganisms, *Pseudomonas fragi*, Solid cellular foodstuffs

## Background

Alternating electric fields stall the planktonic growth of several prokaryotic microorganisms, such as *Escherichia coli, Lactobacillus acidophilus, Staphylococcus aureus, Bacillus cereus, Pseudomonas putida,* and *P. fluorescens* [1–6], leading to its recent application in food preservation [7]. The effectiveness of alternating electric field technology in preventing spoilage can be easily assessed by analyzing the amount of proteins eluted from cellular samples [7]. Alternating electric fields are generally used at either high intensity or high frequencies [1, 3, 4, 7, 8], and the possible applications of moderate electric fields (MEFs, typically <1–100 V/cm) at low frequencies (1–100 Hz) have been far less studied [9]. For example, planktonic *L. acidophilus* exhibits a decreased lag time following treatment with a MEF [6, 9], causing a marked increase in bacteriocin yield [6]. However, few studies have investigated the application of MEF as a food preservative.

Refrigeration is a common food preservation strategy that extends the shelf life of food products by decreasing microbial growth rates and slowing the physical and chemical reactions responsible for spoilage [10–12]. However, microbial species adapted to the temperatures ranging from 0 °C to 22 °C (*e.g.*, psychrotrophs) outcompete other microorganisms and dominate the microbial population present on food products during refrigerated storage [12–14], and thus are the primary contributors to food spoilage at refrigeration temperatures [15–19].

* Correspondence: zhang.h07@hotmail.com
[1]Institute of Food Science and Technology, Chinese Academy of Agricultural Sciences/Key Laboratory of Agro-Products Processing, Ministry of Agriculture, Beijing 100193, People's Republic of China
[2]College of Staple Food Technology, Chinese Academy of Agricultural Sciences, Institute of Food Science and Technology, Harbin 151900, People's Republic of China

*Pseudomonas fragi* is a psychrotrophic species responsible for the majority of spoilage of meat stored aerobically at refrigeration temperatures [17]. *P. fragi* is more widely distributed than other pseudomonads, and has been isolated from water, soil, plant materials, and other natural media [17, 20–22]. This species grows well at temperatures ranging from 2 °C to 35 °C [21]. The prevalence and growth conditions of *P. fragi* contribute to its successful proliferation on foods, especially fresh meat [12, 17, 21, 23, 24], and the strict chill chain applied during fresh meat production from slaughtering to final distribution selectively favors its development [21, 25]. Moreover, *P. fragi* is suggested to promote the growth and survival of several foodborne pathogens, such as *S. aureus* and *Listeria monocytogenes* [26, 27]. Therefore, information about *P. fragi* growth and biological responses to preservation conditions during food storage is urgently needed to improve food safety.

MEF technology may provide a means to control microbial growth on food products where refrigeration fails to do so. However, organisms may develop adaptive responses to continuous MEF treatment, similar to how *P. fragi* has adapted to cooler temperatures [21]. Thus, the aim of this work was to investigate the biological responses of psychrotrophic microorganisms to continuous MEF treatment by characterizing the adaptive response to MEF treatment in *P. fragi*. Additionally, we examined cell structural changes induced by MEF treatment using electron microscopy.

## Methods

### Meat sample preparation and storage

Pork samples were prepared as described by Ercolina et al. [24] with some modifications. Pork muscle (*Longissimus dorsi*) obtained from a slaughterhouse was cut into similar-sized pieces (2 × 2.5 × 3 cm) and packed in sterile stomacher bags (Labplas, Inc., Sainte-Julie, Quebec, Canada). The prepared samples were then stored aerobically at 7 °C for at least 5 days and used as sources for the isolation of psychrotrophs [12, 26].

### Bacterial strain isolation and identification

A meat sample (25 g) was aseptically weighed and homogenized in 225 mL of peptone saline (0.85% NaCl and 0.1% peptone in distilled water) at room temperature (22–25 °C) [27–29]. Serial dilutions were prepared with the homogenized meat sample and 100 μL aliquots of the appropriate dilutions were spread in triplicate on nutrient agar plates (Luqiao Co., Beijing, China) and incubated for 5 days at 7 °C [24]. Colonies present on the countable plates were randomly picked and further purified by repeated streaking. Finally, these purified cultures were maintained as glycerol stock culture and stored at -80 °C [30]. These isolates were also deposited on nutrient agar slants in the dark at 4 °C [31].

The isolates were identified by 16S rDNA sequencing carried out by Sangon Biological Co., Ltd. (Shanghai, China).

### MEF system

The MEF system used in this study was designed and patented by Takayuki et al. [32]. The system consisted of an input controller, transformer, feedback control circuit, output controller, and discharge plate. The transformed voltage was exerted on the discharge plate via the output controller, which could make the discharge plate generate a wide range of electric fields [32]. The waveform of the output voltage measured with an oscilloscope (UTD2102-CEX, Uni-Trend) and a high-voltage prober (P6015A, Tektronix) is a sine wave with a 9.2 kV maximum voltage (peak-to-peak). The frequency of the output voltage is 50 Hz (as described in [6, 9]) and identical to that of the input voltage. Other parameters are identical to those described in the patent [32]. The schematics of this apparatus and the refrigeration system are shown in Fig. 1.

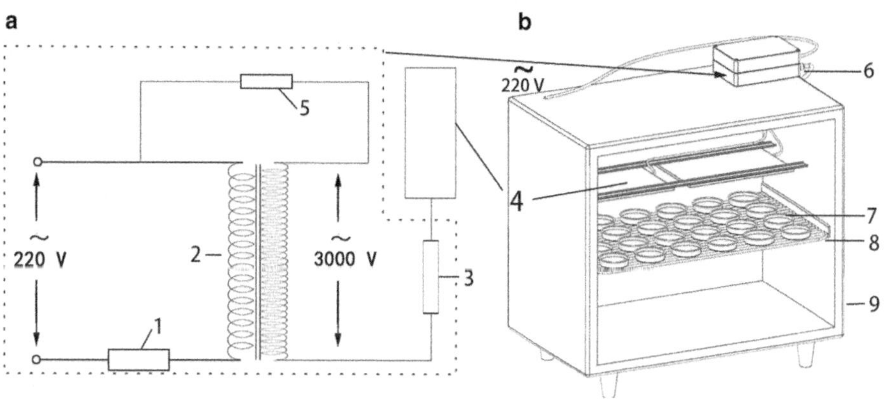


**Fig. 1** Schematic diagrams of the alternating electric field (**a**) and a refrigerator equipped with the alternating electric field (**b**). 1: input controller; 2: transformer; 3: output controller; 4: discharge plate; 5: feedback control circuit; 6: alternating electric field generator; 7: petri dishes; 8: sample holder; 9: refrigerator

**Cultivation on nutrient agar plates**

Microbial isolates were grown on nutrient agar plates because of its extensive use in microbiological studies and relatively definite components. Fresh bacterial cultures were prepared from a glycerol stock culture. After cultivation at 7 °C for 3 days, a single colony on nutrient agar plate was randomly chosen to prepare a bacterial suspension in sterile peptone saline. After serial dilution, 100 μL of the appropriate dilutions (approx. 3,000 CFU/mL) were spread onto nutrient agar plates and were incubated at 7 °C in either a control refrigerator or a refrigerator equipped with the MEF system. MEF-treated samples were placed on holders and continuously exposed to MEF (20 V/cm) throughout the cultivation period (Fig. 1b).

**Microbial growth analysis**

Microbial growth of MEF-treated and control cultures was first evaluated by observing the full biomass development on nutrient agar plates as described by Ercolini et al [24]. Because of the low growth rate at cold temperatures, only cultures incubated for 3, 4, and 5 days were used.

Culture growth was also determined by monitoring the population using plating procedures. Plated cultures were suspended in 4 mL of peptone saline [22], serially diluted, replated on nutrient agar, and then colonies were counted after incubation at 7 °C for 3 days [16, 33]. Three plates were prepared for daily test for each group.

**Dehydrogenase activity assay**

Dehydrogenase activity was assayed by MTT (3-(4, 5-dimethylthiazol-2-yl)-2, 5-diphenyltetrazolium bromide; Beyotime Biotechnology, Jiangsu China) analysis described by Wang et al. [34] with slight modifications. Briefly, 3-, 4-, or 5-day microbial cultures were diluted to an $OD_{600}$ of 0.2 in fresh nutrient broth. A 200 μL aliquot of the cell suspension was then mixed with 20 μL MTT solution (5 mg/mL in sterilized water) and incubated for 20 min at 25 °C. After centrifugation at 12,000 rpm for 1 min, pelleted cell-formazan crystal complexes were transferred into glass test tubes using 2.5 mL of dimethyl sulfoxide (DMSO, Sinopharm, China). The $OD_{550}$ values of the mixtures were measured with a Universal Microplater Reader (Molecular Devices, Sunnyvale, CA) at 10–15 min. Three plates per group were prepared for each test.

Dehydrogenase activity is expressed in MTT reduction units (MRU). One MRU is defined as an $A_{550}$ value of 1.0 produced by the dissolved formazan crystals from cells incubated in LB medium at 7 °C in 20 min [34].

**Intracellular ATP levels determination**

Intracellular ATP levels were measured using a firefly luciferase-based bioluminescence ATP Assay Kit (Beyotime Biotechnology). Cell suspensions prepared from 1 plate was diluted to an $OD_{600}$ of 0.2 and centrifuged at 12,000 rpm for 5 min at 4 °C. The obtained cell pellets were used to prepare intracellular ATP solution according to the supplier's instructions. Cells were mixed with lysis solution and boiled for 2 min to release ATP and to inactivate endogenous ATPase [35, 36]. Then, 20 μL ATP extraction and 100 μL ATP detection solutions were mixed to prepare the luminescent reaction solution.

Luminance (RLU) was recorded using Hidex plate Chameleon™ V. (Finland) and linearly related to the ATP concentration.

**β-Galactosidase activity assay**

β-Galactosidase activity was measured according to Gobinath and Prapulla [37] with some modifications. Cultures grown on 1 agar plate supplemented with 1% lactose were suspended in 4 mL phosphate buffered saline (PBS). The $OD_{600}$ of these cultures was measured and the suspension were centrifuged at 15,000 rpm for 5 min. β-Galactosidase activity in the resulting supernatant was determined by mixing 1 mL supernatant and 0.2 mL ONPG (4 mg/mL, *o*-nitrophenyl-β-D-galactopyraniside). The reaction was terminated after incubation at 25 °C for 20 min by adding 0.5 mL 1 M $Na_2CO_3$. The optical density of the reaction solution was recorded at 420 nm ($OD_{420}$) and 550 nm ($OD_{550}$), as previously described [38]. β-Galactosidase activity was calculated in Miller Units that describe the change in $OD_{420}$/min/mL of cells/$OD_{600}$ [39].

**Microscopic observations**

Control and MEF-treated cultures were observed using a scanning electron microscope (SEM; S570, Hitachi, Japan) and a transmission electron microscope (TEM; H-7500, Hitachi). Cultures grown on nutrient agar plates for 3 days were rinsed 3 times with sterilized distilled water. After centrifugation at 15,000 rpm for 5 min, cell pellets were suspended and fixed in glutaraldehyde (3% aqueous solution). Other steps employed to prepare samples for SEM and TEM were carried out according to the methods described by Racyte and Zituni [1, 40].

**Statistical analysis**

The effect of MEF on *P. fragi* growth on nutrient agar plates at refrigeration temperature was evaluated by three independent experiments. The obtained data were processed with the data module of the Statistical Package for Social Science (IBM SPSS Statistics 19) for Windows and assessed with *t*-tests. A significant difference between the treated and control groups was defined as a $p$ value $< 0.05$.

## Results and Discussion

Isolation and identification of psychrotrophic bacteria from meat samples

After repeated isolation and purification, a total of 25 isolates were obtained from the pork samples. All purified isolates formed visible colonies within 3 days at 7 °C, indicating their adaptation to the refrigeration temperature. These colonies produced the intense fruity odor that is characteristic of pseudomonads, suggesting that these isolates might be *P. fragi* [17, 21].

16S DNA sequencing revealed that the 25 isolates all shared 99% identity with *P. fragi* isolate 7U (Accession No.: LK391544) [41]. The 16S rDNA sequence was deposited in GenBank as KT443784 and the isolate designated as *P. fragi* MC16. These sequencing results indicate that *P. fragi* is primarily responsible for the spoilage of fresh meat stored at refrigeration temperatures. Although many microorganisms are associated with meat spoilage, few species are able to grow at the low temperatures used to preserve perishable foods [12, 15, 16, 18, 22, 24]. Several *Pseudomonas* species (*e.g., P. fragi, P. lundensis, P. fluorescens,* and *P. putida*) have been frequently identified as major spoilage agents in fresh pork, beef, and mutton [12, 17, 21, 23, 24, 42], but only *P. fragi* was isolated from meat samples in this study. Sample storage and processing conditions might have favored *P. fragi* growth by eliminating other spoilage-related species [22]. Additionally, the biotype diversity of *P. fragi* might contribute to this result because different molecular types of *P. fragi* might behave similarly as meat spoilers [21].

### Growth of *P. fragi* MC16 on nutrient agar plates

Control and MEF-treated cultures could be clearly differentiated by observing biomass development. MEF-treated colonies were smaller than untreated colonies after 3 days of growth (Fig. 2). However, the size difference gradually decreased after 4 and 5 days of incubation.

Total viable counts of the control and MEF-treated cultures were determined to quantify the effect of MEF treatment on the growth of *P. fragi* MC16. Total viable counts of the treated and untreated cultures were significantly different ($p < 0.05$) after 3 and 4 days of growth (Fig. 3). Similar to the qualitative biomass development assessments, no significant difference in quantitative biomass between the two groups was found after 5 days of incubation. These results are consistent with a previous study by Loghavi [6], indicating that *P. fragi* MC16 responds differently to MEF exposure during different growth phases.

Changes in the susceptibility of *P. fragi* MC16 on nutrient agar plates to MEF treatment can be attributed to limited nutrient diffusion in the solid matrix and culture age. Previous studies suggested that MEF treatment at lower frequencies could cause temporary, non-lethal membrane permeabilization [9, 43]. For cells grown in liquid medium, increased permeability enhances nutrient diffusion across cell membranes and reduced lag phases are observed [6]. However, nutrient diffusion in solid medium is relatively low. As a result, temporary permeabilization on nutrient agar plates causes sublethal damage rather than rapid nutrient uptake. Therefore, retarded cell growth was observed in this study. Additionally, older

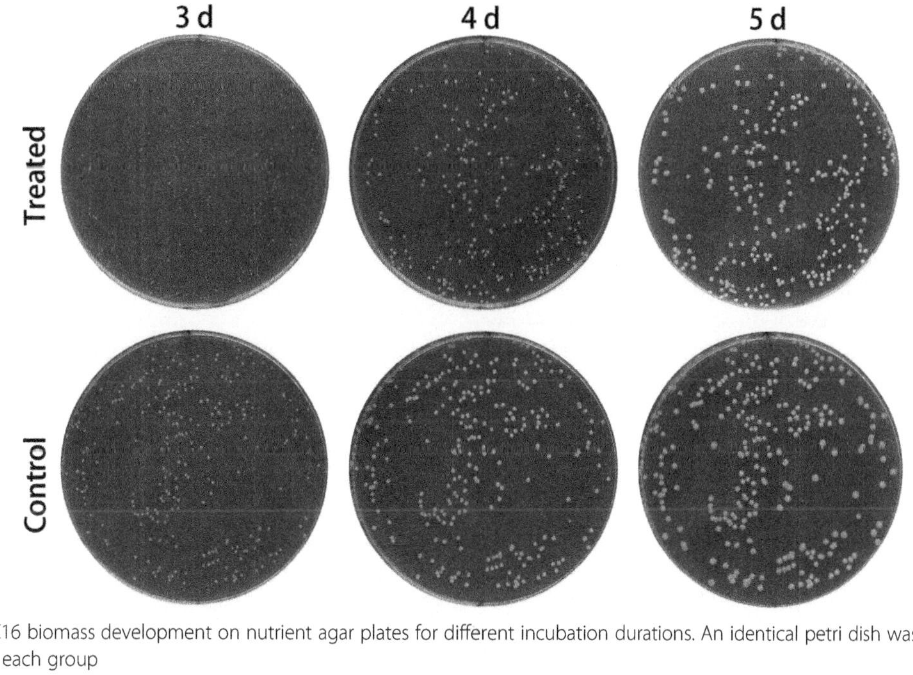


**Fig. 2** *P. fragi* MC16 biomass development on nutrient agar plates for different incubation durations. An identical petri dish was employed on different days for each group

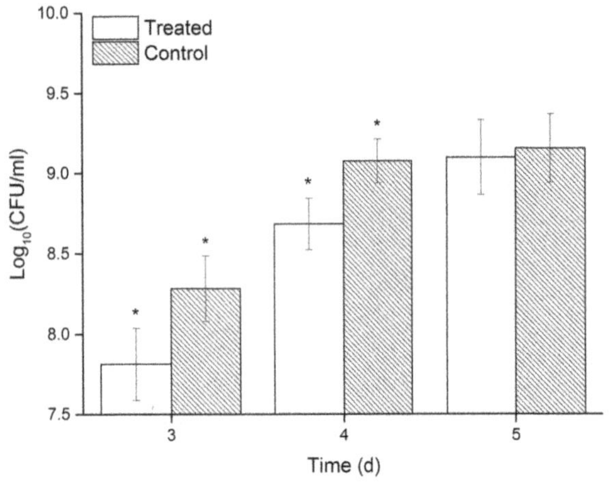


**Fig. 3** Changes in total viable counts of *P. fragi* MC16 grown on nutrient agar plates for different times at 7 °C. * indicates a significant difference from the control group, $p < 0.05$

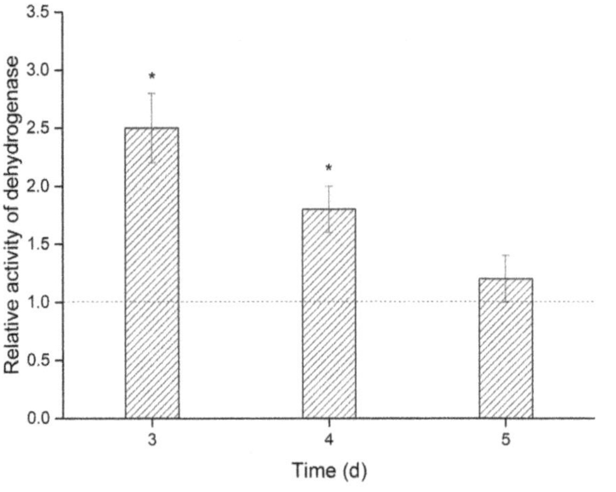


**Fig. 4** The effect of MEF on intracellular dehydrogenase activity. Dehydrogenase activity is expressed in relative units as the ratio of dehydrogenase activity of the treated group to the control group. Error bars represent standard error. The dashed line indicates a ratio of one (where values of the treated group are equal to the control group). * indicates significant difference from the control group, $p < 0.05$

cultures have decreased protein to lipid ratios in membranes [9, 43], causing decreased membrane fluidity and susceptibility to external MEF treatment. Thus, biomass recovered later during incubation.

### The effect of MEF treatment on *P. fragi* MC16 metabolic activity

MEF is an environmental stress to which microorganisms can adapt, similar to cold, heat, and pressure [3, 36]. Microorganisms can modulate their metabolism to cope with sublethal stressors [44, 45] and may enter different growth phases that alter their susceptibility to the stress [6, 9]. Therefore, investigating metabolic changes is one method to characterize microbial stress responses. Dehydrogenases are key enzymes involved in the oxidative activities of various organisms, thus dehydrogenase activity reflects the metabolic state of the organism [36, 46]. We measured intracellular dehydrogenase activity during MEF treatment to characterize the metabolic state of *P. fragi* MC16. MEF-treatment significantly increased intracellular dehydrogenase activity for treated cultures after 3 and 4 days ($p < 0.05$; Fig. 4). Intracellular dehydrogenase activity of MEF-treated cultures after 3 and 4 days was 2.5- and 1.8-times higher than the control groups, respectively. However, dehydrogenase activity decreased after 5 days of treatment such that the enzyme activity in MEF-treated cultures was only 1.2-fold higher than in control cultures. This difference was not significant.

Increased dehydrogenase activity is generally associated with the energy-consuming microbial stress response [44, 46]. This response has been previously observed in *S. aureus* exposed to high osmolarity or high pressure [47]. We found that increased dehydrogenase activity was readily observed when *P. fragi* cultures were treated with MEF. Due to the sublethality of MEF, microorganisms exposed to this field must repair injured membranes [48]. Additionally, MEF treatment may impact the conformation of cellular proteins [9]. To restore normal protein conformations and survive these conditions, microorganisms must modulate gene expression and modify metabolism to engage adaptive cellular mechanisms [49]. All metabolic changes induced by MEF treatment reflected higher metabolic investment, which is generally characterized by increased dehydrogenase activity and ATP level [36]. These metabolic changes may have diminished after 5 days of treatment and incubation because the cells activated adaptive mechanisms and modulated gene expression to manage the stressor.

### The effect of MEF on intracellular ATP concentration

ATP is a direct participant in many cellular metabolic processes [36]. Therefore, changes in ATP concentration may characterize the metabolic state during MEF treatment more specifically than changes in dehydrogenase activity. As shown in Fig. 5, MEF treatment increased intracellular ATP content. ATP concentrations in cultures treated for 3, 4, and 5 days were 1.5-, 1.4-, and 1.1-times higher than control cultures, respectively. Other forms of environmental stress (*e.g.*, heat, pressure, starvation, pH, and irradiation) have been previously reported to increase ATP concentration in microorganisms [50], suggesting that MEF-treatment causes a similar cellular response as these stressors. The concentration of intracellular ATP in MEF-treated cells decreased to the level of the controls after 5 days of exposure, indicating metabolic adaptation to the MEF treatment. This trend in intracellular ATP content reflects similar findings in biomass development and

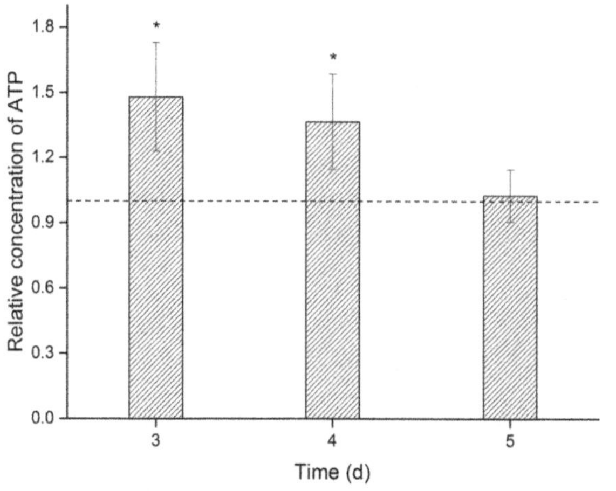


**Fig. 5** The influence of MEF on intracellular ATP concentration in *P. fragi*. Values are presented as ratios of ATP concentration (RLU/OD $_{600}$) in treated cultures relative to control cultures. The dashed line indicates a ratio of one (where means values of treated cultures are equal to control cultures). * indicates a significant difference from the control cultures ($p < 0.05$)

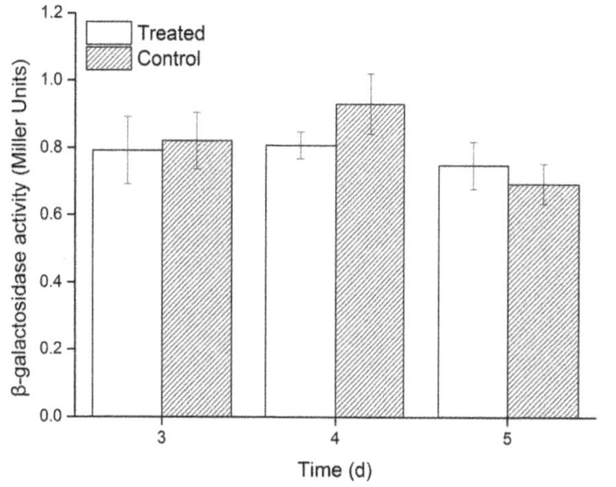


**Fig. 6** Extracellular β-galactosidase activity of the treated and untreated cultures grown on nutrient agar plates for different times

intracellular dehydrogenase activity of the treated cultures over time.

### The effect of MEF on inner membrane permeability

β-Galactosidase is a large, tetrameric protein [39]. Because of its high molecular weight, presence of extracellular β-galactosidase is generally associated with a loss of cell membrane integrity [51]. Thus, the activity of extracellular β-galactosidase produced by MEF-treated and untreated cultures was quantified to indicate cell membrane alterations over time.

Extracellular β-galactosidase activity levels in treated and untreated cultures were 0.79 and 0.80 Miller Units, respectively, after 3 days of cultivation. Extracellular β-galactosidase activity levels were not significantly different between treated and untreated cultures at 3, 4, or 5 days of cultivation (Fig. 6).

These results suggest that MEF treatment did not increase cell membrane permeability, since permeabilized membranes would allow intracellular β-galactosidase to leak into extracellular space. Membrane permeabilization via MEF has been reported in several studies [6, 9, 48]; however, permeabilization is reversible and membrane integrity can be quickly reestablished [48]. Accordingly, the low levels of extracellular β-galactosidase activity observed in this study may indicate that cellular membranes were reestablished too rapidly to detect major changes. In addition, the MEF system used in this study was designed to maintain the freshness of solid cellular foods, such as meat, fish, fruits, and vegetables [32]. The MEF system was not designed to inactivate microorganisms by forming irreversible membrane permeabilization, since inactivation would compromise the freshness and nutrition of the cellular foods that also have cell membranes [7]. Moreover, the intensity of the EMF used in this study was lower than 30 V/cm, ruling out the possibility of electroporation that was reported to occur at field intensities around 1,000 V/cm [3, 52].

### Morphological changes of *P. fragi* MC16 induced by MEF

Intact cell morphology and cytology are essential for the normal growth and multiplication of microorganisms, thus alterations in microbial morphology and cytology reflect cell viability [5, 40, 51]. Microscopic observations of cultures treated for different incubation periods were conducted using SEM and TEM.

SEM observations illustrated changes in cell morphology (Fig. 7). Cultures treated with MEF for 3 days were largely intact, but showed surface indentations (indicated with arrows in Fig. 7a). Cellular debris derived from cell disintegration has been reported by previous literature [5, 40], but was not observed in this study. These results support the findings of the extracellular β-galactosidase activity: MEF treatment appears to have limited impact on microbial cell membranes. Microbial cells treated with MEF for 4 and 5 days could not be differentiated from control cells (Fig. 7e and f), which were characterized by smooth cell surface and uniform cell shapes. The cell morphologies of control and MEF-treated cells after 3 days of treatment may indicate that the treated cells adapted to the MEF exposure. SEM examinations of untreated cultures showed no changes in cell morphology throughout the process (Fig. 7d, e, and f).

TEM observations of MEF-treated cultures showed agglomerations of cytoplasmic materials, vacuole formations, and depressed cell membranes (indicated with arrows in Fig. 8a and b). These cellular observations are consistent with cell damage induced by continuous

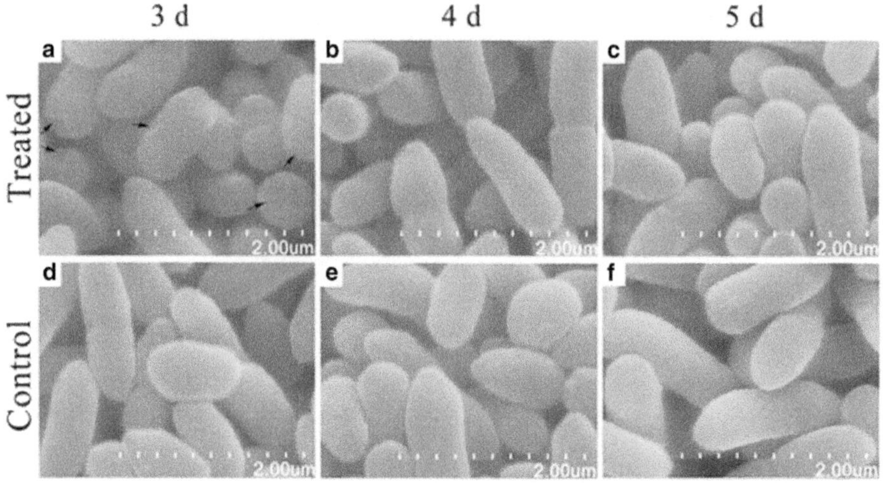


**Fig. 7** Scanning electron microscope observations of *P. fragi* MC16 grown for different times. **a**–**c** Treated samples; **d**–**f** Control samples. Arrows indicate cell surface indentations

ultrasound in *E. coli* ATCC 11775 [53]. Nevertheless, these changes gradually disappeared when the exposure time was extended to 5 days (Fig. 8c). At this point, TEM observations of treated and control cultures (Fig. 8f) were almost identical, suggesting that the cells had completely adapted to the MEF treatment. Control cells showed regular shape and morphology, with the cytoplasm filling the entire cell envelope (Fig. 8d, e, and f).

Summarizing the observations on morphology and cytology, cultures at early growth stages might be susceptible to continuous MEF exposure; however, they subsequently recovered from these injuries as the treatment progressed. These results indicate the microbial adaptation to the MEF treatment and support previous findings that the bacterial stress response supports adaptation to adverse environments [49].

## Conclusions

MEF application is a novel technology for preserving solid cellular foods. The technology can cause adverse morphological and cytological changes to microorganisms growing on a solid matrix during early cell growth stages. However, the impact of MEF application gradually weakens as cells potentially responded to the stress via modified metabolism and gene expression changes. Therefore, future research exploring the adaptive mechanisms to MEF is required before this technology can be improved to further extend the shelf life of solid cellular foodstuffs and reduce food losses caused by microorganisms.

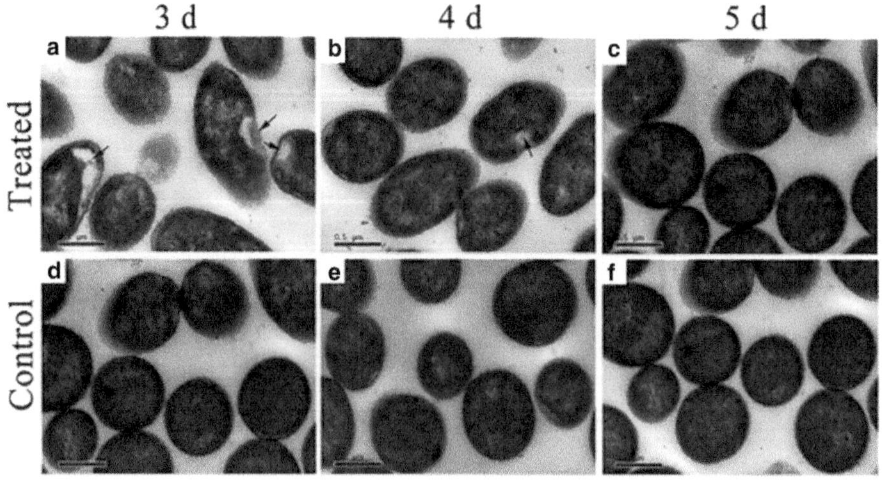


**Fig. 8** Transmission electron microscope observations of *P. fragi* MC16 grown for different times. **a**–**c** Treated samples; **d**–**f** Control samples. Arrows indicate agglomerations of intracellular contents

**Abbreviations**

ATP: Adenosine triphosphate; DNA: Deoxyribonucleic acid; MEF: Moderate electric field; MRU: MTT reduction units; MTT: 3-(4, 5-dimethylthiazol-2-yl)-2, 5-diphenyltetrazolium bromide; OD: Optical density; ONPG: o-nitrophenyl-β-D-galactopyraniside; RLU: Relative luminance unit; SEM: Scanning electron microscope; TEM: Transmission electron microscope

**Acknowledgements**

None.

**Funding**

This research was supported by grants from the National Natural Science Foundation of China (No. 31501549), the Special Fund for Agro-scientific Research in the Public Interest (No. 201303083-1), the Agricultural Science and Technology Innovation Program, CAAS (No. CAAS-XTCX2016005), and the Central Public-interest Scientific Institution Basal Research Fund (No. 0032015016).

**Authors' contributions**

WBC, HHH, and CJZ designed the study and performed the experiments. WBC, FH, and DQZ analyzed the data. WBC and HHH prepared the main manuscript text under the direction of HZ. All authors have read and approved the final manuscript.

**Competing interests**

The authors declare that they have no competing interests.

**References**

1. Racytea J, Bernard S, Paulitsch-Fuchsa AH, Yntema DR, Bruning H, Rijnaarts HHM. Alternating electric fields combined with activated carbon for disinfection of Gram negative and Gram positive bacteria in fluidized bed electrode system. Water Res. 2013;47(16):6395–405.
2. Racyte J, Sharabati J-A-D, Paulitsch-Fuchs AH, Yntema DR, Mayer MJJ, Bruning H, et al. Combining fluidized activated carbon with weak alternating electric fields for disinfection. Carbon. 2011;49(15):5321–8.
3. Giladi M, Porat Y, Blatt A, Wasserman Y, Kirson ED, Dekel E, et al. Microbial growth inhibition by alternating electric fields. Antimicrob Agents Chemother. 2008;52(10):3517–22.
4. Giladi M, Porat Y, Blatt A, Shmueli E, Wasserman Y, Kirson ED, et al. Microbial growth inhibition by alternating electric fields in mice with *Pseudomonas aeruginosa* lung infection. Antimicrob Agents Chemother. 2010;54(8):3212–8.
5. Valle A, Zanardini E, Abbruscato P, Argenzio P, Lustrato G, Ranalli G, et al. Effects of low electric current (LEC) treatment on pure bacterial cultures. J Appl Microbiol. 2007;103(5):1376–85.
6. Loghavi L, Sastry SK, Yousef AE. Effect of moderate electric field frequency on growth kinetics and metabolic activity of *Lactobacillus acidophilus*. Biotechnol Prog. 2008;24:148–53.
7. Ito T, Kawwamura T, Nakagawa A, Yamazaki S, Syuto B, Takaki K. Preservation of fresh food using AC electric field. J Adv Oxid Technol. 2014; 7(2):249–53.
8. Romel R, Mohamed HMH, Sastry SK. Inactivation kinetics of *Bacillus coagulans* spores under ohmic and conventional heating. LWT-Food Sci Technol. 2013;54(1):194–8.
9. Loghavi L, Sastry SK, Yousef AE. Effect of moderate electric field frequency and growth stage on the cell membrane permeability of *Lactobacillus acidophilus*. Biotechnol Prog. 2009;25(1):85–94.
10. Dave D, Ghaly AE. Meat spoilage mechanisms and preservation techniques: a critical review. Am J Agric Biol Sci. 2011;6(4):486–510.
11. Zhou G, Xu X, Liu Y. Preservation technologies for fresh meat - A review. Meat Sci. 2010;86(1):119–28.
12. Doulgeraki AI, Ercolini D, Villani F, Nychas G-JE. Spoilage microbiota associated to the storage of raw meat in different conditions. Int J Food Microbiol. 2012;157(2):130–41.
13. Jääskeläinen E, Hultman J, Parshintsev J, Riekkola M-L, Björkroth J. Development of spoilage bacterial community and volatile compounds in chilled beef under vacuum or high oxygen atmospheres. Int J Food Microbiol. 2016;223:25–32.
14. Jones TH, Vail KM, McMullen LM. Filament formation by foodborne bacteria under sublethal stress. Int J Food Microbiol. 2013;165(2):97–110.
15. Neubeck M, Baurb C, Krewinkel M, Stoeckel M, Kranz B, Stressler T, et al. Biodiversity of refrigerated raw milk microbiota and their enzymatic spoilage potential. Int J Food Microbiol. 2015;211:57–65.
16. Morsy MK, Zór K, Kostesha N, Alstrøm TS, Heiskanen A, El-Tanahi H, et al. Development and validation of a colorimetric sensor array for fish spoilage monitoring. Food Control. 2016;60:346–52.
17. Casaburi A, Piombino P, Nychas G-J, Villani F, Ercolini D. Bacterial populations and the volatilome associated to meat spoilage. Food Microbiol. 2015;45:83–102.
18. Hammond ST, Brown JH, Burger JR, Flanagan TP, Fristoe TS, Mercado-Silva N, et al. Food spoilage, storage, and transport: Implications for a sustainable future. Bioscience. 2015;65(8):758–68.
19. Argyri AA, Mallouchos A, Panagou EZ, Nychas G-JE. The dynamics of the HS/SPME-GC/MS as a tool to assess the spoilage of minced beef stored under different packaging and temperature conditions. Int J Food Microbiol. 2015;193:51–8.
20. Selvakumar G, Joshi P, Nazim S, Mishra P, Bisht J, Gupta H. Phosphate solubilization and growth promotion by *Pseudomonas fragi* CS11RH1 (MTCC 8984), a psychrotolerant bacterium isolated from a high altitude Himalayan rhizosphere. Biologia (Bratisl). 2009;64(2):239–45.
21. Ercolini D, Casaburi A, Nasi A, Ferrocino I, Monaco RD, Ferranti P, et al. Different molecular types of *Pseudomonas fragi* have the same overall behaviour as meat spoilers. Int J Food Microbiol. 2010;142:120–31.
22. Ercolini D, Russo F, Blaiotta G, Pepe O, Mauriello G, Villani F. Simultaneous detection of *Pseudomonas fragi*, *P. lundensis*, and *P. putida* from meat by use of a multiplex PCR assay targeting the *carA* gene. Appl Environ Microbiol. 2007;73(7):2354–9.
23. Storia AL, Ferrocino I, Torrieri E, Monaco RD, Mauriello G, Villani F, et al. A combination of modified atmosphere and antimicrobial packaging to extend the shelf-life of beefsteaks stored at chill temperature. Int J Food Microbiol. 2012;158(3):186–94.
24. Ercolini D, Russo R, Nasi A, Ferranti P, Villani F. Mesophilic and psychrotrophic bacteria from meat and their spoilage potential in vitro and in beef. Appl Environ Microbiol. 2009;75(7):1990–2001.
25. Nychas G-JE, Skandamis PN, Tassou CC, Koutsoumanis KP. Meat spoilage during distribution. Meat Sci. 2008;78(1–2):77–89.
26. Casaburi A, Nasi A, Ferrocino I, Monaco RD, Mauriello G, Villani F, et al. Spoilage-related activity of Carnobacterium maltaromaticum strains in air-stored and vacuum-packed meat. Appl Environ Microbiol. 2011;77(20):7382–93.
27. Zhao F, Zhou G, Ye K, Wang S, Xu X, Li C. Microbial changes in vacuum-packed chilled pork during storage. Meat Sci. 2015;100:145–9.
28. Pacholewicz E, Liakopoulos A, Swart A, Gortemaker B, Dierikx C, Havelaar A, et al. Reduction of extended-spectrum-β-lactamase- and AmpC-β-lactamase-producing *Escherichia coli* through processing in two broiler chicken slaughterhouses. Int J Food Microbiol. 2015;215:57–63.
29. Pacholewicz E, Swart A, Schipper M, Gortemaker GG, Wagenaar JA, Havelaar AH, et al. A comparison of fluctuations of *Campylobacter* and *Escherichia coli* concentrations on broiler chicken carcasses during processing in two slaughterhouses. Int J Food Microbiol. 2015;205:119–27.
30. Cody WL, Wilson JW, Hendrixson DR, McIver KS, Hagman KE, Ott CM, et al. Skim milk enhances the preservation of thawed -80 °C bacterial stocks. J Microbiol Methods. 2008;75(1):135–8.
31. Kermanshahi RK, Sailani MR. Effect of static electric field treatment on multiple antibiotic-resistant pathogenic strains of *Escherichia coli* and *Staphylococcus aureus*. J Microbiol Immunol Infect. 2005;38(6):394–8.
32. Takayuki K, Kanako K. In: Freshness keeping apparatus using space potential generator. JP5683032. Japan; 2015. Available online: http://www.freepatentsonline.com/JP5683032B1.html.

33. Ko W, Yang S, Chang C, Hsieh C. Effect of adjustable parallel high voltage electrostatic field on the freshness of tilapia (*Orechromis niloticus*) during refrigeration. LWT-Food Sci Technol. 2016;66:151–7.
34. Wang H, Cheng H, Wang F, Wei D, Wang X. An improved 3-(4,5-dimethylthiazol-2-yl)-2,5-diphenyl tetrazolium bromide (MTT) reduction assay for evaluating the viability of *Escherichia coli* cells. J Microbiol Methods. 2010;82(3):330–3.
35. Jiang X, Yu X, Fan J, Guo L, Zhu C, Jiang W, et al. RFT2 is overexpressed in esophageal squamous cell carcinoma and promotes tumorigenesis by sustaining cell proliferation and protecting against cell death. Cancer Lett. 2014;353(1):78–86.
36. Filipiča J, Kraigher B, Tepuš B, Kokol V, Mandic-Mulec I. Effects of low-density static magnetic fields on the growth and activities of wastewater bacteria *Escherichia coli* and *Pseudomonas putida*. Bioresour Technol. 2012;120:225–32.
37. Gobinath D, Prapulla SG. Permeabilized probiotic *Lactobacillus plantarum* as a source of β-galactosidase for the synthesis of prebiotic galactooligosaccharides. Biotechnol Lett. 2014;36(1):153–7.
38. Griffith KL, Richar E, Wolf J. Measuring β-galactosidase activity in bacteria: cell growth, permeabilization, and enzyme assays in 96-well arrays. Biochem Biophys Res Commun. 2002;290:397–402.
39. Li W, Zhao X, Zou S, Ma Y, Zhang K, Zhang M. Scanning assay of β-galactosidase activity. Appl Biochem Microbiol. 2012;48(6):603–7.
40. Zituni D, Schutt-Gerowitt H, Kopp M, Kronke M, Addicks K, Hoffmann C, et al. The growth of *Staphylococcus aureus* and *Escherichia coli* in low-direct current electric fields. Int J Oral Sci. 2014;6(1):7–14.
41. Røder HL, Raghupathi PK, Herschend J, Brejnrod A, Knøchel S, Sørensen SJ, et al. Interspecies interactions result in enhanced biofilm formation by co-cultures of bacteria isolated from a food processing environment. Food Microbiol. 2015;51:18–24.
42. Lerma LL, Benomar N, Gálvez A, Abriouel H. Prevalence of bacteria resistant to antibiotics and/or biocides on meat processing plant surfaces throughout meat chain production. Int J Food Microbiol. 2013;161(2):97–106.
43. Loghavi L, Sastry SK, Yousef AE. Effect of moderate electric field on the metabolic activity and growth kinetics of *Lactobacillus acidophilus*. Biotechnol Bioeng. 2007;98(4):872–81.
44. Velasco-Alvarez N, González I, Damian-Matsumura P, Gutiérrez-Rojas M. Enhanced hexadecane degradation and low biomass production by *Aspergillus niger* exposed to an electric current in a model system. Bioresour Technol. 2011;102(2):1509–15.
45. Thrash JC, Coates JD. Review: direct and indirect electrical stimulation of microbial metabolism. Environ Sci Technol. 2008;42(11):3921–31.
46. Danevčič T, Stopar D. Asymmetric response of carbon metabolism at high and low salt stress in *Vibrio* sp. DSM14379. Microb Ecol. 2011;62(1):198–204.
47. Vilhelmsson O, Miller K. Synthesis of pyruvate dehydrogenase in *Staphylococcus aureus* is stimulated by osmotic stress. Appl Environ Microbiol. 2002;68(5):2353–8.
48. Machado LF, Pereira RN, Martins RC, Teixxeira JA, Vicente AA. Moderate electric fields can inactivate *Escherichia coli* at room temperature. J Food Eng. 2010;96(4):520–7.
49. Boor KJ. Bacterial stress responses: what doesn't kill them can make them stronger. PLoS Biol. 2006;4(1), e23.
50. Karu T, Pyatibrat L, Kalendo G. Irradiation with He-Ne laser increase ATP level in cells cultivated in vitro. J Photochem Photobiol B. 1995;27:219–23.
51. Jain S, Sharma A, Basu B. Vertical electric field induced bacterial growth inactivation on amorphous carbon electrodes. Carbon. 2015;81:193–202.
52. Amarjargal A, Tijing LD, Ruelo MTG, Park C-H, Pant HR, Iv FPV, et al. Inactivation of bacteria in batch suspension by fluidized ceramic tourmaline nanoparticles under oscillating radio frequency electric fields. Ceram Int. 2013;39(2):2141–5.
53. Moody A, Marx G, Swanson BG, Bermúdez-Aguirre D. A comprehensive study on the inactivation of *Escherichia coli* under nonthermal technologies: High hydrostatic pressure, pulsed electric fields and ultrasound. Food Control. 2014;37:305–14.

# 10

# Inactivation of MSMEG_0412 gene drastically affects surface related properties of *Mycobacterium smegmatis*

Anna Zanfardino[1], Adriana Migliardi[1], Daniele D'Alonzo[2], Angela Lombardi[2], Mario Varcamonti[1] and Angela Cordone[1*]

**Abstract**

**Background:** The outermost layer of mycobacterial cell wall is rich in lipids and glycolipids, surface molecules which differ among species. *Mycobacterium smegmatis*, an attractive model for the study of both pathogenic and non-pathogenic mycobacteria, presents glycopeptidolipids (GPLs). All the genes necessary for the biosynthesis of such molecules are clustered in a single region of 65 kb and among them, the *msmeg_0412* gene has not been characterized yet. Here we report the isolation and subsequent analysis of a MSMEG_0412 null mutant strain.

**Results:** The inactivation of the *msmeg_0412* gene had a drastic impact on bacterial surface properties which resulted in the lack of sliding motility, altered biofilm formation and enhanced drug susceptibility. The GPLs analysis showed that the observed mutant phenotype was due to GPLs deficiencies on the mycobacterial cell wall. In addition, we report that the expression of the gene is enhanced in the presence of lipidic substrates and that the encoded protein has a membrane localization.

**Conclusion:** *msmeg_0412* plays a crucial role for GPLs production and translocation on *M. smegmatis* surface. Its deletion alters the surface properties and the antibiotic permeability of the mycobacterial cell barrier.

**Keywords:** Mycobacterium smegmatis, msmeg_0412, GPLs, Lipolytic, Cell wall, Biofilm

## Background

Mycobacterial cell wall is a unique well organized barrier which confers low permeability to different chemical agents and common antibiotics [1]. The envelope is waxy, hydrophobic, and thicker compared to that of other bacteria and is organized in two compartments, the inner layer, covalently linked to the plasma membrane and formed by peptidoglycan, arabinogalactan and mycolic acid, and, the outer layer, which is essentially formed by glycolipids and proteins [2, 3]. Glycopeptidolipids (GPLs) are the major surface exposed molecules found in various mycobacterial species including non-pathogenic (e.g *M. smegmatis*) and non–tuberculous pathogenic mycobacteria (e.g. *M. avium, M. intracellulare, M. abscessus*) [4, 5]. GPLs have a common fatty acyl-tetrapeptide core which is glycosylated with 6-deoxy-L-talose and variable *O*-methyl-L-rhamnose residues [6–8]. This is known as non-serovar-specific GPL (nsGPLs) and represents the main product of *M. smegmatis* GPLs. In *M. avium* GPLs present a more complicated structure in which an additional rhamnose and oligosaccharide residues are added to 6-deoxy-L-talose, resulting in serovar-specific GPLs [7, 9, 10]. These molecules, which may represent more than 70 % of total surface lipids [11], are required for cell aggregation, sliding motility and biofilm formation [12–15]. Moreover several studies reported that slight modifications of the GPLs structure are able to affect the mycobacterial ability to stimulate the host immune response, suggesting a role as antigenic molecules [16–19].

The fast growing *M. smegmatis* is frequently used as a model organism to study the molecular and physiological mechanisms used by slow-growing pathogenic mycobacteria. The *M. smegmatis* GPLs biosynthetic

* Correspondence: angelina.cordone@unina.it
[1]Department of Biology, University of Naples "Federico II", Via Cintia, 80126 Naples, Italy
Full list of author information is available at the end of the article

locus has been previously characterized either experimentally or by *in silico* prediction [20–23]. Most genes of the *M. smegmatis* GPL locus are conserved in two clinically significant mycobacterial species such as *M. abscessus and M. avium* subsp. *paratuberculosis* [5, 24].

Among them, the gene *msmeg_0412* is still uncharacterized and its product shares 66 % and 67 % identity with the predicted protein MAB_0402, and MAP2244 from *M. abscessus and M. avium subsp. paratuberculosis*, respectively. Interestingly, *M. tuberculosis*, a mycobacterium which does not produce GPLs on the cell wall, presents Rv1184c which shares 45 % identity with MSMEG_0412. Rv1184c has been recently characterized as the Acyltransferase Chp2, essential for the final biosynthetic steps of polyacyltrehalose (PAT), a pentaacylated, trehalose-based glycolipids only found on the surface of pathogenic mycobacteria [25, 26].

In this study we show that the inactivation of *msmeg_0412* gene alters the lipid profile surface of *M. smegmatis* and enhances the bacterial susceptibility to antibiotics. In addition, we report that *msmeg_0412* gene expression is enhanced in the presence of lipidic substrates and that the encoded protein localizes on the cell wall. Taken together, our results indicate that MSMEG_0412 is an important player in the complex GPLs biosynthetic process in *M. smegmatis* and allow speculating that MSMEG_0412 is translocated to the cell wall where it contributes to GPLs synthesis.

## Results

### Deletion of the *msmeg_0412* gene alters *M. smegmatis* surface properties, enhances suscepitibility to antibiotics and affects GPLs synthesis

The *msmeg_0412* gene (Fig. 1a) belongs to a gene cluster devoted to the biosynthesis and transport of GPLs to mycobacterial cell wall surface. Most of the genes inside the cluster have been previously characterized [5, 15]. In particular, *mps1* and *mps2* genes encode the mycobacterial peptide synthetase required for the formation of tripeptide-amino-alcohol moiety which is then linked to the 3-hydroxy/methoxy C26-C34 acyl chain, the lipid moiety of GPLs. The latter is synthesized by the combined action of Pks, Fad23 and PapA3 gene products. Finally, the action of various glycosyltransferases (Gtf), methyltransferases (Mtf) and acetyltransferases (Atf) leads to the building of the glycosyl side of GPLs. Once synthesized, GPLs are transferred to GPL addressing protein (Gap), a small integral membrane protein, required for the translocation of GPLs on the cell surface. As shown in Fig. 1a, the *msmeg_0412* gene is close to

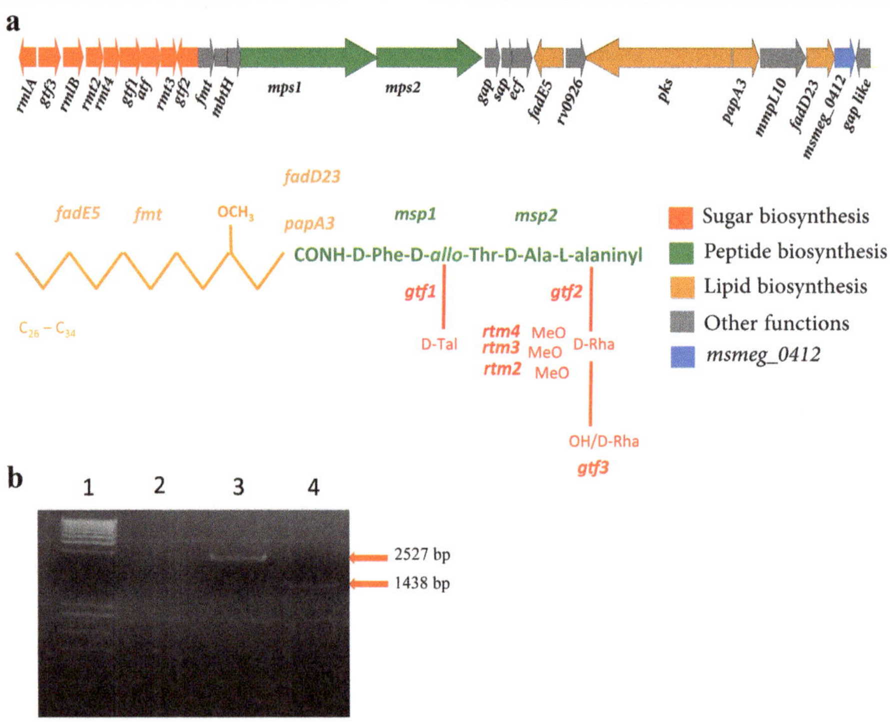


**Fig. 1 a**) Schematic representation of the GPLs structure from *M. smegmatis* and its biosynthetic *locus* on the chromosome. The genes involved in the synthesis of lipid moiety (*orange*), peptide (*green*) and glycosylation (*red*) of GPLs are indicated. **b**) Agarose gel electrophoresis showing PCR-based validation of the *msmeg_0412* gene deletion in the GM1 strain. Lanes: 1) DNA Ladder marker (1Kb); 2) PCR negative control; 3) wt Chromosomal DNA used as template (2527 bp-amplicon expected using 0412-F1 and 0412-R2 as primers); 4) GM1 Chromosomal DNA used as template (1438 bp-amplicon expected using 0412-F1 and 0412-R2 as primers)

the genes involved in the synthesis of lipid portion of GPLs. In order to understand the role of the *msmeg_0412* gene product, we constructed a null mutant strain by using the pGOAL/pNIL procedure as reported in methods. A PCR confirmation of gene deletion in GM1 strain is shown in Fig. 1b. The deletion mutant, GM1, was analyzed for some phenotypes . Compared to its isogenic wild type strain, GM1 displayed altered biofilm formation (Fig. 2a), lack of sliding motility (Fig. 2b), and rough colony phenotype (Fig. 2c).

To confirm that these phenotypes were due to *msmeg_0412* gene inactivation, we performed a complementation analysis. A wild type allele of the *msmeg_0412* gene was PCR-amplified, cloned into pMV10-25 vector [27] without the *gfp* gene (methods), and transformed into strain GM1, yielding strain GM2. As shown in Fig. 2 all phenotypes of strain GM1 were rescued in strain GM2, that appeared identical to the wild type strain.

The drastic impact of *msmeg_0412* deletion on *M. smegmatis* surface properties was also evidenced by a slower growth rate in liquid media with a generation time of the mutant of 4.5 h compared to 3 h of the wt strain (see Additional file 1: Figure S1) and by the enhanced susceptibility to antimicrobials. As shown in Fig. 2d, when streaked on LB agar medium containing Rifampicin at a sub-MIC concentration (10 μg/ml) [28], the GM1 strain was unable to grow compared to the wild type and complemented strains. The same result was obtained using Erythromycin as antibiotic (13 μg/ml) [29]. Since both antibiotics have cytosolic targets, it is possible to hypothesize that the mutation induces changes in the cell wall permeability that lead to an increase in antibiotics penetration.

To better characterize the effects of *msmeg_0412* deletion we extracted and analyzed the GPLs content from the wild type, mutant and complemented strains. TLC analysis of the three organic extracts (Fig. 3a) indicated that GPLs were absent only in GM1 while they were equally abundant in the remaining strains. Conversely, the occurrence of the lipid fraction was detected in both wt, GM1 and GM2 strains (see Additional file 2: Figure S2). An even more accurate picture was provided by ESI-MS analysis (Fig. 3b). In this case, wt and GM2 strains provided identical MS profiles, which could be referred to the occurrence of 6-deoxy-L-talose- and *O*-methyl-L-rhamnose-containing GPLs and their methylated congeners [23]. Conversely, no traces of any sugar-containing compound at the same *m/z* range was found in GM1. The lack of GPLs in the *msmeg_0412* deletion mutant apparently suggests that *msmeg_0412* gene product is involved in the synthesis and assembly of GPLs on the *M. smegmatis* cell wall.

### The *msmeg_0412* gene expression is induced by lipid substrates

In order to evaluate the role played by MSMEG_0412 protein during GPLs production a bioinformatics approach was initially followed. MSMEG_0412 has the canonical penta-peptide sequence motif GXSXG/S and the conserved Ser, Asp and His catalytic triad (indicated in bold in Fig. 4) typical of proteins with lipase/esterase activity and identified as hydrolases involved in lipid metabolism [30, 31].

To experimentally validate the bioinformatic prediction we monitored the effects of two widely used lipid substrates, tributyrin and tween 80, on the transcriptional

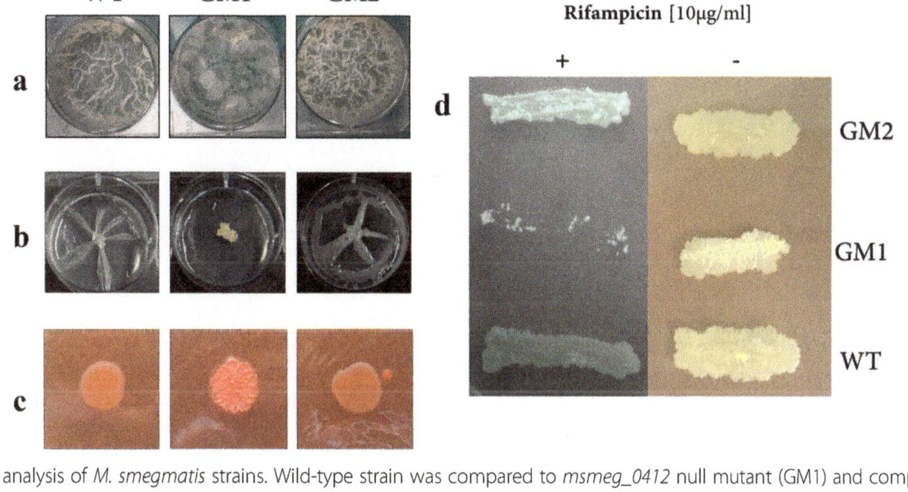


**Fig. 2** Phenotypic analysis of *M. smegmatis* strains. Wild-type strain was compared to *msmeg_0412* null mutant (GM1) and complemented (GM2) strains for the following phenotypes: Biofilm production (**a**), Sliding motility (**b**) Colony morphology (**c**). The same strains were streaked on LB agar plates supplemented with or without Rifampicin (10 μg/ml) to evaluate the Antimicrobial susceptibility towards this antibiotic (**d**)

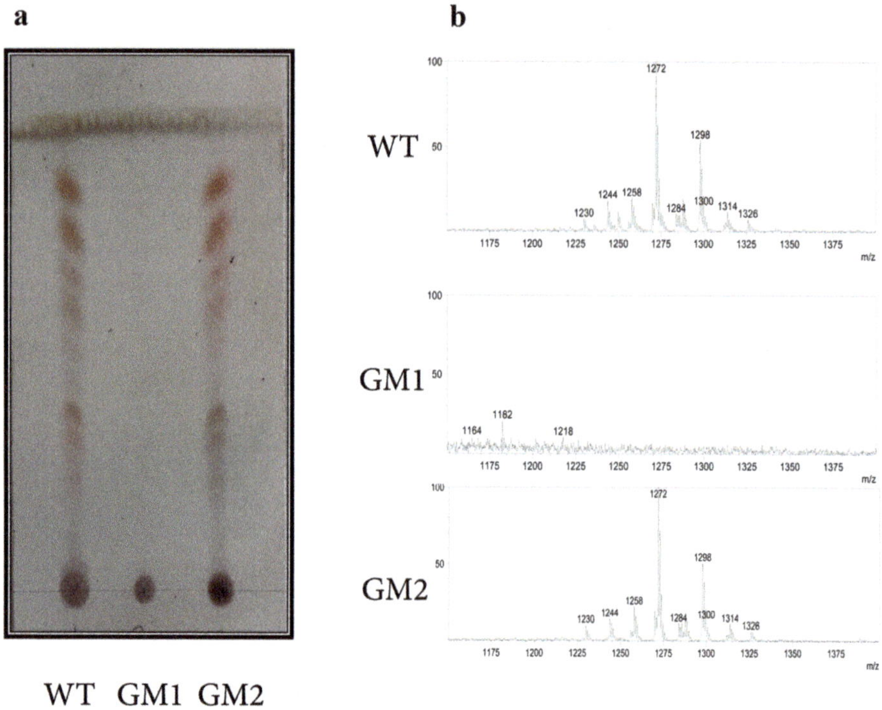


**Fig. 3** Chemical analysis of GPLs. GPLs content from the wild-type, GM1 and GM2 strains as indicated by TLC analysis (**a**) and ESI-MS profile of the organic extracts (**b**)

profile of the *msmeg_0412* gene. To this purpose *M. smegmatis* wild type strain was grown in minimal medium containing glucose, tributyrin or tween 80 as the only carbon source and total RNA was extracted and reverse transcribed as reported in methods. As shown in Fig. 5, the transcriptional level of the gene was induced approximately 50-fold in the presence of tween 80 and 2-fold in the presence of tributyrin.

Based on this result we decided to evaluate the effect of tween 80 on the growth of the mutant strain. To this purpose wt, GM1 and GM2 strains were grown in minimal medium containing 0.2 % (w/v) glucose or 1%t ween 80 as the only carbon source. The result (see Additional file 3: Figure S3) clearly indicated that the mutant was unable to grow in presence of tween 80. Although this result, together with the bioinformatic analysis, allows speculating a lipolytic function for MSMEG_0412, a detailed biochemical characterization is needed to decipher its function.

### The MSMEG_0412 protein is localized on the cell wall

*In silico* analysis suggested that MSMEG_0412 contains a N-terminal signal sequence (underlined in Fig. 4) with a putative cleavage site between residues A29 and D30, most likely involved in targeting the MSMEG_0412 to the cell membrane. To validate this prediction we constructed two gene fusions between *gfp*, the gene encoding for the Green Fluorescent Protein, and the *msmeg_0412* gene with and without the sequence coding for the putative signal peptide (Methods). The gene

MRRRLLAFGTAFTTIGTAGFLGFGVAAADDTKPVDPAPGGAHAEAPSMGTPGRGYALGGA 60

HVLGIPYDEYIMRTGADWFPGLDRQIVDYPAGQVQGHTLERLFPGIGALGERFMPGLGLD 120

GPSYGESIDVGAPNLINAIRQGGPGTVI**GLSEG**ASVLDEVQARLAYDPAAPPPDSLSFAT 180

YGNPVGKHAFGESFLTQMFPVGSVVPSLDYRIPAPVESQYDTYQFVSAY**D**SIADWPDRPD 240

NWISVANAIVGLATG**H**TAVAFTNPSMVPPQNIRTTVNSRGAKDTTIMIPEEHLPLVLPFK 300

YLGVDKDTLNKLDGVLKPYVDAGYSRNDDPLTAPITVDPVNGYDPAAVTAPATQAAFGGG 360

TDPVSQLLAGLQYVVNNQPAPKP 383

**Fig. 4** Amino acid sequence of MSMEG_0412. In bold is reported the conserved pentapeptide sequence motif GXSXG/S. Aminoacid residues potentially involved in the catalytic triad are indicated by an asterisk (*). Underlined is the putative N-terminal signal peptide

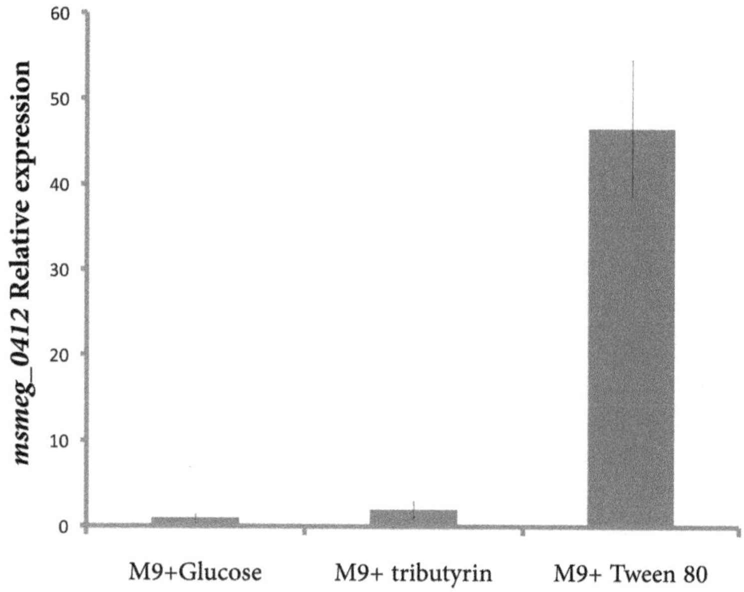


**Fig. 5** Real-Time PCR analysis of *msmeg_0412* gene under different growth conditions. Cells of *M. smegmatis* $mc^2155$ wild type strain were grown in aerated conditions in minimal medium containing glucose, tributyrin or tween 80 to early exponential growth phase. The ΔΔCt method was used to calculate the relative amount of specific RNA present in each sample, and the transcriptional induction determined by comparison to values from cells growing in minimal medium containing glucose

fusions were independently used to transform *M. smegmatis* wild type cells, yielding strains GM7 (carrying the entire *msmeg_0412* gene) and GM8 (carrying the *msmeg_0412* gene without the signal peptide). A fluorescence microscopy analysis showed that the fluorescence signal was localized around the cells of strain GM7 and inside the cells of strain GM8 (Fig. 6). This result indicates that MSMEG_0412 is a membrane protein and that the removal of the signal peptide induces protein aggregation and its retention within the cytosol.

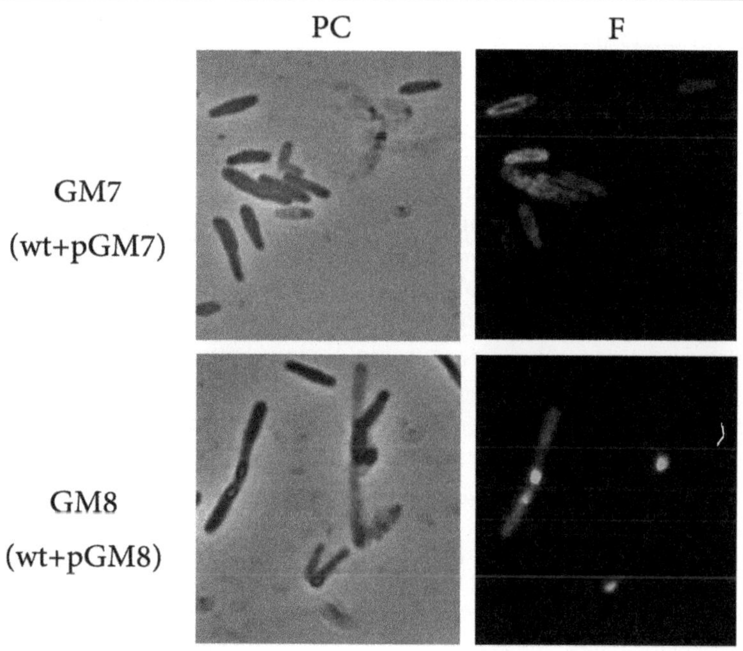


**Fig. 6** Fluorescence analysis. *M. smegmatis* strains carrying gene fusions between *gfp* and *msmeg_0412* gene with (GM7) and without (GM8) the region coding for the putative signal peptide were analysed by phase contrast (PC) and Fluorescent (F) microscopy. Exposure time was 200 ms in both cases

## Discussion

Mycobacteria comprise bacilli with a specialized cell wall divided in two compartments: the inner- and the outer-layer. The latter is rich in lipids and glycolipids which differ among species. *M. smegmatis* shares a richness in glycopeptidolipids (GPLs) with pathogenic non tubercle mycobacteria, such as *M. abscessus and M. avium.*

*Msmeg_0412*, a gene located in the GPLs biosynthetic locus, is highly conserved among mycobacteria, and its product shares 45 % identity with Rv1184c from the human pathogen *M. tuberculosis.* The gene encodes a protein which primary structure has features of proteins annotated as putative lipases or esterases.

Lipase/esterase-like enzymes belong to a family of proteins involved in lipid metabolism, which play a crucial role during mycobacterial infection and persistence inside the host [30, 32]. In fact, it has been reported that such enzymes contribute to the destruction of the host tissues thereby supplying nutrients to the pathogens [33, 34]. Therefore, a better understanding of these proteins will surely help in developing new strategies against mycobacterial infections.

Earlier reports have clearly indicated that the loss of GPLs on *M. smegmatis* surface induce an increased cell wall hydrophobicity [14, 15, 23] which results in the alteration of surface properties such as a change of colony morphotype (from smooth to rough), an altered biofilm formation at liquid/air interface, and the inability to slide on soft agar plates. Here we report the construction and analysis of a *M. smegmatis* strain carrying a null mutation in *msmeg_0412* gene.

Our results show that MSMEG_0412 is required for the correct processing of the GPLs on mycobacterial surface and that its deletion induced a surface perturbation. Furthermore, in line with the bioinformatic predictions, the *msmeg_0412* gene expression is enhanced in the presence of lipid substrates, and the protein encoded localizes on the cell wall. Taken together our results allow to speculate that once synthesized, MSMEG_0412 is translocated to the cell wall, where plays a crucial role for the correct GPLs processing and translocation on *M. smegmatis* surface. A further biochemical approach is needed to better clarify other aspects of the function and the localization of this important protein.

## Conclusions

The analysis of *msmeg_0412* mutant highlighted that this gene is important for *Mycobacterium smegmatis* surface organization and resulted in different important phenotypes, as lack of sliding motility, altered biofilm formation and enhanced drug susceptibility. The analysis of Glycopeptidolipids (GPLs) component of bacterial surface showed that the *msmeg_0412* mutant was strongly impaired for GPLs presence in the mycobacterial cell wall. In addition, based on *in-silico* predictions, we report that the expression of the gene is enhanced by the presence of lipidic substrates and that the encoded protein has a membrane localization.

Finally, generation time of the mutant was shorter than the one of the wild type strain.

Mutations affecting the organization of the cell wall have often pleiotropic effects. This is the first experimental evidence for an involvement of *msmeg_0412* gene in the synthesis of the GPLs.

Taken together, the results presented in this manuscript offer the possibility to consider MSMEG_0412 as a new target to impair the mycobacterial growth and to increase drug susceptibility.

## Methods

### Bacterial strains, media and growth conditions

*M. smegmatis* mc$^2$155 [35] was the parental strain of all the recombinant strains described below. DH5α (*supE44 ΔlacU169 [ϕ80ΔlacZM15] hsdR17 recA1*) [36] was used for all cloning experiments.

The recombinant strain GM1, carrying a *msmeg_0412* gene deletion was engineered using the p2NIL/pGOAL19-based flexible cassette method as previously reported [37].

*M. smegmatis* mc$^2$155 and derivatives were grown in 7H9 medium (Difco) containing 2 % glycerol and 0.05 % tween 80 (Applichem) or in M9 containing 1 mM $Mg_2SO_4$ and 0.2 % glucose. Where indicated, glucose was replaced by 1 % glycerol-tributyrate (tributyrin) or 1 % tween 80.

*E. coli* strains were grown in LB medium. When required, antibiotics were added to the medium at the following final concentrations: ampicillin 100 μg/ml, hygromycin at 200 μg/ml for *E. coli* and 50 μg/ml for *M. smegmatis* mc$^2$155, respectively.

For the antimicrobial susceptibility test, Erythromycin and Rifampicin were added to LB agar medium at the following final concentrations: 13 μg/ml and 10 μg/ml respectively.

For the Microscopy analysis, samples were observed using an Olympus BX51 fluorescence microscope using a FITC filter. The Images were captured using an Olympus DP70 digital camera and processed. Standard acquisition times were 200 ms.

#### *Bioinformatic approach*

A multiple sequence alignment was carried out using PRALINE (http://www.ibi.vu.nl/programs/pralinewww) as program. MSMEG_0412 aminoacid sequence was used as query against the earlier identified *M. tuberculosis* proteins [31] showing the conserved pentapetide sequence motif (GXGXG) and the conserved Ser, Asp and His catalytic triad characteristic of lipase/esterase activities.

The signal peptide server, (http://www.cbs.dtu.dk/services/SignalP), was used to predict the presence and location of the signal peptide cleavage site inside MSMEG_0412.

### Sliding motility assay

5 μl of a liquid culture in stationary phase was dispensed on plate containing 7H9 medium plus 0.3 % agar with no added carbon source. The plate was then incubated at 37 °C for 1 week [15, 23].

### Congo red assays

5 μl of a liquid culture in stationary phase was dispensed on plate containing LB medium supplemented with 2 % agar and 100 μg/ml Congo Red (Sigma) and incubated at 37 °C for 3 days [15, 23].

### Biofilm assay

5 μl of a liquid culture in stationary phase was inoculated into PVC well plates containing 5 ml Sauton liquid medium and incubated in standing condition at 37 °C for 1 week [15, 23].

### DNA manipulation

Plasmid and chromosomal DNA preparation, restriction digestion, ligation, bacterial transformation and agarose gel electrophoresis were performed as described [36].

Plasmid pGM5 carries a gene deletion cassette (Δ0412 cassette) containing 689-bp segment upstream of *msmeg_0412* gene + first 4 codon (5' fragment) followed by last 15 codons + 685-bp segment downstream of the gene (3' fragment). The 5' fragment was PCR amplified using the primer pairs 0412-1 F and 0412-1R listed in Table 1 and the genomic DNA of wild type $mc^2155$ strain as template and the obtained PCR product was cloned in pGEMT-easy vector, giving the plasmid pGM2. The 3' fragment was PCR amplified using the primer pairs 0412-2 F and 0412-2R listed in Table 1 and the genomic DNA of wild type $mc^2155$ strain as template and the obtained PCR product was digested with *KpnI* and cloned in pGM2 vector at the unique *KpnI* restriction site, giving plasmid pGM3. The deletion cassette was then excised from pGM3 vector using the restriction enzymes *BamHI* and *SalI* and transferred into the p2NIL plasmid previously digested by the same restriction enzymes. The resulting plasmid, named pGM4 and the plasmid pGOAL19 were digested with *PacI*, and the *PacI* selectable marker cassette of pGOAL19, carrying the *sacB*, *LacZ* and *Hyg* genes, was ligated to the linearized pGM4 to create the suicide vector pGM5.

Plasmid pGM6, used for complementation analysis, was constructed as follow: the wild type *msmeg_0412* gene from *M. smegmatis* wild type strain was PCR amplified with *Pfu* Turbo high Fidelity DNA polymerase (Stratagene) using chromosomal DNA as template and the oligo 0412-Fw and 0412-Rv (Table 1) as primers. The last ones contained the engineered *NheI* and *BamHI* restriction sites respectively.

The PCR product of expected size was purified with a PCR Purification Kit (Qiagen), digested with *NheI* and *BamHI* restriction enzymes and cloned under the *hsp60* promoter into the shuttle vector pMV10-25 [27], previously digested with the same enzyme pairs which allowed the excision of the *gfp* encoding gene. The resulting plasmid was controlled by sequencing and then used to transform GM1 cells, giving the strain GM2.

Plasmid pGM7 and pGM8 containing the entire version or the truncated version of the *msmeg_0412* gene fused to the *gfp* reporter gene, respectively, were constructed as follow: the *msmeg_0412* gene except the stop codon was PCR amplified with *Pfu* Turbo high Fidelity DNA polymerase (Stratagene) using chromosomal DNA as template and priming the reaction with two different couples of primers: 0412-Fw and 0412-gfpRv or 0412-

**Table 1** Synthetic oligonucleotides

| Name | Sequence (5' - 3')[a] | Position of Annealing[b] |
|---|---|---|
| 0412-S | CATGATCCCCGAGGAGCAC | +857/+875 |
| 0412-A | TTGCCGTTCAAGTACCTCGG | +889/+90 |
| 0412-Fw | AAAGCTAGCATGAGGAGACGACTTCTCGCGTTC | +1/+24 |
| 0412-spFw | AAAGCTAGCGATGACACCAAACCGGTCG | +85/+103 |
| 0412-Rv | AAAGCTAGCTCAGGGCTTGGGCGCCGGTTG | +1132/+1152 |
| 0412-gfpRv | AAAGCTAGCGGGCTTGGGCGCCGGTTG | +1132/+1149 |
| 0412-1F | CTAGAAGCTTGCACCGCACTGGTGAGATA | -690/-672 |
| 0412-1R | CTAGGGTACCCGCTGTAGTCCTCTGCT | -5/+12 |
| 0412-2F | CTAGGGTACCGGACTGCAGTACGTCGTGAACA | +1108/+1129 |
| 0412-2R | CTAGGGTACCACCATGTGGATCCCTCTCCT | +1818/+1837 |

[a]Underlined is an unpaired tail carrying a restriction site

[b]Position of annealing refers to *msmeg_0412* gene sequence, with the first base of the translational initiation codon as +1

spFw and 0412-gfpRv (Table 1). The first couple of primers has allowed to amplify the entire copy of the gene, the other combination of primers has allowed to amplify the gene that misses of 84 bp, encoding a putative N-terminal signal peptide. All primers contained a 5' engineered *NheI* restriction site. The PCR products of expected sizes were purified with the PCR purification Kit (Qiagen) digested with the *NheI* restriction enzyme and cloned in the plasmid pMV10-25 [27], under the control of *hsp60* promoter and *in frame* with GFP coding gene using the *NheI* restriction site located at 5' end of *gfp*.

The resulting plasmids were controlled by sequencing and then used to transform *M. smegmatis* wild type cell, giving the strains GM7 and GM8 respectively.

#### Isolation of Δmsmeg_0412 mutant strain

The *msmeg_0412* null mutant strain, named GM1, was obtained by using the allelic replacement system reported previously [37].

The pGM5 suicide vector was used to transform cells of *M. smegmatis* mc2155 strain and transformants were selected on LB agar plates supplemented with hygromicin (200 mg/ml), kanamycin (25 mg/ml) and X-Gal (50 mg/ml). Blue colonies were restreaked onto LB agar plates; a loopful of cells was then resuspended in LB broth, serially diluted and plated on LB agar containing 2 % sucrose (Suc) and X-gal. White $Kan^S$, $Hyg^S$ and $Suc^R$ colonies were isolated and identified by PCR analysis. One $Kan^S$, $Hyg^S$ and $Suc^R$ colony with a deleted *msmeg_0412* gene was identified as GM1 strain.

#### RNA extraction

Total RNA was extracted by using the RNeasy Mini Kit (QIAGEN) according to manufacturer's protocol. RNAs were resuspended in 100 ml of RNAse-free water and treated with DNAse I (Fermentas, 1U/ml) for 30 minutes at 37 °C. The quality of RNA samples were estimated using the RNA nanochip on the Thermoscientific Nanodrop 1000. The concentration of RNA was determined by measuring the absorbance at 260 nm.

#### Real Time PCR analysis

cDNA was synthesized by using PrimeScript Enzyme mix (Takara) and random hexamers as primers. Primers 0412-S and 0412-A (Table 1), and those for the normalizing gene 16S, 16S-fw (5′-GGCGAACGGGTGAGT AACA-3′) and 16S-Rv (5′-GCCCTGCACTTTGGG ATAAG-3′), were designed with ABI PRISM Primer Express software (PE Applied Biosystems). Real-Time PCR was performed by using SYBR green RT-PCR (Takara). Reactions were performed as previously described [38]. Values reported here were the average of at least three independent experiments and were expressed as the mean ± SEM (Standard Error of the Mean). The fluorescence signal due to SYBR Green intercalation was monitored to quantify the double-stranded DNA product formed in each PCR cycle. Statistical significance was determined by Student's unpaired *t*-test and the significance levels were reported in the text.

#### Extraction and analysis of mycobacterial lipids

Total lipids were extracted from whole cells as previously reported [16]. In brief, cells from wt, mutant and complemented strains were grown in LB medium until stationary phase ($OD_{600}$: 2.5), harvested 10 minutes at 7000 rpm and resuspended in PBS, pH 7.4. Lipids were extracted with $CHCl_3$:$CH_3OH$ (1:2 v/v) for 24 hours, separately dried under vacuum and partitioned between water and chloroform (1:1 v/v).

The organic phases were further washed with distilled water and evaporated to dryness. For analytical thin-layer chromatography (TLC), the crude residues (1.0 mg) were dissolved in $CHCl_3$:$CH_3OH$ (0.75 mL, 9:1 v/v), spotted on silica gel 60-precoated plates (0.25-mm thickness, Merck) and eluted with $CHCl_3$:$CH_3OH$ (9:1 v/v). Sugar-containing compounds were visualized by spraying TLC plates with a chromosulfuric acid staining solution and charring them at 200 °C for 2 min. The GPLs were detected as brownish-red spots. Conversely, the lipid content in the three strains was detected by extracting the crudes with hexane (0.3 mL). The resulting solutions were spotted on TLC plates which were eluted in toluene:acetone (99:1 v/v). In this case, lipids were visualized by spraying the plates with a cerium molybdate staining solution and charring them at 200 °C for 2 min. The lipids were detected as dark blue spots.

ESI-MS analysis of the organic extracts was performed using a Shimadzu LC-MS-2010EV system with ESI interface, Q-array-octapole-quadrupole mass analyser and a Shimadzu LC-MS solution Workstation software for data processing. The optimized MS parameters were selected as follows: CDL (curved desolvation line) temperature 250 °C; heat block temperature 250 °C; probe temperature 250 °C; detector gain 1.6 kV; probe voltage +4.5 kV; CDL voltage -15 V. Nitrogen served as nebulizing (flow rate: 1.5 L/min) and drying gas (flow rate: 15 L/min). Each sample (0.5 mg) was dissolved in 1 mL of acetonitrile (ACN)/0.1 % trifluoroacetic acid (TFA); 2 μL aliquots of the resulting solutions were injected using ACN/0.1 % TFA as solvent (flow rate: 0.2 mL/min). MS spectra were acquired in positive ion mode.

### Additional files

**Additional file 1: Figure S1.** Growth curve of *M. smegmatis* strains in LB medium. *M. smegmatis* wild type and GM1 strains were grown in LB medium containg 0,05 % tween 80 and $OD_{600nm}$ determined every

3 hours. For each strains the data reported in graph are the mean of three independent experiments.

**Additional file 2: Figure S2.** Chemical analysis of total lipids. Total lipids were extracted from whole wt, GM1 and GM2 cells and analyzed by TLC.

**Additional file 3: Figure S3.** Growth curve of *M. smegmatis* strains in Minimal medium. *M. smegmatis* wild type, GM1, GM2 strains were grown in minimal medium containing 0.2 % (w/v) glucose or 1%t ween 80 as the only carbon source. and $OD_{600nm}$ determined every 3 hours. For each strains the data reported in graph are the mean of three independent experiments.

### Abbreviations
ESI-MS: Electrospray Ionization Mass Spectometer; GPLs: Glycopeptidolipids; TLC: Thin Layer chromatography

### Acknowledgements
We would like to acknowledge Prof. Ezio Ricca for critical reading of the manuscript.

### Funding
This work was financially supported by the Department of Biology of the University of Naples "Federico II".

### Authors' contributions
AC conceived the study; AZ, AM, and AC performed mutant isolation, strains construction and phenotypic analysis, Real-Time PCR experiments and Fluorescence microscopy analysis; DD and AL carried out lipid extraction and analysis; AZ, MV and AC analyzed data and wrote the manuscript. All authors read and approved the final manuscript.

### Competing interest
The authors declare that they have no competing interests.

### Author details
[1]Department of Biology, University of Naples "Federico II", Via Cintia, 80126 Naples, Italy. [2]Department of Chemical Sciences, University of Naples "Federico II", Via Cintia, 80126 Naples, Italy.

### References
1. Hett EC, Rubin J. Bacterial growth and cell division. Microbiol Mol Biol Rev. 2008;72:126–56.
2. Brennan PJ. Structure, function, and biogenesis of the cell wall of *Mycobacterium tuberculosis*. Tuberculosis (Edinburgh). 2003;83:91–7.
3. Brennan PJ, Nikaido H. The envelope of mycobacteria. Annu Rev Biochem. 1995;64:29–63.
4. Howard ST, Byrd TF. The rapidly growing mycobacteria: Saprophytes and parasites. Microbes Infect. 2000;2:1845–53.
5. Ripoll F, Deshayes C, Pasek S, Laval F, Beretti JL, Biet F, Risler JL, et al. Genomics of glycopeptidolipid biosynthesis in *Mycobacterium abscessus* and *M. chelonae*. BMC Genomics. 2007;8:114.
6. Daffè M, Laneelle MA, Puzo G. Structural elucidation by field desorption and electron-impact mass spectrometry of the C-mycosides isolated from *Mycobacterium smegmatis*. Biochim Biophys Acta. 1983;751:439–43.
7. Aspinall GO, Chatterjee D, Brennan PJ. The variable surface glycolipids of mycobacteria: structures, synthesis of epitopes, and biological properties. Adv Carbohydr Chem Biochem. 1995;51:169–242.
8. Miyamoto Y, Mukai T, Maeda Y, Kai M, Naka T, Yano I, Makino M. Characterization of the Fucosylation Pathway in the Biosynthesis of Glycopeptidolipids from Mycobacterium avium Complex. J Bacteriol. 2007;189:5515–22.
9. Vergne I, Daffe M. Interaction of mycobacterial glycolipids with host cells. Front Biosci. 1998;3:865–76.
10. Chatterjee D, Khoo KH. The surface glycopeptidolipids of mycobacteria: structures and biological properties. Cell Mol Life Sci. 2001;58:2018–42.
11. Billman-Jacobe H. Glycopeptidolipid synthesis in Mycobacteria. Curr Sci. 2004;86:11–114.
12. Recht J, Martinez A, Torello S, Kolter R. Genetic analysis of sliding motility in *Mycobacterium smegmatis*. J Bacteriol. 2000;182:4348–51.
13. Recht J, Kolter R. Glycopeptidolipid acetylation affects sliding motility and biofilm formation in *Mycobacetrium smegmatis*. J Bacteriol. 2001;183:5718–24.
14. Etienne G, Villeneuve C, Billman-Jacobe H, Astarie-Dequeker C, Dupont MA, Daffè M. The impact of the absence of glycopeptidolipids on the ultrastructure, cell surface and cell wall properties, and phagocytosis of *Mycobacterium smegmatis*. Microbiology. 2002;148:3089–100.
15. Deshayes C, Laval F, Montrozier H, Daffè M, Etienne G, Reyrat JM. A Glycosyltransferase Involved in Biosynthesis of Triglycosylated Glycopeptidolipids in *Mycobacterium smegmatis*: Impact on Surface Properties. J Bacteriol. 2005;187:7283–91.
16. Villeneuve C, Etienne G, Abadie V, Montrozier H, Bordier C, Laval F, Daffè M, et al. Surface-exposed glycopeptidolipids of *Mycobacterium smegmatis* specifically inhibit the phagocytosis of mycobacteria by human macrophages. Identification of a novel family of glycopeptidolipids. JBC. 2003;278:51291–300.
17. Sweet L, Schorey JS. Glycopeptidolipids from *Mycobacterium avium* promote macrophage activation in a TLR2- and MyD88-dependent manner. J Leukoc Biol. 2006;80:415–23.
18. Sweet L, Zhang W, Torres-Fewell H, Serianni A, Boggess W, Schorey J. *Mycobacterium avium* glycopeptidolipids require specific acetylation and methylation patterns for signaling through Toll-like Receptor 2. JBC. 2008;238:33221–31.
19. Nessar R, Reyrat J, Davidson LB, Byrd TF. Deletion of the *mmpL4b* gene in the *Mycobacterium abscessus* glycopeptidolipid biosynthetic pathway results in loss of surface colonization capability, but enhanced ability to replicate in human macrophages and stimulate their innate immune response. Microbiology. 2011;157:1187–95.
20. Patterson JH, McConville MJ, Haites RE, Coppel RL, Billman-Jacobe H. Identification of a methyltransferase from *Mycobacterium smegmatis* involved in glycopeptidolipid synthesis. JBC. 2000;275:24900–6.
21. Jeevarajah D, Patterson JH, McConville MJ, Billman-Jacobe H. Modification of glycopeptidolipids by an O-methyltransferase of *Mycobacterium smegmatis*. Microbiology. 2002;148:3079–87.
22. Trivedi OA, Arora P, Sridharan V, Tickoo R, Mohanty D, Gokhale RS. Enzymic activation and transfer of fatty acids as acyl-adenylates in mycobacteria. Nature. 2004;428:441–5.
23. Sonden B, Kocincova D, Deshayes C, Euphrasie D, Rhayat L, Laval F, Frehel C, et al. Gap, a mycobacterial specific integral membrane protein, is required for glycolipid transport to the cell surface. Mol Microbiol. 2005;54:426–40.
24. Biet F, Bay S, Thibault VC, Euphrasie D, Grayon M, Ganneau C, Lanotte P, et al. Lipopentapeptide induces a strong host humoral response and distinguishes *Mycobacterium avium* subsp. *paratuberculosis* from *M. avium* subsp. *avium*. Vaccine. 2008;26:257–68.
25. Touchette MH, Holsclaw CM, Previti ML, Solomon VC, Leary JA, Bertozzi CR, Seeliger JC. The *rv1184c* Locus Encodes Chp2, an Acyltransferase in *Mycobacterium tuberculosis* Polyacyltrehalose Lipid Biosynthesis. J Bacteriol. 2015;197:201–10.
26. Belardinelli JM, Larrouy-Maumus G, Jones V, de Sorio de Carvalho LP, McNeil MR, Jackson M. Biosynthesis and Translocation of Unsulfated Acyltrehaloses in Mycobacterium tuberculosis. JBC. 2014;289:27952–65.
27. Delogu G, Pusceddu C, Bua A, Fadda G, Brennan MJ, Zanetti S. Rv1818c-encoded PE_PGRS protein of *Mycobacterium tuberculosis* is surface exposed and influences bacterial cell structure. Mol Microbiol. 2004;52:725–33.
28. Larrouy-Maumus G, Kovierová HS, Dhouib R, Angala SK, Zuberogoitia S, Pham H, et al. A Small Multidrug Resistance-like Transporter Involved in the

Arabinosylation of Arabinogalactan and Lipoarabinomannan in Mycobacteria. JBC. 2012;47:39933–41.
29. Lu T, Zhao X, Li X, Hansen G, Blondeau J, Drlica K. Effect of chloramphenicol, erythromycin, moxifloxacin, penicillin and tetracycline concentration on the recovery of resistant mutants of *Mycobacterium smegmatis* and *Staphylococcus aureus*. J Antimicrobial Chemother. 2003;52:61–4.
30. Singh G, Singh G, Jadeja D, Kaur J. Lipid hydrolizing enzymes in virulence: *Mycobacterium tuberculosis* as a model system. Crit Rev Microbiol. 2010;36:259–69.
31. Sultana R, Tanneeru K, Guruprasad L. The PE-PPE domain in mycobacterium reveals a Serine α/β Hydrolase fold and function: an in-silico analysis. PLoS One. 2011;6, e16745.
32. Shen G, Singh K, Chandra D, Serveau-Avesque C. maurin D, Canaan S, Singla R, Behera D, Laal S. LipC (Rv0220) is an immunogenic cell surface esterase of *Mycobacterium tuberculosis*. Infect Immun. 2012;80:243–53.
33. Deb C, Daniel J, Sirakova TD, Abomoelak B, Dubey VS, Kolattukudy PE. A novel lipase belonging to the Hormone-sensitive Lipase Family induced under starvation to utilize stored triacylglycerol in *Mycobacterium tuberculosis*. JBC. 2006;281:3866–75.
34. Cotes K, Bakala N'goma CJ, Dhouib R, Douchet I, Maurin D, Carrière F, Canaan S. Lipolytic enzymes in *Mycobacterium tuberculosis*. Appl Microbiol Biotechnol. 2008;78:741–9.
35. Snapper SB, Melton RE, Mustafa S, Kieser T, Jacobs Jr WR. Isolation and characterization of efficient plasmid transformation mutants of *Mycobacterium smegmatis*. Mol Microbiol. 1990;4:1911–9.
36. Sambrook J, Fitsch EF, Maniatis T. Molecular Cloning: A Laboratory Manual. New York: Cold Spring Harbor Press; 1989.
37. Parish T, Soker NG. Use of a flexible cassette method to generate a double unmarked *Mycobacterium tuberculosis tlyA plcABC* mutant by gene replacement. Microbiology. 2000;146:1969–75.
38. Cordone A, Mauriello EMF, Pickard DJ, Dougan DJ, De Felice M, Ricca E. The *lrp* Gene and Its Role in Type I Fimbration in *Citrobacter rodentium*. J Bacteriol. 2005;187:7009–17.

# 11

# *Bacillus subtilis* 5′-nucleotidases with various functions and substrate specificities

Ayako Terakawa[1], Ayane Natsume[1], Atsushi Okada[1], Shogo Nishihata[1], Junko Kuse[1], Kosei Tanaka[2], Shinji Takenaka[1,2], Shu Ishikawa[3] and Ken-ichi Yoshida[2,3*]

**Abstract**

**Background:** In *Escherichia coli*, *nagD*, *yrfG*, *yjjG*, *yieH*, *yigL*, *surE*, and *yfbR* encode 5′-nucleotidases that hydrolyze the phosphate group of 5′-nucleotides. In *Bacillus subtilis*, genes encoding 5′-nucleotidase have remained to be identified.

**Results:** We found that *B. subtilis ycsE*, *araL*, *yutF*, *ysaA*, and *yqeG* show suggestive similarities to *nagD*. Here, we expressed them in *E. coli* to purify the respective $His_6$-tagged proteins. YcsE exhibited significant 5′-nucleotidase activity with a broader specificity, whereas the other four enzymes had rather weak but suggestive activities with various capacities and substrate specificities. In contrast, *B. subtilis yktC* shares high similarity with *E. coli suhB* encoding an inositol monophosphatase. YktC exhibited inositol monophosphatase activity as well as 5′-nucleotidase activity preferential for GMP and IMP. The *ycsE*, *yktC*, and *yqeG* genes are induced by oxidative stress and were dispensable, although *yqeG* was required to maintain normal growth on solid medium. In the presence of diamide, only mutants lacking *yktC* exhibited enhanced growth defects, whereas the other mutants without *ycsE* or *yqeG* did not.

**Conclusions:** Accordingly, in *B. subtilis*, at least YcsE and YktC acted as major 5′-nucleotidases and the four minor enzymes might function when the intracellular concentrations of substrates are sufficiently high. In addition, YktC is involved in resistance to oxidative stress caused by diamide, while YqeG is necessary for normal colony formation on solid medium.

**Keywords:** *Bacillus subtilis*, Haloacid dehalogenase superfamily, Inositol monophosphatase, Inositol phosphate, Nucleoside/nucleotide metabolism, 5′-nucleotidase, Oxidative stress, Phosphatase, Protein motif

## Background

The pool sizes of nucleotides and nucleosides are balanced to enable the efficient synthesis of DNA and RNA [1]. Numerous enzymes involved in the biosynthesis and catabolism of nucleic acids are controlled to regulate the appropriate pool size of each compound. For example, 5′-nucleotidases hydrolyze 5′-nucleotides to generate nucleosides and inorganic phosphate, which participate in the regulatory mechanism that opposes the generation of nucleotides with the phosphorylation of nucleosides catalyzed by kinases.

* Correspondence: kenyoshi@kobe-u.ac.jp
[2]Organization of Advanced Science and Technology, Kobe University, 1-1 Rokkodai, Nada, Kobe, Hyogo 657-8501, Japan
[3]Department of Science, Technology and Innovation, Kobe University, 1-1 Rokkodai, Nada, Kobe, Hyogo 657-8501, Japan
Full list of author information is available at the end of the article

5′-Nucleotidases are ubiquitous among species and reside in different subcellular locations. Extracellular 5′-nucleotidases are produced by certain bacteria. For example, *Vibrio parahaemolyticus* NutA is a 5′-nucleotidase bound to the membrane by a lipid anchor [2]. *Escherichia coli* UshA is a periplasmic enzyme with 5′-nucleotidase and UDP-sugar hydrolase activities [3]. These 5′-nucleotidases degrade extracellular nucleotides to satisfy the cell's nutritional requirements.

In contrast, bacteria produce numerous intracellular 5′-nucleotidases that belong to various enzyme families. In *E. coli*, the substrates of NagD are UMP, GMP, AMP, and CMP [4]. NagD belongs to the haloacid dehalogenase superfamily (HADSF) characterized by a specific protein motif [5]. The HADSF family comprises numerous proteins in organisms ranging from prokaryotes to higher eukaryotes, including humans. The vast majority of enzymes of the HADSF family are phosphoryl transferases, although the superfamily was named after 2-haloacid dehalogenase, because it is the first structurally characterized member. When the amino acid sequences of their entire coding regions are compared, similarities among the HADSF members are not usually very high (15–30 % identical), although their central regions involved in catalytic activity are relatively conserved. Further, correlation between structure and catalytic activity is frequently observed and interrelates with similarities among the structures of substrates [6–8]. *E. coli* produces other HADSF-family enzymes such as YrfG, YjjG, YieH, and YigL, and each exhibits 5′-nucleotide phosphatase activity [9, 10].

In addition, *E. coli* produces SurE and YfbR, which exhibit 5′-nucleotidase activity and do not belong to the HADSF family [10]. SurE shows broad substrate specificity, by dephosphorylating 5′-nucleotides as well as 3′-nucleotides, with highest affinity for 3′-AMP. SurE hydrolyzes polyphosphate with preference for short-chain substrates. Homologs of *E. coli surE* are present in numerous eubacteria and archaea, and SurE represents a family of metal-dependent phosphatases [11]. YfbR belongs to the HD domain superfamily of metal-dependent phosphatases and phosphodiesterases, and the HD domain was named after a protein motif with predicted catalytic residues containing the conserved doublet His–Asp [12]. YfbR specifically hydrolyzes 5′-deoxyribonucleotides.

Inositol monophosphatase denotes phosphatases that liberate inorganic phosphate from *myo*-inositol 1-monophosphate (MIMP). The mammalian and plant enzymes are involved in the metabolism of inositol phospholipids, including the generation and degradation of inositol phosphates and phosphatidylinositols that mediate cellular signal transduction [13]. Inositol monophosphatases are present in diverse organisms such as bacteria and higher eukaryotes, and evolved from a common ancestral gene [14, 15]. Mammalian inositol monophosphatases exhibit relatively broad substrate specificity for phosphate-containing compounds [16, 17]. For example, the enzyme isolated from rat testis hydrolyzes adenosine 2′-monophosphate [16] and the bovine brain enzyme hydrolyzes β-glycerophosphate and adenosine 2′-monophosphate [17]. *E. coli suhB* encodes a protein homologous to eukaryotic inositol monophosphatases with equivalent activities [18].

However, phosphatidylinositol is not present in *E. coli* usually [19], and its physiological function in cells is unknown. In contrast, *E. coli* SuhB exhibits wider substrate specificity and hydrolyzes β-glycerophosphate and adenosine 2′-monophosphate as well [18]. Thus, we hypothesized that bacterial inositol monophosphatases might also function as 5′-nucleotidase. Moreover, various *E. coli* enzymes likely dephosphorylate 5′-nucleotides with different specificities and functions and may therefore act together to maintain the sizes of the intracellular pools of nucleotides and nucleosides.

In *Bacillus subtilis*, the partial nucleotide limitation induces 5′-nucleotidase activity [20], and *B. subtilis* is used for fermentative inosine production, which may involve the dephosphorylation of IMP catalyzed by an unidentified 5′-nucleotidase [21]. However, *B. subtilis* genes encoding 5′-nucleotidase have remained to be identified.

Here we selected *B. subtilis* genes potentially encoding 5′-nucleotidases [22] and expressed them in *E. coli* to characterize their function. We identified two major and four minor genes encoding 5′-nucleotidases, with various functions and substrate specificities. A major 5′-nucleotidase was involved in resistance to oxidative stress, and a minor enzyme was required for normal growth on solid medium.

## Results

### Selection of *B. subtilis* genes similar to *E. coli* genes encoding 5′-nucleotidase

In *E. coli*, at least *nagD*, *yrfG*, *yjjG*, *yieH*, *yigL*, *surE*, and *yfbR* encode 5′-nucleotidases, and their gene products are classified in the families as follows: HADSF (NagD, YrfG, YjjG, YieH, and YigL), SurE, and HD domain (YfbR). In the *B. subtilis* genome, no gene shares significant homology with *surE* or *yfbR*. However, gene products of *araL*, *yutF*, *yqeG*, *ysaA*, *ycsE*, *gapB*, *ftsA*, and *hprP* share some suggestive similarities with that of *E. coli nagD* (Fig. 1); AraL (65.8 % similarity/272 aa overlap),

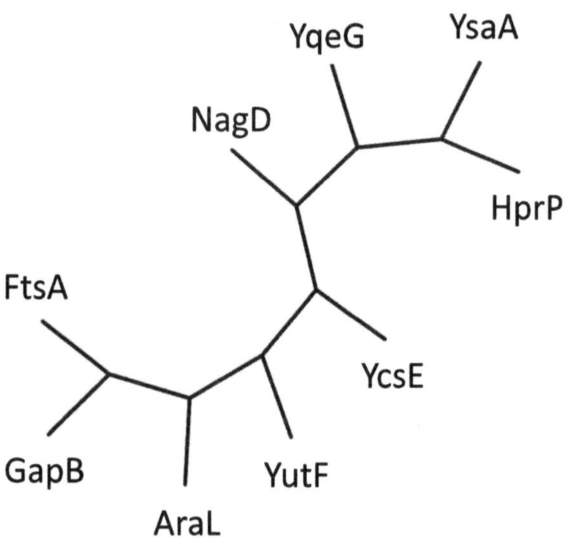


**Fig. 1** Similarities among the amino acid sequences of the proteins encoded by *E. coli nagD* homologs in the *B. subtilis* genome. The amino acid sequences of the proteins encoded by *B. subtilis araL, yutF, yqeG, ysaA, ycsE, gapB, ftsA*, and *hprP*, and *E. coli nagD* were analyzed using CLUSTALW (default fast/approximate settings [http://www.genome.jp/tools/clustalw/]) to generate the phylogenic tree

YutF (73.0 %/256 aa), YqeG (15.7 %/172 aa), YsaA (16.9 %/260 aa), YcsE (51.4 %/249 aa), GapB (6.8 %/340 aa), FtsA (9.5 %/440 aa), and HprP (15.7 %/216 aa). Among them, AraL, YutF, YqeG, YsaA, YcsE, and HprP were supposed to belong to HADSF, since their amino acid sequences contain the HADSF motif (http://pfam.xfam.org/family/PF00702.24).

The *araL* gene resides within the L-arabinose operon, which encodes enzymes required for the degradation of L-arabinose and is under the control of AraR repressor to be induced in the presence of L-arabinose [23]. AraL was predicted to be a phosphatase that hydrolyzes certain sugar-phosphates [23].

The *yutF* gene encodes a protein that was predicted to be involved in N-acetyl-glucosamine catabolism and similar to 4-nitrophenyl phosphatase [24]. The condition-dependent transcriptome analysis revealed that *yutF* was almost constitutively expressed at low levels [25].

The *yqeG*, *ysaA*, and *ycsE* genes encode a putative HADSF-family enzyme of unknown function. The *yqeG* gene is the first gene of the long operon comprising *yqeG*, *yqeH*, *aroD*, *yqeI*, *nadD*, *yqeK*, *yqeL*, and *yqeM*. This operon encodes essential genes [26], which are expressed constitutively at significant levels that increase during germination and under conditions of oxidative stress [25]. The *ysaA* and *ycsE* genes are constitutively expressed and are induced, respectively, during the nutritional shift from malate to malate plus glucose and in the presence of ethanol or diamide [25].

The products of *gapB*, *ftsA*, and *hprP* show relatively lower similarities to NagD (Fig. 1). GapB is NADP-dependent glyceraldehyde-3-phosphate dehydrogenase, which is involved in gluconeogenesis [27]. FtsA is an actin-like ATPase involved in cell division [28]. And HprP is a P-Ser-HPr phosphatase involved in carbon catabolite repression [29].

### 5′-Nucleotidase activity of *B. subtilis* enzymes sharing some similarities with NagD

Of the eight *B. subtilis* genes that share some similarities with *nagD* as described above, *gapB*, *ftsA*, and *hprP* have respective and specific functions other than as 5′-nucleotidases [27–29]. Therefore, these genes were excluded from the present study. We cloned and expressed *araL*, *yutF*, *yqeG*, *ysaA*, and *ycsE* in *E. coli* as $His_6$-tag fusion proteins.

The $His_6$-tag fusion proteins were purified to form respective single bands in sodium dodecyl sulfate polyacrylamide gel electrophoresis (SDS-PAGE, data not shown), and were subjected to phosphatase assays using the substrates as follows: AMP, CMP, GMP, IMP, UMP, and glucose 6-phosphate (G6P) (Fig. 2). G6P was included because it is the substrate of certain HADSF enzymes [9]. YcsE appeared to be the most efficient phosphatase acting on all six substrates at a lower concentration (2 mM), indicating its broad substrate specificity with preference for IMP, CMP, and G6P. Its activity increased in the presence of 10 mM substrates, particularly AMP, GMP, and UMP. The other four enzymes were relatively less active even at the higher substrate concentration. All 5′-nucleotides served almost equally as substrates for AraL at lower and higher concentrations, and AraL activity was enhanced particularly against G6P at the higher concentration. YutF exhibited purine 5′-nucleotidase activity with the substrates GMP and IMP, but only at the higher concentration, and did not hydrolyze pyrimidine nucleotides and G6P. YqeG hydrolyzed GMP and G6P only at the higher concentration. YsaA activity was weakest with purine-nucleotide substrates at the higher concentration.

The catalytic properties of YcsE are shown in Table 1. The substrate specificity of YcsE was broad with higher $K_M$ values, suggesting that it efficiently dephosphorylated these substrates at higher concentrations. In contrast, the properties of the other four enzymes were not determined because of their lower activities. Nevertheless, the data do not exclude their physiological function as 5′-nucleotidases, because the activities of two of the 5′-nucleotidases YieH and YigL identified previously in *E. coli* were too low to determine their enzymatic properties [9].

### 5′-Nucleotidase activity of SuhB homolog in *B. subtilis*

As described above, *E. coli* SuhB is an inositol monophosphatase, which exhibits wider substrate specificity, as it hydrolyzes β-glycerophosphate and adenosine 2′-monophosphate as well [18]. Thus, we hypothesized that bacterial inositol monophosphatases might also function as 5′-nucleotidase.

The *B. subtilis yktC* gene, which is the homolog of *E. coli suhB* (31.7 % similarity of 265 overlapping amino acid residues), was cloned and expressed as a $His_6$-tagged protein in *E. coli*. $His_6$-tagged YktC was purified to form a single band in SDS-PAGE (data not shown) and was subjected to phosphatase assays in the presence of various substrates (Fig. 2, Table 2). The inositol monophosphatase activity of YktC was indeed demonstrated by its efficient ability to hydrolyze MIMP, and it also turned out to exhibit 5′-nucleotidase activity preferentially against IMP and GMP as substrates. To our knowledge, this is the first identification of an inositol monophosphatase with 5′-nucleotidase activity. Compared with *E. coli* SuhB, YktC hydrolyzed G6P less efficiently, while β-glycerophosphate more efficiently [18]. Expression of *yktC* is mainly constitutive and is markedly enhanced by stressors such as diamide, ethanol, or high salt as well as at higher and lower temperatures [25].

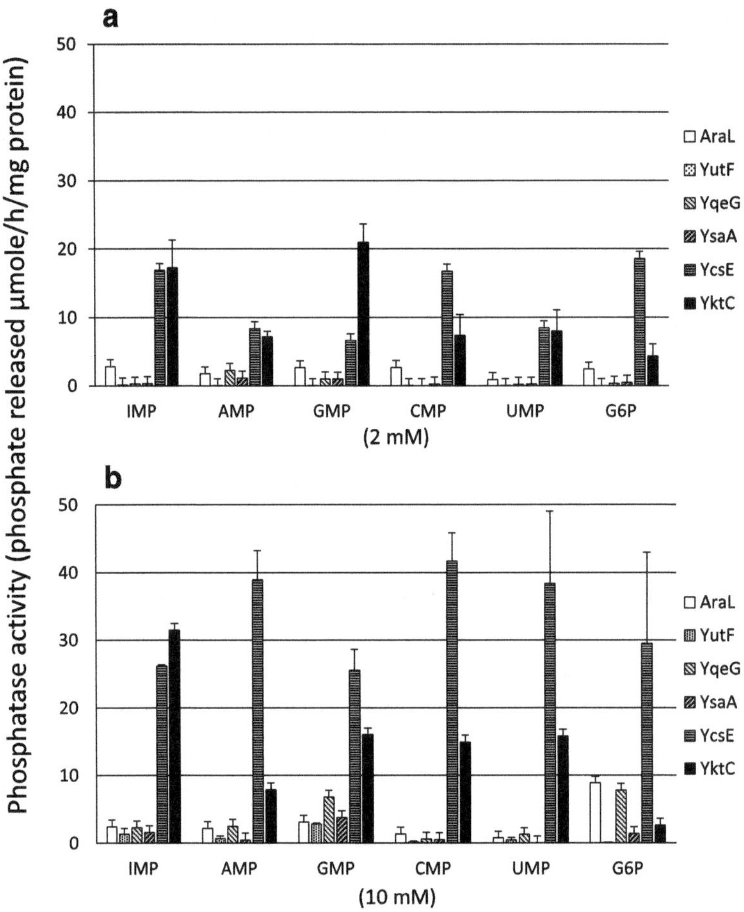


**Fig. 2** Phosphatase activities of *B. subtilis* proteins homologous to *E. coli* NagD. Each purified protein (0.25 mg/ml, indicated on the right) was incubated with 2 mM (**a**) or 10 mM (**b**) of various substrates for 12 h and assayed for phosphatase activity as described in Methods. All presented data are the mean values of three independent experiments ± SD

**Physiological functions of YqeG, YcsE, and YktC**

As mentioned above, *yqeG* expression is enhanced by oxidative stress, *ycsE* is induced in the presence of diamide or ethanol, and *yktC* is markedly enhanced by stressors such as diamide, ethanol, or high salt, as well as at higher and lower temperatures [25]. These findings inspired us to investigate the three genes encoding 5′-nucleotidase for their possible functional association with the response to oxidative stress.

We used conventional marker replacement to inactivate *ycsE* and *yktC*. The *yqeG* gene is the first of the long operon comprising *yqeG, yqeH, aroD, yqeI, nadD, yqeK, yqeL,* and *yqeM* [26]. At least *yqeH, yqeI,* and *nadD* are essential genes [26]. Therefore, we tried to introduce an in-frame deletion of *yqeG* to maintain the expression of the downstream essential genes. However, numerous attempts failed. Therefore, once an additional copy of *yqeG* was introduced into the *amyE* locus under the

**Table 1** Kinetic constants of YcsE

| Substrate | $K_M$ (mM)[a] | $V_{max}$ (µmol min$^{-1}$ mg protein$^{-1}$)[a] | $k_{cat}$ (s$^{-1}$)[a] |
|---|---|---|---|
| IMP | 3.2 ± 0.036 | 0.13 ± 0.060 | $3.7 \pm 1.4 \times 10^2$ |
| AMP | 8.6 ± 0.80 | 0.50 ± 0.029 | $2.3 \pm 0.13 \times 10^3$ |
| GMP | 10.3 ± 0.12 | 0.45 ± 0.12 | $2.0 \pm 0.55 \times 10^3$ |
| CMP | 4.7 ± 0.23 | 0.53 ± 0.18 | $3.4 \pm 0.011 \times 10^3$ |
| UMP | 8.5 ± 0.69 | 0.20 ± 0.072 | $9.6 \pm 3.5 \times 10^2$ |
| G6P | 22.2 ± 0.96 | 2.3 ± 0.48 | $1.06 \pm 0.22 \times 10^4$ |

[a]All presented data are the mean values of three independent experiments ± SD

**Table 2** Kinetic constants of YktC

| Substrate | $K_M$ (mM)[a] | $V_{max}$ (μmol $min^{-1}$ mg $protein^{-1}$)[a] | $k_{cat}$ ($s^{-1}$)[a] |
|---|---|---|---|
| MIMP | 0.076 ± 0.006 | 0.82 ± 0.16 | 2.9 ± 0.33 × $10^3$ |
| IMP | 1.1 ± 0.083 | 0.078 ± 0.0084 | 1.8 ± 0.59 × $10^2$ |
| AMP | 1.8 ± 0.13 | 0.030 ± 0.00022 | 5.3 ± 0.039 × $10^1$ |
| GMP | 1.9 ± 0.026 | 0.169 ± 0.028 | 3.0 ± 0.34 × $10^2$ |
| CMP | 1.6 ± 0.017 | 0.049 ± 0.014 | 1.5 ± 0.52 × $10^2$ |
| UMP | 2.4 ± 0.052 | 0.11 ± 0.0099 | 9.8 ± 4.6 × $10^2$ |
| G6P | 2.6 ± 0.51 | 0.021 ± 0.010 | 3.7 ± 1.3 × $10^1$ |
| β-glycerophosphate | 0.49 ± 0.035 | 1.2 ± 0.32 | 5.7 ± 1.2 × $10^2$ |

[a]All presented data are the mean values of three independent experiments ± SD

control of *spac* promoter, which is conditionally induced by IPTG. And then, in the presence of IPTG, the existing *yqeG* gene was deleted using marker replacement so that the genes downstream were constitutively expressed by read-through from the kanamycin-resistance cassette (strain NON05). Regardless of the presence and absence of IPTG, strain NON05 grew in liquid medium (Fig. 3), suggesting that *yqeG* might be dispensable in planktonic growth. However, the *spac* promoter is somewhat leaky; a significant basal level of expression still exists in the absence of IPTG [30]. Therefore, in the absence of IPTG, there might be some production of YqeG, and this might be sufficient for planktonic growth. Nevertheless, NON05 formed fewer colonies in the absence of IPTG (Fig. 4), suggesting that *yqeG* was required to maintain normal growth particularly on solid medium. Similarly, less efficient colony-formation was exhibited by the triple mutant NON06 in the absence of IPTG, and colony formation was slightly decreased when *ycsE* and *yktC* were inactivated additionally.

The addition of diamide to liquid medium arrested the growth of the parental strain 168 (Fig. 3). The growth characteristics of the mutant NON01 lacking *ycsE* were comparable to those of the parental strain in the presence of diamide. Conversely, the mutants NON02 and NON03 lacking *yktC* exhibited the enhanced growth defect in the presence of the higher concentration of diamide (4 mM), as the optical density for cells decreased significantly possibly because of cell lysis. The results suggest that *yktC*, but not *ycsE*, contributed to resistance to oxidative stress induced by diamide. In contrast, NON05 did not exhibit significantly increased sensitivity to diamide, depending on the repression of *yqeG* in the absence of IPTG (Fig. 3). The triple-mutant NON06 exhibited the growth defect similar to that of NON02 or NON03 in the presence of diamide (Fig. 3), and the growth of NON06 did not depend on IPTG. Therefore, the conditional expression of *yqeG* was not associated with the susceptibility to diamide. Taken together, the data indicate that only YktC was involved in the resistance to oxidative stress.

## Discussion

In eukaryotes, multiple nucleotidases and nucleoside kinases control the intracellular concentration of nucleotides [31–33]. Nucleotidases play a role in the maintenance of a constant nucleoside supply, which is required to maintain RNA metabolism and biosynthesis of cell wall [1]. In bacteria, numerous nucleotidases with different substrate specificities control the intracellular nucleotide pool [34], and it is believed that the accumulation of intracellular nucleotides is potentially toxic and therefore must be prevented [31–33].

Here we show that *B. subtilis* YktC and YcsE functioned as 5′-nucleotidases, suggesting that they may represent two of the main enzymes involved in intracellular nucleoside/nucleotide metabolism. The other enzymes similar to *E. coli* NagD, which belong to the HADSF family (YutF, YsaA, YqeG, and AlaL), exhibited lower activity, although the data suggest they may function as minor 5′-nucleotidases. HADSF phosphatases often have broader substrate specificity and hydrolyze various phosphorylated compounds [5]. Thus, 5′-nucleotidases with lower substrate affinity might contribute to the balance of intracellular nucleotide concentrations as well as that of other phosphorylated metabolites. The $K_M$ values of *E. coli* NagD for UMP, CMP, GMP, AMP, and G6P are 0.16 ± 0.038, 1.47 ± 0.044, 0.40 ± 0.130, 0.84 ± 0.250, and 5.90 ± 0.750 mM, respectively [4], which are smaller than those of YcsE and YktC (Tables 1 and 2, respectively). Therefore, *B. subtilis* YktC and YcsE might be less efficient as 5′-nucleotidase than NagD, but it is still suggestive that they may be able to function as two of the 5′-nucleotidases that are the first to respond to the accumulation of nucleotides, and the other enzymes may function when the concentration of nucleotides is in far excess.

Some of the genes encoding 5′-nucleotidases might be induced upon partial nucleotide limitation as previously reported [20]. Interestingly, the expression of *yqeG* is enhanced under oxidative stress, that of *ycsE* is induced by diamide or ethanol, and that of *yktC* is markedly enhanced in the presence of the various stressors, including

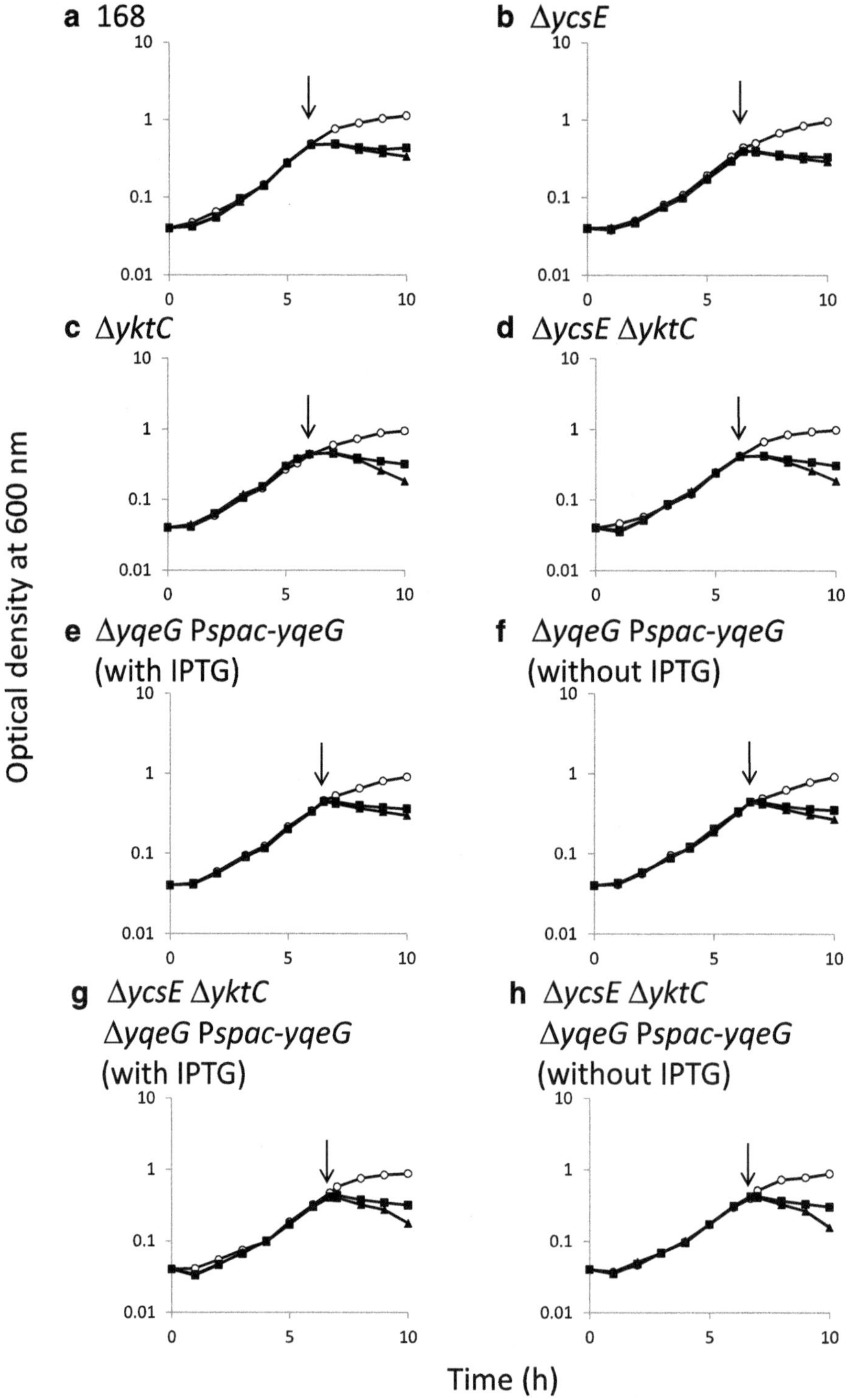


**Fig. 3** Growth curves of *B. subtilis* strains. Strains 168 (**a**), NON01 (**b** *ΔycsE*), NON02 (**c** *ΔyktC*), NON03 (**d** *ΔycsE ΔyktC*), NON05 (**e** and **f**, *ΔyqeG Pspac-yqeG*), and NON06 (**g** and **h**, *ΔycsE ΔyktC ΔyqeG Pspac-yqeG*) were inoculated into liquid medium and their growth was monitored. At the times indicated by the arrowheads, diamide was added to final concentrations of 0 mM (open circle), 1 mM (solid square), and 4 mM (solid triangle). Strains NON05 and NON06 were grown in the presence (**e** and **g**) and absence (**f** and **h**) of 1 mM IPTG. All experiments were repeated more than three times and similar results were observed

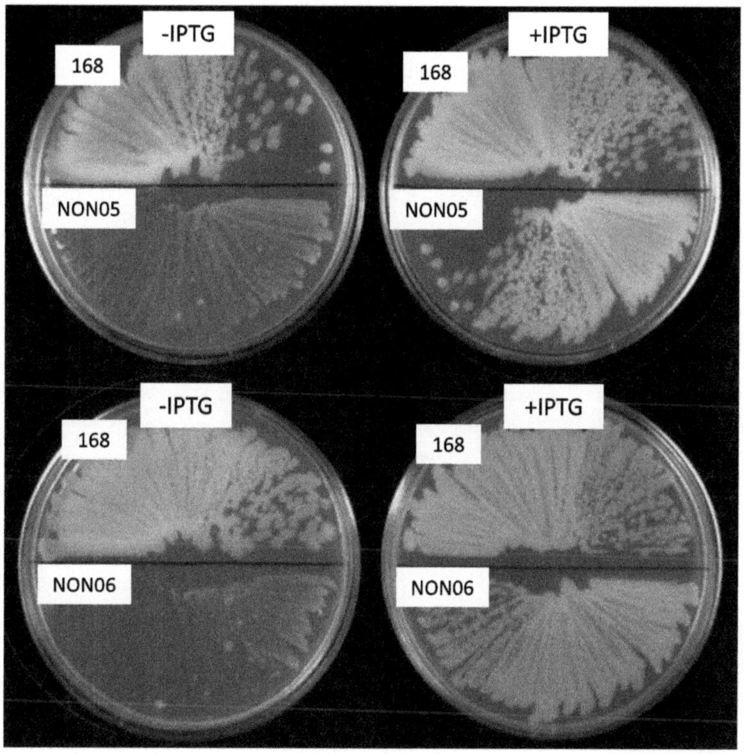


**Fig. 4** Colony formation by *B. subtilis* strains. Suspensions of strains 168, NON05, and NON06 were streaked on LB plates with (right) or without (left) 1 mM IPTG and incubated overnight at 37 °C. All experiments were repeated more than three times and similar results were observed

diamide, ethanol, and high salt as well as at higher and lower temperatures [25]. These findings indicate that the intracellular levels of phosphorylated metabolites and nucleotides including 5′-nucleotides may be altered in the presence of these stressors, in particular oxidative stress. Our results show that inactivation of *yktC* led to increased sensitivity to diamide (Fig. 4). When the expression of *yqeG* was artificially modified, the cells grew normally in liquid medium and did not exhibit increased sensitivity to diamide, depending on repression of *yqeG* in the absence of IPTG (Fig. 4). Further, in the presence of diamide, the triple mutant NON06, lacking *ycsE* and *yktC* and with repressed *yqeG*, exhibited a growth defect similar to that of NON02 or NON03, which lacks *yktC* (Fig. 4). Therefore, we conclude that only YktC was involved in resistance to diamide-induced oxidative stress. However, further studies are required to elucidate how its enzymatic functions, including its 5′-nucleotidase activity, are related to resistance to the oxidative stress.

*B. subtilis* induces S-cysteinylation to protect protein thiols after exposure to diamide, because this bacterium does not produce thiols such as glutathione [35]. To our knowledge, enzymes with 5′-nucleotidase activity have not been demonstrated to be involved in resistance to oxidative stress. In response to oxidative stress, *Pseudomonas fluorescens* evokes a metabolic adaptation to increase NADPH synthesis and to decrease NADH production, accompanied by the induction and repression, respectively, of $NAD^+$ kinase and $NADP^+$ phosphatase, which regulate the levels of $NAD^+$ and $NADP^+$ [36]. Namely, oxidative stress represses the $NADP^+$ phosphatase in *P. fluorescens* [36], whereas it induces *yktC* in *B. subtilis* [25]. If YktC might function as $NADP^+$ phosphatase, it could decrease NADPH synthesis under oxidative stress, which is contradictory to the physiological demand. Therefore, YktC is unlikely to act as a $NADP^+$ phosphatase, and we did not determine if $NADP^+$ was a substrate of the phosphatase activity of YktC. On the other hand, *yqeG* is the first gene of a long operon comprising essential genes such as *nadD* that encodes nicotinamide-nucleotide adenylyltransferase required for *de novo* biosynthesis of $NAD^+$ and $NADP^+$ [37]. The induction of *yqeG* by oxidative stress, together with the other members of the operon, might be relevant to the enhanced NADPH synthesis as found in *P. fluorescens*. However, our results indicate that *yqeG* was dispensable, except for normal colony formation on solid medium (Fig. 4). We were therefore unable to determine the physiological function of *yqeG*.

In mammalian and plant cells, inositol monophosphatases play important roles in the metabolism of inositol polyphosphates and phosphatidylinositols [13, 38]. *De novo* phosphatidylinositol synthesis in mammals and *Saccharomyces cerevisiae* involves a reaction that combines CDP-diacylglycerol with *myo*-inositol. *myo*-Inositol is produced through the isomerization of glucose 6-phosphate to form MIMP [39] and the subsequent dephosphorylation of MIMP by inositol monophosphatase [16]. Further, MIMP is produced by the dephosphorylation of the second messenger inositol polyphosphates [40]. In *E. coli* and *B. subtilis*, however, phosphatidylinositols are not components of phospholipids usually present in the cell membrane [41, 42], and inositol 1-phosphate synthase has not been identified. Therefore, it is unlikely that MIMP occurs naturally in these bacteria and might serve as the physiological substrate of *E. coli* SuhB and *B. subtilis* YktC. In *E. coli*, a decrease in intracellular SuhB levels compensated the defects in *secY* and *rpoH* [43, 44]. The *secY* gene encodes a membrane component involved in protein secretion [45], and *rpoH* encodes $\sigma^{32}$, a component of the RNA polymerase involved in heat-shock induction [46]. Therefore, defects in *secY* and *rpoH* impair secretion and the heat-shock response, respectively, which are involved in posttranslational quality control of proteins [18]. Therefore, SuhB may control the rates of peptide chain elongation and protein folding [42]. In contrast, the SuhB homolog of *P. aeruginosa* plays an important role in pathogenesis to control the genes required for acute and chronic infection [47]. *Burkholderia cepacia* SuhB is required for the secretion of proteins associated with motility and biofilm formation [48], suggesting that the physiological functions of SuhB homologs are complex, which may be true for *B. subtilis* YktC. Nevertheless, it is intriguing that MIMP is the best substrate for these enzymes. Kozloff and the colleagues detected a small amount of phosphatidylinositol from *E. coli* cells, implying the possibility of a weak phosphatidylinositol biosynthesis under some specific conditions [49]. *E. coli* SuhB and *B. subtilis* YktC might be involved in bacterial metabolism of phosphatidylinositol, which has not been characterized yet.

The broad substrate specificity of YktC revealed here implies the association of this enzyme with sugar-phosphate stress. Although sugars serve as energy and carbon sources, sugar phosphates are produced during their metabolism. Excess accumulation of sugar phosphates impairs cell growth [50] and may trigger cell death [51, 52] through an unknown mechanism. Similarly, accumulation of nucleotides is detrimental to the cell [34]. Therefore, by hydrolyzing sugar phosphates as well as nucleotides, YktC may play a role in correcting imbalances of metabolites. For example, we found that the $K_M$ value of YktC for G6P was $2.6 \pm 0.51$ mM, which is greater than the reported intracellular concentrations of glucose 6-phosphate in *E. coli* (0.8–2.0 mM) [53]. However, under certain conditions, the intracellular concentration of G6P in *Lactococcus lactis* reaches 20–50 mM [54], indicating that G6P accumulates to higher levels in bacterial cells. Accordingly, in the presence of excess intracellular concentrations of G6P, YktC may hydrolyze it to intervene in glucose metabolism. This may be true as well for YcsE, because it hydrolyzed G6P in vitro more efficiently than YktC (Fig. 2). Similarly, YktC may hydrolyze β-glycerophosphate, which is a biosynthetic precursor of phospholipids [55], to interfere with the maintenance of the cell membrane. Therefore, their phosphatase activities with broad-specificity may exert pleiotropic effects on metabolism and other cellular functions. It would be worthwhile to perform metabolomic analyses on the mutants lacking *yktC* and *ycsE* to determine concentrations of the various metabolites, including 5′-nucleotides and other sugar phosphates, which would reinforce the hypothesis concerning the role of these enzymes.

Together with sodium glutamate, IMP and GMP are generally used as food additives to enhance taste. Industrial production of IMP is achieved using enzymatic conversion of inosine into IMP. *B. subtilis* strains producing large amounts of inosine in the fermentation medium were generated [56–60], and 5′-nucleotidase activity of these strains is remarkably increased, which may convert IMP into inosine [59]. Thus, optimizing the expression of genes encoding 5′-nucleotidases such as those studied here may further improve the fermentation of inosine production.

## Conclusions

*B. subtilis ycsE*, *araL*, *yutF*, *ysaA*, and *yqeG* show suggestive similarities to *E. coli nagD* encoding 5′-nucleotidase. Among the five, only YcsE exhibited significant 5′-nucleotidase activity with a broader specificity, whereas the other four enzymes had rather weak but suggestive activities with various capacities and substrate specificities. Interestingly, YqeG was required to maintain normal growth on solid medium. On the other hand, *B. subtilis yktC* shares high similarity with *E. coli suhB* encoding inositol monophosphatase, the gene product of which had 5′-nucleotidase activity preferential for GMP and IMP, and turned out to be involved in resistance to oxidative stress.

## Methods

### Bacterial strains, plasmids, oligonucleotide primers, and culture conditions

Bacterial strains and plasmids are listed in Table 3. Bacterial strains were usually grown aerobically at 37 °C and maintained in Luria-Bertani (LB) medium (Becton Dickinson, NJ, USA) containing 50 mg/l ampicillin and 50 mg/l kanamycin as required. Oligonucleotide primers are listed in Table 4. To assess the resistance to oxidative stress of *B. subtilis* strains, cells were inoculated into LB and allowed to grow. The oxidizing agent diamide was

**Table 3** Bacterial strains and plasmids

| Strain and plasmid | Description | Source or reference |
|---|---|---|
| *B. subtilis* | | |
| 168 | *trpC2* | Laboratory stock |
| NON01 | *trpC2 ycsE::spc* | This study |
| NON02 | *trpC2 yktC::cat* | This study |
| NON03 | *trpC2 ycsE::spc yktC::cat* | This study |
| NON04 | *trpC2 amyE::(Pspac-yqeG erm)* | This study |
| NON05 | *trpC2 amyE::(Pspac-yqeG erm)* Δ*yqeG::kan* | This study |
| NON06 | *trpC2 ycsE::spc yktC::cat amyE::(Pspac-yqeG lacI erm)* Δ*yqeG::kan* | This study |
| TMO310 | *trpC2 aprE::(spec lacI* Pspac-*mazF)* | [62] |
| TMO311 | *trpC2 aprE::(kan lacI* Pspac-*mazF)* | [62] |
| *E. coli* | | |
| DH5α | $F^-$ Φ80*lacZ*ΔM15 Δ(*lacZYA-argF*) *U169 recA1 endA1 hsdR17*($rK^-$, $mK^+$) *phoA supE44* $\lambda^-$ *thi-1 gyrA96 relA1* | Laboratory stock |
| BL21(DE3) | $F^-$ *ompT gal dcm lon* $hsdS_B(r_B^-, m_B^-)$ λ(DE3 [*lacI lacUV5-T7p07 ind1 sam7 nin5*]) $[malB^+]_{K\text{-}12}(\lambda^S)$ | Laboratory stock |
| Plasmids | | |
| pCRE-test | *cat* | [63] |
| pET28b | *kan* | Takara Bio |
| pET28b-araL | *kan, araL* | This work |
| pET28b-yutF | *kan, yutF* | This work |
| pET28b-yqeG | *kan, yqeG* | This work |
| pET28b-ysaA | *kan, ysaA* | This work |
| pET28b-ycsE | *kan, ycsE* | This work |
| pET28b-yktC | *kan, yktC* | This work |
| pMD20-araL | *amp* | This work |
| pMD20-yutF | *amp* | This work |
| pMD20-yqeG | *amp* | This work |
| pMD20-ysaA | *amp* | This work |
| pMD20-ycsE | *amp* | This work |
| pMD20-yktC | *amp* | This work |
| pMutin2 | *erm lacI* | [64] |

added to the exponentially growing culture at various concentrations, and growth was continuously monitored.

### Construction of bacterial strains

*E. coli* strains expressing *B. subtilis* genes were constructed as follows: DNA fragments corresponding to the open reading frames of genes of interest were PCR-amplified using *B. subtilis* strain 168 chromosomal DNA (Table 3) as template and pairs of respective oligonucleotide primers with generation of unique restriction sites at their 5′-termini (araL-F/araL-R for *araL*, ycsE-F/ycsE-R for *ycsE*, yktC-F/yktC-R for *yktC*, yqeG-F/yqeG-R for *yqeG*, ysaA-F/ysaA-R for *ysaA*, and yutF-F/yutF-R for *yutF*) (Table 4). Each fragment was cloned into the pMD20-T plasmid using the Mighty TA-cloning kit (Takara Bio, Otsu, Japan) to yield pMD20-araL, pMD20-ycsE, pMD20-yktC, pMD20-yqeG, pMD20-ysaA, and pMD20-yutF, which were used to transform *E. coli* strain DH5α. After performing nucleotide sequencing to confirm the constructs, each recombinant plasmid was digested with the appropriate restriction enzyme(s) to prepare a DNA fragment corresponding to its respective open reading frame and then ligated to the arms of pET28b (Takara Bio) digested with the same enzyme(s). The plasmids pET28b-araL, pET28B-ycsE, pET28B-yktC, pET28B-yqeG, pET28B-ysaA, and pET28B-yutF were constructed to express each protein as a C-terminal $His_6$-fusion, and expression was controlled by the T7 promoter in *E. coli* strain BL21(DE3) (Takara Bio).

Mutant strains of *B. subtilis* lacking *ycsE* and *yktC* were constructed using conventional marker replacement as follows (Additional file 1: Figure S1): To delete

**Table 4** Oligonucleotide primers

| Name | Sequence (5′–3′)[a] | Restriction site |
|---|---|---|
| araL-F | gggaattc<u>catatg</u>cgtattatggccagtcatgat | NdeI |
| araL-R | ccg<u>ctcgag</u>tatcagaatcccctcctcagc | XhoI |
| yutF-F | cgc<u>ggatcc</u>atgaaaacatataaagggtattta | BamHI |
| yutF-R | ccc<u>aagctt</u>aatgtatggaatccattcagt | HindIII |
| ysaA-F | cgc<u>ggatcc</u>atgaaagccgtatttttgattta | BamHI |
| ysaA-R | ccg<u>ctcgag</u>tttctctaatataggaaacaa | XhoI |
| YqeG-F | gggaatc<u>catatg</u>ttaaaaaagtttttt | NdeI |
| YqeG-R | ccc<u>aagctt</u>ttactcctcccactgaat | HindIII |
| ycsE-F | gggaattc<u>catatg</u>tctgtccaaagagaagatgta | NdeI |
| ycsE-R | ccg<u>ctcgag</u>tagtacccaatggcgaatcgcttta | XhoI |
| yktC-F | cgc<u>ggatcc</u>atgacaaattggacggaaatcgat | BamHI |
| yktC-R | ccc<u>aagctt</u>cttccgagcatgaagatactc | HindIII |
| ycsE-1 | gtttattgatggcgtgac | |
| ycsE-2 | ctgtgtagccttgagagtgatgaacagacatatatgtacctctc | |
| ycsE-3 | tcactctcaaggctacacag | |
| ycsE-4 | cgcaagcttcaaaaattatatggag | |
| ycsE-5 | ctccatataatttttgaagcttgcggggtactataaaaaaagagagtcc | |
| ycsE-6 | aacacttcatttgcggtc | |
| yktC-1 | gttcgcctagagctgtaagcttc | |
| yktC –2 | gccgatgataagctgtcaaagcaatctcatcgatttccgtcc | |
| yktC –3 | tttgacagcttatcatcggc | |
| yktC –4 | ttataaaagccagtcattaggc | |
| yktC –5 | gcctaatgactggcttttataagtgctagccggaaacccatc | |
| yktC –6 | gaacatctatcaccagacgacatc | |
| yqeG-d1 | gacataatgcgccatttgatggttg | |
| yqeG-d2 | gcttgagtcaattccgctgtcgctcaaacgaaaagggcacattcag | |
| yqeG-d3 | cgacagcggaattgactcaagc | |
| yqeG-d4 | cgcaagcttacgataaacccagc | |
| yqeG-d5 | gctgggtttatcgtaagcttgcgcaaaacccgcacccttccc | |
| yqeG-d6 | caggcagagcatatggatacg | |
| yqeG-s1 | ccttccagggtatgtttctc | |
| yqeG-s2 | gatactgcactatcaacacactctttcgacatggatgagcgatgatg | |
| yqeG-s3 | aagagtgtgttgatagtgcagtatc | |
| yqeG-s4 | caacaagctggggatccg | |
| yqeG-s5 | cggatccccagcttgttgcgacaatcattttgaaagaaagaaaaaggg | |
| yqeG-s6 | gatgacctcgtttccaccggcaaccttttccatttcttactcctcc | |
| yqeG-s7 | ccggtggaaacgaggtcatc | |
| yqeG-s8 | cacaaattaaaaactggtctgatcggcctaactcacattaattgcgttg | |
| yqeG-s9 | cgatcagaccagtttttaatttgtg | |
| yqeG-s10 | ttaacaaaattctccagtcttcacatcg | |

[a]Restriction enzyme cleavage sites are underlined

*ycsE*, two PCR fragments (each approximately 700-bp) corresponding to the upstream and downstream regions of *ycsE* were amplified from the DNA of *B. subtilis* strain 168 using the primer pairs ycsE-1/ycsE-2 and ycsE-5/ycsE-6, respectively (Table 4). A PCR fragment containing the spectinomycin-resistance gene cassette was amplified from the DNA of strain TMO310 (Table 3) using the primer pair ycsE-3/ycsE-4 (Table 4). The three PCR fragments were ligated using recombinant PCR with the primer pair ycsE-1/ycsE-6 to generate a single PCR fragment comprising the spectinomycin-resistance gene cassette flanked by the upstream and downstream regions of *ycsE*. *B. subtilis* strain 168 was transformed with the recombinant PCR fragment to confer resistance to spectinomycin, and a transformant with correct marker replacement was selected and designated strain NON01 (Table 3). To delete *yktC*, upstream and downstream regions of *yktC* were amplified using the primer pairs yktC-1/yktC-2 and yktC-5/yktC-6, respectively (Table 4). The chloramphenicol-resistance cassette was amplified from pCRE-test using the primer pair yktC-3/yktC-4 (Table 4). The three PCR fragments were ligated using recombinant PCR with the primer pair of yktC-1/yktC-6 to generate a single PCR fragment, which was integrated into chromosome of *B. subtilis* strain 168 to yield strain NON02 (Table 3). Strain NON01 was further transformed with NON02 DNA to confer resistance to chloramphenicol and generate strain NON03 lacking both *ycsE* and *yktC* (Table 3).

Another mutant that conditionally expressed *yqeG* was constructed as follows (Additional file 1: Figure S1). Five DNA fragments, corresponding to the C-terminal region of *amyE*, the stretch containing the erythromycin-resistance cassette and the *spac* promoter of pMutin2 (Table 3), the coding region of *yqeG*, the *lacI* gene of pMutin2, and the N-terminal region of *amyE*, were PCR-amplified using the primer pairs yqeG-s1/yqeG-s2, yqeG-s3/yqeG-s4, yqeG-s5/yqeG-s6, yqeG-s7/yqeG-s8, and yqeG-s9/yqeG-s10, respectively (Table 4). Using recombinant PCR with the primer pair yqeG-s1/yqeG-s10, all these five fragments were ligated in the order as described above and then integrated into the *amyE* locus of strain 168 to generate the erythromycin-resistant strain NON04 harboring an additional IPTG-inducible copy of *yqeG*.

Next, the endogenous *yqeG* gene of NON04 was deleted using marker replacement. Two DNA fragments comprising the upstream and downstream regions of *yqeG* (approximately 700 bp each) were amplified using the primer pairs yqeG-d1/yqeG-d2 and yqeG-d5/yqeG-d6, respectively (Table 4). The kanamycin-resistance cassette was amplified from the DNA of strain TMO311 using the primer pair yqeG-d3/yqeG-d4 (Table 4). Recombinant PCR was used to ligate together the three fragments using the primer pair of yqeG-d1/yqeG-d6t, and the ligation product was integrated into the chromosome of NON04 to yield strain NON05 in which expression of the additional *yqeG* was induced with 1 mM IPTG (Table 3). In NON05, *yqeG* was replaced with the kanamycin-resistance cassette so that the rest of operon comprising *yqeH, aroD, yqeI, nadD, yqeK, yqeL*, and *yqeM*, which reside downstream of *yqeG*, are constitutively expressed by read-through from the kanamycin-resistance gene. NON05 was transformed further with NON03 DNA to confer resistance to both chloramphenicol and spectinomycin inactivating *yktC* and *ycsE*, respectively (strain NON06).

Chromosomal DNAs of mutant strains of *B. subtilis* were subjected to diagnostic PCR experiments with the respective primer pairs used for recombinant PCR, and their correct construction was confirmed by appearance of the respective PCR fragments with altered length (Additional file 1: Figure S1).

### Purification of gene products produced in *E. coli*

*E. coli* strain BL21(DE3) bearing the pET28b derivatives were inoculated into 600 ml of LB medium containing kanamycin and were cultured at 37 °C with shaking at 180 rpm. When the optical density of the cultures reached 0.4–0.6, IPTG was added to a final concentration of 1 mM, the cells were cultivated further for 5 h, harvested using centrifugation, and resuspended into 10.8 ml of NP buffer (50 mM $NaH_2PO_4$, pH 8.0, and 300 mM NaCl). The suspension was chilled in an ice bath, mixed with 1.2 ml of 10 mg/ml lysozyme in NP buffer, incubated for 30 min, and sonicated to disrupt the cells. After centrifugation, the supernatant was subjected to $His_6$-tag affinity purification using TALON Metal Affinity Resins (Takara Bio). The $His_6$-tagged proteins were purified in a reaction buffer containing 1 mM $NH_4Cl$, 250 mM KCl, 10 mM $MgCl_2$, and 50 mM Tris–HCl (pH 7.8), and subjected to SDS-PAGE followed by Coomassie brilliant blue staining.

### Enzyme assay

The phosphatase assay was performed as previously described [61]. Each purified enzyme (0.25 mg/ml) was incubated with substrate in 50 μl of reaction buffer containing 1 mM $NH_4Cl$, 250 mM KCl, 10 mM $MgCl_2$, and 50 mM Tris–HCl (pH 7.8) at 37 °C, and the reaction was terminated using 50 μl of 10 % trichloroacetic acid. After centrifugation, the supernatant was mixed with 20 μl of solution A (10 % ascorbic acid in 2.25 M $H_2SO_4$) and 20 μl of solution B (2.4 % ammonium molybdate and 1 mg/ml potassium antimonyl tartrate in 2.25 M $H_2SO_4$). The mixture was incubated at room temperature for 10 min, and absorbance at 820 nm was measured to calculate the concentration of the released phosphate ion with reference to a standard curve made with various concentrations of inorganic phosphate. Trace of phosphate released in the reaction mixtures in the absence of purified enzyme was subtracted as background.

**Abbreviations**

G6P: Glucose 6-phosphate; HADSF: Haloalkane acid dehalogenase superfamily; IPTG: Isopropyl β-D-1-thiogalactopyranoside; MIMP: *Myo*-inositol 1-monophosphate; SDS-PAGE: Sodium dodecyl sulfate polyacrylamide gel electrophoresis

**Acknowledgements**

We thank Mr. Kentaro Kojima for his technical contributions in the earlier stage of this work.

**Funding**

This work was supported by the Ministry of Education, Culture, Sports, Science and Technology, Japan and in part by the Special Coordination Funds for Promoting Science and Technology, Creation of Innovative Centers for Advanced Interdisciplinary Research Areas; the Advanced Low-Carbon Technology Research and Development Program; KAKENHI (26660067); and Grants-in-Aid from the NC-CARP project.

**Authors' contributions**

AT conducted most of the experiments and analyzed the results. The enzymatic experiments were performed under the supervision of KT and ST. KT, AN, AO, SN, and JK conducted experiments with the mutant strains of *B. subtilis*. KY conceived the idea for the project and wrote the final manuscript with SI. All authors read and approved the final manuscript.

**Competing interests**

The authors declare that the research was conducted in the absence of any commercial or financial relationships that could be construed as a potential conflict of interest.

**Author details**

[1]Department of Agrobioscience, Kobe University, 1-1 Rokkodai, Nada, Kobe, Hyogo 657-8501, Japan. [2]Organization of Advanced Science and Technology, Kobe University, 1-1 Rokkodai, Nada, Kobe, Hyogo 657-8501, Japan. [3]Department of Science, Technology and Innovation, Kobe University, 1-1 Rokkodai, Nada, Kobe, Hyogo 657-8501, Japan.

**References**

1. Rampazzo C, Ferraro P, Pontarin G, Fabris S, Reichard P, Bianchi V. Mitochondrial deoxyribonucleotides, pool sizes, synthesis, and regulation. J Biol Chem. 2004;279:17019–26.
2. Tamao Y, Noguchi K, Sakai-Tomita Y, Hama H, Shimamoto T, Kanazawa H, Tsuda M, Tsuchiya T. Sequence analysis of *nutA* gene encoding membrane-bound Cl(–)-dependent 5′-nucleotidase of *Vibrio parahaemolyticus*. J Biochem. 1991;109:24–9.
3. Burns DM, Beacham IR. Nucleotide sequence and transcriptional analysis of the *E. coli ushA* gene, encoding periplasmic UDP-sugar hydrolase (5′-nucleotidase): regulation of the *ushA* gene, and the signal sequence of its encoded protein product. Nucleic Acids Res. 1986;14:4325–42.
4. Tremblay LW, Dunaway-Mariano D, Allen KN. Structure and activity analyses of *Escherichia coli* K-12 NagD provide insight into the evolution of biochemical function in the haloalkanoic acid dehalogenase superfamily. Biochemistry. 2006;45:1183–93.
5. Koonin EV, Tatusov RL. Computer analysis of bacterial haloacid dehalogenases defines a large superfamily of hydrolases with diverse specificity: Application of an iterative approach to database search. J Mol Biol. 1994;244:125–32.
6. Allen KN, Dunaway-Mariano D. Phosphoryl group transfer: evolution of a catalytic scaffold. Trends Biochem Sci. 2004;29:495–503.
7. Lu Z, Dunaway-Mariano D, Allen KN. HAD superfamily phosphotransferase substrate diversification: structure and function analysis of HAD subclass IIB sugar phosphatase BT4131. Biochemistry. 2005;44:8684–96.
8. Shin DH, Roberts A, Jancarik J, Yokota H, Kim R, Wemmer DE, Kim SH. Crystal structure of a phosphatase with a unique substrate binding domain from *Thermotoga maritime*. Protein Sci. 2003;12:1464–72.
9. Kuznetsova E, Proudfoot M, Gonzalez CF, Brown G, Omelchenko MV, Borozan I, Carmel L, Wolf YI, Mori H, Savchenko AV, Arrowsmith CH, Koonin EV, Edwards AM, Yakunin AF. Genome-wide analysis of substrate specificities of the *Escherichia coli* haloacid dehalogenase-like phosphatase family. J Biol Chem. 2006;281:36149–61.
10. Proudfoot M, Kuznetsova E, Brown G, Rao NN, Kitagawa M, Mori H, Savchenko A, Yakunin AF. General enzymatic screens identify three new nucleotidases in *Escherichia coli*. Biochemical characterization of SurE, YfbR, and YjjG. J Biol Chem. 2004;279:54687–94.
11. Lee JY, Kwak JE, Moon J, Eom SH, Liong EC, Pedelacq JD, Berendzen J, Suh SW. Crystal structure and functional analysis of the SurE protein identify a novel phosphatase family. Nat Struct Biol. 2001;8:789–94.
12. Aravind L, Koonin EV. The HD domain defines a new superfamily of metal-dependent phosphohydrolases. Trends Biochem Sci. 1998;23:469–72.
13. Drøbak BK. The plant phosphoinositide system. Biochem J. 1992;288:697–712.
14. Gläser HU, Thomas D, Gaxiola R, Montrichard F, Surdin-Kerjan Y, Serrano R. Salt tolerance and methionine biosynthesis in *Saccharomyces cerevisiae* involve a putative phosphatase gene. EMBO J. 1993;12:3105–10.
15. Neuwald AF, York JD, Majerus PW. Diverse proteins homologous to inositol monophosphatase. FEBS Lett. 1991;294:16–8.
16. Eisenberg Jr F. D-Myoinositol 1-phosphate as product of cyclization of glucose 6-phosphate and substrate for a specific phosphatase in rat testis. J Biol Chem. 1967;242:1375–82.
17. Gee NS, Ragan CI, Watling KJ, Aspley S, Jackson RG, Reid GG, Gani D, Shute JK. The purification and properties of *myo*-inositol monophosphatase from bovine brain. Biochemistry. 1988;249:883–9.
18. Matsuhisa A, Suzuki N, Noda T, Shiba K. Inositol monophosphatase activity from the *Escherichia coli suhB* gene product. J Bacteriol. 1995;177:200–5.
19. Kates M. Bacterial lipids. Adv Lipid Res. 1964;2:17–90.
20. Dhariwal KR, Vasantha N, Freese E. Partial nucleotide limitation induces phosphodiesterase I and 5′-nucleotidase in *Bacillus subtilis*. J Bacteriol. 1982;149:1146–9.
21. Asahara T, Mori Y, Zakataeva NP, Livshits VA, Yoshida K, Matsuno K. Accumulation of gene-targeted *Bacillus subtilis* mutations that enhance fermentative inosine production. Appl Microbiol Biotechnol. 2010;87:2195–207.
22. Kunst F, Ogasawara N, Moszer I, Albertini AM, Alloni G, Azevedo V, Bertero MG, Bessières P, Bolotin A, Borchert S, Borriss R, Boursier L, Brans A, Braun M, Brignell SC, Bron S, Brouillet S, Bruschi CV, Caldwell B, Capuano V, Carter NM, Choi S-K, Codani JJ, Connerton IF, Cummings NJ, Daniel RA, Denizot F, Devine KM, Düsterhöft A, Ehrlich SD, Emmerson PT, Entian KD, Errington J, Fabret C, Ferrari E, Foulger D, Fritz C, Fujita M, Fujita Y, Fuma S, Galizzi A, Galleron N, Ghim S-Y, Glaser P, Goffeau A, Golightly EJ, Grandi G, Guiseppi G, Guy BJ, Haga K, Haiech J, Harwood CR, Hénaut A, Hilbert H, Holsappel S, Hosono S, Hullo M-F, Itaya M, Jones L, Joris B, Karamata D, Kasahara Y, Klaerr-Blanchard M, Klein C, Kobayashi Y, Koetter P, Koningstein G, Krogh S, Kumano M, Kurita K, Lapidus A, Lardinois S, Lauber J, Lazarevic V, Lee S-M, Levine A, Liu H, Masuda S, Mauël C, Médigue C, Medina N, Mellado RP, Mizuno M, Moestl D, Nakai S, Noback M, Noone D, O'Reilly M, Ogawa K, Ogiwara A, Oudega B, Park S-H, Parro V, Pohl TM, Portetelle D, Porwollik S, Prescott AM, Presecan E, Pujic P, Purnelle B, Rapoport G, Rey M, Reynolds S, Rieger M, Rivolta C, Rocha E, Roche B, Rose M, Sadaie Y, Sato T, Scanlan E, Schleich S, Schroeter R, Scoffone F, Sekiguchi J, Sekowska A, Seror SJ, Serror P, Shin B-S, Soldo B, Sorokin A, Tacconi E, Takagi T, Takahashi H, Takemaru K, Takeuchi M, Tamakoshi A, Tanaka T, Terpstra P, Tognoni A, Tosato V, Uchiyama

S, Vandenbol M, Vannier F, Vassarotti A, Viari A, Wambutt R, Wedler E, Wedler H, Weitzenegger T, Winters P, Wipat A, Yamamoto H, Yamane K, Yasumoto K, Yata K, Yoshida K, Yoshikawa H-F, Zumstein E, Yoshikawa H, Danchin A. The complete genome sequence of the Gram-positive bacterium *Bacillus subtilis*. Nature. 1997;390:249–56.

23. Sa-Nogueira I, Nogueira TV, Soares S, de Lencastre H. The *Bacillus subtilis* L-arabinose (*ara*) operon: nucleotide sequence, genetic organization and expression. Microbiology. 1997;143:957–69.
24. Guo Z, Wang F, Shen T, Huang J, Wang Y, Ji C. Crystal structure of thermostable p-nitrophenylphosphatase from *Bacillus Stearothermophilus* (Bs-TpNPPase). Protein Pept Lett. 2014;21:483–9.
25. Nicolas P, Mäder U, Dervyn E, Rochat T, Leduc A, Pigeonneau N, Bidnenko E, Marchadier E, Hoebeke M, Aymerich S, Becher D, Bisicchia P, Botella E, Delumeau O, Doherty G, Denham EL, Fogg MJ, Fromion V, Goelzer A, Hansen A, Härtig E, Harwood CR, Homuth G, Jarmer H, Jules M, Klipp E, Le Chat L, Lecointe F, Lewis P, Liebermeister W, March A, Mars RA, Nannapaneni P, Noone D, Pohl S, Rinn B, Rügheimer F, Sappa PK, Samson F, Schaffer M, Schwikowski B, Steil L, Stülke J, Wiegert T, Devine KM, Wilkinson AJ, van Dijl JM, Hecker M, Völker U, Bessières P, Noirot P. Condition-dependent transcriptome reveals high-level regulatory architecture in *Bacillus subtilis*. Science. 2012;335:1103–6.
26. Morimoto T, Loh PC, Hirai T, Asai K, Kobayashi K, Moriya S, Ogasawara N. Six GTP-binding proteins of the Era/Obg family are essential for cell growth in *Bacillus subtilis*. Microbiology. 2002;148:3539–52.
27. Fillinger S, Boschi-Muller S, Azza S, Dervyn E, Branlant G, Aymerich S. Two glyceraldehyde-3-phosphate dehydrogenases with opposite physiological roles in a nonphotosynthetic bacterium. J Biol Chem. 2000;275:14031–7.
28. Feucht A, Lucet I, Yudkin MD, Errington J. Cytological and biochemical characterization of the FtsA cell division protein of *Bacillus subtilis*. Mol Microbiol. 2001;40:115–25.
29. Galinier A, Kravanja M, Engelmann R, Hengstenberg W, Kilhoffer MC, Deutscher J, Haiech J. New protein kinase and protein phosphatase families mediate signal transduction in bacterial catabolite repression. Proc Natl Acad Sci U S A. 1998;95:1823–8.
30. Yansura DG, Henner DJ. Use of the *Escherichia coli lac* repressor and operator to control gene expression in *Bacillus subtilis*. Proc Natl Acad Sci U S A. 1984;81:439–43.
31. Bianchi V, Spychala J. Mammalian 5′-nucleotidases. J Biol Chem. 2003;278:46195–8.
32. Koshland Jr DE. Control of enzyme activity and metabolic pathways. Trends Biochem Sci. 1984;9:155–9.
33. Newsholme EA, Challiss RAJ, Crabtree B. Substrate cycles: their role in improving sensitivity in metabolic control. Trends Biochem Sci. 1984;9:277–80.
34. Breton A, Pinson B, Coulpier F, Giraud MF, Dautant A, Daignan-Fornier B. Lethal Accumulation of Guanylic Nucleotides in *Saccharomyces cerevisiae* HPT1-Deregulated Mutants. Genetics. 2008;178:815–24.
35. Pöther DC, Liebeke M, Hochgräfe F, Antelmann H, Becher D, Lalk M, Lindequist U, Borovok I, Cohen G, Aharonowitz Y, Hecker M. Diamide triggers mainly S thiolations in the cytoplasmic proteomes of *Bacillus subtilis* and *Staphylococcus aureus*. J Bacteriol. 2009;191:7520–30.
36. Singh R, Mailloux RJ, Puiseux-Dao S, Appanna VD. Oxidative stress evokes a metabolic adaptation that favors increased NADPH synthesis and decreased NADH production in *Pseudomonas fluorescens*. J Bacteriol. 2007;189:6665–75.
37. Kobayashi K, Ehrlich SD, Albertini A, Amati G, Andersen KK, Arnaud M, Asai K, Ashikaga S, Aymerich S, Bessieres P, Boland F, Brignell SC, Bron S, Bunai K, Chapuis J, Christiansen LC, Danchin A, Débarbouille M, Dervyn E, Deuerling E, Devine K, Devine SK, Dreesen O, Errington J, Fillinger S, Foster SJ, Fujita Y, Galizzi A, Gardan R, Eschevins C, Fukushima T, Haga K, Harwood CR, Hecker M, Hosoya D, Hullo MF, Kakeshita H, Karamata D, Kasahara Y, Kawamura F, Koga K, Koski P, Kuwana R, Imamura D, Ishimaru M, Ishikawa S, Ishio I, Le Coq D, Masson A, Mauël C, Meima R, Mellado RP, Moir A, Moriya S, Nagakawa E, Nanamiya H, Nakai S, Nygaard P, Ogura M, Ohanan T, O'Reilly M, O'Rourke M, Pragai Z, Pooley HM, Rapoport G, Rawlins JP, Rivas LA, Rivolta C, Sadaie A, Sadaie Y, Sarvas M, Sato T, Saxild HH, Scanlan E, Schumann W, Seegers JF, Sekiguchi J, Sekowska A, Séror SJ, Simon M, Stragier P, Studer R, Takamatsu H, Tanaka T, Takeuchi M, Thomaides HB, Vagner V, van Dijl JM, Watabe K, Wipat A, Yamamoto H, Yamamoto M, Yamamoto Y, Yamane K, Yata K, Yoshida K, Yoshikawa H, Zuber U, Ogasawara N. Essential *Bacillus subtilis* genes. Proc Natl Acad Sci U S A. 2003;100:4678–83.
38. Hawkins AR, Lamb HK, Smith M, Keyte JW, Roberts CF. Molecular organisation of the quinic acid utilization (QUT) gene cluster in *Aspergillus nidulans*. Mol Gen Genet. 1988;214:224–31.
39. Chen IW, Charalampous FC. A soluble enzyme system from yeast which catalyzes the biosynthesis of inositol from glucose. Biochem Biophys Res Commun. 1963;12:62–7.
40. Majerus PW. Inositol phosphate biochemistry. Annu Rev Biochem. 1992;61:225–50.
41. Cronan JE. Bacterial membrane lipids: where do we stand? Annu Rev Microbiol. 2003;57:203–24.
42. den Kamp JA, Redai I, van Deenen LL. Phospholipid composition of *Bacillus subtilis*. J Bacteriol. 1969;99:298–303.
43. Shiba K, Ito K, Yura T. Mutation that suppresses the protein export defect of the *secY* mutation and causes cold-sensitive growth of *Escherichia coli*. J Bacteriol. 1984;160:696–701.
44. Yano R, Nagai H, Shiba K, Yura T. A mutation that enhances synthesis of sigma 32 and suppresses temperature-sensitive growth of the *rpoH15* mutant of *Escherichia coli*. J Bacteriol. 1990;172:2124–30.
45. Ito K, Wittekind M, Nomura M, Shiba K, Yura T, Miura A, Nashimoto H. A temperature-sensitive mutant of *E. coli* exhibiting slow processing of exported proteins. Cell. 1983;32:789–97.
46. Grossman AD, Erickson JW, Gross CA. The *htpR* gene product of *E. coli* is a sigma factor for heat-shock promoters. Cell. 1984;38:383–90.
47. Li K, Xu C, Jin Y, Sun Z, Liu C, Shi J, Chen G, Chen R, Jin S, Wu W. SuhB is a regulator of multiple virulence genes and essential for pathogenesis of *Pseudomonas aeruginosa*. MBio. 2013;4:e00419–13.
48. Rosales-Reyes R, Saldías MS, Aubert DF, El-Halfawy OM, Valvano MA. The *suhB* gene of *Burkholderia cenocepacia* is required for protein secretion, biofilm formation, motility and polymyxin B resistance. Microbiology. 2012;158:2315–24.
49. Kozloff LM, Turner MA, Arellano F, Lute M. Phosphatidylinositol, a phospholipid of ice-nucleating bacteria. J Bacteriol. 1991;173:2053–60.
50. Englesberg E, Anderson RL, Weinberg R, Lee N, Hoffee P, Huttenhauer G, Boyer H. L-Arabinose-sensitive, L-ribulose 5-phosphate 4-epimerase-deficient mutants of *Escherichia coli*. J Bacteriol. 1962;84:137–46.
51. Irani MH, Maitra PK. Properties of *Escherichia coli* mutants deficient in enzymes of glycolysis. J Bacteriol. 1977;132:398–410.
52. Yarmolinsky MB, Wiesmeyer H, Kalckar HM, Jordan E. Hereditary defects in galactose metabolism in *Escherichia coli* mutants. II. Galactose-induced sensitivity. Proc Natl Acad Sci U S A. 1959;45:1786–91.
53. Lowry OH, Carter J, Ward JB, Glaser L. The effect of carbon and nitrogen sources on the level of metabolic intermediates in *Escherichia coli*. J Biol Chem. 1971;246:6511–21.
54. Papagianni M, Avramidis N, Filiousis G. Glycolysis and the regulation of glucose transport in *Lactococcus lactis* spp. *lactis* in batch and fed-batch culture. Microb Cell Fact. 2007;6:16.
55. Paoletti L, Lu YJ, Schujman GE, de Mendoza D, Rock CO. Coupling of fatty acid and phospholipid synthesis in *Bacillus subtilis*. J Bacteriol. 2007;189:5816–24.
56. Ishii K, Shiio I. Improved inosine production and derepression of purine nucleotide biosynthetic enzymes in 8-azaguanine resistant mutants of *Bacillus subtilis*. Agric Biol Chem. 1972;36:1511–22.
57. Matsui H, Sato K, Enei H, Hirose Y. Mutation of an inosine producing strain of *Bacillus subtilis* to DL-methionine sulfoxide resistant for guanosine production. Appl Environ Microbiol. 1977;34:337–41.
58. Kotani Y, Yamaguchi K, Kato F, Furuya A. Inosine accumulation by mutants of *Brevibacterium ammoniagenes*: strain improvement and culture conditions. Agric Biol Chem. 1978;42:399–405.
59. Matsui H, Sato K, Enei H, Takinami K. 5′-Nucleotidase activity in improved inosine-producing mutants of *Bacillus subtilis*. Agric Biol Chem. 1982;46:2347–52.
60. Chen S, Chu J, Zhuang Y, Zhang S. Enhancement of inosine production by *Bacillus subtilis* through suppression of carbon overflow by sodium citrate. Biotechnol Lett. 2005;27:689–92.
61. Barnett JE, Brice RE, Corina DL. A colorimetric determination of inositol monophosphates as an assay for D-glucose 6-phosphate-1 L-myoinositol 1-phosphate cyclase. Biochem J. 1970;119:183–6.
62. Morimoto T, Ara K, Ozaki K, Ogasawara N. A new simple method to introduce marker-free deletions in the *Bacillus subtilis* genome. Genes Genet Syst. 2009;84:315–8.
63. Miwa Y, Nakata A, Ogiwara A, Yamamoto M, Fujita Y. Evaluation and characterization of catabolite-responsive elements (*cre*) of *Bacillus subtilis*. Nucleic Acids Res. 2000;28:1206–10.
64. Vagner V, Dervyn E, Ehrlich SD. A vector for systematic gene inactivation in *Bacillus subtilis*. Microbiology. 1998;144:3097–104.

# 12

# Bacterial and fungal core microbiomes associated with small grain silages during ensiling and aerobic spoilage

Lysiane Duniere[1], Shanwei Xu[1], Jin Long[1], Chijioke Elekwachi[1], Yuxi Wang[1], Kelly Turkington[2], Robert Forster[1] and Tim A. McAllister[1*]

**Abstract**

**Background:** Describing the microbial populations present in small grain silage and understanding their changes during ensiling is of interest for improving the nutrient value of these important forage crops. Barley, oat and triticale forages as well as an intercropped mixture of the 3 crops were harvested and ensiled in mini silos for a period of 90 days, followed by 14 days of aerobic exposure. Changes in fermentation characteristics and nutritive value were assessed in terminal silages and bacterial and fungal communities during ensiling and aerobic exposure were described using 16S and 18S rDNA sequencing, respectively.

**Results:** All small grain silages exhibited chemical traits that were associated with well ensiled forages, such as low pH value (4.09 ± 0.28) and high levels of lactic acid (59.8 ± 14.59 mg/g DM). The number of microbial core genome operational taxonomic units (OTUs) decreased with time of ensiling. Taxonomic bacterial community profiles were dominated by the Lactobacillales after fermentation, with a notable increase in Bacillales as a result of aerobic exposure. Diversity of the fungal core microbiome was shown to also be reduced during ensiling. Operational taxonomic units assigned to filamentous fungi were found in the core microbiome at ensiling and after aerobic exposure, whereas the Saccharomycetales were the dominate yeast population after 90 days of ensiling and aerobic exposure. Bacterial and fungal orders typically associated with silage spoilage were identified in the core microbiome after aerobic exposure.

**Conclusion:** Next Generation Sequencing was successfully used to describe bacterial communities and the first record of fungal communities throughout the process of ensiling and utilization. Adequately describing the microbial ecology of silages could lead to improved ensiling practices and the selection of silage inoculants that act synergistically with the natural forage microbiome.

**Keywords:** Barley, Oats, Triticale, Silage, Microbiome, Bacteria, Fungi, Aerobic exposure

## Background

Silage production is of great economic importance in the world. The land area devoted to worldwide forage production was estimated at 135.7 million ha in 2000 [1]. Corn and alfalfa are the most important crops used for ensiling worldwide [2], but small grain cereals (e.g., barley, triticale, oats) and various legumes are often ensiled in Northern climates [3–5]. The type of crop selected for forage and silage production depends on their growth requirements, yield and their nutritive value [6].

Silage conservation of moist crops depends on microbial fermentation [7] with lactic acid bacteria (LAB) being the main producers of the organic acids essential for silage conservation [8, 9]. Inoculants have been selected for their ability to improve fermentation during ensiling of the forage, increase aerobic stability and enhance silage digestibility [10]. Among the many microbial species found in silage, some are considered undesirable as their presence can negatively impact silage quality. Undesirable microorganisms can lower the dry matter

* Correspondence: tim.mcallister@agr.gc.ca
[1]Agriculture and Agri-Food Canada Research Centre, Lethbridge T1J 4B1, AB, Canada
Full list of author information is available at the end of the article

(DM) and water-soluble carbohydrates (WSC) content of silage [11] and cause undesirable fermentation profiles such as the production of butyric acid [12]. Among undesirable microorganisms, some such as *Listeria* sp. or mycotoxigenic fungi can be pathogenic or produce toxins that have adverse effects on the health of both livestock and humans [13]. Yeasts are the main microorganisms involved in the spoilage of silage as they metabolize lactic acid, increasing silage pH and creating conditions that are conducive for the growth of other spoilage or pathogenic microorganisms [14]. A better understanding of the microbial communities actively involved in the ensiling process could provide additional insight into approaches to improve the conservation of silages.

Many of the microorganisms found in silage such as *Lactobacillus* sp. or *Saccharomyces* sp. have been proven to enter a viable, but nonculturable state in the face of environmental stress [15, 16]. Microbial communities described solely on the basis of culturing are often incomplete as many species are unculturable or poorly represented by the culturing process. Metagenomic deep sequencing of microbial DNA is a culture-independent technique that allows microbial diversity to be described without the need to culture isolates. Ribosomal DNA (rDNA) gene sequencing has been successfully used in many studies to describe microbiomes in complex environments including the rhizosphere [17, 18], soil [19, 20], compost [21] and rumen [22]. Eikmeyer et al. [23] were the first to study the bacterial microbiome of grass silage during ensiling with and without inoculants. However, these authors focused only on bacterial communities and did not describe the nature of the fungal microbiome.

This study aimed to characterize the bacterial and fungal core microbiomes associated with small grain cereals (i.e., barley, oats, triticale) during ensiling and upon aerobic exposure.

## Methods

### Forage

Whole-crop barley (*Hordeum vulgare*, L. Variety "Sundre" [24]), oats (*Avena sativa* L. Variety "AC Morgan", [25]) and triticale (*Triticosecale Wittm* Variety "Bunker", [26]) or an intercropped mixture of all 3 crops were planted on 12 June 2013 at the Lacombe Research Centre, Agriculture and Agri-Food Canada (113.7° W, 52.5°N) and harvested on 4 September 2013. Individual crop species were seeded at 300 seeds $m^{-2}$ and an intercropped mixture of the three cereals was seeded at 100 seeds $m^{-2}$. Seeding was conducted with a 3.7 m seed drill (Conserva-Pak™, John Deere, Moline, IL, USA) configured with knife openers spaced 23 cm apart and a plot size of 3.7 × 7.6 m. All plots received N, P and K fertilizers applied at the time of seeding, alone or as a blend, typically in the form of urea (46-0-0), monoammonium phosphate (11-51-0), and potassium chloride (0-0-60) at rates according to soil test recommendations. Prior to seeding, the plot areas were sprayed with glyphosate [N-(phosphonomethyl)glycine] across the entirety of each site 24 to 48 h prior to seeding using label recommended rates and application parameters for the Canadian Prairies [27]. In-crop herbicides were used for weed control depending on the weed spectrum present. At harvest, crops were ensiled at DM levels of 48.1% (firm dough), 30.2% (early dough), 41.7 and 38.5% (medium dough) in barley, oats, triticale and the intercropped mixture, respectively. Forage was chopped to a theoretical chop length of 9.5 mm using a self-propelled forage harvester (Harvester 6610, John Deere, Moline, IL, USA).

### Mini silo experiment

Forages were packed into mini PVC silos (2.5 to 3 kg of fresh forage) with a hydraulic press to a density of approximately 240 kg/$m^3$ as previously described [10]. The silos were weighed prior to filling and immediately after sealing, and stored at ambient temperature (22 °C). Each crop was harvested without wilting from 3 replicate field plots. Triplicate silos for each crop (one from each plot) were prepared and opened after 90 day of ensiling. Prior to ensiling (day 0), samples of each forage type from each plot were collected for chemical and microbial analyses. Silos were weighed prior to opening to calculate DM loss. At sampling, triplicate mini silos were opened and the contents were thoroughly mixed by hand. Subsamples were then collected for chemical, microbial and molecular analyses.

### Aerobic stability

After 90 days of ensiling, silos were opened and approximately 7.5% of the wet weight in each silo was subsampled. These samples were placed into separate 4-L insulated containers (3 replicates per treatment), covered with two layers of cheesecloth and stored at 20 °C for 14 days. Two Dallas Thermochron iButtons (Embedded Data Systems, Lawrenceburg, KY, USA) were embedded in the silage near the bottom and centre of each container with the temperature in the storage room recorded every 15 min. Ambient temperature and the temperature in each container were simultaneously monitored for 14 days. The contents of each container were thoroughly mixed and sampled after 14 days of aerobic exposure for chemical (15 g) and microbial analyses (10 g) and for DNA extraction (30 g). Aerobic stability was calculated according to Teller et al. [28] as the number of hours before the temperature of aerobically exposed silage exceeded the baseline ambient temperature by 2 °C.

### Chemical analysis

Forages collected at harvest and silage samples collected on d 90 from each silo were analyzed for water-soluble carbohydrates (WSC), ammonia nitrogen ($NH_3$-N) and starch, as described by Zahiroddini et al. [29]. Volatile fatty acids and lactate were determined by the methods of Kudo et al. [30] on a Hewlett Packard model 5890A Series Plus II gas liquid chromatograph (column: 30 m FFAP fused silica capillary, 0.32 mm i.d., 1.0 m film thickness, Phenomenex, Torrance, CA, USA). Total nitrogen (N) was determined by elemental analysis (Dumas Nitrogen) using a NA1500 Nitrogen/Carbon analyzer (Carlo Erba Instruments, Milan, Italy). Crude protein was calculated as N × 6.25. The DM of forages and silage samples was determined by drying at 105 °C in forced-draft oven for 24 h. Organic matter (OM) was estimated by ashing 1 g of dried sample in a muffle furnace at 550 °C for 5 h. Neutral detergent fibre (NDF) and acid detergent fibre (ADF) were analyzed using an Ankom 200 system following the manufacturer's instructions (Ankom Technology Corporation, Fairport, NY, USA) with sodium sulfite and α-amylase used for NDF analysis. Nitrogen in ADF residues (ADF insoluble nitrogen, ADIN) was measured by combustion as described above. Samples (15 g) from each mini silo at each sampling time were mixed with 135 mL of deionized water, blended for 30 s and filtered through two layers of cheesecloth and pH of the filtrate was measured with a Symphony pH meter (VWR, Mississauga, ON, Canada).

### Microbial analysis

For microbial analyses, forage or silage samples (10 g) from each mini silo were treated as described previously by Addah et al. [31] with lactobacilli counts estimated using Man Rogosa Sharpe (MRS, [32]) plates amended with 200 μg/mL of cycloheximide (MRS; Dalynn Biologicals, Calgary, AB, Canada), total bacteria using nutrient agar (NA) amended with 200 μg/mL of cycloheximide (Dalynn Biologicals, Calgary, AB, Canada) and yeast and molds using Sabouraud's dextrose agar (SDA; Dalynn Biologicals, Calgary, AB, Canada) containing 100 μg/mL of tetracycline and 100 μg/mL of chloramphenicol. Lactobacilli MRS agar and NA plates were incubated at 37 °C for 24–48 h whereas SDA plates were incubated at ambient temperature for 48–72 h. Numbers of yeasts and filamentous fungi (i.e., molds) were differentiated based on colony appearance and morphology on SDA plates.

### Molecular analysis

#### *DNA extraction*

For each sampling, forage and silage samples (30 g) from each mini silo were frozen (−80 °C), lyophilized and ground through a 4 mm screen. Subsamples (5 g) were then ball milled for 1 min at ambient temperature and DNA extraction was performed according to the procedure of Yu and Morrison (2004) [33]. Briefly, DNA was extracted from 0.3 g of ground silage using a bead beating step (FastPrep-24, MP Biomedicals, Santa Ana, CA, USA; 3 min, maximum speed) with zirconia silica beads (0.3 g of 0.1 mm and 0.1 g of 0.5 mm). Nucleic acids were precipitated with ammonium acetate and isopropanol, washed in ethanol and re-suspended in Tris-EDTA buffer. All samples were purified through a QiaAmp DNA Stool kit column (Qiagen Sciences, Germantown, MD, USA) and eluted in nuclease free water. Yield and purity of extracted DNA was measured using a Pico-Green® dsDNA quantitation assay (Invitrogen Canada Inc., Burlington, ON, Canada). High molecular weight DNA samples at a minimal concentration of 20 ng/μl were used for sequencing.

#### *Sequencing*

Extracted DNA samples were pair-end sequenced using Illumina MiSeq (San Fransico, CA, USA) at Genome Quebec (McGill University, Génome Québec Innovation Centre, Montreal, QC, Canada). For bacteria, the V3-V4 region of 16S was targeted using universal primers 347 F-CS1 (ACACTGACGACATGGTTCTACAGGAGGCAGC AGTRRGGAAT) and 803R-CS2 (TACGGTAGCAGA-GACTTGGTCTCTACCRGGGTATCTAATCC). Primers nu-SSU-0817 (TTAGCATGGAATAATRRAATAGGA) and nu-SS-1196 (TCTGGACCTGGTGAGTTTCC) originally designed by Borneman and Hartin (2000) [34] were used to target around 400 bp of the genomic region containing part of the V4 and total V5 variable domains of the SSU rDNA gene from all four major phyla of fungi. Library construction and the Illumina MiSeq paired-end sequencing was performed by Genome Quebec according to manufacturer's recommendations.

### Bioinformatic analysis

Bacterial sequences were scanned for contaminants and unpaired reads, trimmed to a fixed length of 250 bp and paired-end assembled using FLASH software [35]. Read counts per sample were checked to confirm no sample failed to be amplified or sequenced. Reads were clustered at 97% identity using DNACLUST [36] and clusters showing abundances higher than 3 were then scanned for chimeras with UCHIME denovo and UCHIME reference in order to estimate final operational taxonomic units (OTUs) [37]. OTUs were analyzed for taxonomic distribution with the RDP classifier software using the complete Greengenes database (http://greengenes.secondgenome.com/downloads) supplemented with eukaryotic sequences from the Silva databases and a customized set of mitochondria and chloroplasts 16S sequences. Each OTUs taxonomic depth with a RDP classifier score below 0.5 was kept to reconstruct the final lineage. Diversity metrics were obtained by aligning OTU sequences to a Greengenes core reference alignment [38] using the PyNAST aligner [39]. Alpha diversity index (Chao1, Shannon-Weiner indexes, and rarefaction curves), and taxonomic classifications were then computed using QIIME software [40, 41].

Fungal sequences were analyzed following the MiSeq standard operating procedure for MOTHUR analysis [42] using a process adapted for 18S rDNA gene sequences. Reference and taxonomy alignments were performed using the updated SILVA reference file v119 [43], containing the full length sequences and taxonomic references for 51,533 SSU sequences from Eukaryota, 464,618 sequences from Bacteria and 18,797 sequences from Archaea. Briefly, the reads from each sample were randomly subsampled down to 15,000 using HTSeq and Python [44]. The sequences were then assembled, trimmed and singletons were removed. Reads with homopolymers greater than eight bases and sequences with one or more ambiguous bases were removed from the data set. Remaining sequences were aligned and run against the SILVA reference database. Before being classified into OTUs, sequences were pre-clustered at 97% identity and chimeras were scanned and removed using UCHIME [37]. Final OTUs were obtained and analyzed for taxonomic distribution with the SILVA database. Alpha diversity indexes and taxonomic classifications were then computed in a manner similar to that of bacterial sequences.

Determination of the core microbiome was accomplished by comparing samples from all crops across time. For each time, any taxa found to be ubiquitous across all samples were then defined to be part of the core microbiome of small grain cereal silages.

### Calculations and statistical analysis

Microbial populations were estimated as colony forming unit (CFU)/g of forage or silage DM and were log transformed prior to statistical analysis. Ensiling parameters, aerobic stability and microbial data were assessed using a randomized complete block design considering 4 crops (barley, oats, triticale and intercrop) at 3 times of sampling (0, 90 and 104 days). An analysis of variance using the crops and time was performed to determine significant differences in diversity indexes. Significance was defined as a $P$ value of less than 0.05 using SAS [45].

## Results

### Silages characteristics

At ensiling, barley forage exhibited the highest ($P < 0.001$) initial pH of all forages (Table 1). A higher level of DM was observed in barley silage compared to oat and triticale silages with the DM of intercropped silage being close to the average DM of the 3 crops. With the exception of barley, the pH of all silages rapidly declined to below 4.0 within 7 days (data not shown) of ensiling and remained stable until sampling of terminal silage (90 days). In this study, barley silage was the exception to this pattern as it gradually declined to a pH of 4.5 over 90 days of ensiling. After ensiling, DM was higher for barley silage and lower for oat silage as compared to triticale or intercropped silage ($P < 0.001$, Table 1). DM loss was significantly higher for triticale than other silages ($P < 0.05$)

The nutrient composition and fermentation profile of all silages differed ($P < 0.05$) after 90 days of fermentation (Table 1). At harvest, the OM of barley and triticale forages was higher than oat and intercropped forages, but after 90 days of ensiling OM levels differed ($P < 0.001$) among all small grain silages. Barley forage exhibited the highest level of CP, whereas after 90 days of ensiling oat silage exhibited the lowest ($P < 0.05$) CP of the silages and a higher ADIN ($P < 0.05$) content than triticale silage. Both barley and triticale forage exhibited lower ($P < 0.05$) levels of NDF than oats or intercropped forages, but after 90 days of ensiling only barley silage exhibited lower ($P < 0.05$) levels of NDF and ADF than other silages. Levels of WSC were noticeably lower ($P < 0.001$) in barley than other forages, but still differed among all forages. A higher ($P < 0.05$) WSC content was observed for triticale silage, while the WSC of barley silage remained low. The level of starch in barley was higher ($P < 0.001$) than other forages and silages, but declined in all silages by about 2 mg/g DM during ensiling. Lactic acid levels differed ($P < 0.05$) among all silage and were the highest for triticale silage and the lowest for barley silage. Acetic acid levels were highest ($P < 0.05$) in barley silage and the lowest in triticale silage. Propionic and butyric acids were either absent or only detected at trace levels.

Total bacterial populations were lower ($P < 0.05$) in oat forage as compared to triticale and intercropped forages (Table 1). Populations of LAB, molds and yeasts did not differ among fresh small grain forages. Compared to other silages, the total bacterial population (TB) was higher ($P < 0.05$) in barley and lower ($P < 0.05$) in triticale, a pattern that was also observed for LAB populations. No molds were detected in any of the silages after 90 days of ensiling, with yeast populations ranging between 5.84 and 6.75 $\log_{10}$ CFU/g DM in oat, triticale and intercropped silages and were virtually absent in barley silage.

After aerobic exposure, the pH of oat, triticale and intercropped silages increased ($P < 0.001$) while the pH of barley silage remained largely unchanged (Table 1). The aerobic stability of barley was > 336 h, with no measurable increase in temperature over 14 days of aerobic exposure (Fig. 1). Oat and intercropped silage exhibited lower ($P < 0.05$) aerobic stability than triticale silage (Table 1), with temperature increases in these

**Table 1** Physico-chemical characteristics and microbial populations of barley, oat, triticale and intercropped forages, 90day silages and aerobically exposed silages ($n = 3$)

| Item | Forage | | | | SEM | *P*-value* | Silage | | | | SEM | *P*-value* | Aerobically exposed Silage | | | | SEM | *P*-value* |
|---|---|---|---|---|---|---|---|---|---|---|---|---|---|---|---|---|---|---|
| | Barley | Oat | Triticale | Intercropped | | | Barley | Oat | Triticale | Intercropped | | | Barley | Oat | Triticale | Intercropped | | |
| pH | $7.57^{b}$ | $6.28^{d}$ | $6.52^{c}$ | $6.57^{c}$ | 0.143 | <0.001 | $4.50^{b}$ | $3.98^{c}$ | $3.93^{c}$ | $3.94^{c}$ | 0.070 | <0.001 | $4.39^{d}$ | $8.71^{b}$ | $6.29^{c}$ | $8.43^{b}$ | 0.548 | <0.001 |
| DM | $48.08^{b}$ | $30.21^{e}$ | $41.65^{c}$ | $38.48^{d}$ | 23.390 | <0.001 | $47.02^{b}$ | $29.87^{e}$ | $39.06^{c}$ | $37.85^{d}$ | 22.060 | <0.001 | $57.88^{b}$ | $35.63^{d}$ | $53.80^{b}$ | $46.95^{c}$ | 2.671 | <0.001 |
| DM loss (%) | NM | | | | | | $3.10^{c}$ | $1.65^{c}$ | $7.07^{b}$ | $2.11^{c}$ | 0.879 | 0.009 | NM | | | | | |
| Nutrient composition, % DM basis | | | | | | | | | | | | | | | | | | |
| OM | $93.61^{b}$ | $92.99^{c}$ | $93.92^{b}$ | $93.11^{c}$ | 0.114 | 0.002 | $93.38^{b}$ | $92.58^{d}$ | $92.98^{c}$ | $92.12^{e}$ | 0.144 | <0.001 | NM | | | | | |
| CP | $14.20^{b}$ | $11.71^{c}$ | $11.60^{c}$ | $11.91^{c}$ | 0.059 | 0.003 | $13.41^{b}$ | $11.48^{c}$ | $13.63^{b}$ | $12.43^{bc}$ | 0.049 | 0.018 | NM | | | | | |
| NDF | $53.10^{d}$ | $58.73^{b}$ | $51.28^{d}$ | $55.59^{c}$ | 0.878 | <0.001 | $49.34^{c}$ | $57.12^{b}$ | $56.06^{b}$ | $58.00^{b}$ | 1.121 | 0.003 | NM | | | | | |
| ADF | $27.04^{d}$ | $36.89^{b}$ | $30.51^{c}$ | $30.89^{c}$ | 1.116 | 0.001 | $24.89^{d}$ | $34.89^{b}$ | $33.30^{c}$ | $33.14^{c}$ | 1.189 | <0.001 | NM | | | | | |
| ADIN (% total N) | NM | | | | | | $4.43^{bc}$ | $5.19^{b}$ | $3.74^{c}$ | $4.78^{b}$ | 0.188 | 0.013 | NM | | | | | |
| Starch | $21.54^{b}$ | $9.14^{d}$ | $7.41^{e}$ | $10.76^{c}$ | 1.709 | <0.001 | $19.20^{b}$ | $7.04^{cd}$ | $5.54^{d}$ | $8.30^{c}$ | 1.651 | <0.001 | NM | | | | | |
| WSC (mg/g DM) | $4.66^{e}$ | $31.51^{d}$ | $74.60^{b}$ | $46.33^{c}$ | 7.305 | <0.001 | $0.99^{c}$ | $2.52^{c}$ | $29.20^{b}$ | $12.01^{bc}$ | 7.821 | 0.031 | NM | | | | | |
| Fermentation products, mg/g DM | | | | | | | | | | | | | | | | | | |
| Lactic acid | | | NM | | | | $44.64^{e}$ | $77.87^{b}$ | $64.56^{c}$ | $52.12^{d}$ | 3.913 | <0.001 | $43.69^{b}$ | $6.60^{d}$ | $28.27^{c}$ | $5.86^{d}$ | 4.935 | <0.001 |
| Acetic acid | | | NM | | | | $28.92^{b}$ | $17.06^{c}$ | $16.85^{c}$ | $10.32^{d}$ | 1.895 | <0.001 | $26.63^{b}$ | $1.19^{c}$ | $1.33^{c}$ | $0.87^{c}$ | 3.331 | <0.001 |
| Propionic acid | | | NM | | | | 0.09 | ND | ND | ND | NA | NA | 0.11 | ND | ND | ND | 0.015 | NA |
| Butyric acid | | | NM | | | | ND | 0.01 | ND | 0.03 | 0.031 | 0.156 | ND | $0.03^{b}$ | $0.02^{b}$ | $0.02^{b}$ | 0.004 | 0.021 |
| La:Ac ratio | | | NM | | | | $1.54^{d}$ | $4.56^{b}$ | $3.83^{c}$ | $5.05^{b}$ | 0.404 | <0.001 | $1.64^{c}$ | $5.38^{c}$ | $21.42^{b}$ | $6.69^{c}$ | 2.487 | 0.002 |
| Total VFAs | | | NM | | | | $29.09^{b}$ | $17.21^{c}$ | $17.08^{c}$ | $10.42^{d}$ | 1.892 | <0.001 | $26.82^{b}$ | $1.35^{cd}$ | $1.93^{c}$ | $1.07^{d}$ | 3.314 | <0.001 |
| $NH_3$-N | $0.13^{c}$ | $0.10^{d}$ | $0.19^{b}$ | $0.11^{cd}$ | 0.011 | 0.001 | $2.34^{b}$ | $2.27^{d}$ | $2.13^{c}$ | $1.43^{d}$ | 0.109 | <0.001 | $1.27^{b}$ | $0.22^{d}$ | $1.13^{b}$ | $0.61^{c}$ | 0.134 | 0.001 |
| Microbiology, $\log_{10}$ CFU/g DM | | | | | | | | | | | | | | | | | | |
| Total bacteria | $7.71^{bc}$ | $7.47^{c}$ | $7.97^{b}$ | $7.85^{b}$ | 0.085 | 0.0288 | $8.92^{b}$ | $8.40^{c}$ | $7.22^{d}$ | $8.05^{c}$ | 0.198 | 0.0011 | $7.75^{d}$ | $9.50^{b}$ | $8.78^{c}$ | $9.56^{b}$ | 0.223 | <0.001 |
| LAB | 6.23 | 5.82 | 6.68 | 6.98 | 0.193 | 0.1267 | $8.98^{b}$ | $8.52^{c}$ | $7.83^{d}$ | $8.21^{cd}$ | 0.143 | 0.0017 | $7.78^{c}$ | $9.37^{b}$ | $9.28^{b}$ | $9.09^{b}$ | 0.216 | 0.003 |
| Molds | 6.22 | 6.06 | 5.99 | 6.12 | 0.051 | 0.4966 | ND | ND | ND | ND | NA | NA | ND | $8.00^{b}$ | ND | $7.18^{c}$ | 7.180 | <0.001 |
| Yeasts | 7.1 | 7.04 | 7.09 | 7.03 | 0.021 | 0.4055 | $<1.00^{c}$ | $5.84^{b}$ | $6.34^{b}$ | $6.75^{b}$ | 0.481 | 0.0002 | $4.28^{c}$ | $8.82^{b}$ | $9.19^{b}$ | $8.63^{b}$ | 0.631 | 0.001 |
| Aerobic Stability[a] | NM | | | | | | NM | | | | | | $>336.00^{d}$ | $51.14^{c}$ | $142.91^{b}$ | $39.10^{c}$ | 17.379 | 0.001 |

[*] Within a row, means with a different superscript (b,c,d,e) differ ($P < 0.05$)

[a] Hours of aerobic exposure before a 2 °C rise in temperature was recorded in the silage mass after exposure to air

*Abbreviations*: *CP* crude protein, *NDF* neutral detergent fiber, *ADF* acid detergent fiber, *ADIN* acid detergent insoluble nitrogen, *WSC* water soluble carbohydrates, *NM* not measured, *ND* not determined, *NA* not available

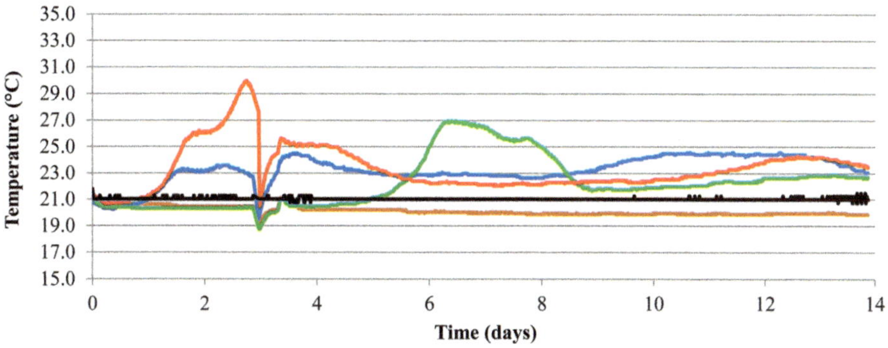


**Fig. 1** Silage temperature during aerobic exposure. Average temperature (°C) in barley (*orange*), oat (*blue*), triticale (*green*) and intercropped (*red*) silages ($n = 3$) and ambient temperature (*black*) ($n = 2$) recorded during 14 days of aerobic exposure

silages noticeable after 1 day of aerobic exposure (Fig. 1). With the exception of barley, levels of lactic acid declined ($P < 0.05$) in all silages, being the lowest in oat and intercropped silage. After 14 days of aerobic exposure, TB and LAB were lower ($P < 0.05$) in barley silage than in other small grain silages. Yeast populations in aerobically exposed barley silage remained at least 4 $\log_{10}$ CFU/g DM lower than ($P < 0.001$) in other silages. Molds were only detected in oat and intercropped silages after 14 days of aerobic exposure.

## Sequencing

A total of 8,937,418 and 3,959,537 reads were obtained for the bacterial and fungal microbiomes in silages, respectively. After bioinformatics analysis, a total of 526,210 bacterial sequences and 70,219 fungal sequences were classified.

### *Microbial silage diversity*

Rarefaction curves for bacterial and fungal OTUs of silage samples at 97% identity are shown in Additional file 1: Figures S1a and S1b, respectively. Sequencing depth was insufficient to fully describe the diversity of the microbial populations in silage as rarefaction curves did not reach a clear plateau for either bacterial or fungal sequences. Bacterial diversity of the samples was observed to increase over the period of ensiling and aerobic exposure (i.e., 0, 90 and 104 days) as shown by differences in Chao1 and Shannon-Weiner indexes ($P < 0.001$, Additional file 2: Table S1). Barley silage was the exception as its bacterial diversity was similar to other silages after ensiling. This observation is supported by Additional file 3: Figure S2a which showed that the nature of the silage microbiome was more influenced by sampling time (Additional file 3: Figure S2a) than by silage type (Additional file 3: Figure S2c). Fungal Chao1 index did not differ ($P > 0.05$) among sampling times and exhibited similar richness among all silages. In contrast to bacteria, fungal communities appeared to be less influenced by sampling time or ensiling, although fungal sequences obtained prior to ensiling did tend to cluster together (Additional file 3: Figure S2b).

A higher number of bacterial core microbiome OTUs were identified in fresh small grain forage (Fig. 2a) than in terminal (Fig. 2b) or aerobically exposed silage (Fig. 2c). The bacterial core microbiome of fresh forage was composed of 167 unique OTUs with the number of OTUs exclusively associated with each forage type ranging from 12 for triticale silage to 33 for oat silage (Fig. 2a). The diversity of the bacterial core microbiome declined after ensiling, being composed of 79 OTUs in terminal silage (Fig. 2b) and only 49 OTUs in aerobically exposed silage (Fig. 2c). After aerobic exposure, the number of bacterial OTUs in barley silage increased dramatically in spite of this silage being aerobically stable.

The core fungal microbiome of silage consisted of 36 OTUs in fresh forage (Fig. 3a), and declined to 21 OTUs in terminal silage (Fig. 3b) and 14 OTUs in aerobically exposed silage (Fig. 3c). The highest number of shared OTUs occurred between intercropped and triticale forages (Fig. 3a) and between triticale and oat terminal silages (Fig. 3b) and between aerobically exposed intercropped and oat silages (Fig. 3c). With the exception of barley silage, the number of fungal OTUs for each crop decreased after ensiling, with aerobic exposure increasing the number of crop specific OTUs in intercropped, oat and barley silage.

### *Bacterial core microbiome during forage ensiling*

Taxonomic profile of the bacterial core microbiome varied among fresh forage (Fig. 4a), terminal silage (Fig. 4b) and aerobically exposed silage (Fig. 4c). Up to 23 bacterial orders exhibited changes in relative abundance during ensiling, of which 12 orders accounted for less than 1.2% relative abundance variation in all silages (Fig. 5a). Among all silages, relative abundance of all bacterial orders, with the exception of the Lactobacillales, declined during the ensiling process (Fig. 4b). The

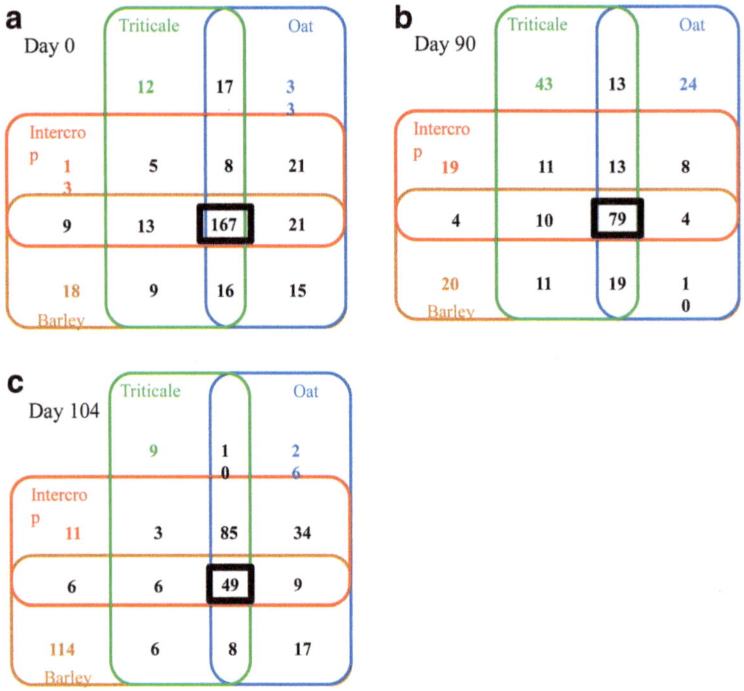


**Fig. 2** Four-way Venn diagram depicting unique bacterial OTUs in barley (*orange*), oat (*blue*), triticale (*green*) and intercropped (*red*) silages, or shared bacterial OTUs among silages at 0 days **a** 90 days **b** and 104 days **c**. Bacterial core microbiome OTUs number is displayed with *bold black frame*

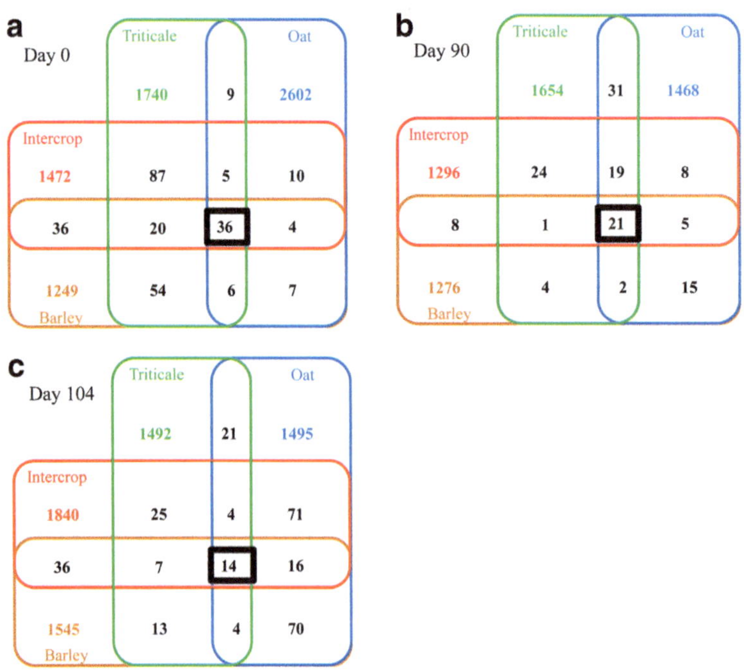


**Fig. 3** Four-way Venn diagram depicting unique fungal OTUs in barley (*orange*), oat (*blue*), triticale (*green*) and intercropped (*red*) silages, or shared bacterial OTUs among silages at 0 days **a** 90 days **b** and 104 days **c**. Fungal core microbiome OTUs number is displayed with *bold black frame*

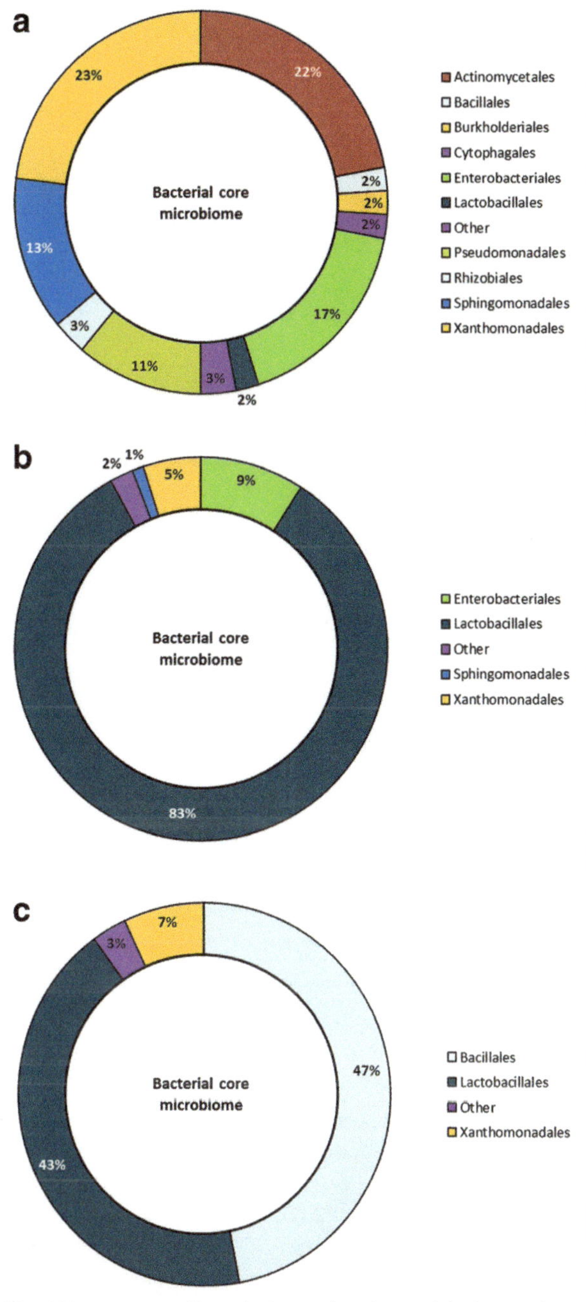


**Fig. 4** Taxonomic profile and relative abundance of the bacterial core microbiome after 0 days, fresh forage **a** 90 days, terminal silage **b** and terminal silage after 14 days of aerobic exposure **c**. OTUs were assigned at the order level

decline in relative abundance of OTUs was the highest for the Actinomycetales as a decrease ranging from −11.2 to −32% was observed among silages, with the Sphingomonadales also showing a consistent decline in all silages (Fig. 5a). A strong reduction in the Xanthomonadales was also noted to occur during the ensiling of barley and intercropped silage, whereas Enterobacteriales strongly decreased in triticale silage (−24.1%). Most often the reduction in bacterial diversity during ensiling was also accompanied by a reduction in abundance. The most notable decreases in OTUs and relative abundance were observed in the Actinomycetales, Shingomonadales, Pseudomonadales and Xanthomonadales. Ten of the bacterial orders associated with core bacterial microbiome of fresh forage were no longer detected in terminal silage.

#### *Bacterial core microbiome in aerobically exposed silage*

The diversity of the bacterial microbiome continued to decline after aerobic exposure (Fig. 4c). During aerobic exposure, up to 13 orders from bacterial core microbiome exhibited differential abundance, of which 4 orders accounted for less than 0.05% of this variation among silages. The abundance of OTUs associated with Bacillales notably increased in oat, triticale and intercropped silages, but only a slight increase was observed in barley silage (Fig. 5b). The relative abundance of OTUs associated with Lactobacillales, also increased as a result of aerobic exposure of barley silage, whereas this order declined considerably in oat, triticale and intercropped silages (−49.3, −61.9 and −47.6%, respectively). The abundance of Xanthomonadales also increased in the core bacterial microbiome during aerobic exposure and this was linked to a higher diversity in all silages (Fig. 5b). With the exception of the Sphingomonadales and the Actinomycetales in intercropped silage (+0.07%), the abundance of all other bacterial orders declined after aerobic exposure. Bacillales represented almost half of the relative abundance of the core microbiome after aerobic exposure in all silages except barley (Fig. 4c).

Shared OTUs among aerobically exposed oat, triticale and intercropped silages were mainly members of the Bacillales order (98.82% of the number of OTUs observed) with only a single OTU belonging to the Streptococcaceae order being associated with these silages (Table 2). Among the Bacillales, the Bacillaceae represented 70.6% of the shared OTUs, whereas the Planococcaceae accounted for 22.4%. The OTUs found in aerobically exposed barley silage were mostly members of the Lactobacillales (27.2%).

#### *Fungal core microbiome during forage ensiling*

The diversity of fungi also tended to decline during the ensiling process (Fig. 6a and b). In total, 19 fungal orders were altered by the ensiling process, of which 9 orders accounted for less than 1% variation in relative abundance in all silages (Fig. 7a). The changes in the fungal population observed during the ensiling of barley differed from that of other silages as the relative abundance of OTUs associated with members of the Sporidiobolales, Pucciniales, Tremellomycetes and Cystofilobasidiales increased (Fig. 7a). During ensiling, numbers of

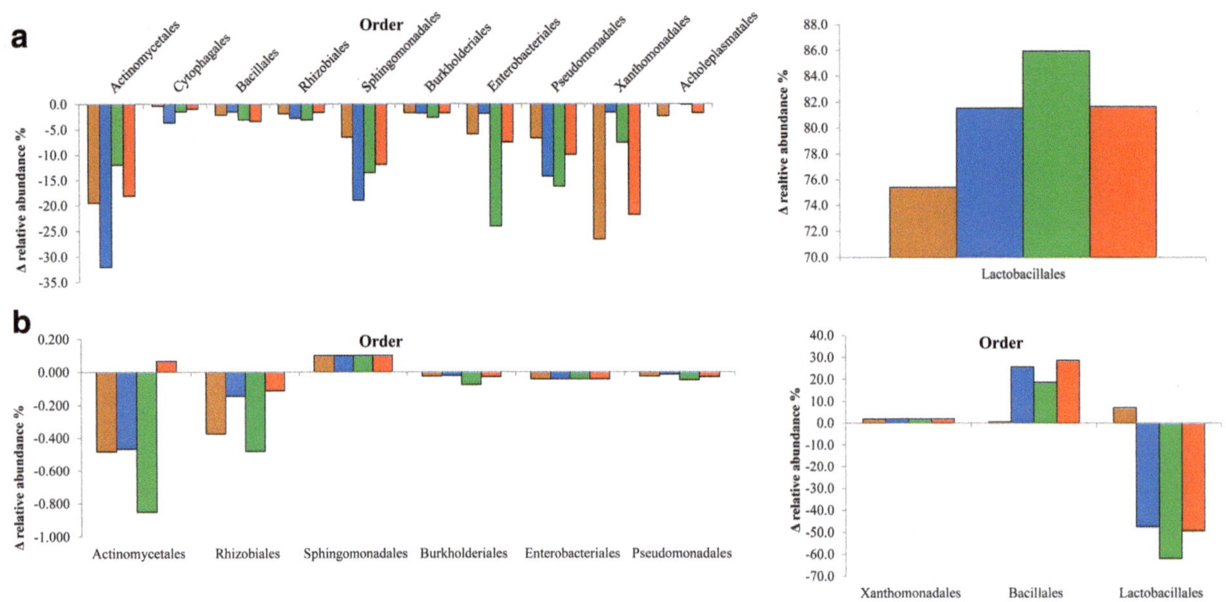


**Fig. 5** Bacterial core microbiome. **a** Average difference in the relative abundance of OTUs observed in bacterial core microbiome for barley (*orange*), oat (*blue*), triticale (*green*) and intercropped (*red*) fresh forages as compared to terminal silages (0 to 90 day) ($n = 3$). Bacterial orders that exhibited variation ≤ to 1% are not shown. Members of the Lactobacillales are presented separately to ease interpretation. **b** Average difference in the relative abundance of OTUs observed in bacterial core microbiome for barley (*orange*), oat (*blue*), triticale (*green*) and intercropped (*red*) silages after 14 days of aerobic exposure (90 to 104 days) ($n = 3$). Bacterial orders that exhibited variation ≤ to 0.05% in all silages are not shown. Xanthomonadales, Bacillales and Lactobacillales are presented separately to ease interpretation

OTUs in the Tremellales and Agaricales increased in barley and oat silages. The number of OTUs associated with Capnodiales, identified as members of the *Cladosporium* (Additional files 4 and 5: Figures S3 and S4) also declined, representing only 8% of the fungal core microbiome after ensiling (Fig. 6b). Saccharomycetales represented almost 70% of the fungal core microbiome in terminal silage, with most of the OTUs being associated with *Kazachtania* (61%) and *Pichia* (8%). (Additional files 4 and 5: Figures S3 and S4). The fungal core microbiome was not consistent among terminal silages as the relative abundance of Saccharomycetales increased by 1.52% in barley silage as compared to an increase of 73.95 to 96.37% OTUs in other silages (Fig. 7a). In contrast, Pleosporales slightly decreased in relative abundance in all silages. Orders such as Eurotiales and Chaetothyriales represented a small proportion of the fungal core microbiome with few differences between fresh small grain forages and terminal silages (data not shown).

#### *Fungal core microbiome in aerobically exposed silage*

In general, the diversity of the fungal core microbiome decreased during aerobic exposure (Fig. 6c). As many as 7 fungal orders that were detected in terminal silage were no longer detected in aerobically exposed silage. Fifteen orders of the fungal core microbiome underwent changes in relative abundance after aerobic exposure of which 4 orders accounted for less than 0.7% variation in all silages (Fig. 7b). The greatest decline was observed in barley and oat silages, but the Sporidiobolales, Pleosporales, Puccinales, Tremellomycetes, Agaricales and Cystofilobasidiales also declined in all silages after aerobic exposure (Fig. 7b). An increase in diversity and OTUs associated with the Microascales, occurred because of an increase in the *Corollospora* (Additional file 6: Figure S5), mainly in barley, oat and intercropped silages (Fig. 7b). The relative abundance of OTUs belonging to Saccharomycetales also increased by 55.33 and 2.85% in barley and triticale silages respectively, whereas these populations decreased in oat and intercropped silages (−29.35 and −27.51%). Saccharomycetales still represented 70% of the fungal core genome after aerobic exposure, primarily due to the high abundance of *Kazachstania* (62%) and *Pichia* (8%) (Additional file 6: Figure S5). An increase of 12.92% in relative abundance of Hypocreales was also observed in intercropped silage (Fig. 7b).

## Discussion

Present knowledge of the silage microbiome has been mainly garnered through culture dependant methods which are known to underestimate microbial diversity. Microbial epiphytic population of plants may affect the quality and productivity of agricultural crops [46]. In this study, rDNA sequencing was used to elucidate the dynamics of the microbial ecology of small grain silages

**Table 2** Taxonomic assignment, number and percent of shared and unique OTUs in oat, triticale and intercropped silages and barley silage respectively after 14 days of aerobic exposure

| Item | Phylum | Class | Order | Family | OTU number | OTU % |
|---|---|---|---|---|---|---|
| Shared OTUs between oat, triticale and intercropped silages | | | | | 85 | 100 |
| | Firmicutes | Bacilli | Bacillales | Bacillaceae | 60 | 70.59 |
| | Firmicutes | Bacilli | Bacillales | NM | 5 | 5.88 |
| | Firmicutes | Bacilli | Bacillales | Planococcaceae | 19 | 22.36 |
| | Firmicutes | Bacilli | Lactobacillales | Streptococcaceae | 1 | 1.18 |
| Unique barley silage OTUs | | | | | 114 | 100 |
| | Actinobacteria | Actinobacteria | Actinomycetales | NS | 14 | 12.28 |
| | Actinobacteria | Thermoleophilia | Solirubrobacterales | NS | 2 | 1.75 |
| | Bacteroidetes | Cytophagia | Cytophagales | NS | 2 | 1.75 |
| | Bacteroidetes | Flavobacteriia | Flavobacteriales | NS | 7 | 6.14 |
| | Bacteroidetes | Sphingobacteriia | Sphingobacteriales | NS | 7 | 6.14 |
| | Firmicutes | Bacilli | Bacillales | NS | 2 | 1.75 |
| | Firmicutes | Bacilli | Lactobacillales | NS | 31 | 27.19 |
| | Proteobacteria | α-proteobacteria | Rhizobiales | NS | 11 | 9.65 |
| | Proteobacteria | α-proteobacteria | Rhodobacterales | NS | 3 | 2.63 |
| | Proteobacteria | α-proteobacteria | Sphingomonadales | NS | 6 | 5.26 |
| | Proteobacteria | β-proteobacteria | Burkholderiales | NS | 9 | 7.89 |
| | Proteobacteria | γ-proteobacteria | Alteromonadales | NS | 1 | 0.88 |
| | Proteobacteria | γ-proteobacteria | Enterobacteriales | NS | 6 | 5.26 |
| | Proteobacteria | γ-proteobacteria | Pseudomonadales | NS | 9 | 7.89 |
| | Proteobacteria | γ-proteobacteria | Xanthomonadales | NS | 3 | 2.63 |
| | Tenericutes | Mollicutes | Acholeplasmatales | NS | 1 | 0.88 |

*Abbreviations*: *NM* not measured, *NS* non shown

by studying the variability over time of the silage microbial core genome. While silages can have their own distinct microbial profile, the true 'core microbiome' can be defined as those microbes that are common to all small grain cereals.

### Silage characteristics

As observed by McCartney and Vaage [6], differences in stage of maturity among forages at harvest influences the composition and quality of ensiled crops. In this study, all small grain cereal silages, with the exception of barley silage, obtained a pH of < 4.0 after 90 days of ensiling. Barley silage was harvested at the firm-dough stage and exhibited a high DM and low WSC content, suggesting it was overly mature at harvest [47]. Compared to the other silages, levels of lactic acid were notably lower and acetic acid higher in barley silage. This indicates a strong homofermentative acid fermentation in oat, triticale and intercropped silages, whereas a switch from homo- to heterofermentation likely occurred during the ensiling of barley. Only traces of butyric acid were found in oat and intercropped silages, indicating that all silages were well conserved and that growth of spoilage microorganisms was limited. These observations were in accordance with the microbial counts observed, as no molds were detected at opening, yeasts were only cultured from oat, triticale and intercropped silages and numbers of LAB exceeded 7.83 $\log_{10}$ CFU/g DM in all small grain silages.

### Core microbiome during ensiling of small grain cereal forages

The number of different OTUs identified were higher for fungal than for the bacterial core microbiome, a result that can be explained by differences in the databases and analyzes used for taxonomic assignments. Bacterial taxonomic assignments relied on the Greengenes database and a RDP classifier with a depth taxonomic OTU threshold and similarity cut off for taxonomic clustering. As the RDP classifier is not designed to process 18S rDNA sequences, the Silva database was used for fungal identification and only a similarity cut off was applied for taxonomic assignment. It might be possible that OTUs belonging to the same lineage (i.e., one hit identified at the family level and the other at the genus level for example) were identified as different OTUs through

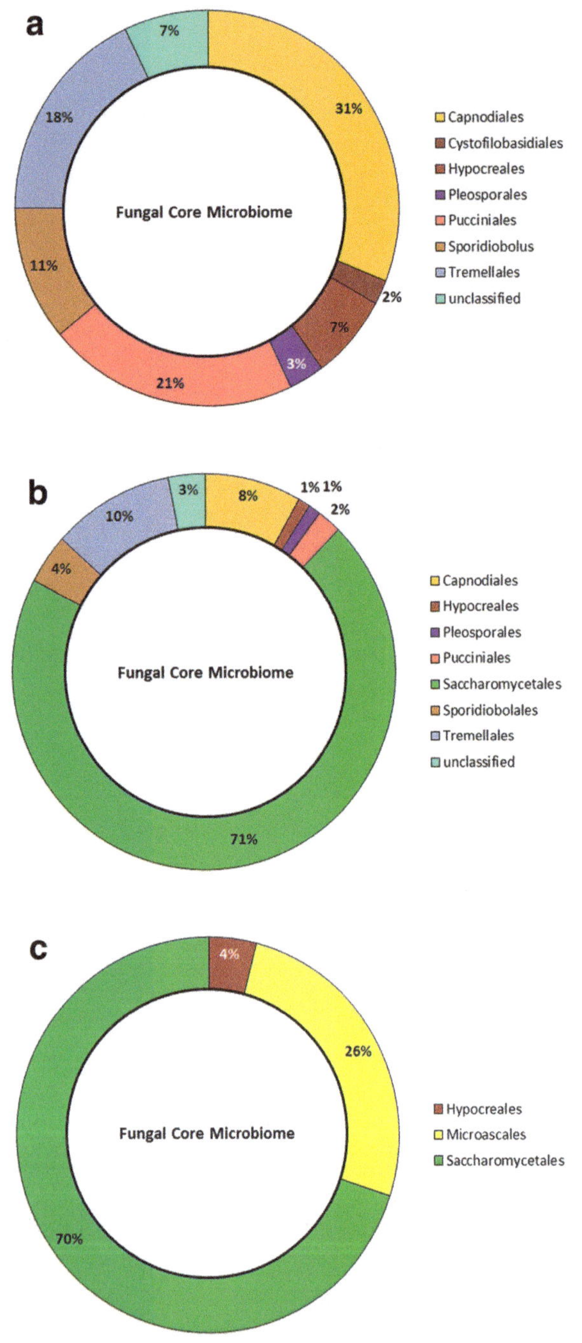


**Fig. 6** Taxonomic profile and relative abundance of the fungal core microbiome after 0 days, fresh forage **a** 90 days, terminal silage **b** and terminal silage after 14 days of aerobic exposure **c**. OTUs were assigned at the order level

18S rDNA analysis, increasing the final number of estimated fungal OTUs, a bias that likely did not occur during bacterial analysis. The validity of these taxonomic assignments was still acceptable as bacterial and fungal communities were never compared directly. Bacterial communities in silage varied over time as shown by PCoA. The greatest microbial diversity in the core bacterial microbiome of barley, oat, triticale and intercropped silages was observed to be associated with the fresh forage. The bacterial population observed at that time was dominated by Proteobacteria, mainly members of the Xanthomonadales, Pseudomonadales Enterobacteriales and Sphingomonadales. Actinobacteria also accounted for a significant portion of the core bacterial microbiome at ensiling. These results are in accordance with literature as α- and γ-Proteobacteria were found to be dominant members of the bacterial microbiome on maize leaves [48]. Leaves of field-grown barley (cultivar Sumo) were found to be largely colonized by Pseudomonadales, although Enterobacteriales such as *Erwinia* were also detected [49]. *Azospirillum* and *Agrobacterium* belonging to α-proteobacteria have also been shown to be associated with oat leaves [50]. It is noteworthy that the initial proportion of Lactobacillales in the core bacterial microbiome of small grain forages was low. This is a reflection of the limited LAB population observed in the field (only 26 different OTUs belonging to Lactobacillales were identified in fresh forages), but also the diversity of epiphytic LAB found among the different crops, precluding some of the OTUs as contributors to the core microbiome of fresh small grain forages. Others have found that LAB constitute a small proportion of the epiphytic bacterial population in most forages as Mogodiniyai Kasmaei et al. [51] observed a relative abundance of initial LAB population in green grass, red clover and maize ranging from 0.18 to 3.77%.

Little information is available with regard to the epiphytic fungal communities in silage, although the presence and growth of mycotoxigenic fungi has been extensively studied [52]. Barley and oat forages can be heavily colonized by mycotoxigenic fungi such as *Fusarium* which belongs to the Hypocreales [53]. *Fusarium, Aspergillus, Penicillium, Mucor, Rhizopus, Cladosporium,* and *Absidia* were identified in a 4-year survey of small grain cereals including barley, oats and triticale in eastern Romania [54]. We also identified the presence of potential mycotoxigenic fungi within the core microbiome of small grain silages such as members of the *Cladosporium* of the Capnodiales. Although we did not specifically identify *Fusarium,* fungi belonging to the Hypocreales were found in high quantities in barley, intercropped and triticale silages.

Yeasts are considered one of the main fungal colonizers of wheat leaves [55] and several genera have been identified in barley including *Sporobolomyces* and *Cryptococcus* [56]. The fungal core microbiome at ensiling was dominated by members of the Basidiomycota phylum, mainly members of the Pucciniales and Tremellales, such as *Cryptococcus* or Sporodiobolales, such as *Sporidiobolus.* Pucciniales OTUs were identified primarily as *Puccinia,* which was more prominent in oat silage. *Puccinia* are known to be responsible for rust, a

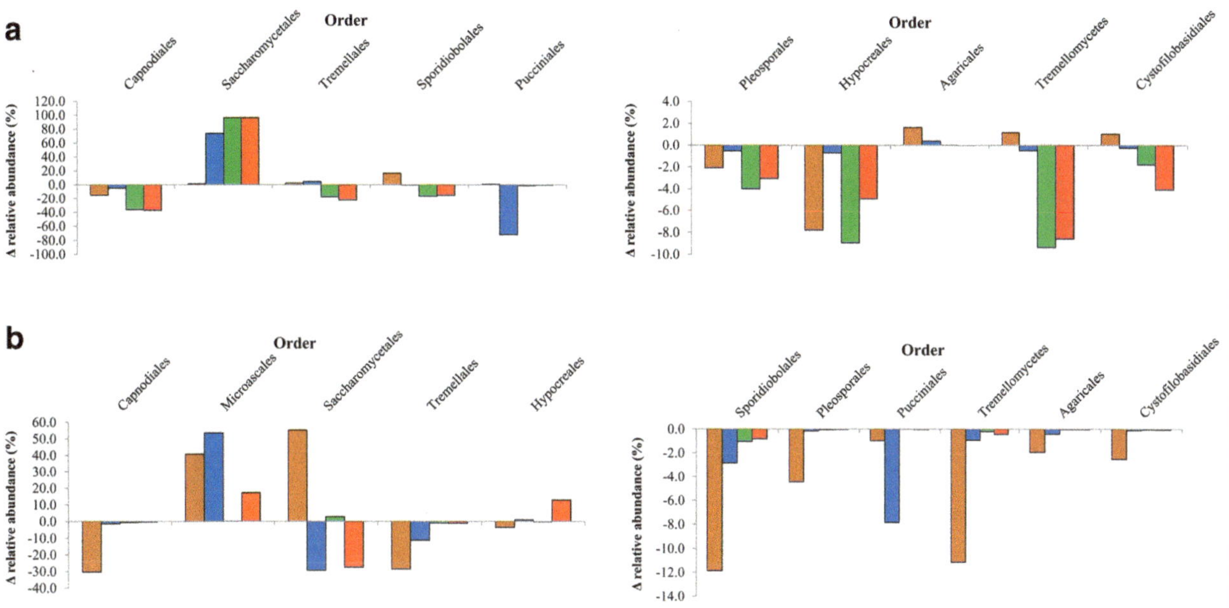


**Fig. 7** Fungal core microbiome during ensiling. **a** Average difference in the relative abundance of OTUs observed in fungal core microbiome for barley (*orange*), oat (*blue*), triticale (*green*) and intercropped (*red*) forages as compared to terminal silages (0 to 90 day) ($n = 3$). Fungal orders that exhibited variation ≤ 1% are not shown. **b** Average difference in the number of OTUs observed in fungal core microbiome for barley (*orange*), oat (*blue*), triticale (*green*) and intercropped (*red*) silages after aerobic exposure for 14 days (from 90 to 104 days) ($n = 3$). Fungal orders that exhibited variation ≤ 0.7% are not shown

disease that results in significant losses in small grain cereal crops [57]. Stripe rust caused by *Puccinia striiformis* has been frequently found in wheat, barley and triticale in Alberta [58].

The main microbial changes that occur during ensiling are well known from a culturable perspective [59] and have been recently reviewed [13]. Aerobic microflora remain metabolically active in silage for a few days until oxygen is depleted and acidification inhibits microbial metabolism. In properly ensiled forage, facultative anaerobic bacteria gradually decrease silage pH, promoting the growth of acid-tolerant LAB, which typically dominate the bacterial microbiome of terminal silage. Numbers of viable LAB tend to decrease during storage with the exception of some heterofermentative species such as *Lactobacillus buchneri,* which convert lactic acid into acetic acid and have been identified in corn silage a month after fermentation [60]. Culture independent techniques such as DGGE, T-RFLP or LH-PCR have been used to characterized silage bacterial communities and shown a predominance of *Lactobacillus, Lactococcus* and *Pediococcus* during ensiling [61, 62]. These observations are in accordance with our results as the core bacterial microbiome after 90 day of fermentation in silage showed reduced diversity and was largely dominated by members of the Lactobacillales. Using 16S pyrosequencing and metagenomic analysis, Eikmeyer et al. [23] observed a predominance of bacteria from the phylum Firmicutes in grass silage after 58 days of ensiling. Members of the Lactobacillales, including *Lactococcus, Lactobacillus* and *Weissella* accounted for 80% of relative bacterial abundance, a result similar to our observations. Bacterial population sequencing in ryegrass, whole crop corn and alfalfa silages also illustrated the intensive selection process that occurs during ensiling as bacterial diversity in initial forages dramatically declined with Leuconostocaceae and Lactobacillaceae representing about 70 to 95% of the bacterial population in terminal silages [63]. Bao et al. [64] used single molecule real-time sequencing technology to study the bacterial population in inoculated alfalfa silages. They observed an expansion in the LAB population during ensiling with inoculated LAB species dominating, but the extent of this domination could not be ascertained as uninoculated control silage was not included in the study. The shift in bacterial microbiome from Proteobacteria and Actinobacteria to Firmicutes is key to ensuring the conservation of small grain silages. Members of the Proteobacteria, Xanthomonadales, Sphingomonadales, Enterobacteriales and Actinomycetales comprised the remainder of the microbiome in small grain silages. McEniry et al. [65], observed that the population of Enterobacteria associated with grass silage declined within 48 h after ensiling.

Butyric acid producing bacteria (BAB) such as Clostridia and *Bacillus* are considered undesirable in silage. These bacteria promote the growth of less acid-tolerant spoilage microorganisms and can contaminate milk and milk products with BAB spores, resulting in significant economic losses [12]. Clostridia was not detected within the core microbiome of terminal silage in this study and only trace amounts of butyric acid were observed in oat and intercropped silages. Our study used mini-silos, which can create an ensiling environment that is likely more homogeneous than that achieved in field scale bunker or upright silos. These differences in ensiling environment may also influence the nature of the microbiome that establishes in terminal silage. Kraut-Cohen et al. [66] defined bacterial populations in bunker silos using Ion Torrent PGM sequencing technology, as being "preserved" regions in the center and "spoiled" regions along the edges, with the composition of the bacterial microbiome clearly differing between these sampling sites. Although numbers of total bacteria and LAB were the lowest in triticale silages after ensiling, a higher number of specific OTUs were observed in this silage. This suggests that some of the microorganisms associated with triticale silage were nonculturable using plating techniques within the laboratory, an observation that may also account for the slightly higher numbers of LAB as compared to total bacteria in this silage.

The limitations of using culture-based approaches to document the impact of fungal populations in small grain silages is illustrated by our inability to culture molds and to only detect yeast in oat, triticale and intercropped silages. Use of sequencing to characterize the fungal microbiome identified 21 OTUs within the core microbiome. The number of fungal OTUs did decline with ensiling, reflecting the acidification and transition from an aerobic to an anaerobic environment. Although most fungi are strict aerobes, a few filamentous fungi and several yeasts are capable of fermentative growth and can survive the ensiling process. The strongest decrease in members of the fungal core microbiome was observed for *Cladosporium* a member of the Capnodiales. Saccharomycetales dominated the core fungal microbiome in all silages, with the exception of barley silage. The initial low level of WSC in barley silage (4.66 mg/g DM) may have limited the growth of Saccharomycetales as compared to the other silages. Saccharomycetales including *Candida* and *Issatchenkia* have been shown to be associated with barley silage using culture techniques, but no filamentous fungi were observed [67]. Members of the Sporidiobolales and Tremellales were also found to be minor components of the fungal core microbiome in small grain silages. As in our study, others have used molecular techniques to confirm that members of the Saccharomycetales including *Candida*, *Pichia* or *Saccharomyces* are the predominant fungi associated with terminal silage [68–70].

#### Core microbiome in aerobically exposed silage

The diversity of the core bacterial microbiome decreased during aerobic exposure. Oat, triticale and intercropped silages exhibited similar bacterial diversity, whereas a high number of specific OTUs were observed in barley silage. These observations are in accordance with the significant lower counts of total bacterial and LAB populations in barley silage as compared to other silages. However, the number of OTUs associated with Lactobacillales increased in aerobically exposed barley. It can be hypothesized that some of the Lactobacillales species were in a viable, but nonculturable state, with these populations increasing during aerobic exposure. This is in accordance with the high DM content observed at opening in barley silage. This low level of moisture likely prevented a rapid growth of many microorganisms and limited further silage spoilage during aerobic exposure as indicated by its low temperature profile and a lack of pH increase. In contrast, the temperature of oat, triticale and intercropped silages increased and high pH values were reached in these silages after aerobic exposure. Exposure to oxygen is known to promote yeast growth and lactic acid consumption. Contrary to the observations in other silages, diversity of Lactobacillales and Xanthomonadales increased in barley silage during aerobic exposure and far fewer OTUs were assigned to Bacillales as compared to other silages. The Lactobacillales present in barley silage during ensiling may have proliferated at the beginning of the aerobic exposure despite the high DM content observed. Such a scenario in aerobically exposed silage could lead to a slower increase in silage pH and thus limit the growth of spoilage microorganisms.

*Listeria* is considered as one of the main pathogenic bacteria in silage [71] as it is responsible for listeriosis which can result in encephalitis, abortion, septicemia and even death in animals and humans [72]. Although *Listeria* belongs to the Bacillales family, which increased in most silages during aerobic exposure, none of the observed OTUs were identified as *Listeria* at the sequencing depth used in this study. Production of butyric acid in silage arises primarily from *Clostridium* and *Bacillus* and can result in reductions in silage intake [12, 73]. Although these undesirable bacteria were not formally identified at the species level, many OTUs were assigned to *Bacillus* after 14 days of aerobic exposure in all silages. These observations confirm the significant role of Bacillales in the deterioration of small grain silage during aerobic exposure.

Yeasts belonging to the Saccharomycetales such as *Issatchenkia* sp., *Candida* sp. and *Saccharomyces* sp. have been shown to be associated with barley silage during aerobic exposure [67]. In this study, the Saccharomycetales order dominated the core fungal microbiome of silage before and after aerobic exposure. Most of these OTUs were assigned to *Kazachstania* and *Pichia*. Depending on the teleomorph or anamorph state of the cell, some *Kazachstania* can also be identified as *Candida* species [74]. It can thus be hypothesized that the high proportion of Saccharomycetales observed in the core microbiome after aerobic exposure is likely reflective of the main spoilage genera, *Candida* and *Pichia* [75]. Intercropped and oat silages contained the lowest concentration of acetic acid, which is known to improve aerobic stability of silage by inhibiting the growth of fungi [9, 76]. These two silages exhibited similar temperature profiles during aerobic exposure and the poorest aerobic stability. Aerobic exposure in these silages was linked with a decrease in the diversity and abundance of the Saccharomycetales and an increase in the diversity of Microascales and Hypocreales. This is in accordance with the literature as the succession of fungal species during aerobic exposure of silage is typically initiated by yeasts with the increase in pH allowing less acid-tolerant spoilage microorganisms to proliferate [77]. According to the temperature profile, spoilage was delayed in triticale silage as the switch from Saccharomycetales to other spoilage fungi did not appear to occur even after 14 days of aerobic exposure. A clear link cannot be made between Saccharomycetales abundance and temperature in barley silage as there was no evidence of deterioration during aerobic exposure. As a result, Saccharomycetales diversity and relative abundance increased as well as for the Microascales. Microascales have been found in composted silage using Illumina Miseq sequencing [23] and the filamentous fungal species *Microascus brevicaulis* has been identified in barley silage after exposure to air [67]. Fungi such as *Aspergillus flavus* and *Penicillium verrucosum*, belonging to the Eurotiales are known to be among the main post-harvest mycotoxin producers and are thus considered undesirable in silage [78]. Their presence has been previously observed in barley silage several decades ago [79]. OTUs assigned to Eurotiales were found in the fungal core microbiome although their relative abundance was only of 0.2%, and they were associated solely with oat silage.

## Conclusion

To our knowledge, this study is the first to use next-generation sequencing to define the core bacterial and fungal microbiome of small grain silages. Using rDNA gene sequencing through Illumina technology highlighted the role of the main bacterial and fungal orders as well as the microbial succession that occured during the ensiling process. As expected, a high dominance of Lactobacillales species was observed during storage and growth of spoilage microorganisms was noted during aerobic exposure. This work is also the first to study the core fungal microbiome depicting fungal communities succession during the complex process of ensiling (from fresh forage to aerobically exposed silage). This paper has highligted the dominance of yeasts belonging to the Saccharomycetale order in the core microbiome of small grain silages after ensiling. Aerobic exposure was characterized by an increase of OTUs belonging to the Hypocreales, frequently associated with saprophytic fungi, indicating a decaying process in silage, or with plant pathogenic fungi such as *Sarocladium* sp.. Using next generation sequencing to define the microbial ecology of silage could facilitate the development of additives that could act synsergistically with defined populations to improve the quality and aerobic stability of small grain cereal silages.

## Additional files

**Additional file 1: Figure S1.** Rarefaction curves. Rarefaction curves depicting the effect of 3% dissimilarity on the number of bacterial (A) or fungal (B) OTUs observed for barley (orange), oat (blue), triticale (green) and intercropped (red) silages.

**Additional file 2: Table S1.** Bacterial and Fungal Diversity Index. **Table S2.** Mapping File used for MOTHUR pipeline.

**Additional file 3: Figure S2.** PCo analysis. Principal coordinates analysis for bacterial (left) and fungal (right) communities according to sampling time; fresh forage, terminal silage, aerobically exposed silage (A and B) and silage type (C and D).

**Additional file 4: Figure S3.** Taxonomic profile and relative abundance of the fungal core microbiome of fresh forage. OTUs were assigned at the genus level.

**Additional file 5: Figure S4.** Taxonomic profile and relative abundance of the fungal core microbiome after ensiling (90 day). OTUs were assigned at the genus level.

**Additional file 6: Figure S5.** Taxonomic profile and relative abundance of the fungal core microbiome after aerobic exposure (14 days). OTUs were assigned at the genus level.

### Abbreviations

ADF: Acid detergent fibre; ADIN: Acid detergent insoluble nitrogen; BAB: Butyric acid bacteria; CFU: Colony forming units; CP: Crude protein; DDGE: Denaturing gradient gel electropheriss; DM: Dry matter; DNA: Deoxyribonucleic acid; EDTA: Ethylenediaminetetraacetate; K: Potassium; LAB: Lactic acid bacteria; LH-PCR: Length heterogeneity polymerase chain reaction; MRS: Man Rogosa Sharpe; N: Nitrogen; NA: Nutrient agar; NDF: Neutral detergent fibre; OM: Organic matter; OTU: Operational taxonomic unit; P: Phosphorus; PCoA: Principal component analysis; rDNA: Ribosomal DNA; RDP: Ribosomal database project; SDA: Sabouraud dextrose agar; SSU: Small sub unit; TB: Total bacteria; T-RFLP: Terminal restriction fragment length polymorphism; WSC: Water soluble carbohydrate

### Acknowledgments

The authors thank C. Barkley, Z. Madic, D. Vedres, F. Van Herk, R. Brandt, L. Schneider, N. Rauhala, D. Clark and J. Busaan for their invaluable technical assistance on this project.

### Funding

Financial support for this study from Barley Research Cluster of Agriculture and Agri-Food Canada is gratefully acknowledged. The funder had no role in the design, collection of samples, analysis, interpretation or writing of the manuscript.

## Authors' contributions

Conceived and designed the experiments: LD, KT, TAM. Performed the experiments: LD, SX, LJ, YW. Analyzed data: LD, CE, RF. Wrote the paper: LD, TAM. All authors have read and approve of the final version of the manuscript.

## Competing interests

The authors declare that they have no competing interests.

## Author details

[1]Agriculture and Agri-Food Canada Research Centre, Lethbridge T1J 4B1, AB, Canada. [2]Agriculture and Agri-Food Canada (AAFC), Lacombe T4L 1 W1, AB, Canada.

## References

1. Monfreda C, Ramankutty N, Foley JA. Geographical distribution of crop areas, yields, physiological types, and net primary production in the year 2000. Global Biochem Cycles. 2008;22:1–19.
2. Weinberg ZG, Ashbell G. Engineering aspects of ensiling. Biochem Eng J. 2003;13:181–8.
3. Alberta Agriculture Food and Rural Development. Triticale production and utilization manual 2005: spring and winter triticale for grain, forage and value-added. Edmonton: Alberta Agriculture, Food and Rural Development; 2005.
4. Kennelly J, Okine EK, Khorasani R. Barley as a grain and forage source for ruminants. Edmonton: Alberta Agriculture, Food and Rural Development; 1995.
5. Kirilov A. Fooder oats in Europe. In: Suttie, J.M., Reynolds, S.G., editors. Fodder oats: a world overview. 2004. FAO Plant production and protection Series N°33. Rome: FAO.
6. McCartney DH, Vaage AS. Comparative yields and feeding value of barley, oat and triticale silages. Can J Anim Sci. 1994;74:91–6.
7. Gollop N, Zakin V, Weinberg ZG. Antibacterial activity of lactic acid bacteria included in inoculants for silage and in silages treated with these inoculants. J Appl Microbiol. 2005;98:662–6.
8. Baah J, McAllister TA, Bos L, Van Herk F, Charley RC. Effect of *Lactobacillus buchneri* 40788 and buffered propionic acid on preservation and nutritive value of alfalfa and timothy high-moisture hay. Asian-Autralas J Anim Sci. 2005;18:649–60.
9. Filya I, Sucu E. The effect of bacterial inoculants and a chemical preservative on the fermentation and aerobic stability of whole-crop cereal silages. Asian-Australas J Anim Sci. 2007;20:378–84.
10. Addah W, Baah J, Groenewegen P, Okine EK, McAllister TA. Comparison of the fermentation characteristics, aerobic stability and nutritive value of barley and corn silages ensiled with or without a mixed bacterial inoculant. Can J Anim Sci. 2011;91:133–46.
11. Pedroso AF, Nussio LG, Paziani SF, Loures RDS, Igarasi MS, Coelho RM, et al. Fermentation and epiphytic microflora dynamics in sugar cane silage. Sci Agric. 2005;62:427–32.
12. Visser MMM, Driehuis F, Te Giffel MC, De Jong P, Lankveld JMG. Concentration of butyric acid bacteria spores in silage and relationship with aerobic deterioration. J Dairy Sci. 2007;90:928–36.
13. Dunière L, Sindou J, Chaucheyras-Durand F, Chevallier I, Thévenot-Sergentet D. Silage processing and strategies to prevent persistence of undesirable microorganisms. Anim Feed Sci Technol. 2013;182:1–15.
14. Woolford MK. The detrimental effects of air on silage. J Appl Bacteriol. 1990;68:101–16.
15. Oliver JD. The viable but nonculturable state in bacteria. J Microbiol. 2005;43:93–100.
16. Salma M, Rousseaux S, Sequeira-Le Grand A, Divol B, Alexandre H. Characterization of the viable but nonculturable state in *Saccharomyces cerevi*siae. Plos One. 2013;8:e77600.
17. Blaalid R, Carlsen TOR, Kumar S, Halvorsen R, Ugland KI, Fontana G, et al. Changes in the root-associated fungal communities along a primary succession gradient analysed by 454 pyrosequencing. Mol Ecol. 2012;21:1897–908.
18. Bulgarelli D, Rott M, Schlaeppi K, Ver Loren Van Themaat E, Ahmadinejad N, Assenza F, et al. Revealing structure and assembly cues for Arabidopsis root-inhabiting bacterial microbiota. Nat Lett. 2012;488:91–5.
19. Keshri J, Mishra A, Jha B. Microbial population index and community structure in saline–alkaline soil using gene targeted metagenomics. Microbiol Res. 2013;168:165–73.
20. Lim YW, Kim BK, Kim C, Jung HS, Kim B-S, Lee HJ, et al. Assessment of soil fungal communities using pyrosequencing. J Microbiol. 2010;48:284–9.
21. Neher DA, Weich TR, Bates ST, Leff JW, Fierer N. Changes in bacterial and fungal communities across compost recipes, preparation methods, and composting times. Plos One. 2013;8:e79512.
22. Fouts DE, Szpakowski S, Purushe J, Torralba M, Waterman RC, MacNeil MD, et al. Next generation sequencing to define prokaryotic and fungal diversity in the bovine rumen2012. Plos One. 2012;7:e48289.
23. Eikmeyer FG, Köfinger P, Poschenel A, Jünemann S, Zakrzewski M, Heini S, et al. Metagenome analyses reveal the influence of the inoculant *Lactobacillus buch*neri CD034 on the microbial community involved in grass ensiling. J Biotechnol. 2013;167:334–43.
24. Nyachiro JM, Helm JH, Oro M, Juskiw PE, Salmon DF. Registration of 'Sundre' barley. J Plant Reg. 2007;1:35–6.
25. Kibite S, Menzies JG. AC Morgan oat. Can J Plant Sci. 2001;81:85–7.
26. Salmon DF, Helm JH, Juskiw PE, Nyachiro J, Albers S, Bergen C, Oro M. Registration of 'Bunker' triticale. J Plant Reg. 2007;1:101.
27. Alberta Agriculture and Rural Development. Crop Protection 2013. Edited by H. Brook and M. Cutts. Edmonton: Alberta Agriculture and Rural Development; 2013.
28. Teller RS, Schmidt RJ, Whitlow LW, Kung Jr L. Effect of physical damage to ears of corn before harvest and treatment with various additives on the concentration of mycotoxins, silage fermentation, and aerobic stability of corn silage. J Dairy Sci. 2012;95:1428–36.
29. Zahiroddini H, Baah J, Absalom W, McAllister TA. Effect of an inoculant and hydrolytic enzymes on fermentation and nutritive value of whole crop barley silage. Anim Feed Sci Technol. 2004;117:317–30.
30. Kudo H, Cheng KJ, Costerton JW. Interactions between *Treponema bryantii* and cellulolytic bacteria in the in vitro degradation of straw cellulose. Can J Microbiol. 1987;33:244–8.
31. Addah W, Baah J, Okine EK, McAllister TA. A third-generation esterase inoculant alters fermentation pattern and improves aerobic stability of barley silage and the efficiency of body weight gain of growing feedlot cattle. J Anim Sci. 2011;90:1541–52.
32. De Man JC, Rogosa M, Sharpe ME. A medium for the cultivation of Lactobacilli. J Appl Bacteriol. 1960;23:130–5.
33. Yu Z, Morrison M. Improved extraction of PCR-quality community DNA from digesta and fecal samples. Biotechniques. 2004;36:808–12.
34. Borneman J, Hartin RJ. PCR primers that amplify fungal rRNA genes from environmental samples. Environ Microbiol. 2000;66(10):4356–60.
35. Magoc T, Salzberg SL. FLASH: fast length adjustment of short reads to improve genome assemblies. Bioinformatics. 2011;27:2957–63.
36. Ghodsi M, Liu B, Pop M. DNACLUST: accurate and efficient clustering of phylogenetic marker genes. BMC Bioinformatics. 2011;12:271.
37. Edgar RC, Haas BJ, Clemente JC, Quince C, Knight R. UCHIME improves sensitivity and speed of chimera detection. Bioinformatics. 2011;27:2194–200.
38. Desantis TZ, Hugenholtz P, Larsen N, Rojas M, Brodie EL, Keller K, et al. Greengenes, a chimera-checked 16S rRNA gene database and workbench compatible with ARB. Appl Environ Microbiol. 2006;72:5069–72.
39. Caporaso JG, Bittinger K, Bushman FD, Desantis TZ, Andersen GL, Knight R. PyNAST: a flexible tool for aligning sequences to a template alignment. Bioinformatics. 2010;26:266–7.

40. Caporaso JG, Kuczynski J, Stombaugh J, Bittinger K, Bushman FD, Costello EK, et al. QIIME allows analysis of high-throughput community sequencing data. Nat Methods. 2010;7:335–6.
41. Kuczynski J, Stombaugh J, Walters WA, Gonzalez A, Caporaso JG, Knight R. Using QIIME to analyze 16S rRNA gene sequences from microbial communities. 2011. Curr Protoc Bioinformatics Chapter 10: Unit 10.17. doi:10.1002/0471250953.bi1007s36.
42. Kozich JJ, Westcott SL, Baxter NT, Highlander SK, Schloss PD. Development of a dual-index sequencing strategy and curation pipeline for analyzing amplicon sequence data on the MiSeq Illumina sequencing platform. Environ Microbiol. 2013;79(17):5112–20.
43. Quast C, Pruesse E, Yilmaz P, Gerken J, Schweer T, Yarza P, et al. The SILVA ribosomal RNA gene database project: improved data processing and web-based tools. Nucleic Acids Res. 2013;41(D1):D590–6.
44. Anders S, Pyl PT, Huber W. HTSeq - a python framework to work with high-throughput sequencing data. Bioinformatics. 2014;31(2):166–9.
45. SAS Institute Inc. SAS OnlineDoc®9.3.1. Cary: SAS Institute Inc; 2012.
46. Whipps JM, Hand P, Pink D, Bending GD. Phyllosphere microbiology with special reference to diversity and plant genotype. J Appl Microbiol. 2008; 105:1744–55.
47. Schroeder JW. Haylage and other fermented forages: NDSU extension service Fargo. North Dakota: Quality Forage AS1252; 2004.
48. Kadivar H, Stapleton AE. Ultraviolet radiation alters maize phyllosphere bacterial diversity. Microb Ecol. 2003;45:353–61.
49. Gravouil C, Fountaine JM, Dickinson MJ, Newton AC. Ecological tolerance: changing our approach to crop protection. Proc Crop ProtN Br. 2012;2012:95–100.
50. Rekosz-Burlaga H, Garbolinska M. Characterization of selected groups of microorganisms occurring in soil rhizosphere and phyllosphere of oats. Pol J Microbiol. 2006;55:227–35.
51. Mogodiniyai Kasmaei K, Dicksved J, Spörndly R and Udén P. Separating the effects of forage source and field microbiota on silage fermentation quality and aerobic stability. Grass Forage Sci. 2016; doi: 10.1111/gfs012238
52. Alonso VA, Pereyra CM, Keller LAM, Dalcero AM, Rosa CAR, Chiacchiera SM, et al. Fungi and mycotoxins in silage: an overview. J Appl Microbiol. 2013;115:637–43.
53. Streit E, Schatzmayr G, Tassis P, Tzika E, Marin D, Taranu I, et al. Current situation of mycotoxin contamination and co-occurrence in animal feed—focus on Europe. Toxins. 2012;4:788–809.
54. Tabuc C, Taranu I, Calin L. Survey of moulds and mycotoxin contamination of cereals in South-Eastern Romania in 2008–2010. Arch Zootech. 2011;14:25–38.
55. Karlsson I, Friberg H, Steinberg CPP. Fungicide effects on fungal community composition in the wheat phyllosphere. Plos One. 2014;9(11):e111786.
56. Buck JW, Andrews JH. Role of adhesion in the colonization of barley leaves by the yeast Rhodosporidium toruloides. Can J Microbiol. 1999;45:433–40.
57. Dean R, Van Kan JA, Pretorius ZA, Hammond-Kosack KE, Di Pietro A, Spanu PD, et al. The top 10 fungal pathogens in molecular plant pathology. Mol Plant Pathol. 2012;13:414–30.
58. Xi K, Turkington TK, Salmon D, McCallum BD, Navabi A. Stripe rust 101: what is it, why do we have it, what can be done about it. Alberta: Agriculture and Forestry; 2007. http://www1.agric.gov.ab.ca/$department/deptdocs.nsf/all/prm11383. Accessed 1 Nov 2015.
59. Woolford MK, Pahlow, G. The silage fermentation. Microbiology of fermented foods. Wood, BJB, ed. USA: Springer; 1997. p. 73-102.
60. Driehuis F, Elferink SJWHO, Spoelstra SF. Anaerobic lactic acid degradation during ensilage of whole crop maize inoculated with Lactobacillus buchneri inhibits yeast growth and improves aerobic stability. J Appl Microbiol. 1999;87:583–94.
61. Mceniry J, O'Kiely P, Clipson NJW, Forristal PD, Doyle EM. Assessing the impact of various ensilage factors on the fermentation of grass silage using conventional culture and bacterial community analysis techniques. J Appl Microbiol. 2010;108:1584–93.
62. Muck RE. Microbiology of ensiling. Finland: Proc XVI Int Silage Conf Hameenlinna; 2012. p. 75–86.
63. Ni K, Minh TT, Minh Tu TT, Tsuruta T, Pang H and Nishino N. Comparative microbiota assessment of wilted Italian ryegrass, whole crop corn, and wilted alfalfa silage using denaturing gradient gel electrophoresis and next-generation sequencing. Appl Microbiol Biotech. 2016; doi: 10.1007/s00253-016-7900-2
64. Bao W, Mi Z, Xu H, Zheng Y, Kwok LY, Zhang H, Zhang W. Assessing quality of Medicago sativa silage by monitoring bacterial composition with single molecule, real-time sequencing technology and various physiological parameters. Sci Rep. 2016;6:28358. doi:10.1038/srep28358.
65. Mceniry J, O'Kiely P, Clipson NJ, Forristal PD, Doyle EM. Bacterial community dynamics during the ensilage of wilted grass. J Appl Microbiol. 2008;105:359–71.
66. Kraut-Cohen J, Tripathi V, Chen Y, Gatica J, Volchinski V, Sela S, Weinberg Z, Cytryn E. Temporal and spatial assessment of microbila communities in commercial silages from bunker silos. Appl Microbiol Biotechnol. 2016;100: 6827–35.
67. Inglis GD, Yanke LJ, Kawchuk LM, McAllister TA. The influence of bacterial inoculants on the microbial ecology of aerobic spoilage of barley silage. Can J Microbiol. 1999;45:77–87.
68. Dolci P, Tabacco E, Cocolin L, Borreani G. Microbial dynamics during aerobic exposure of corn silage stored under oxygen barrier or polyethylene films. Appl Environ Microbiol. 2011;77:7499–507.
69. May LA, Smiley B, Schmidt MG. Comparative denaturing gradient gel electrophoresis analysis of fungal communities associated with whole plant corn silage. Can J Microbiol. 2001;47:829–41.
70. Rossi F, Dellaglio F. Quality of silages from Italian farms as attested by number and identity of microbial indicators. J Appl Microbiol. 2007; 103:1707–15.
71. Wiedmann M. ADSA foundation scholar award—an integrated science-based approach to dairy food safety: listeria monocytogenes as a model system. J Dairy Sci. 2003;86:1865–75.
72. Low J, Donachie W. A review of Listeria monocytogenes and listeriosis. Vet J. 1997;153:9–29.
73. Driehuis F, Oude Elferink SJ. The impact of the quality of silage on animal health and food safety: a review. Vet Q. 2000;22:212–6.
74. Kurtzman CP, Robnett CJ. Phylogenetic relationships among yeasts of the 'Saccharomyces complex' determined from multigene sequence analyses. FEMS Yeast Res. 2003;3:417–32.
75. Jonsson A, Pahlow G. Systematic classification and biochemical characterization of yeasts growing in grass silage inoculated with Lactobacillus cultures. Anim Res Dev. 1984;20:7–22.
76. Reich LJ, Kung Jr L. Effects of combining Lactobacillus buchneri 40788 with various lactic acid bacteria on the fermentation and aerobic stability of corn silage. Anim Feed Sci Technol. 2010;159:105–9.
77. McAllister TA, Selinger LB, Mcmahon LR, Bae HD, Lysyk TJ, Oosting J, et al. Intake, digestibility and aerobic stability of barley silage inoculated with mixtures of Lactobacillus plantarum and Enterococcus faecium. Can J Anim Sci. 1995;75:425–32.
78. Cheli F, Campagnoli A, Dell'orto V. Fungal populations and mycotoxins in silages: from occurrence to analysis. Anim Feed Sci Technol. 2013;183:1–16.
79. Niles EV. Fungi in stored barley treated with systemic fungicides. Pestic Sci. 1980;11:458–66. 24.

# 13

# Illumina sequencing-based community analysis of bacteria associated with different bryophytes collected from Tibet, China

Jing Yan Tang, Jing Ma, Xue Dong Li and Yan Hong Li*

**Abstract**

**Background:** Previous studies on the bacteria associated with the bryophytes showed that there were abundant bacteria inhabited in/on these hosts. However, the type of bacteria and whether these discriminate between different bryophytes based on a particular factor remains largely unknown.

**Results:** This study was designed to analyze the biodiversity and community of the bacteria associated with ten liverworts and ten mosses using Illumina-sequencing techniques based on bacterial 16S rRNA gene. A total of 125,762 high quality sequences and 437 OTUs were obtained from twenty bryophytes. Generally, there were no obvious differences between the richness of bacteria associated with liverworts and mosses; however, the diversity was significantly higher in liverworts than in mosses. The taxonomic analyses showed that there were abundant bacteria inhabited with each bryophyte and those primarily detected in all samples were within the phyla *Proteobacteria*, *Actinobacteria*, *Acidobacteria*, *Bacteroidetes*, *Armatimonadetes* and *Planctomycetes*. In addition, bacteria assigned to *Chloroflexi, Fibrobacteres*, *Gemmatimonadetes*, *Chlamydiae*, group of TM6 and WCHB1-60 also appeared in part of the bryophytes. The assigned bacteria included those adapted to aquatic, anaerobic and even extreme drought environments, which is consistent with the bryophyte transition from aquatic to terrestrial conditions. Of them, approximately 10 recognized genera were shared by all the samples in a higher proportion, such as *Burkholderia*, *Novosphingobium*, *Mucilaginibacter*, *Sorangium*, *Frankia*, *Frondihatitans*, *Haliangium*, *Rhizobacter*, *Granulicella* and *Hafnia,* and 11 unclassified genera were also detected in all samples, which exhibited that large amounts of unclassified bacteria could interact with the bryophytes. The Heatmap and Principle Coordinate Analyses showed that bacteria associated with six mosses displayed a higher community similarity. Notably, the bacteria associated with another four mosses exhibited higher similarity with the ten liverworts.

**Conclusions:** The result of further analysis of the bacterial community in different bryophytes revealed that the phylogeny of hosts might portray a strong influence on the associated bacterial community and that niche also played important roles when the hosts were phylogenetically more similar. Further studies are needed to confirm the role of phylogeny on bacterial communities and determine the level of influence on predicting which bacteria is associated with the host.

**Keywords:** Illumina-sequencing, Bacterial diversity, Community, Bryophytes, Phylogeny status

* Correspondence: liyh@cnu.edu.cn
College of Life Science, Capital Normal University, Xisanhuan North Road 105#, Haidian District, Beijing 100048, China

## Background

The presence of microbes is essential to the overall health of all plants in our environment [1]. Many of the bacteria associated with host plants work to degrade organic pollutants, improve plant growth [2–4] or enhance hosts the capability of adapting extreme environment [5]. Of the nearly all the plant species that exist on the earth, each individual plant is host to many bacteria [6]. Despite this widespread dependence, the majority of related research has concentrated on only the common higher plants and few reports have incorporated effects on community structure or the diversity of bacteria associated with plants such as bryophytes [7–9].

Bryophytes are the simplest land plants and form the basal clade of land plants. They are considered to be the ancestors of pteridophytes and all other tracheophytes. They represent the first green plants to colonize the terrestrial environment [10], and have evolved numerous important adaptations, including the alternation of gametophytic and sporophytic generations, elaboration of gametophytes, specialization of gametangia, and the adaption of desiccation-resistant spore walls [11]. Thus, bryophytes adapt to various environments ranging from harsh Antarctic conditions to extremely drought niches by their ability to preserve both water and many nutrients in unfavorable environments [12]. Importantly, bryophytes play important roles in nutrient cycling and can act as bio-indicators of air pollutants or heavy metals, making them crucial to the environmental health of many ecosystems [11].

A recent study of *Grimmia montana* [13] and other bryophytes [8, 14, 15] showed that there were abundant bacteria associated with these hosts. However, the type of bacteria that inhabits in/on bryophytes and whether these discriminate between different bryophytes based on a particular factor remains largely unknown. Thus, we studied 20 different types of bryophyte samples, including ten liverworts and ten mosses collected from Tibet, China, in order to determine their associated bacteria based on the Illumina-sequencing of 16S rRNA gene amplicons. The aims of this study were to find out which bacteria were dominant in/on the different bryophytes, what was the differences between the bacteria associated with the bryophytes and common higher plants, and which factor made a strong influence on the bacteria distribution in/on these bryophytes. The results of this study would be helpful to improve our understanding of the composition of bacteria associated with these hosts, and also could provide valuable insight into the mechanism of interactions between microbes and bryophytes.

## Methods

### Plant material and surface treatment

The investigation concentrated on the biodiversity and community structure of bacteria associated with 20 kinds of bryophytes, which were collected from four sites (shown in Table 1) in Tibet, China at Oct 13-15th, 2014. These bryophytes were not in the list of national key protected wild plants (http://guoqing.china.com.cn/2012-11/01/content_26977329.htm) and allowed to free collection for this survey (http://rep.iplant.cn/news/27). After the bryophytes were sampled, they were stored at 4 °C and transferred into the lab as soon as possible. In total, the samples included 10 types of liverworts and 10 types of mosses, which were labeled by _T or _X, respectively (shown in Table 1). Samples were first cleaned by rinsing several times with tap water to remove the attached matrix, and then washed by sterilized water for three times. Subsequently, the samples were surface treated by 70% ethanol for 3 min, which was followed by washing five times with sterile distilled water. Finally, the water was absorbed by the sterilized filter paper and the surface treated samples were placed into the sterile glass dish for use.

### DNA extraction and PCR amplification of the bacterial 16S rRNA gene

Approximately 1 g (including about 50 ~ 200 individuals) of each bryophyte sample was homogenized in liquid nitrogen, and then total DNA was extracted using the FastDNA SPIN Kit for Soil (MPBio) according to the instructions provided by the manufacturer. The first 16S rRNA gene amplification of bacteria associated with 20 bryophytes was conducted by using the Taq Mix kit (2 × Taq PCR Master Mix, Biomed) in a total volume of 50 μl containing 25 μl Taq Mix and 20 μl $ddH_2O$, 2 μl of each primer, and 1 μl template DNA based on the primers 799 F (5′-AACAGGATTAGATACCCTG-3′) and 1492R (5′-GGTTACCTTGTTACGACTT-3′) [16]. The thermal cycling conditions were as follows: an initial denaturation at 94 °Cfor 3 min, then 30 cycles of 94 °C for 30 s, 55 °C for 30 s, and 72 °C for 1 min, with a final extension step of 72 °C for 5 min. PCR products were separated by the electrophoresis technique using a 1% agarose gel. The bands of approximately 730 bp in size were excised and purified by the TIAN gel Mini Purification Kit (TIAN GEN Co.) as described by the manufacturer.

### Nested-PCR amplification and Illumina sequencing

Purified PCR products of the bacterial 16S rRNA gene were used for a second amplification step based on the primers 926 F (5′-AAACTYAAAKGAATTGACGG-3′) and 1392R (5′-ACGGGCGGTGTGTRC-3′). To distinguish the different samples, a barcoded-tag with eight nucleotide bases was randomly added to the upstream of the universal primers. The second PCR amplification was carried out by using the TransStart Fastpfu DNA Polymerase in a total volume of 20 μl containing 4 μl

**Table 1** Information of all samples

| Sample ID[a] | Taxonomic status | Scientific Name | Locality | Longitude/Latitude | Altitude (m) |
|---|---|---|---|---|---|
| JM1_T | Jungermanniales Plagiochilaceae | *Plagiochilion mayebarae* S. Hatt. | Zhamo roadside | N29° 48′ 07.17″ E95° 41′ 58.70″ | 4100 |
| JM2_T | Jungermanniales Lepidoziaceae | *Lepidozia reptans* (L.) Dum. | Zhamo roadside | N29° 48′ 07.17″ E95° 41′ 58.70″ | 4100 |
| JM28_T | Jungermanniales Jungermanniaceae | *Jamesoniella elongella* (Tayl.) Steph. | Zhamo roadside | N29° 45′ 16.57″ E95° 42′ 38.39″ | 4175 |
| JM30_T | Jungermanniales Jungermanniaceae | *Jungermannia parviperiantha* Gao et Bai | Zhamo roadside | N29° 45′ 16.57″ E95° 42′ 38.39″ | 4175 |
| GL3_T | Jungermanniales Jungermanniaceae | *Jungermannia leiantha* Grolle. | Galongla glacier | N29° 48′ 07.17″ E95° 41′ 58.70″ | 3993 |
| GL7_T | Jungermanniales Jungermanniaceae | *Jungermannia leiantha* Grolle. | Galongla glacier | N29° 48′ 07.17″ E95° 41′ 58.70″ | 3993 |
| GL9_T | Marchantiales Gymnomitriaceae | *Apomarsupella revolute* (Nees) Schust. | Galongla glacier | N29° 48′ 07.17″ E95° 41′ 58.70″ | 3993 |
| GL10_T | Jungermanniales Jungermanniaceae | *Jungermannia parviperiantha* Gao et Bai | Galongla glacier | N29° 48′ 07.17″ E95° 41′ 58.70″ | 3993 |
| GWL5_T | Jungermanniales Jungermanniaceae | *Horikawaella rotundifolia* Gao et Yi | Gawalong glacier | N29° 45′ 10.75″ E95° 42′ 23.61″ | 4337 |
| GWL7_T | Marchantiales Gymnomitriaceae | *Apomarsupella revolute* (Nees) Schust. | Gawalong glacier | N29° 45′ 10.75″ E95° 42′ 23.61″ | 4337 |
| JM3_X | Polytrichales Polytrichaceae | *Pogonatum urnigerum* (Hedw.) Beauv. | Zhamo roadside | N29° 48′ 07.17″ E95° 41′ 58.70″ | 4100 |
| JM5_X | Hypnales Hypnaceae | *Hypnum revolutum* (Mitt.) Lindb. | Zhamo roadside | N29° 48′ 07.17″ E95° 41′ 58.70″ | 4100 |
| JM25_X | Hypnales Amblystegiaceae | *Sanionia uncinata* (Hedw.) Loeske | Zhamo roadside | N29° 45′ 16.57″ E95° 42′ 38.39″ | 4175 |
| JM29_X | Grimmiales Grimmiaceae | *Racomitrium himalayanum* (Mitt.) Jaeg. | Zhamo roadside | N29° 45′ 16.57″ E95° 42′ 38.39″ | 4175 |
| TM2_X | Hypnales Hypnaceae | *Hypnum plumaeforme* Wils. | Tongmai Roadside | N30° 3′46.56″ E95° 9′35.45″ | 2047 |
| GL1_X | Grimmiales Grimmiaceae | *Racomitrium barbuloides* Card. | Galongla glacier | N29° 48′ 07.17″ E95° 41′ 58.70″ | 3993 |
| GL4_X | Polytrichales Polytrichaceae | *Pogonatum perichaetiale* (Mont.) Jaeg. | Galongla glacier | N29° 48′ 07.17″ E95° 41′ 58.70″ | 3993 |
| GL5_X | Grimmiales Grimmiaceae | *Racomitrium sudeticum* (Funck.) Bruch. et Schimp. | Gawalong glacier | N29° 45′ 10.75″ E95° 42′ 23.61″ | 4337 |
| GWL2_X | Hypnales SematopHyllaceae | *Wijkia deflexifolia* Crum | Galongla glacier | N29° 48′ 07.17″ E95° 41′ 58.70″ | 3993 |
| GWL9_X | Hypnales Brachytheciaceae | *Brachythecium rutabulum* (Hedw.) B. S. G. | Gawalong glacier | N29° 45′ 10.75″ E95° 42′ 23.61″ | 4337 |

[a]T stands for liverwort and X stands for moss

5 × FastPfu Buffer, 2 μl $ddH_2O$, 2 μl 2.5 mM dNTPs, 0.8 μl Forward Primer (5 μM), 0.8 μl Reverse Primer (5 μM), 0.4 μl FastPfu Polymerase and 10 ng template DNA. The thermal cycling conditions were as follows: an initial denaturation of 95 °C for 3 min, and 27 cycles of 95 °C for 30 s, 55 °C for 30 s and 72 °C for 45 s and a final extension step of 72 °C for 10 min. Amplicons were extracted from 2% agarose gels and purified using the AxyPrep DNA Gel Extraction Kit (Axygen Biosciences, Union City, CA, U.S.) according to the manufacturer's instructions and quantified using QuantiFluor™ -ST (Promega, U.S.). Purified amplicons were pooled in equimolar concentrations and sequenced using 300 bp paired-end model with the MiSeq system (Illumina, USA) in Majorbio (Shanghai).

**Bioinformatic analysis of the sequences**

The lengths of the short reads were extended by identifying the overlap between paired-end reads by the FLASH software [17]. The singleton sequences were removed and the low quality sequences were filtered out using QIIME software (version 1.17) [18]. Reads that were not assembled were also discarded. The reads were sorted according to barcode sequences and the sample sources. The sequences number of each sample was counted. Sequences covered the V6-V8 region of bacterial 16S rRNA gene were clustered into operational taxonomic units (OTUs) at 97% sequence similarity by using UPARSE software (version 7.1 http://drive5.com/uparse/) [19] and chimeric sequences were identified and removed using UCHIME [20].

To further calculate the Alpha diversity, species richness (Chao), species coverage (Coverage), species diversity (Shannon-Wiener Index and Simpson's diversity index) and rarefaction analyses were calculated using the software of Mothur version v.1.30.1 [21]. The richness index Chao estimator, was used to estimate the richness of the bacteria [22]. Shannon diversity and Simpson indexes were used to estimate the biodiversity of the bacterial communities. These alpha diversity indexes were compared between sample groups using Two Independent Sample tests of nonparametric analysis in SPSS version 16.0 for Windows (SPSS Inc., Chicago, IL). We analyzed the taxonomy of each 16S rRNA gene sequence with RDP Classifier [23] (http://rdp.cme.msu.edu/) against the Silva 16S rRNA database using a confidence threshold of 70% [24]. Community structure analyses were based on the phylum and genus taxonomy levels.

Heatmaps were generated on the basis of the relative abundance of phyla and genera, respectively, using R (version 2.15; The R Project for Statistical Computing, http://www.R-project.org). For phylogeny-based cluster comparisons, the composition of the microbial communities present in the samples of ten liverworts, ten mosses and all these twenty bryophytes were analyzed based on the Bray-Curtis distance and principal coordinate analysis (PCoA) plots were generated.

The two groups of bryophytes (obtained by the analysis of PCoA plots) specific to different bacteria types was performed using the linear discriminant analysis (LDA) effect size (LEfSe) method (http://huttenhower.sph.harvard.edu/lefse/) for biomarker discovery, which emphasizes both statistical significance and biological relevance. With a normalized relative abundance matrix, LEfSe uses the Kruskal-Wallis rank sum test to detect features with significantly different abundances between assigned taxa and performs LDA to estimate the effect size of each feature. A significance alpha of 0.05 and an effect size threshold of 4 were used for all biomarkers discussed in this study. All tests for significance were two-sided, and $p$ values below 0.05 were considered statistically significant.

## Results

### Amplicon analyses by illumina sequencing

The statistical results showed that 305,486 sequences were obtained from the 730 bp amplified products derived from the total DNA extracted from twenty bryophytes. After removing the low quality data, the singleton sequences and the chimeric sequences, about 125,762 high quality sequences were obtained. Totally more than ninety-nine percent of these sequences were classified as bacteria and their average length was 479 bp.

### Bacterial species richness and diversity in different bryophytes

The overall number of OTUs detected by the analysis reached 437, based on ≥97% nucleotide sequence identity between sequences. To assess whether our sampling effort provided sufficient OTU coverage of the bacterial composition associated with each bryophyte, rarefaction curves and the library coverage were generated for each sample. The rarefaction curves and >98.5% coverage in all samples showed that the libraries could reflect the main bacterial information in each sample (Fig. 1). Of 437 OTUs detected overall, 387 OTUs could be detected in both liverwort and moss species. Thirty-one special OTUs were only detected in liverworts, while only 19 special OTUs were found in mosses. In addition, the OTU number was varied from 143 (*Hypnum plumaeforme*, TM2_X) to 308 (*Jungermannia parviperiantha*, GL10_T) in different samples. In general, the average OTU number was 240 in liverworts, which was slightly higher than the average OTU number in mosses but was not significantly different ($p > 0.05$). There was no relationship between the OTU numbers and the sampling sites (Table 2).

The Chao estimator showed that the richness of the bacteria associated with these bryophytes ranged from 191 to 357. The average Chao estimator was 294 in liverworts, which was also slightly higher but not significantly ($p > 0.05$) different than that detected in mosses

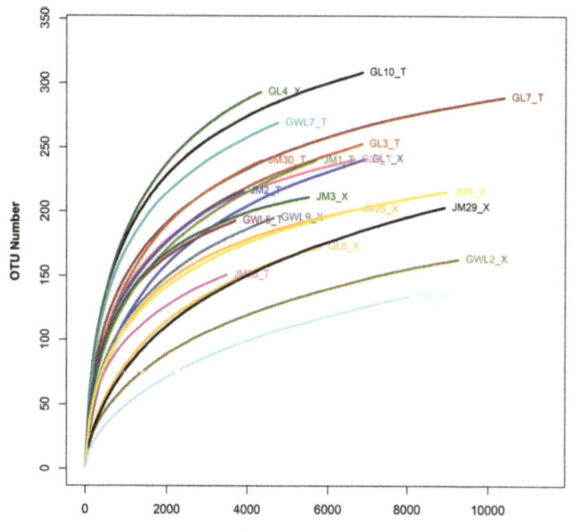


Fig. 1 Rarefaction curves of bacterial 16S rDNA gene sequences

**Table 2** The diversity indexes of bacteria associated with all the bryophytes

| Sample ID | Reads | OTUs | Chao estimator | Coverage (%) | Shannon diversity | Simpson |
|---|---|---|---|---|---|---|
| JM1_T | 5693 | 240 | 320 ± 51.5 | 98.8 | 3.95 ± 0.04 | 0.039 ± 0.002 |
| JM2_T | 3896 | 216 | 267 ± 35.5 | 98.5 | 4.15 ± 0.05 | 0.029 ± 0.002 |
| JM28_T | 3475 | 151 | 198 ± 39.5 | 98.7 | 3.44 ± 0.05 | 0.07 ± 0.005 |
| JM30_T | 4350 | 240 | 285 ± 30.5 | 98.6 | 3.7 ± 0.06 | 0.071 ± 0.004 |
| GL3_T | 6857 | 253 | 310 ± 39.5 | 99.1 | 4.1 ± 0.04 | 0.032 ± 0.001 |
| GL7_T | 10382 | 289 | 334 ± 32.5 | 99.5 | 4.1 ± 0.04 | 0.041 ± 0.002 |
| GL9_T | 6653 | 241 | 288 ± 36.5 | 99.3 | 4.12 ± 0.04 | 0.032 ± 0.001 |
| GL10_T | 6869 | 308 | 357 ± 34.5 | 99.1 | 4.58 ± 0.04 | 0.020 ± 0.001 |
| GWL5_T | 3697 | 193 | 231 ± 31 | 98.8 | 3.95 ± 0.05 | 0.042 ± 0.003 |
| GWL7_T | 4771 | 269 | 346 ± 51.5 | 98.6 | 4.44 ± 0.04 | 0.024 ± 0.001 |
| JM3_X | 5521 | 211 | 233 ± 20.5 | 99.4 | 3.95 ± 0.04 | 0.042 ± 0.002 |
| JM5_X | 8966 | 216 | 256 ± 31.5 | 99.5 | 3.47 ± 0.04 | 0.083 ± 0.004 |
| JM25_X | 6662 | 202 | 262 ± 49.5 | 99.3 | 3.01 ± 0.06 | 0.18 ± 0.009 |
| JM29_X | 8919 | 203 | 270 ± 43.5 | 99.2 | 2.22 ± 0.04 | 0.25 ± 0.008 |
| TM2_X | 8035 | 134 | 191 ± 42.5 | 99.3 | 1.89 ± 0.04 | 0.27 ± 0.007 |
| GL1_X | 6929 | 241 | 317 ± 50 | 99.0 | 2.81 ± 0.05 | 0.19 ± 0.008 |
| GL4_X | 4349 | 293 | 328 ± 24 | 98.6 | 4.41 ± 0.05 | 0.032 ± 0.002 |
| GL5_X | 9263 | 163 | 207 ± 33 | 99.4 | 2.37 ± 0.04 | 0.20 ± 0.006 |
| GWL2_X | 5808 | 173 | 214 ± 30 | 99.1 | 2.34 ± 0.06 | 0.23 ± 0.009 |
| GWL9_X | 4667 | 195 | 250 ± 40.5 | 98.8 | 3.57 ± 0.05 | 0.064 ± 0.004 |

(253). However, the Shannon diversity indexes of bacteria associated with liverworts exhibited a much greater value than those with mosses ($p < 0.05$). The average Shannon diversity index was 4.053 in liverwort samples, whereas it was only 3.004 in mosses. Interestingly, the Shannon diversity indexes were nearly similar between different liverwort samples, but relatively bigger changes among different moss species (Table 2).

#### Community structure of bacteria associated with the bryophytes at phylum level

According to the sequencing results, the phylum of less than 2% OTUs could not be identified (Table 3). For the other 98% of OTUs, bacteria could be classified into a phylum and for those associated with ten liverworts, 14 phyla were identified. Of them, the phyla *Proteobacteria, Actinobacteria, Acidobacteria, Bacteroidetes, Armatimonadetes, Chloroflexi, Planctomycetes* and *Firmicutes* were detected in all ten liverworts, but bacteria assigned to *Fibrobacteres, Gemmatimonadetes, Chlamydiae,* TM6 and WCHB1-60 also appeared in some of the liverwort samples (Fig. 2a). Of the bacteria associated with mosses, seven phyla could be detected in all samples, including *Proteobacteria, Actinobacteria, Acidobacteria, Bacteroidetes, Armatimonadetes, Chlamydiae* and *Planctomycetes*. In addition, the five phyla *Fibrobacteres, Gemmatimonadetes, Firmicutes,* TM6 and WCHB1-60 were also

**Table 3** Ratio of unclassified sequences at different taxonomic levels

| | Relative abundance (%) | | | | |
|---|---|---|---|---|---|
| Sample ID | Phylum | Class | Order | Family | Genus |
| JM1_T | 0.00 | 0.12 | 4.53 | 8.59 | 26.25 |
| JM2_T | 0.00 | 0.08 | 6.57 | 15.25 | 30.06 |
| JM28_T | 0.00 | 0.32 | 0.83 | 8.39 | 18.11 |
| JM30_T | 0.00 | 0.07 | 2.37 | 4.73 | 31.05 |
| GL3_T | 0.04 | 1.71 | 7.16 | 9.85 | 33.26 |
| GL7_T | 0.01 | 0.10 | 1.18 | 5.70 | 26.63 |
| GL9_T | 0.12 | 0.24 | 2.39 | 6.50 | 18.54 |
| GL10_T | 0.31 | 0.93 | 6.35 | 9.28 | 18.59 |
| GWL5_T | 0.03 | 0.08 | 3.41 | 6.86 | 17.85 |
| GWL7_T | 0.13 | 0.19 | 7.84 | 12.03 | 28.38 |
| JM3_X | 0.00 | 0.14 | 3.97 | 5.10 | 23.47 |
| JM5_X | 0.00 | 0.01 | 1.25 | 4.17 | 37.94 |
| JM25_X | 0.00 | 0.23 | 3.17 | 6.54 | 27.37 |
| JM29_X | 0.00 | 0.02 | 12.66 | 14.07 | 62.47 |
| TM2_X | 0.00 | 0.01 | 15.41 | 15.52 | 61.79 |
| GL1_X | 0.01 | 0.07 | 10.58 | 12.85 | 54.00 |
| GL4_X | 0.02 | 0.90 | 5.10 | 7.10 | 26.50 |
| GL5_X | 0.01 | 0.22 | 6.71 | 7.08 | 16.99 |
| GWL2_X | 0.03 | 0.10 | 15.05 | 16.03 | 60.62 |
| GWL9_X | 0.00 | 0.00 | 7.71 | 8.86 | 45.01 |

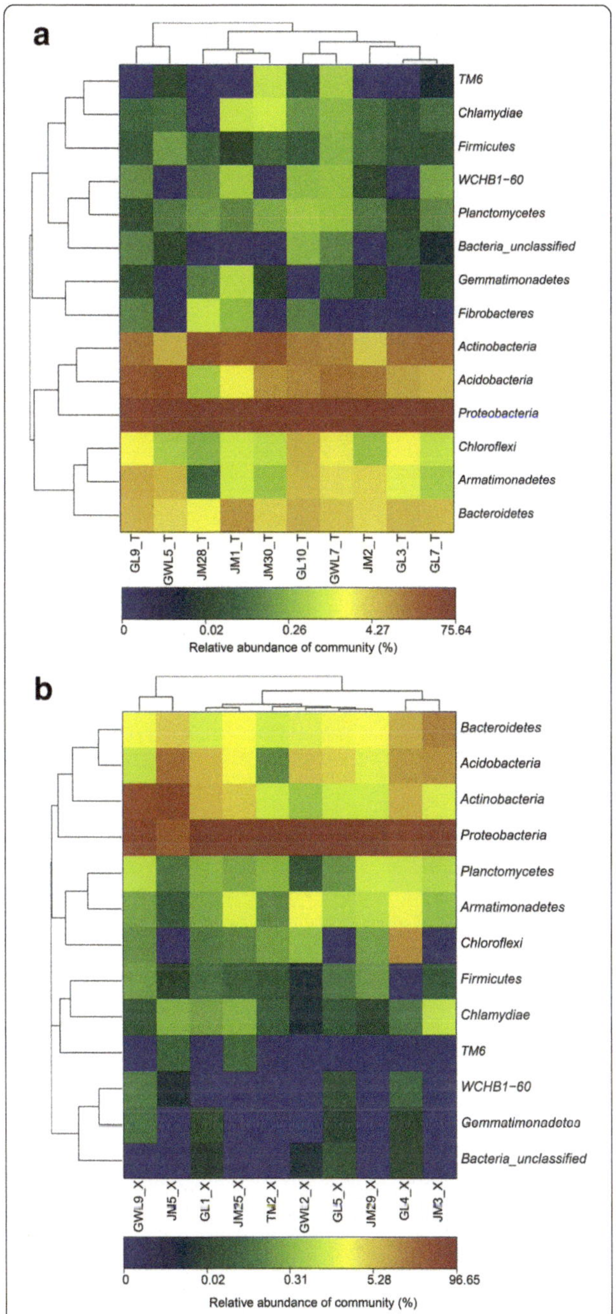


**Fig. 2** The distribution of all sequences from each bryophyte sample on the phyla (**a** and **b**) level

detected in some of the mosses (Fig. 2b). Thus, the common phyla to all the bryophytes were *Proteobacteria*, *Actinobacteria*, *Acidobacteria*, *Bacteroidetes*, *Armatimonadetes* and *Planctomycetes*.

Further analysis showed that *Proteobacteria* was the most predominant phylum inhabited in the nineteen samples, with the exception of the sample of *Hypnum revolutum* (JM5_X). In consideration of the dominant phyla associated with liverworts and mosses, there were no big differences among ten liverworts, and *Actinobacteria*, *Acidobacteria*, *Bacteroidetes* and *Armatimonadetes* were the subsequent dominant phyla associated with all of them. However, there were big differences among the bacteria associated with the ten mosses. For example, *Actinobacteria* was the most dominant phylum in *Hypnum revolutum* (JM5_X), and it was also the abundant phylum in the samples of *Racomitrium barbuloides* (GL1_X), *Pogonatum perichaetiale* (GL4_X), *Brachythecium rutabulum* (GWL9_X) and *Sanionia uncinata* (JM25_X). In contrast, it was infrequently detected in *Wijkia deflexifolia* (GL5_X), *Racomitrium sudeticum* (GWL2_X), *Pogonatum urnigerum* (JM3_X) and *Hypnum plumaeforme* (TM2_X). *Acidobacteria* was also abundant in seven of the mosses, but only a small quantity could be detected in GWL9_X, JM29_X and TM2_X.

The phylum *Proteobacteria* that was detected in the libraries could be subdivided into four classes: *gamma*, *beta*, *alpha* and *delta*. Although the proportions of each class were different in different liverwort samples, all four classes were present in all but the abundance of *gamma* and *beta* classes were usually greater. In mosses, there were bigger differences among different samples. In four samples (JM29_X, TM2_X, GL5_X and GWL2_X), the *Proteobacteria* consisted of primarily the *gamma* class, while *beta*, *alpha* and *delta* classes were only present in a small proportion (Fig. 3). Thus, there were similar compositions of bacteria associated with the ten liverworts at both the phylum and class level of *Proteobacteria*, while large differences in bacterial composition existed among different moss species.

#### Community structures of bacteria associated with the bryophytes at the genus level

According to the taxonomic analysis, a relatively large number of OTUs could not be assigned into any genus with a confidence level higher than 70%, suggesting that the presence of many unclassified sequences were detected in these bryophytes (Table 3). In total there were at least 176 genera of bacteria associated with the twenty bryophytes, and the first 100 abundant genera were reflected by the Heatmap diagram (Fig. 4). Of them, approximately 21 genera were shared by all the samples in a higher proportion, including 10 recognized genera such as *Burkholderia*, *Novosphingobium*, *Mucilaginibacter*, *Sorangium*, *Frankia*, *Frondihatitans*, *Haliangium*, *Rhizobacter*, *Granulicella* and *Hafnia*, and another 11 unclassified genera in family *Xanthomonadaceae*, *Oxalobacteraceae*, *Microbacteriaceae*, *Comamonadaceae* and *Enterobacteriaceae*, and some in the order *Myxococcales*, *Burkholderiales* and *Xanthomonadales*. But those in the genera of *Acidiphilum*, *Bradyrhizobium*, *Flexibacter*, *Bryobacter*, *Acidothermus*, *Acidobacterium*, and family of *Acetobacteraceae* and *Chitinophagaceae*, order of

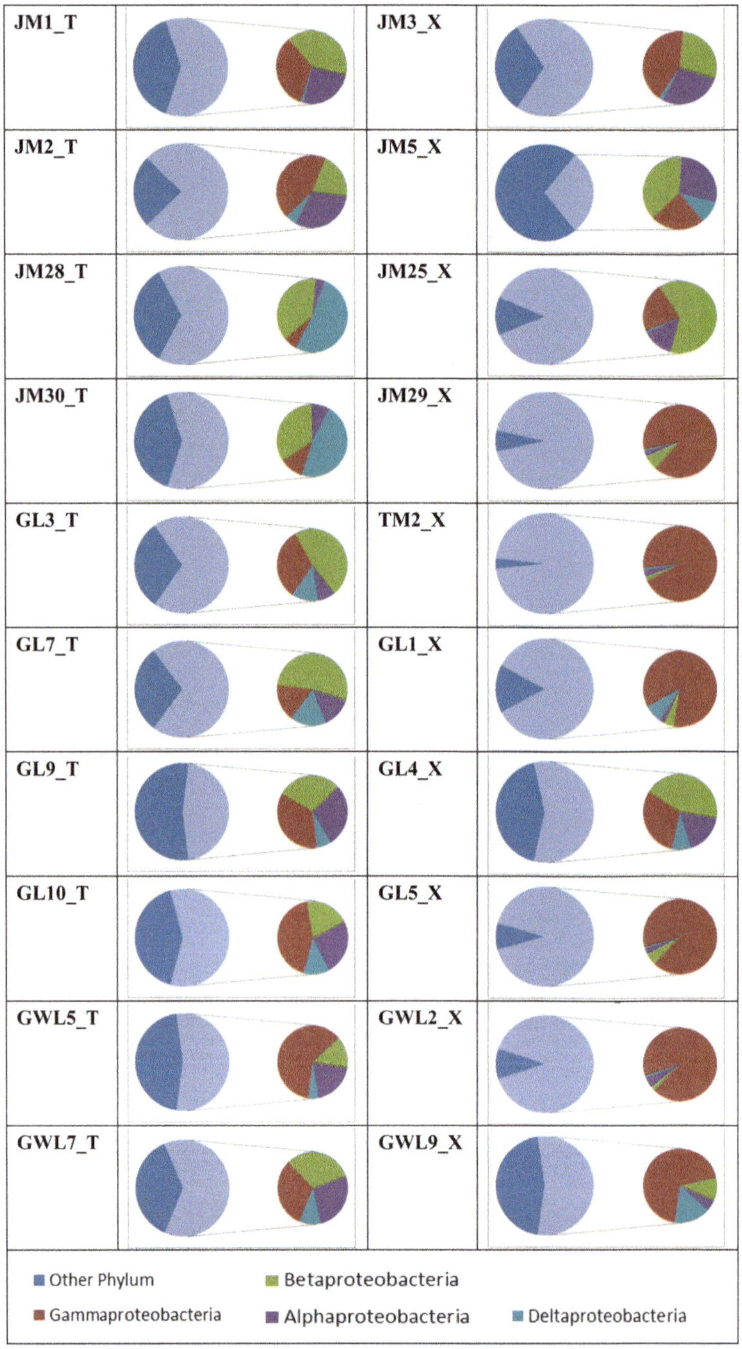


**Fig. 3** The distribution of all sequences from each bryophyte sample on the class level of phylum *Proteobacteria*

*Armatimonadales* and *Acidimicrobiales*, even some in the alphal_cluster of *Ktedonobacteria* and *Armatimonadetes*, were mainly associated with the liverworts. The bacterial community reflected by the heatmap that was based on the genus level showed that bacteria associated with six mosses (GWL2_X, GWL9_X, GL1_X, GL5_X, JM29_X and TM2_X) displayed a higher community similarity; whereas, the bacteria associated with the other four mosses exhibited a higher similarity with the liverworts.

### The bacterial community similarities among different bryophytes

In the PCoA diagram based on ten liverworts generated based on 50.80% variance (34.09% first axis and 16.71% second axis, Fig. 5a), although there was not obvious laws for the distribution of the samples in the plots, totally the bacterial community in four liverworts sampled from Galongla glacier (GL: GL7_T, GL3_T, GL9_T and GL10_T) and two from Gawalong glacier

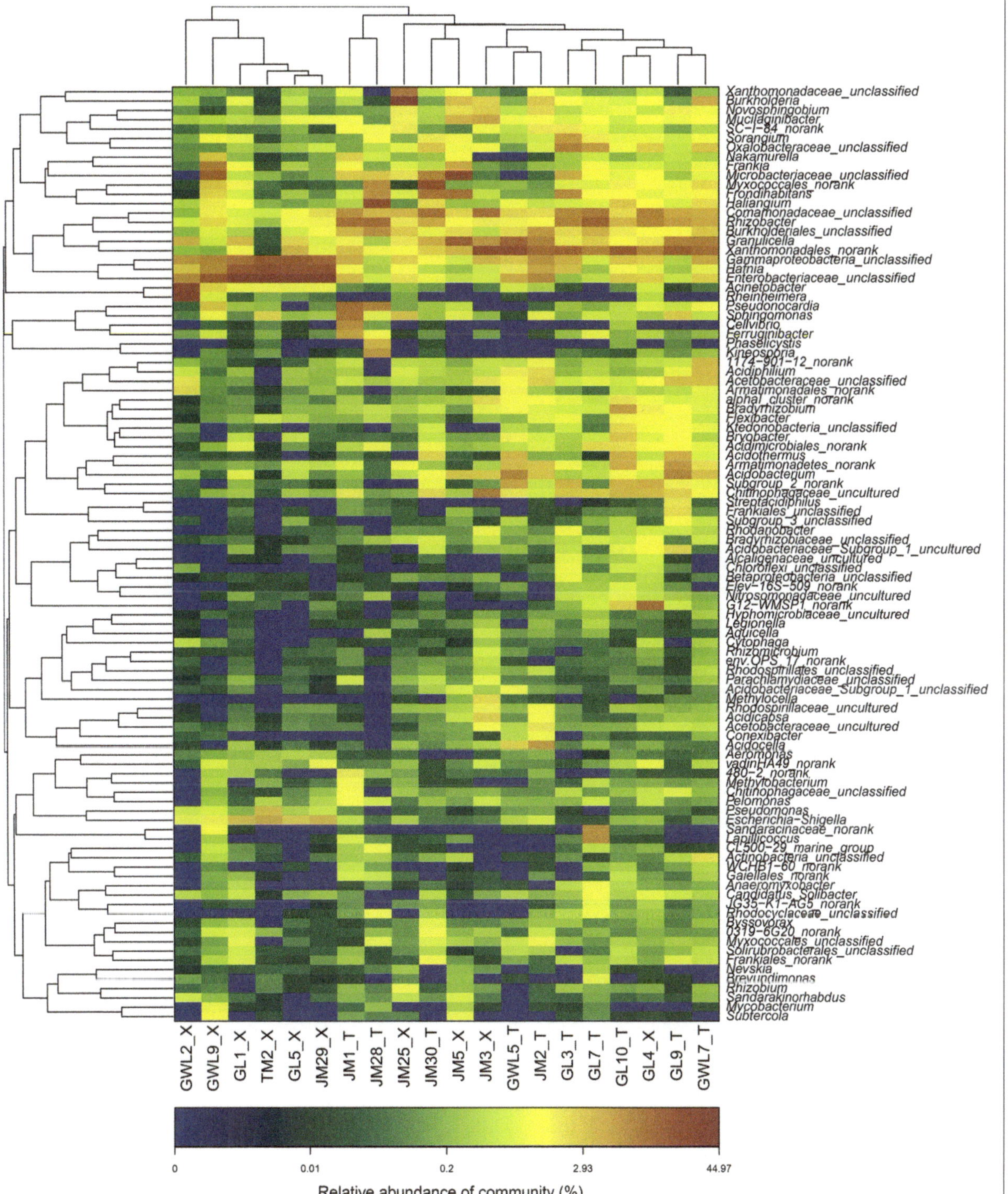


**Fig. 4** The Heatmap of bacterial composition in all bryophytes at genus level

(GWL: GWL7_T and GWL5_T) displayed the relatively closer distances, respectively, while for another four liverworts collected from Zhamo roadside (JM), their associated bacterial community displayed a relatively bigger differences. For the PCoA analysis based on ten mosses, the bacterial community associated with them also exhibited that samples from the same site of JM (JM3_X, JM5_X, JM25_X), GL (GL1_X and GL5_X) and GWL (GWL2_X and GWL9_X) had a relatively closer distances with the exception of GL4_X, JM29_X and TM2_X based

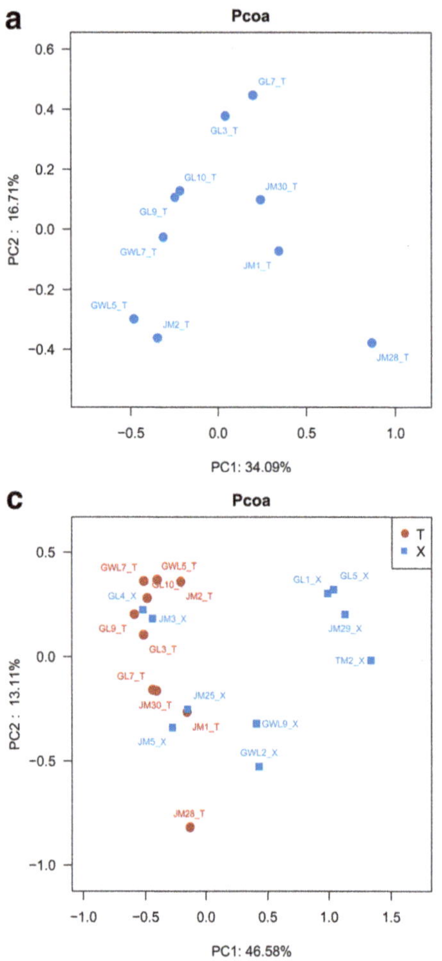

**Fig. 5** The PCoA analysis of the bacterial community in ten liverworts (**a**), ten mosses (**b**) and all twenty bryophytes (**c**)

on 70.56% variance (57.42% first axis and 13.14% second axis, Fig. 5b). Interestingly, when the PCoA analysis was based on all twenty bryophytes, the bacteria associated with all the samples could be divided into two groups: (1) the bacteria community associated with the four specific mosses (JM3_X, JM5_X, JM25_X, GL4_X,) and all ten liverworts displayed a closer relationship. (2) bacteria associated with another six moss species (JM29_X, TM2_X, GL1_X, GL5_X, GWL2_X and GWL9_X) were more similar. The cumulative percentage variance of species was explained by 59.69% variance (Fig. 5c).

The LEfSe analysis further identified the specific bacterial taxa that were differentially present or abundant in above two groups (six mosses made up group 1, and other 14 bryophytes were group 2). The results showed that bacteria in the order *Ardenticatenales* and family *Bacillaceae* were significant abundant in six of the mosses (Fig. 6a and b). In addition, four orders (*Enterobacteriales, Aeromonadales, Chromatiales* and *Pseudomonadales*) of bacteria in the *gamma* class of *Proteobacteria* were also significantly abundant in these 6 samples, and they were significantly differentiated from the other 14 bryophytes. In contrast, the bacteria that were more abundant in the 14 other samples compared to group 1 were those in: *Sphingobacteriales* and *Chitinophagaceae* of phylum *Bacteroidetes*; in *Conexibacteraceae, Microbacteriaceae, Acidothermaceae* and *Acidimicrobiales* of phylum *Actinobacteria*; in TA18 group; in *Xanthomonadales, Legionellales* and NKB5 group of the *gamma* class of *Proteobacteria*; in *Comamonadaceae*, SC_1_84, and the unclassified family of *Burkholderiales* in *betaproteobacteria*; and finally those in the order *Rhizobiales* and *Rhodospirillales* of *alphaproteobacteria*.

## Discussion

### Advantage of the combination of the nested-PCR system and illumina sequencing technique

The objective of this study was to evaluate the composition of the bacterial communities that are associated with different bryophytes using a high throughput-sequencing technique. Since the bacteria associated with the bryophytes either resided inside the plant or was attached to

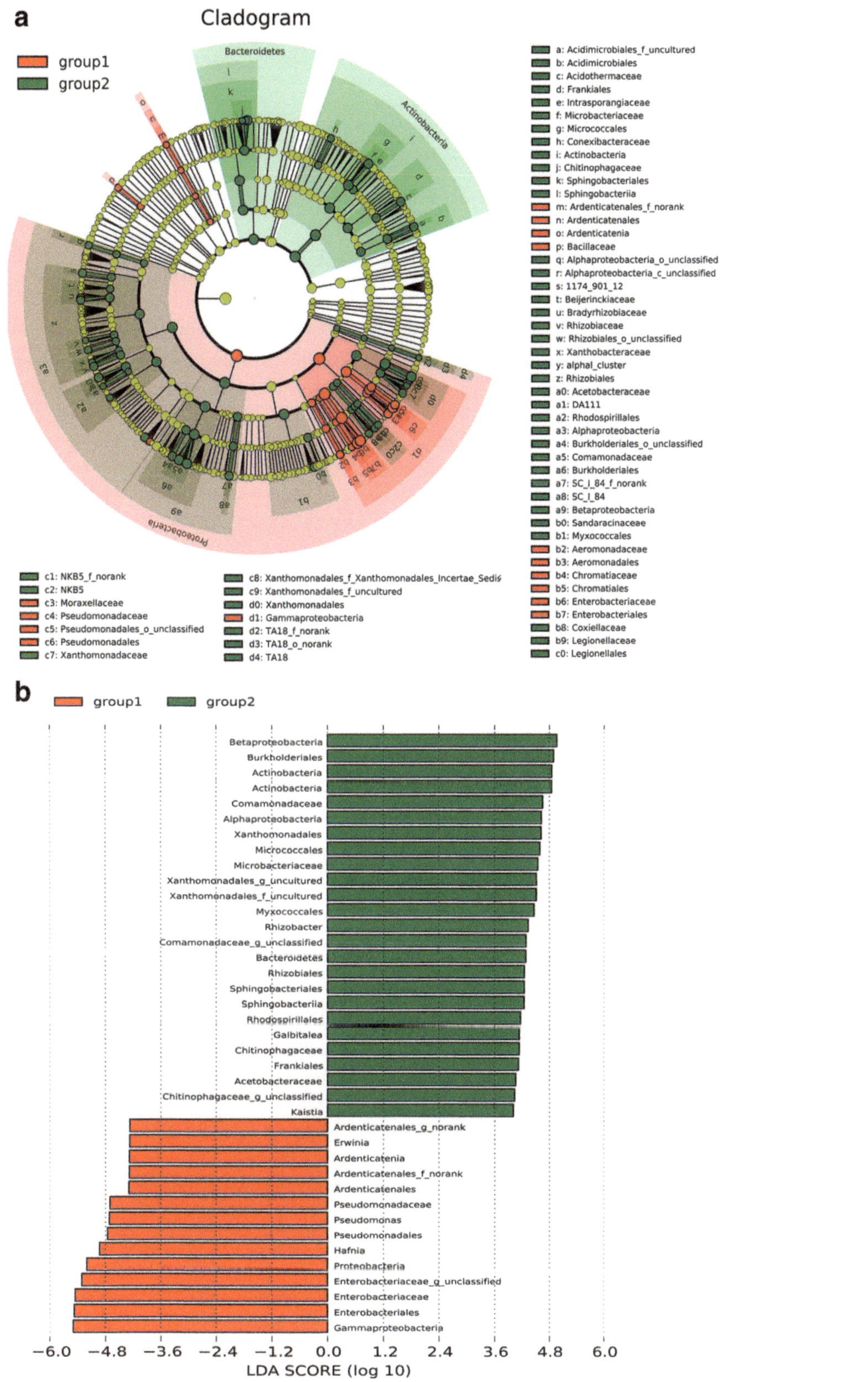


**Fig. 6** (See legend on next page.)

(See figure on previous page.)

**Fig. 6** The LEfSe analysis of bacteria associated with part bryophytes of group 1 (samples JM29_X, TM2_X, GWL9_X, GWL2_X, GL1_X and GL5_X) and group 2 (samples JM3_X, JM5_X, JM25_X, GL4_X and ten liverworts). **a** Cladogram representing the taxonomic hierarchical structure of the identified phylotype biomarkers generated using LEfSe. Phylotype biomarkers were identified comparing samples from group 1 and group 2. Each filled circle represents one biomarker. Red, phylotypes statistically overrepresented under the condition of group 1; green, phylotypes overrepresented under the condition of group 2; yellow, phylotypes for which relative abundance is not significantly different between the two conditions. The diameter of each circle is proportional to the phylotype's effect size, phylum is indicated in their names on the cladogram and the class, order, family, or genera are given in the key. **b** Identified phylotype biomarkers ranked by effect size in group1 and group 2. The phylotype biomarkers were identified as being significantly abundant when samples from group 1 and group 2 were compared and the alpha value was <0.05

the surface, we used the total DNA that was extracted from the sampled bryophytes as the template. In order to control for the DNA of plant chloroplasts or mitochondria, a nested PCR technique was used in this study. Primers used in the first run were 799 F and 1492R, which have been used successfully to filter out chloroplast DNA and separate the bacterial 16S rRNA gene fragment from the mitochondria DNA of hosts [16]. The second run was based on the primer pairs of 926 F and 1392R and was used to further amplify the V6-V8 region of bacterial 16S rRNA gene. According to the data obtained in this survey, ≥ 99.9% of the reads obtained were bacterial 16S rRNA gene fragments. This confirmed that the nested PCR technique based on these two primer pairs could be successfully used for the high throughput sequencing of bacteria on plant hosts.

In line with previous investigations of bacterial communities and diversity inhabited in *Grimmia Montana* using 16S rDNA clone library techniques [13], the dominant bacterial phyla were *Proteobacteria, Actinobacteria, Firmicutes* and *Bacteriodetes*; while another bacterial-biota dynamics research made by Koua (2015) only detected two phyla of *Proteobacteria* and *Firmicutes* in the phyllosphere of the eight bryophytes from different ecosystems when using the PCR-DGGE technique based on the bacterial V3 region of 16S rRNA gene [15]. However, those in phyla *Acidobacteria, Armatimonadetes, Chlamydiae* and *Planctomycetes* were also detected in all the bryophytes within this study, and the bacteria affiliated to phyla *Firmicutes, Gemmatimonadetes*, TM6 and WCHB1-60 were also found in some of moss species, which showed that the nested-PCR combined with the Illumina-sequencing technique used here is a more sensitive and effective method for the identification of bacteria associated with plant hosts.

#### The bacterial richness and diversity in different bryophytes

Here, we studied the bacteria associated with twenty bryophytes mainly derived from four places including four different niches, the roadside of Zhamo (JM) and Tongmai (TM) and the glaciers of Galongla (GL) and Gawalong (GWL). According to the OTU numbers, Chao estimators, Shannon diversity indexes and Simpson indexes, the richness and diversity of the bacteria associated with TM2_X were the lowest on the scale. This may be due to the altitude (2000 m) of the sampling site of TM at Tongmai, which was far lower than those collected from the higher sampling locations (4100 ~ 4337 m). Equally, there was no relationship between the bacterial richness or diversity in samples collected from all other sampling sites, JM, GL or GWL, which are at similar altitudes. Thus, higher altitudes might be helpful to stimulate bacteria inhabitance in or on hosts as a way to help them adapt to extreme environments.

#### Comparison of bacteria community between those associated with common higher plants and bryophytes

Sequencing taxonomy analysis showed that the bacteria that inhabit bryophytes are primarily members of eight to twelve specific phyla. Of them, *Proteobacteria, Actinobacteria, Acidobacteria* and *Bacteriodetes* were the dominant phyla in all samples. Consistent with other reports, the members of phylum *Proteobacteria* were also detected as the most dominant bacteria in many plant hosts, and the proportion of bacteria in other phyla were differentially varied across the different hosts [15, 16, 25–27]. Compared with the dominant bacteria associated with other common higher plants, *Acidobacteria* and *Actinobacteria* were still in high proportions, and those in *Armatimonadetes, Planctomycetes, Chloroflexi, Gemmatimonadetes, Firmicutes* and group TM6 and WCHB1-60, which were rarely detected in higher plants, were found in the majority of the bryophytes examined in this survey.

*Acidobacteria* is the second most dominant phylum inhabited in soil, sediments or benthonic species and its persistence could be based on the general mechanism of trace gas oxidation [28]. In this study, we detected it in nearly all the bryophytes at varied levels (0.15 ~ 30.54%). Although we don't understand its role on the bryophytes, the bryophytes appeared to be good hosts and could provide suitable growth conditions for many *Acidobacteria*.

The bacterium of *Armatimonadetes* was originally described solely on the basis of environmental 16S rRNA gene clone sequences, and a bacterial strain *Armatimonas rosea* of this phylum was first isolated from an aquatic plant in Japan in 2011 [29]. The report of the first *Armatimonadetes* genome from the thermophile *Chthonomonas calidirosea* T49 (T) showed that it could act as a saccharide scavenger in a geothermal steam-affected soil environment. Its predicted genes encode for carbohydrate transporters and carbohydrate-metabolizing enzymes, which would help the plant hosts utilize a wide range of carbohydrates [30]. Small amounts of *Armatimonadetes* bacteria detected in all bryophytes showed that they are important to the micro-ecosystem of hosts by helping them utilize a wide range of carbohydrates to prompt their growth. Likewise *Planctomycetes*, which are aquatic bacteria that often inhabit brackish and fresh water, were identified by a stable-isotope probing technique to degrade complex heteropolysaccharides in soil [31]. In this survey, we demonstrated that they also inhabit bryophytes, which are a type of lower group derived from water and depend on nutrient resources and rainwater.

*Chloroflexi* is a phylum of bacteria containing isolates with a diversity of phenotypes, including members that are aerobic thermophiles and use oxygen and grow well in high temperatures; anoxygenic phototrophs that use light for photosynthesis; and anaerobic halorespirers that act as energy sources. Barton and colleagues (2014) reported that the microbial diversity in a Venezuelan orthoquartzite cave was dominated by the *Chloroflexi* (Class *Ktedonobacterales*) and postulated that the poor buffering capacity of quartzite or the low pH of the environment selected for this unusual community structure [32]. Interestingly, some of these were also detected in the majority of bryophytes and could be related to the extreme niches of the bryophyte locations.

*Gemmatimonadetes* have been found in a variety of arid soils, such as grassland, prairie, and pasture soil, as well as eutrophic lake sediments and alpine soils. The phylum *Gemmatimonadetes* diverged in early microbial evolution at least 3 billion years ago [33]. As the living fossil of plants, bryophytes might be good hosts for their growth and production, and thus, it was not particularly surprising that this phylum was also detected in this study.

In addition to those listed above, bacteria within the phylum TM6 are those chiefly found in the biofilm of sink drains and are considered to be the primary low abundance members in a wide range of habitats, as identified by culture-independent rRNA surveys. Several genomes of these bacteria were analyzed by two independent research groups, McLean (2013) and Yoeh (2016). Both studies found that these bacteria are small and lack complete biosynthetic pathways for various essential cellular building blocks including amino acids, lipids, and nucleotides. Other features identified in the TM6 genomes included a degenerated cell envelope, the expression of ATP/ADP translocases for parasitizing host ATP pools, and protein motifs that facilitate eukaryotic host interactions [34, 35]. Another uncultured phylum, the WCHB1-60, was detected in pond samples, although the possible functions of these phyla in the liverworts and mosses are not clear and should be considered in future research.

Thus, there were many differences between the bacterial communities associated with bryophytes compared to those in the common higher plants. The bacteria associated with bryophytes included those adapted to aquatic, drought and even anaerobic environments, which is consistent with the bryophyte transition from aquatic to terrestrial conditions. In addition, there were a small proportion of sequences assigned to unclassified phylum and many of them assigned to unclassified genera, suggesting that there are large amounts of new symbiotic bacteria associated with bryophytes.

### The factors involved in the bacterial community similarities among different bryophytes

Combined analyses of the bacteria associated with ten liverworts via Heatmap and PCoA patterns (Fig. 4a) showed that samples from each same sampling site of Gawalong glacier (GWL), Galongla glacier (GL) and Zhamo roadside (JM) generally displayed more similar bacterial communities, although the bacterial community associated with the liverworts from JM displayed more scattered distribution. Thus, we inferred that the bacterial community similarity was primarily predictive of the sampling site itself. Each sampling site represented a specific environmental niche; therefore, the niche was considered to be a determining factor of bacterial community in the liverworts to some extent. It was deeply confirmed by the relatively closer distribution of samples from two different glacier regions, GL and GWL, which also showed that the similar bacterial community compositions inhabited in/on the liverworts from similar niches might be more similar. In addition, although the samples JM30_T and GL10_T were both *Jungermannia parviperiantha* of order *Jungermanniales*, and collected from Zhaomo roadside and Galongla glacier respectively, their bacterial compositions were shown in a lower similarity by the relatively far distance in the first axis of PCoA plot (Fig. 4a). Thus, it further suggested that the niches were likely the most important factor to determine which bacteria could inhabit in/on liverworts in this survey.

Furthermore, the bacteria associated with ten mosses were divided into three groups (Fig. 4b) and totally displayed that the same sampling site derived samples nearly had the higher bacterial community similarity, but with the exception of GL4_X, JM29_X and TM2_X. The exception of the bacteria associated with these three mosses suggested that there must be other factors influenced on the bacterial composition. Further analysis the phylogenetic status of sample GL4_X and found that it was a species of genus *Pogonatum*, family *Polytrichaceae* and belonged to the same genus and with very close relationship to sample JM3_X (shown in Table 1); while JM29_X were the species *Racomitrium himalayanum*, and belonged to the same genus with samples GL1_X and GL5_X., which meant JM29_X was phylogenetically closely related to GL1_X and GL5_X. According to the PCoA pattern (Fig. 4b), the bacteria associated with GL4_X showed the higher similarity with those inhabited in/on JM3_X, JM5_X and JM25_X, while bacteria inhabited in /on JM29_X showed the higher similarity with those in GL1_X and GL5_X, which demonstrated that the phylogenetic status of the hosts might be another important factor (besides sampling site) that could influence the bacterial community in mosses. For TM2_X, although it was identified as the same genus of *Hypum* with sample JM5_X, the reason why they were not clustered together probably related to the characters of genus *Hypum* in order *Hypnales*. As reporting on the phylogenetic research of different mosses by the molecular methods [36], order *Hypnales* formed a large clade with the inclusion of exemplars of pleurocarpous mosses, some families such as *Brachytheciaceae, Amblystegiaceae* and *Sematophyllaceae* appeared as monophyletic entities, whereas *Hypnaceae,* as another family of *Hypnales,* was polyphyletic and the genus *Hypnum* was also placed in several distinct clades. Thus, it might be the reason why the bacteria in sample TM2_X and JM5_X were not clustered together.

However, when the PCoA profile was made based on all 20 bryophytes (Fig. 4c), the bacterial communities of four mosses (JM3_X, JM5_X, JM25_X, GL4_X,) and ten liverworts were clustered together, showing that their associated bacterial communities were more similar to each other than to the other six mosses. Further analysis of the genetic characteristics of these bryophytes also confirmed that the phylogenetic relationships of these bryophytes might play an important role in determining which bacteria inhabitance with the hosts. For example, ten samples of the liverworts belonged to the order of *Jungermanniales* or *Marchantiales*, respectively, while the ten mosses were from three different orders, *Hypnales, Polytrichales* and *Grimmiales*. According to the phylogeny status of bryophytes previously described, bryophytes represent the first green plants to colonize the terrestrial environment. This plant type includes liverworts, hornworts and mosses. Some analyses that have combined morphological characters with rRNA sequences and that of the chloroplast gene rbcL [37, 38] have generally supported a liverworts-basal topology (LBT) and placed either mosses or hornworts as sister to vascular plants. Other reports based on extensive cladistics analysis of morphological, ultrastructural data and mitochondrial small-subunit (mtSSU) rDNA [39, 40] yielded a monophyletic moss-liverwort clade, with hornworts sister to all land plants (hornworts-basal topology, HBT). Thus, whether the phylogeny of bryophytes was supported by LBT or HBT hypothesis, liverworts had little differentiation on organs such as stem or leaf, and were lower than mosses in the taxonomic status. The four mosses that were clustered together with liverworts, the JM3_X, GL4_X JM5_X and JM25_X, belonged to the families *Polytrichaceae* of order *polytrichales*, and *Hypnaceae, Amblystegiacae* of order *Hypnales*, respectively, whereas another three mosses GL1_X, GL5_X and JM29_X belonged to *Grimmiaceae* of order *Grimmiales*, and GWL9_X, GWL2_X and TM2_X belonged to the family of *Brachytheciaceae*, *Sematophyllaceae* and *Hypnaceae* of *Hypnales*, respectively (Shown in Table 1). According to the molecular phylogenetics and ordinal relationships based on analyses of a large-scale data set of 600 rbcL sequences of mosses [36], *Polytrichaceae* is the representative of *Polytrichales* and formed the basal placements clade in these mosses, while the order *Grimmmiales* and *Hypnales* were considered as the sister clade in a relatively higher phylogenetic status of the mosses [41]. In addition, order *Hypnales* formed a large clade, some involved families such as *Amblystegiaceae, Sematophyllaceae* and *Brachytheciaceae* appeared as monophyletic entities [36], but *Hypnaceae* was polyphyletic, and one of its genus *Hypnum* was also placed in different clades. Thus, the reason why the four mosses of JM5_X, JM25_X, JM3_X and GL4_X had similar bacterial communities with ten liverworts was likely due to the phylogenetic status of *Polytrichaceae* (JM3_X and GL4_X) were closer to the liverworts. In addition, although *Amblystegiaceae* (JM25_X) and some species in genus of *Hypnum* of *Hypnaceae* (JM5_X) were assigned to a higher phylogenetic status of order *Hypnales*, their differentiate extent might be more close to *Polytrichales* and liverworts. While the other six mosses, in which three of them belonged to *Grimmiaceae* in order *Grimmiales* (GL1_X, GL5_X and JM29_X), one *Brachytheciaceae* (GWL9_X) and one *Sematophyllaceae* species in *Hypnales* (GWL2_X) and another *Hypnum* species (TM2_X), were grouped together because it might be related to they (family *Grimmiaceae, Brachytheciaceae, Sematophyllaceae* and some *Hypnum* species) were in a higher and closer phylogenetic position in bryopsida and further away from the liverworts.

## Conclusions

In this survey, the bacterial community similarity among different bryophytes seemed related to their phylogenetic position and niches of hosts inhabited. For the hosts with the closer phylogenetic status, the bacterium associated with them was more similar. For those with a very similar phylogenetic status like ten liverworts and some mosses samples, their bacterial community was mainly related to the niches of samples. This hypothesis was basically consistent with reports by Dynesius [42], who concluded that tolerance to ash treatment of different bryophytes was more related to phylogeny than ecology when examining the responses of bryophytes to wood-ash recycling. Certainly, further studies with more bryophytes genotypes will still be needed to determine if this is consistent for all bryophytes or limited to those studied here. In addition, further isolation and identification of more potential useful bacterial resources (especially abundant in different groups) and enhancing the work on their possible biological function in the hosts would be very necessary to make a good understanding on the interaction mechanism between bryophytes and microbes.

#### Acknowledgements
We would like to thank Alison Beamish at the University of British Columbia for her assistance with English language and grammatical editing of the manuscript.

#### Funding
The work was funded by the Scientific Research Program of National Natural Science Foundation of China (No. 31470136).

#### Authors' contributions
JYT and JM performed the experiments and JYT analyzed the data and wrote the paper; XDL collected the bryophyte samples from Tibet, China and identified their taxonomy status; YHL designed the experiment, revised and edited the manuscript. All authors read approved the final manuscript.

#### Competing interests
The authors declare that they have no competing interests.

#### New clinical tools and procedures
Not Applicable.

#### Research authorization
Research was authorized by College of Life Science, Capital Normal University.

#### References
1. Hallmann J, Quadt-Hallmann A, Mahaffee WF, Kloepper JW. Bacterial endophytes in agricultural crops. Can J Microbiol. 1997;43:895–914.
2. Strobel G, Daisy B, Castillo U, Harper J. Natural products from endophytic microorganisms. J Nat Prod. 2004;67:257–68.
3. Mastretta C, Barac T, Vangronsveld J, Newman L, Taghavi S, Van der Lelie D. Endophytic bacteria and their potential application to improve the phytoremediation of contaminated environments. Biotech Gen Eng Rev. 2006;23:175–207.
4. Afzal M, Khan QM, Sessitsch A. Endophytic bacteria: prospects and applications for the phytoremediation of organic pollutants. Chemosphere. 2014;117:232–42.
5. Raymond JA. Dependence on epiphytic bacteria for freezing protection in an Antarctic moss, *Bryum argenteum*. Environ Microbiol Rep. 2016;8(1):14–9.
6. Brock T, Bregman R. Periodicity in growth, productivity, nutrient content and decomposition of *Sphagnum recurvum* var. *mucronatum* in a fen woodland. Oecologia. 1989;80:44–52.
7. Opelt K, Berg G. Diversity and antagonistic potential of bacteria associated with bryophytes from nutrient-poor habitats of the Baltic Sea coast. Appl Environ Microbiol. 2004;70(11):6569–79.
8. Opelt K, Berg C, Schoenmann S, Eberl L, Berg G. High specificity but contrasting biodiversity of *Sphagnum*-associated bacterial and plant communities in bog ecosystems independent of the geographical region. ISME J. 2007;1(6):502–16.
9. Opelt K, Chobot V, Hadacek F, Schoenmann S, Eberl L, Berg G. Investigations of the structure and function of bacterial communities associated with Sphagnum mosses. Environ Microbiol. 2007;9(11):2795–809.
10. Nickrent DL, Parkinson CL, Palmer JD, Duff RJ. Multigene phylogeny of land plants with special reference to bryophytes and the earliest land plants. Mol Biol Evol. 2000;17(12):1885–95.
11. Ye J, Hao ZQ, Yu DY, Yan HB, Feng DQ. Research advances in bryophyte ecological function. Chin J Appl Ecol. 2004;15:1939–42.
12. Turetsky MR. The role of bryophytes in carbon and nitrogen cycling. Bryologist. 2003;106:395–409.
13. Liu XL, Liu SL, Liu M, Kong BH, Liu L, Li YH. A primary assessment of the endophytic bacterial community in a xerophilous moss (*Grimmia montana*) using molecular method and cultivated isolates. Brazn J Microbiol. 2014;45(1): 165–73.
14. Opelt K, Berg C, Berg G. The bryophyte genus *Sphagnum* is a reservoir for powerful and extraordinary antagonists and potentially facultative human pathogens. FEMS Microbiol Ecol. 2007;61(1):38–53.
15. Koua FHM, Kimbara K, Tani A. Bacterial-biota dynamics of eight bryophyte species from different ecosystems. Saudi J Biol Sci. 2015;22(2):204–10.
16. Chelius M, Triplett E. The diversity of archaea and bacteria in association with the roots of *Zea mays* L. Microb Ecol. 2001;41:252–63.
17. Magoc T, Salzberg SL. FLASH: fast length adjustment of short reads to improve genome assemblies. Bioinformatics. 2011;27(21):2957–63.
18. Caporaso JG, Kuczynski J, Stombaugh J, Bittinger K, Bushman FD, Costello EK, Fierer N, Pena AG, Goodrich JK, Gordon JI, et al. QIIME allows analysis of high-throughput community sequencing data. Nat Methods. 2010;7(5):335–6.
19. Edgar RC. UPARSE: highly accurate OTU sequences from microbial amplicon reads. Nat Methods. 2013;10(10):996–8.
20. Edgar RC, Haas BJ, Clemente JC, Quince C, Knight R. UCHIME improves sensitivity and speed of chimera detection. Bioinformatics. 2011;27(16):2194–200.
21. Schloss PD, Westcott SL, Ryabin T, Hall JR, Hartmann M, Hollister EB, Lesniewski RA, Oakley BB, Parks DH, Robinson CJ, et al. Introducing mothur: Open-source, platform-independent, community-supported software for describing and comparing microbial communities. Appl Environ Microbiol. 2009;75(23):7537–41.
22. Chao A. Nonparameteric-estimation of the number of classes in a population. Scand J Stat. 1984;11(4):265–70.
23. Cole JR, Wang Q, Cardenas E, Fish J, Chai B, Farris RJ, Kulam-Syed-Mohideen AS, McGarrell DM, Marsh T, Garrity GM, et al. The ribosomal database project: improved alignments and new tools for rRNA analysis. Nucleic Acids Res. 2009;37:D141–5.
24. Quast C, Pruesse E, Yilmaz P, Gerken J, Schweer T, Yarza P, Peplies J, Gloeckner FO. The SILVA ribosomal RNA gene database project: improved data processing and web-based tools. Nucleic Acids Res. 2013; 41(D1):D590–6.
25. Li YH, Liu QF, Liu Y, Zhu JN, Zhang QA. Endophytic bacterial diversity in roots of *Typha angustifolia* L. in the constructed Beijing Cuihu Wetland (China). Res Microbiol. 2011;162(2):124–31.
26. Faria DC, Dias AC, Melo IS, de Carvalho Costa FE. Endophytic bacteria isolated from orchid and their potential to promote plant growth. World J Microbiol Biotechnol. 2013;29(2):217–21.
27. Koiv V. Microbial population dynamics in response to *Pectobacterium atrosepticum* infection in potato tubers. Sci Rep. 2015;5:167–72.
28. O'Connor-Sanchez A, Rivera-Dominguez AJ, De los Santos-Briones C, Lopez-Aguiar LK, Pena-Ramirez YJ, Prieto-Davo A. Acidobacteria appear to dominate the microbiome of two sympatric Caribbean Sponges and one Zoanthid. Biolog Res. 2014;47:1–6.
29. Tamaki H, Tanaka Y, Matsuzawa H, Muramatsu M, Meng X-Y, Hanada S, Mori K, Kamagata Y. *Armatimonas rosea* gen. nov., sp nov., of a novel bacterial phylum,

Armatimonadetes phyl. nov., formally called the candidate phylum OP10. Int J Syst Evol Microbiol. 2011;61:1442–7.
30. Lee KCY, Morgan XC, Dunfield PF, Tamas I, McDonald IR, Stott MB. Genomic analysis of *Chthonomonas calidirosea*, the first sequenced isolate of the phylum Armatimonadetes. ISME J. 2014;8(7):1522–33.
31. Wang XQ, Sharp CE, Jones GM, Grasby SE, Brady AL, Dunfield PF. Stable-isotope probing identifies uncultured Planctomycetes as primary degraders of a complex heteropolysaccharide in soil. Appl Environ Microbiol. 2015; 81(14):4607–15.
32. Barton HA, Giarrizzo JG, Suarez P, Robertson CE, Broering MJ, Banks ED, Vaishampayan PA, Venkateswaran K. Microbial diversity in a Venezuelan orthoquartzite cave is dominated by the Chloroflexi (Class Ktedonobacterales) and Thaumarchaeota Group I.1c. Front Microbiol. 2014;5:14.
33. Takaichi S, Maoka T, Takasaki K, Hanada S. Carotenoids of *Gemmatimonas aurantiaca* (Gemmatimonadetes): identification of a novel carotenoid, deoxyoscillol 2-rhamnoside, and proposed biosynthetic pathway of oscillol 2,2′-dirhamnoside. Microbiology-SGM. 2010;156:757–63.
34. McLean JS, Lombardo MJ, Badger JH, Edlund A, Novotny M, Yee-Greenbaum J, Vyahhi N, Hall AP, Yang Y, Dupont CL, et al. Candidate phylum TM6 genome recovered from a hospital sink biofilm provides genomic insights into this uncultivated phylum. Proc Nat Acad Sci USA. 2013;110(26):E2390–9.
35. Yeoh YK, Sekiguchi Y, Parks DH, Hugenholtz P. Comparative genomics of candidate phylum TM6 suggests that parasitism is widespread and ancestral in this lineage. Mol Biol Evol. 2016;33(4):915–27.
36. Tsubota H, Luna ED, Gonzalez D, Ignatov MS, Deguchi H. Molecular phylogenetics and ordinal relationships based on analyses of a large-scale data set of 600 rbcL sequences of mosses. Hikobia. 2004;14(2):149–69.
37. Shanker A, Sharma V, Daniell H. Phylogenomic evidence of bryophytes' monophyly using complete and incomplete data sets from chloroplast proteomes. J Plant Biochem Biotech. 2011;20(2):288–92.
38. Zhang L, Zhang A, Chen J, Liu Y. Phylogenetic relationship in some bryophytes based on rbcL-a gene of their chloroplast. Biotechnology. 2012;22(1):50–2.
39. Beckert S, Steinhauser S, Muhle H, Knoop V. A molecular phylogeny of bryophytes based on nucleotide sequences of the mitochondrial *nad5* gene. Plant Syst Evol. 1999;218:179–92.
40. Nishiyama T, Kato M. Molecular phylogenetic analysis among bryophytes and tracheophytes based on combined data of plastid coded genes and the 18S rRNA gene. Mol Biol Evol. 1999;16(8):1027–36.
41. Wahrmund U, Rein T, Muller KF, Groth-Malonek M, Knoop V. Fifty mosses on five trees: comparing phylogenetic information in three types of non-coding mitochondrial DNA and two chloroplast loci. Plant Syst Evol. 2009;282(3–4):241–55.
42. Dynesius M. Responses of bryophytes to wood-ash recycling are related to their phylogeny and pH ecology. Per Plant Ecol Evol Syst. 2012;14(1):21–31.

# 14

# Reclassification of the taxonomic status of SEMIA3007 isolated in Mexico B-11A Mex as *Rhizobium leguminosarum* bv. *viceae* by bioinformatic tools

Luciano Takeshi Kishi[†], Camila Cesário Fernandes[†], Wellington Pine Omori, João Carlos Campanharo and Eliana Gertrudes de Macedo Lemos[*]

**Abstract**

**Background:** Evidence based on genomic sequences is extremely important to confirm the phylogenetic relationships within the *Rhizobium* group. SEMIA3007 was analyzed within the *Mesorhizobium* groups to define the underlying causes of taxonomic identification. We previously used biochemical tests and phenotypic taxonomic methods to identify bacteria, which can lead to erroneous classification. An improved understanding of bacterial strains such as the *Mesorhizobium* genus would increase our knowledge of classification and evolution of these species.

**Results:** In this study, we sequenced the complete genome of SEMIA3007 and compared it with five other *Mesorhizobium* and two *Rhizobium* genomes. The genomes of isolated SEMIA3007 showed several orthologs with *M. huakuii*, *M. erdmanii* and *M. loti*. We identified SEMIA3007 as a *Mesorhizobium* by comparing the 16S rRNA gene and the complete genome.

**Conclusion:** Our ortholog, 16S rRNA gene and average nucleotide identity values (ANI) analysis all demonstrate SEMIA3007 is not *Rhizobium leguminosarum* bv. *viceae*. The results of the phylogenetic analysis clearly show SEMIA3007 is part of the *Mesorhizobium* group and suggest a reclassification is warranted.

**Keywords:** Genome sequencing, Core genome, Comparative analysis, Ortholog genes, Phylogenetic analysis

## Background

Rhizobia is the collective name of the genera *Rhizobium*, *Sinorhizobium* and *Mesorhizobium*, which are soil and rhizosphere bacteria of agronomic importance because they form nitrogen-fixing symbioses with leguminous plants [1, 2]. Thus, rhizobia are considered bio-fertilizers and have been used as inoculants for over 120 years. Rhizobial genetic diversity and the plant-bacteria molecular interactions have been well-studied [3]. The growth rate of *Mesorhizobium* is intermediate between the genera *Rhizobium* and *Bradyrhizobium* and is one of the largest genera. Additionally, the *Mesorhizobium* genera consists of 24 species found in Asia, Europe, the Mediterranean region and Africa [4, 5].

Jarvins et al. [6] were the first to request the creation of the *Mesorhizobium* genus and reclassified several genera identified as *Rhizobium* into *Mesorhizobium*. The correct phylogenetic identification of a species requires an accurate technical characterization [7, 8].

The taxonomy of *Mesorhizobium* requires the reclassification of species because there is a need for studies to avoid classification problems. Taxonomic information provides access to basic trait information such as physiology, epidemiology and evolutionary history [9]. The correct taxonomic assignment of bacterial genomes is a primary and challenging task [10–13].

* Correspondence: egerle@fcav.unesp.br
[†]Equal contributors
Departamento de Tecnologia, Laboratório de Bioquímica de Microrganismos e Planta – LBMP, UNESP - Universidade Estadual Paulista, Faculdade de Ciências Agrárias e Veterinárias, Via de Acesso Prof. Paulo Donato Castellane s/n, 14884-900 Jaboticabal, SP, Brazil

The partial 16S ribosomal RNA gene (16S rRNA) is a molecular marker widely used in the taxonomy of bacteria. However, this gene has no consensus sequence to correctly classify microorganisms at the species level [14–16]. Thus, DNA-DNA hybridization (DDH) has been used as the gold standard for defining prokaryotic species at the genomic level. DDH is the only taxonomic method that offers a numerical and relatively stable species. Therefore, DDH influences how the current classification system has been constructed [17].

DDH is an expensive and laborious method that is available in only a few laboratories worldwide, since it requires the hybridization of hundreds of strains and often does not resolve the taxonomic problems. However, it is an important limiting factor for the description of new species, particularly in countries with the greatest biodiversity. Prokaryotic species continue to be a group of strains due to DNA-DNA re-association values greater than 70 % [14, 18].

The recent development of sequencing technologies has enabled us to carefully assess microbial communities by generating many nucleotide sequences at lower costs. Next generation sequencing (NGS) technologies have revolutionized the field of microbial ecology and allows researchers to determine the level of diversity more closely using in-depth sequencing. There are various applications using these NGS platforms, which range from single-gene targeted sequencing to whole-genome sequencing and shotgun metagenome sequencing [19]. With the availability of whole genome sequences, the gene content based approaches appear promising in inferring the bacterial taxonomy. The complete genome sequencing of a bacterial genome often reveals a substantial number of unique genes present only in that genome which can be used for its taxonomic classification [11, 12].

The recent improved access to various new gene sequences and the definition of prokaryote species has led to doubts regarding the suitability of the DNA-DNA hybridization method [20]. The new proposals include the analysis of several genes or the entire genome. One proposed analysis method is to analyze common genes between two strains and determine the average nucleotide identity values (ANI). An ANI value exceeding 94 % corresponds to 70 % traditional DNA-DNA hybridization [21, 22].

This analysis method also considers genes with ecological functions. Other ANI values suggested replacing 70 % hybridization with 95 % ANI and 69 % conserved DNA. In the protein coding portion of the genome, these values would suggest 85 % conserved genes [23]. The most recent proposals recommend > 95–96 % ANI to delineate species and would replace the traditional 70 % cut off threshold used for DDH sequences [17].

The aim of this study was to evaluate SEMIA3007 isolated in Mexico as B-11A Mex and is classified by phenotypic taxonomic methods such as *Rhizobium leguminosarum* bv. *viceae* by different groups of researchers. We used a combination of complete 16S rRNA sequencing and complete genome analysis to reclassify B-11A as *Mesorhizobium* sp.

## Results and discussion

### Bacterial growth curve

The bacterial growth curve of SEMIA3007 is shown in Fig. 1. SEMIA3007 grew similar to the median strains of *Rhizobium*, *Mesorhizobium* and *Bradyrhizobium*. These findings phenotypically characterize SEMIA3007 as part of the genus *Mesorhizobium*. This strain was originally isolated in Mexico (B-11A Mex) and classified taxonomically as *Rhizobium leguminosarum* bv. *viceae* SEMIA3007 by a combination of phenotypic methods, biochemical tests and partial sequencing of the 16S rRNA gene.

### Genome assembly of SEMIA3007 and its features

The sequencing result shows that strain SEMIA3007 has the following characteristics: one contig of 6,990,002 bp, G + C content 63 %, 6,814 coding sequences (CDS) and a total of 55 RNAs. In the SEMIA3007 genome, there are two clusters encoding nitrite reductase (nirV and nirK) and four clusters related to denitrification processes that reduce nitrate to nitrogen gas. It is postulated that after host infection this cluster is responsible for allowing *Brucella suis* to survive low oxygen concentrations because the cells can use nitrogen oxides as final electron acceptors [24, 25]. The presence of this pathway enables SEMIA3007 to use this mechanism of intracellular survival during host infection.

We also found the following other genes were present in the genome of SEMIA3007: *nif*A, *nif*S, *nif*U, IscA-like,

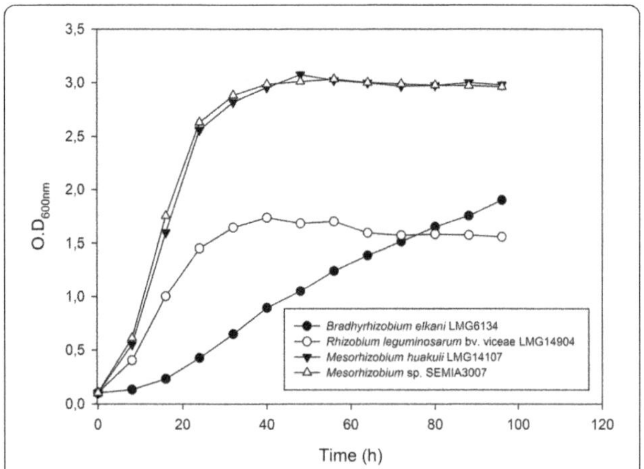


**Fig. 1** Bacterial growth curve. *Bradyrhizobium elkanii* LMG6134, *Rhizobium leguminosarum* bv. *viceae* LMG14904, *Mesorhizobium huakuii* LMG14107 and *Mesorhizobium* sp. SEMIA3007 strains

*nifB*, *frdN*, *nifX*, *nifX2*, *nifE*, *nifN*, *nifQ*, *nifW*, *nifH*, *nifD*, *nifK*, *nifZ* and *nifT*. These results suggest mechanisms for denitrification processes. The *nif* genes function in the transformation of ammonia nitrogen, nitrate, and nitrite ammonification and code for proteins such as nitrate reductase (EC 1.7.99.4), nitrite reductase [NAD(P)H] (EC 1.7.1.4), ferredoxin-nitrite reductase (EC 1.7.7.1) and nitrite reductase (cytochrome; ammonia-forming) (EC 1.7.2.2).

The SEMIA3007 genome also contains a subsystem for assimilation of ammonia and the bacteria can use ammonia assimilated for metabolism of amino acids (glutamate). The system uses glutamate ammonia ligase (EC 6.3.1.2), glutamate synthase (NADPH) (EC 1.4.1.13), glutamate synthase (NADH) (EC 1.4.1.14) and glutamate synthase (ferredoxin) (EC 1.4.7.1).

System secretions of type I, type II, type IV and type VI were identified. These system components are common among rhizobia. The type IV secretion systems are identified in microorganisms associated with plants and are usually composed of Vir proteins [26]. The operon of the type IV SEMIA3007 system features 12 genes encoding the following proteins: VirB1-VirB4, VirB6, VirB8-VirB9, VirB11, VirD4 and VirG. The virB region is responsible for coding key virulence factors in the symbiosis species *Mesorhizobium* [26, 27]. This operon may assist in inducing acidification of the phagosome in the cells after phagocytosis. The acidification may lead to the segregation of unknown effector molecules and create changes in the host cell endosome that generate a new intracellular compartment in which the attacker can replicate [28]. The secretion systems of types III, IV and VI and the nodulation factors are considered responsible for lead host specificity in *Mesorhizobium huakuii* [4].

#### Genome comparisons of SEMIA3007 and *Rhizobium*

Our analysis of the similarity between genomes can be used to differentiate microorganisms. We used the genomes of SEMIA3007, *Rhizobium leguminosarum* bv. *viceae* (gi 115254414), *Rhizobium leguminosarum* bv. *trifolii* (gi 240861949), *Mesorhizobium erdmanii* (gi 548692182), *Mesorhizobium ciceri* (gi 317165637), *Mesorhizobium huakuii* (gi 657121522), *Mesorhizobium loti* (gi 47118328) and *Mesorhizobium opportunistum* (gi 336024847) to construct a progressive alignment using the program Mauve. We found there was a high degree of similarity (block synteny and direction) between SEMIA3007 and the *Mesorhizobium* group and a limited number of blocks collinear between *Rhizobium* (Fig. 2).

ANI [23] is one method that has replaced the DDH [29], and it is the best *in silico* parameter representing DDH that has been experimentally demonstrated [22, 29]. Our genome comparisons for taxonomic purposes were based on BLAST calculations [30]. An ANI value of 95 % ± 0.5 % identity corresponds to 70 % DDH [23], which is a value often recommended to delimit species when used in conjunction with other criteria, such as phenotypic traits [31].

Richter and Rossello-Mora [17] describe a software tool (JSpecies) designed to easily allow the calculation of ANI based on the BLAST algorithm [30] and the MUMmer ultra-rapid aligning tool [32]. We also calculated the tetranucleotide frequencies, which are alignment-free parameters that have been successfully applied to phylogenetically sort metagenome inserts [33]. Therefore, the 95–96 % ANI threshold can be readily used as an objective boundary for species circumscription if it is reinforced by high TETRA correlation values [17]. Our results demonstrate that SEMIA3007 is more genetically similar to *Mesorhizobium huakuii* than *Rhizobium* (Table 1).

#### Phylogenetic analysis using 16S rRNA

The results of sequencing the 16S rRNA gene SEMIA3007 were subjected to a membership analysis taxonomy in RDPII bank. We utilized the classifier tool with a threshold of 95 %. The result showed the identity was 100 % *Mesorhizobium*. Additionally, there was 100 % identity with the 16S Ribosomal RNA database using the Blast program (June 2006).

A phylogenetic analysis was performed using data available on the NCBI database to assess whether SEMIA3007 should be identified and cataloged as *Rhizobium leguminosarum* bv. *viceae* within *Rhizobium* or be reclassified as part of the *Mesorhizobium* group (Additional file 1: Table S1).

The results of the phylogenetic analysis clearly show SEMIA3007 is a member of *Mesorhizobium* and is separate from the *Rhizobium* group, which suggests a reclassification of SEMIA3007 is warranted (Fig. 3).

#### Comparison of gene orthologs

Previous studies have compared genes to differentiate organisms. We used OrthoMCL clustering to identify "core genes", which are the number of unique and shared orthologs of SEMIA3007 and *Mesorhizobium* (Fig. 4). A total of 32,604 proteins from SEMIA3007 (6,814 proteins), *M. huakuii* (5,838 proteins), *M. erdmanii* (6,491 proteins), *M. loti* (7,043 proteins) and *M. opportunistum* (6,418 proteins) were evaluated. We used an inflation index of 1.5 to complete genes and identified 3,075 ortholog groups within the five genomes.

The clusters of orthologs in Fig. 4 show there are 3,075 ortholog groups in SEMIA3007 representing 69.5 % of the total CDS in the genome. However, SEMIA3007 and *M. huakuii* showed 3,951 (79.1 %) common ortholog groups. We found that SEMIA3007

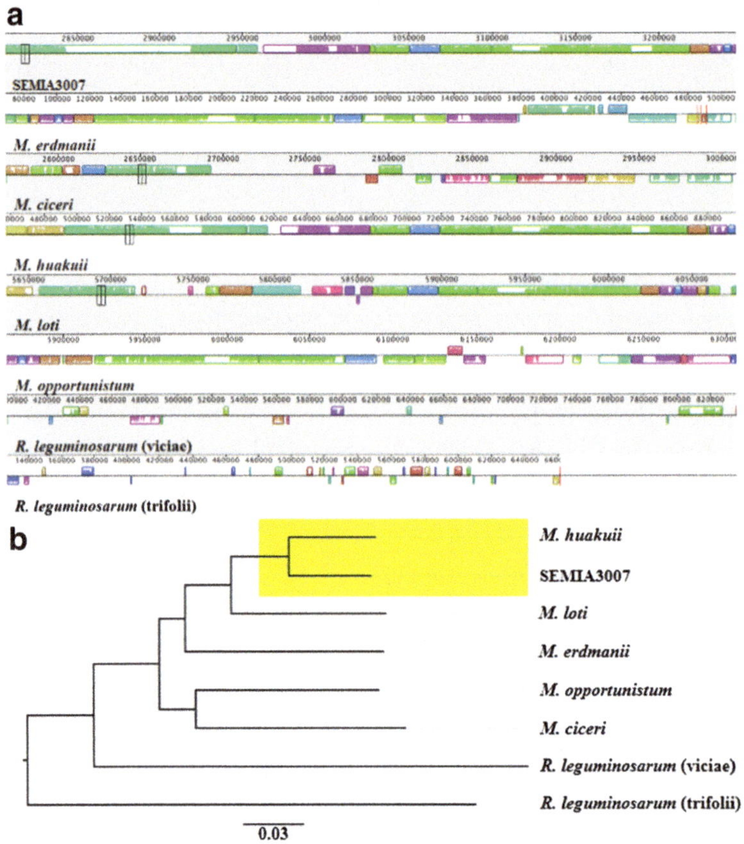


**Fig. 2** Genomes comparison between *Mesorhizobium* and *Rhizobium*. **a** A genome alignment of eight genomes using Mauve reveals collinear blocks conserved (LCB) among all genomes. Each chromosome is shown horizontally and homologous blocks in each genome are shown as identically colored regions. **b** Similarity between genomes aligned by Mauve program showing the phylogenetic relationships between genomes

and *M. erdmanii* shared 4,392 (87.9 %) orthologs. There were 4,197 (84 %) orthologs in common between SEMIA3007 and *M. loti*. There were also 3,984 (79 %) orthologs shared between SEMIA3007 and *M. opportunistum*. Therefore, isolated SEMIA3007 shows a large number of *Mesorhizobium* gene orthologs. These findings suggest that SEMIA3007 is a *Mesorhizobium* strain.

Therefore, the results for growth curve of SEMIA3007, comparative analysis of the genome, ANI,

**Table 1** Probability of pairwise comparison

| | Genome (bp) | %GC | Gene | %ANIb | %ANIm | %Tetra |
|---|---|---|---|---|---|---|
| SEMIA3007 | 6990002 | 62.9 | 7173 | - | - | - |
| *M. huakuii* | 6364365 | 63.2 | 5838 | 98.45 | 98.94 | 99.97 |
| *M. loti* | 7036071 | 62.7 | 7043 | 93.46 | 94.64 | 99.96 |
| *M. opportunistum* | 6884444 | 62.9 | 6418 | 87.31 | 89.17 | 99.85 |
| *M. ciceri* | 6264489 | 62.5 | 6100 | 85.70 | 87.25 | 99.85 |
| *M. erdmanii* | 7018265 | 62.7 | 6491 | 88.89 | 90.40 | 99.86 |
| *R. leg_viciae* | 5057142 | 61.1 | 4797 | 70.72 | 83.54 | 95.75 |
| *R. leguminosarum* | 7418122 | 60.7 | 7293 | 70.27 | 82.79 | 96.01 |

Nucleotide identity values (ANI) and correlation indexes of their Tetra-nucleotide signatures between SEMIA3007 with *Mesorhizobium* and *Rhizobium*

gene orthologs and phylogenetic analysis using 16S rRNA show that SEMIA3007 is not *Rhizobium leguminosarum* bv. *viceae* suggesting its reclassification for Mesorhizobium group [10–13].

## Conclusions

NGS technologies have proven their utility in genomic and metagenomics areas since their earliest application appeared in 2006. Identifying each individual sequence is important in microbial community analysis because the taxonomic information provides access to basic trait information such as physiology, epidemiology and evolutionary history. The taxonomic information also permits indirect inference of their ecological roles in a given environment [19].

Whole-genome sequencing has proven to be valuable and critical for refining the phylogenetic positions and correct taxonomic classification of rhizobial strains [10, 11, 34]. In this study, we sequenced, assembled and annotated the SEMIA3007 genome. We used this genome sequence to examine the phylogenetic relationship between *Mesorhizobium* and *Rhizobium* genus. SEMIA3007 was classified by phenotypic

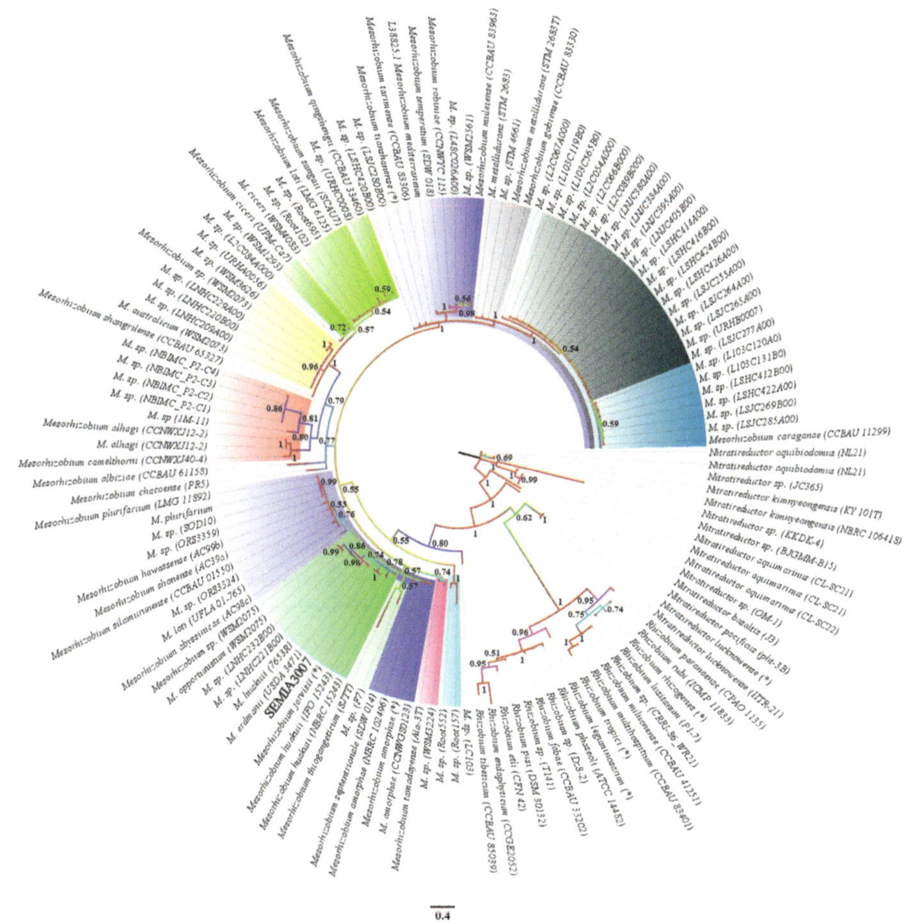

**Fig. 3** Phylogenetic tree showing the taxonomic position of SEMIA3007 strain between groups of *Mesorhizobium* and *Rhizobium*. Genetic differences between bacteria of 0.4 %. The numbers in the branches show the probability calculated by MrBayes with the colors

taxonomic methods and biochemical tests as *Rhizobium leguminosarum* bv. *viceae*. However, our results strongly suggest that SEMIA3007 belongs to the *Mesorhizobium* genus. The placement of SEMIA3007 in a *Mesorhizobium* genus is supported by our analysis of ANI, ortholog genes and phylogenetic analysis.

We can see a high degree of similarity and block synteny and direction between SEMIA3007 and the *Mesorhizobium* group. Our results demonstrated there were a limited number of blocks collinear between *Rhizobium*. Additionally, the ANI based on a pairwise genome comparison of all shared ortholog protein coding genes is 98 % with *Mesorhizobium huakuii*. Our phylogenetic analysis demonstrated that SEMIA3007 is not part of the *Rhizobium* genus, and the ortholog genes revealed sufficient ability to identify SEMIA3007 as *Mesorhizobium*.

The concepts of orthology originated from the field of molecular systematics [35] and have recently been applied to functional characterizations and classifications on the scale of whole-genome comparisons [36–38]. In comparative genomics, the clustering of orthologous genes provides a framework for integrating information from multiple genomes by highlighting the divergence and conservation of gene families and biological processes.

The identification of orthologous groups in prokaryotic genomes has permitted cross-referencing of genes from multiple species and has facilitated genome annotation, protein family classification, studies on bacterial evolution and the identification of strains. The ultimate goal of taxonomy is to construct a classification that is operative and predictive for any discipline in microbiology. The classification is also essentially stable for old and new strain such as Rhizobia and the collective names of the genera *Rhizobium*, *Sinorhizobium*, *Mesorhizobium*.

## Methods

### Bacterial growth curve

The strains of *Bradyrhizobium elkanii* LMG6134, *Rhizobium leguminosarum* bv. *viceae* LMG14904, *Mesorhizobium huakuii* LMG14107 and *Mesorhizobium* sp. SEMIA3007 were cultured for 96 h with shaking

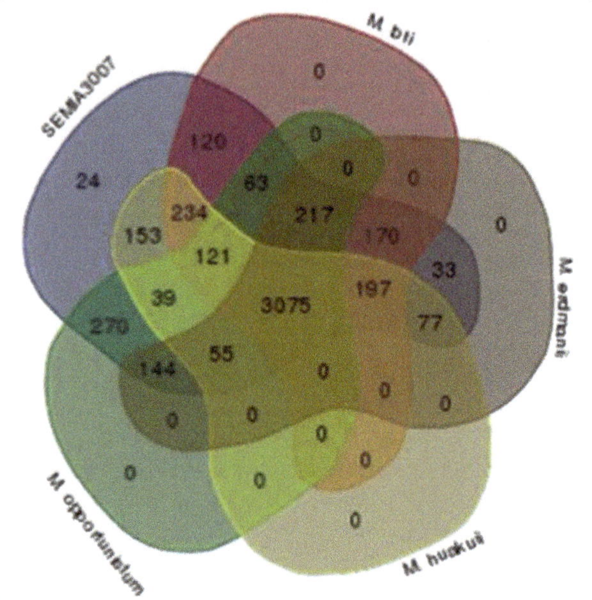


**Fig. 4** Venn diagram showing core genome analyses of *Mesorhizobium* strains. The number of protein-coding gene ortholog sharing among five *Mesorhizobium*. SEMIA3007; *M. huakuii* (CP006581.1); *M. loti* (NC_002678.2); *M. opportunistum* (NC_015675.1); *M. erdmanii* (NZ_AXAE01000048.1)

(150 rpm) at 30 °C in TY medium [39] in triplicate. To obtain the bacterial growth curve, the OD reading was collected every 8 h.

**Bacterial strain and DNA preparation**

SEMIA3007 was cultured for 48 h at 28 °C with 145 rpm shaking in TY medium [39]. The SEMIA3007 cells were harvested by centrifugation, and the total DNA was prepared using a Wizard® Genomic DNA Purification Kit (Promega).

**Sequencing and annotation of the genome**

The *de novo* sequencing of the SEMIA3007 genome used a combined strategy involving Illumina – HiscanSQ. The libraries were constructed using a TruSeq® DNA Sample Prep kit and Nextera Mate Pair Sample Preparation kit (Illumina®). The cluster formation of library templates was performed with the TruSeq PE Cluster kit v3 (Illumina®) and the Illumina cBot workstation using conditions recommended by the manufacturer. Paired end 100 base pair (2x100bp) sequencing by synthesis was performed with TruSeq SBS kit v3 (Illumina®) on an Illumina HiscanSQ using protocols defined by the manufacturer. The base call conversion to sequence reads was performed using CASAVA 1.8.3 (Illumina®). As a result, paired-end and mate pair fastq files were trimmed using Scythe 0.991 (https://github.com/vsbuffalo/scythe), Cutadapt 1.7.1 [40] and the quality of data was filtered by Prinseq program [41] with Phred ≥20. The sequence assembly was performed using the Spades 3.6.1 program [42]. The prediction of ORFs and annotation were performed using the Rast system [43].

**Genome comparisons and average nucleotide identity (ANI)**

For comparing the genome of SEMIA3007 to others genomes we compute an alignment of the six genomes, we used the Progressive Mauve algorithm [44]. An alignment of the four *Mesorhizobium* and *Rhizobium* genomes was constructed using the default mauveAligner parameters. The resulting LCBs were inspected using the Mauve alignment viewer, and the minimum LCB weight was adjusted to eliminate LCBs consisting of only repetitive elements (LCB Weight 600).

Reference genomes for comparison purposes were retrieved from the GenBank database (http://www.ncbi.nlm.nih.gov/genbank/). Sequences were uploaded into the JSpecies software package (http://www.imedea.uib.es/jspecies) to perform pairwise genome calculations of the average nucleotide identity (ANI) [17, 23] and support the proposed cut-off level of 95 % as a species delineation threshold [22].

**Ortholog analysis**

The ortholog groups in multiple genomes can be useful for annotation and revealing the patterns of phylogenetic proteins from different strains. The groups also provide insights into the evolutionary conservation and diverse cellular functions in different species.

Four coding sequences (CDS) from genomes/drafts of *Mesorhizobium loti, Mesorhizobium huakuii, Mesorhizobium erdmanii* and *Mesorhizobium sp.* were extracted from GenBank files (Additional file 1: Table S1), representing four species (five with SEMIA3007 CDS). The pan and core genome analysis was conducted by determining shared (homologous) and species-specific protein-coding genes using OrthoMCL [36] with e-value cutoff $1 \times 10^{-20}$, protein percent identity ≥50 % and MCL inflation of 1.5. OrthoMCL computes families of homologous genes for pan and core genome analyses. The families in which two or more genomes participate were used to determine numbers plotted. OrthoMCL was run with blast e-value cut-off of 1e-5 and an inflation parameter of 1.5. The table with orthologs was used to plot Venn diagrams (http://bioinformatics.psb.ugent.be/webtools/Venn/) [36].

**16S rRNA gene sequencing**

The amplification of the 16S rRNA gene of the SEMIA3007 was performed with FD1 and RD1 primers [45]. The PCR reaction mixture consisted of 30 ng of DNA, 7.5 pmol of each primer, 0.2 mM of dNTPs, 1.5 mM of $MgCl_2$, Buffer 1X and 2.5 U Taq DNA polymerase (Ludwig Biotec). A thermocycler model PTC-100 ™ Programmable Thermal Controller (MJ Research, Inc.) was used with a thermal profile of 96 °C for 2 min, 40 cycles of 96 °C for 30 s, 53 °C for 1 min and 60 °C for 4 min. After the PCR reaction, the products were purified with a Wizard® SV Gel and PCR Clean-Up System (Promega). The amplicon was sequenced with 1 μl of BigDye Terminator v3.1, buffer 0.75X (Tris-HCl 200 mM, pH 9.0 and MgCl2 5 mM), 10 pmoles of primer FD1, 50 ng of DNA and sterile Milli-Q distilled water (10 μL q.s.p). Sequencing was performed on Sequencer ABI PRISM 3130xl DNA Analyzer (Applied Biosystems) following the manufacturer's instructions.

### Downloading the sequences 16S rRNA in GenBank

The National Center for Biotechnology Information (NCBI) was used to search the genome for species *Mesorhizobium* (March 15, 2016). All complete gene sequences for *16S rRNA* (16S ribosomal RNA) were downloaded from GenBank (Additional file 1: Table S1) [46].

### Phylogenetic analysis of 16S rRNA gene

The 16S rRNA gene set were aligned using the MAFFT v7.215 program [47]. The search for the best nucleotide substitution matrix was performed with the Phangorn package [48] in R [49] and the feature modelTest. The construction of a phylogenetic tree was performed with the Mrbayes v3.2.2 program [50] using the matrix replacement General Time Reversible (GTR) with gamma variation (G) and invariable sites (I) with 10.000.000 generations. The best evolutionary model was chosen based on Akaike information criterion with correction (AICc).

### Nucleotide sequence accession number

The data sets results of this article are available in the NCBI BioProject SRR3703040.

#### Abbreviations

16S rRNA: 16S ribosomal RNA gene; AICc: Akaike information criterion with correction; ANI: Average nucleotide identity values; CDS: Coding sequences; DDH: DNA-DNA hybridization; DNA: Deoxyribonucleic acid; GTR: General time reversible; LCB: Collinear blocks conserved; LCB: Locally collinear blocks; NAD(P)H: Nitrite reductase; NADH: Glutamate synthase; NADPH: Glutamate synthase; NCBI: National Center for Biotechnology Information; NGS: Next generation sequencing; OD: Optical density; ORF: Open read frame; PCR: Polymerase chain reaction; TY: Triptone, yeast medium

#### Acknowledgements

We gratefully acknowledge Universidade Estadual Paulista - UNESP, the Programa de Pós Graduação em Microbiologia Agropecuária, UNESP, Jaboticabal, São Paulo State, Brazil. We also thank FAPESP (grants: 2009/539842; 2014/14234-6) and CNPq.

#### Funding

This work was supported by FAPESP (grants: 2009/539842; 2014/14234-6) and CNPq.

#### Authors' contributions

Conceived and designed the experiments: LTK CCF EGML. Performed the experiments: LTK CCF JCC EML. Analyzed the data: LTK WPO. Contributed reagents/materials/analysis tools: EGML. Wrote the paper: CCF LTK. All authors read and approved the final manuscript.

#### Competing interests

The authors declare that they have no competing interests.

#### References

1. Kaneko T, Nakamura Y, Sato S, Asamizu E, Kato T, Sasamoto S, et al. Complete genome structure of the nitrogen-fixing symbiotic bacterium Mesorhizobium loti. DNA Res. Int J Rapid Publ Rep Genes Genomes. 2000;7:331–8.
2. Velázquez E, Peix A, Zurdo-Piñiro JL, Palomo JL, Mateos PF, Rivas R, et al. The coexistence of symbiosis and pathogenicity-determining genes in *Rhizobium rhizogenes* strains enables them to induce nodules and tumors or hairy roots in plants. Mol Plant Microbe Interact. 2005;18:1325–32.
3. Graham PH, Sadowsky MJ, Keyser HH, Barnet YM, Bradley RS, Cooper JE, et al. Proposed minimal standards for the description of new genera and species of root- and stem-nodulating bacteria. Int J Syst Bacteriol. 1991;41:582–7.
4. Wang S, Hao B, Li J, Gu H, Peng J, Xie F, et al. Whole-genome sequencing of Mesorhizobium huakuii 7653R provides molecular insights into host specificity and symbiosis island dynamics. BMC Genomics. 2014;15:440.
5. Degefu T, Wolde-meskel E, Liu B, Cleenwerck I, Willems A, Frostegard A. Mesorhizobium shonense sp. nov., Mesorhizobium hawassense sp. nov. and Mesorhizobium abyssinicae sp. nov., isolated from root nodules of different agroforestry legume trees. Int J Syst Evol Microbiol. 2013;63:1746–53.
6. Jarvis BDW, Van Berkum P, Chen WX, Nour SM, Fernandez MP, Cleyet-Marel JC, et al. Transfer of rhizobium loti, rhizobium huakuii, rhizobium ciceri, rhizobium mediterraneum, and rhizobium tianshanense to mesorhizobium gen. nov. Int J Syst Bacteriol. 1997;47:895–8.
7. Mousavi SA, Willems A, Nesme X, de Lajudie P, Lindström K. Revised phylogeny of rhizobiaceae: proposal of the delineation of pararhizobium gen. nov., and 13 new species combinations. Syst Appl Microbiol. 2015;38:84–90.
8. Mousavi SA, Österman J, Wahlberg N, Nesme X, Lavire C, Vial L, et al. Phylogeny of the Rhizobium–Allorhizobium–Agrobacterium clade supports the delineation of Neorhizobium gen. nov. Syst Appl Microbiol. 2014;37:208–15.
9. Martínez-Hidalgo P, Martínez-Molina E, Mateos PF, Velázquez E, Peix Á, Flores-Félix JD, et al. Revision of the taxonomic status of type strains of Mesorhizobium loti and reclassification of strain USDA 3471 T as the type strain of Mesorhizobium erdmanii sp. nov. and ATCC 33669 T as the type strain of Mesorhizobium jarvisii sp. nov. Int. J. Syst. Evol. Microbiol. 2015;65:1703–8.
10. Ormeño-Orrillo E, Servín-Garcidueñas LE, Rogel MA, González V, Peralta H, Mora J, et al. Taxonomy of rhizobia and agrobacteria from the Rhizobiaceae family in light of genomics. Syst Appl Microbiol. 2015;38:287–91.
11. Gupta A, Sharma VK. Using the taxon-specific genes for the taxonomic classification of bacterial genomes. BMC Genomics [Internet]. 2015 [cited 2016 Oct 4];16. Available from: http://www.biomedcentral.com/1471-2164/16/396. Accessed 5 Oct 2016.

12. Teng JLL, Tang Y, Huang Y, Guo F-B, Wei W, Chen JHK, et al. Phylogenomic Analyses and Reclassification of Species within the Genus Tsukamurella: Insights to Species Definition in the Post-genomic Era. Front. Microbiol. [Internet]. 2016 [cited 2016 Oct 4];7. Available from: http://journal.frontiersin.org/Article/10.3389/fmicb.2016.01137/abstract. Accessed 5 Oct 2016.
13. McIlroy SJ, Lapidus A, Thomsen TR, Han J, Haynes M, Lobos E, et al. High quality draft genome sequence of Meganema perideroedes str. Gr1T and a proposal for its reclassification to the family Meganemaceae fam. nov. Stand. Genomic Sci. [Internet]. 2015 [cited 2016 Oct 4];10. Available from: http://www.standardsingenomics.com/content/10/1/23. Accessed 5 Oct 2016.
14. Vandamme P, Pot B, Gillis M, de Vos P, Kersters K, Swings J. Polyphasic taxonomy, a consensus approach to bacterial systematics. Microbiol Rev. 1996;60:407–38.
15. Kim B-Y, Weon H-Y, Cousin S, Yoo S-H, Kwon S-W, Go S-J, et al. Flavobacterium daejeonense sp. nov. and Flavobacterium suncheonense sp. nov., isolated from greenhouse soils in Korea. Int J Syst Evol Microbiol. 2006;56:1645–9.
16. Menna P, Hungria M, Barcellos FG, Bangel EV, Hess PN, Martínez-Romero E. Molecular phylogeny based on the 16S rRNA gene of elite rhizobial strains used in Brazilian commercial inoculants. Syst Appl Microbiol. 2006;29:315–32.
17. Richter M, Rossello-Mora R. Shifting the genomic gold standard for the prokaryotic species definition. Proc Natl Acad Sci. 2009;106:19126–31.
18. Coenye T, Gevers D, Van de Peer Y, Vandamme P, Swings J. Towards a prokaryotic genomic taxonomy. FEMS Microbiol Rev. 2005;29:147–67.
19. Kim M, Lee K-H, Yoon S-W, Kim B-S, Chun J, Yi H. Analytical tools and databases for metagenomics in the next-generation sequencing era. Genomics Inform. 2013;11:102.
20. Achtman M, Wagner M. Microbial diversity and the genetic nature of microbial species. Nat. Rev. Microbiol. [Internet]. 2008 [cited 2016 Jun 22]; Available from: http://www.nature.com/nrmicro/journal/v6/n6/abs/nrmicro1872.html. Accessed 5 Oct 2016.
21. Konstantinidis KT, Tiedje JM. Genomic insights that advance the species definition for prokaryotes. Proc Natl Acad Sci. 2005;102:2567–72.
22. Konstantinidis KT, Tiedje JM. Trends between gene content and genome size in prokaryotic species with larger genomes. Proc Natl Acad Sci U S A. 2004;101:3160–5.
23. Goris J, Klappenbach JA, Vandamme P, Coenye T, Konstantinidis KT, Tiedje JM. DNA–DNA hybridization values and their relationship to whole-genome sequence similarities. Int J Syst Evol Microbiol. 2007;57:81–91.
24. Kohler S, Foulongne V, Ouahrani-Bettache S, Bourg G, Teyssier J, Ramuz M, et al. Nonlinear partial differential equations and applications: the analysis of the intramacrophagic virulome of Brucella suis deciphers the environment encountered by the pathogen inside the macrophage host cell. Proc Natl Acad Sci. 2002;99:15711–6.
25. Haine V, Dozot M, Dornand J, Letesson J-J, De Bolle X. NnrA is required for full virulence and regulates several brucella melitensis denitrification genes. J Bacteriol. 2006;188:1615–9.
26. Hubber AM, Sullivan JT, Ronson CW. Symbiosis-induced cascade regulation of the *Mesorhizobium loti* R7A VirB/D4 type IV secretion system. Mol Plant Microbe Interact. 2007;20:255–61.
27. Soto MJ, Sanjuan J, Olivares J. Rhizobia and plant-pathogenic bacteria: common infection weapons. Microbiology. 2006;152:3167–74.
28. Boschiroli ML, Ouahrani-Bettache S, Foulongne V, Michaux-Charachon S, Bourg G, Allardet-Servent A, et al. Type IV secretion and Brucella virulence. Vet Microbiol. 2002;90:341–8.
29. Garrity GM, Trüper HG, Whitman WB, Grimont PAD, Nesme X, Frederiksen W, et al. Report of the ad hoc committee for the re-evaluation of the species definition in bacteriology. Int J Syst Evol Microbiol. 2002;52:1043–7.
30. Altschul SF, Gish W, Miller W, Myers EW, Lipman DJ. Basic local alignment search tool. J Mol Biol. 1990;215:403–10.
31. Rosselló-Mora R, Amann R. The species concept for prokaryotes. FEMS Microbiol Rev. 2001;25:39–67.
32. Delcher AL, Kasif S, Fleischmann RD, Peterson J, White O, Salzberg SL. Alignment of whole genomes. Nucleic Acids Res. 1999;27:2369–76.
33. Teeling H, Meyerdierks A, Bauer M, Amann R, Glockner FO. Application of tetranucleotide frequencies for the assignment of genomic fragments. Environ Microbiol. 2004;6:938–47.
34. Schuldes J, Rodriguez Orbegoso M, Schmeisser C, Krishnan HB, Daniel R, Streit WR. Complete genome sequence of the broad-host-range strain sinorhizobium fredii USDA257. J Bacteriol. 2012;194:4483.
35. Fitch WM. Distinguishing homologous from analogous proteins. Syst Zool. 1970;19:99.
36. Li L. OrthoMCL: identification of ortholog groups for eukaryotic genomes. Genome Res. 2003;13:2178–89.
37. Tatusov RL, Koonin EV, Lipman DJ. A genomic perspective on protein families. Science. 1997;278:631–7.
38. Chervitz SA, Aravind L, Sherlock G, Ball CA, Koonin EV, Dwight SS, et al. Comparison of the complete protein sets of worm and yeast: orthology and divergence. Science. 1998;282:2022–8.
39. Beringer JE. R factor transfer in Rhizobium leguminosarum. J Gen Microbiol. 1974;84:188–98.
40. Martin M. Cutadapt removes adapter sequences from high-throughput sequencing reads. EMBnet J. 2011;17:10.
41. Schmieder R, Edwards R. Quality control and preprocessing of metagenomic datasets. Bioinformatics. 2011;27:863–4.
42. Bankevich A, Nurk S, Antipov D, Gurevich AA, Dvorkin M, Kulikov AS, et al. SPAdes: a new genome assembly algorithm and its applications to single-cell sequencing. J Comput Biol. 2012;19:455–77.
43. Aziz RK, Bartels D, Best AA, DeJongh M, Disz T, Edwards RA, et al. The RAST server: rapid annotations using subsystems technology. BMC Genomics. 2008;9:75.
44. Darling ACE. Mauve: multiple alignment of conserved genomic sequence with rearrangements. Genome Res. 2004;14:1394–403.
45. Weisburg WG, Barns SM, Pelletier DA, Lane DJ. 16S ribosomal DNA amplification for phylogenetic study. J Bacteriol. 1991;173:697–703.
46. Laranjo M, Alexandre A, Oliveira S. Legume growth-promoting rhizobia: an overview on the Mesorhizobium genus. Microbiol Res. 2014;169:2–17.
47. Katoh K. MAFFT: a novel method for rapid multiple sequence alignment based on fast Fourier transform. Nucleic Acids Res. 2002;30:3059–66.
48. Schliep KP. phangorn: phylogenetic analysis in R. Bioinformatics. 2011;27:592–3.
49. Ihaka R, Gentleman R. R: A language for data analysis and graphics. J Comput Graph Stat. 1995;5:299–314.
50. Ronquist F, Teslenko M, van der Mark P, Ayres DL, Darling A, Hohna S, et al. MrBayes 3.2: efficient bayesian phylogenetic inference and model choice across a large model space. Syst Biol. 2012;61:539–42.

# 15

# Detailed analysis of c-di-GMP mediated regulation of *csgD* expression in *Salmonella typhimurium*

Irfan Ahmad[1,3,2], Annika Cimdins[1], Timo Beske[1,4] and Ute Römling[1*]

**Abstract**

**Background:** The secondary messenger cyclic di-GMP promotes biofilm formation by up regulating the expression of *csgD*, encoding the major regulator of rdar biofilm formation in *Salmonella typhimurium*. The GGDEF/EAL domain proteins regulate the c-di-GMP turnover. There are twenty- two GGDEF/EAL domain proteins in the genome of *S. typhimurium*. In this study, we dissect the role of individual GGDEF/EAL proteins for *csgD* expression and rdar biofilm development.

**Results:** Among twelve GGDEF domains, two proteins upregulate and among fifteen EAL domains, four proteins down regulate *csgD* expression. We identified two additional GGDEF proteins required to promote optimal *csgD* expression. With the exception of the EAL domain of STM1703, solely, diguanylate cyclase and phosphodiesterase activities are required to regulate *csgD* mediated rdar biofilm formation. Identification of corresponding phosphodiesterases and diguanylate cyclases interacting in the *csgD* regulatory network indicates various levels of regulation by c-di-GMP. The phosphodiesterase STM1703 represses transcription of *csgD* via a distinct promoter upstream region.

**Conclusion:** The enzymatic activity and the protein scaffold of GGDEF/EAL domain proteins regulate *csgD* expression. Thereby, c-di-GMP adjusts *csgD* expression at multiple levels presumably using a multitude of input signals.

**Keywords:** c-di-GMP, CsgD, GGDEF/EAL domain proteins, rdar morphotype, biofilm formation, *Salmonella typhimurium*

## Background

Bacteria have the ability to live as free-swimming planktonic cells or in the form of multicellular communities known as biofilms, the life style that confers tolerance towards challenging environmental conditions [1, 2]. The ubiquitous second messenger bis- (3′–5′)-cyclic dimeric GMP (c-di-GMP), plays a major role in the transition from the motile to the sessile life style on the single cell level [3–6]. C-di-GMP is synthesized by diguanylate cyclases (DGCs), GGDEF domain proteins [7–10], and degraded by c-di-GMP phosphodiesterases (PDEs), EAL or HD-GYP domain proteins [11–13]. These cyclic di-GMP metabolizing proteins occur almost ubiquitously in bacterial genomes with single bacterial genomes to possess from a single GGDEF/EAL domain protein to more than hundred [6, 14–16]. Elucidating the precise role of individual GGDEF/EAL domain proteins will contribute to the understanding of the complex regulation of bacterial physiology by the c-di-GMP signalling system.

A variety of phenotypes such as motility, cell cycle and differentiation and virulence are regulated by c-di-GMP signalling, however, biofilm formation is studied most extensively [17–19]. In the model organism *Salmonella enterica* serovar Typhimurium UMR1, c-di-GMP promotes a rdar (**r**ed, **d**ry **a**nd **r**ough) biofilm formation by stimulating the production of the extracellular matrix components, the exopolysaccharide cellulose and proteinaceous curli fimbriae [20, 21].

Expression of the rdar morphotype is regulated by the LuxR family transcriptional activator CsgD, a major hub in rdar biofilm formation in *S. typhimurium* [22, 23]. CsgD is central in regulating the transition between biofilm formation and virulence. *csgD* expression is usually regulated by environmental stimuli such as temperature

* Correspondence: ute.romling@ki.se
[1]Department of Microbiology, Tumor and Cell Biology, Karolinska Institutet, Stockholm, Sweden
Full list of author information is available at the end of the article

and growth phase from the transcriptional to the post-transcriptional level [24]. Global transcriptional regulators such as RpoS, OmpR, H-NS and IHF regulate the transcription of *csgD* in *S. typhimurium* [25]. *CsgD* expression is also adjusted post-transcriptionally by several small sRNAs and the RNA chaperone Hfq [26–28] and is a major target of c-di-GMP signalling [20, 29].

The genome of *S. typhimurium* contains twenty-two GGDEF/EAL domain proteins, not all are bona fide c-di-GMP metabolizing proteins [20, 30]. Task distribution is shown as distinct panels of proteins are associated with specific phenotypes such as *csgD* expression, cellulose biosynthesis, motility, invasion of epithelial cells, stimulation of a pro-inflammatory immune response and colonization of the gastrointestinal tract of mice [20, 30].

In rdar biofilm formation, two GGDEF-EAL proteins, STM3388 and STM2123 promote, while the four EAL domain proteins STM1703, STM4264, STM3611 and STM1827 suppress *csgD* expression [20, 31]. The transcriptional regulator CsgD activates the expression of *csgBA*, encoding the minor and major subunit of curli and *adrA*, encoding the diguanylate cyclase AdrA. C-di-GMP produced by AdrA stimulates the cellulose synthase in order to activate cellulose biosynthesis [20, 32].

To further dissect the network of GGDEF and EAL domain proteins, we identified two novel GGDEF domain proteins to regulate *csgD* expression. Deletion of major phosphodiesterases indicates that elevated c-di-GMP levels regulate *csgD* expression and rdar morphotype by multiple pathways. Identification of corresponding diguanylate cyclases and phosphodiesterases points to local and global regulation of *csgD* expression by c-di-GMP signalling.

## Methods

### Bacterial strains, plasmids, and growth conditions

Bacterial strains and plasmids are listed in Additional file 1. For cloning purposes, *E. coli* TOP10 and *S. typhimurium* were grown on Luria-Bertani (LB) agar plates supplemented with appropriate antibiotics. Otherwise, bacteria were pre cultured on LB agar plates at 37°C overnight and directly inoculated on LB agar plates without salt. Antibiotics were ampicillin (100 μg $ml^{-1}$), chloramphenicol (20 μg $ml^{-1}$), kanamycin (30 μg $ml^{-1}$), and tetracycline (20 μg $ml^{-1}$). For expression of genes, 0.1% arabinose and 1 mM IPTG was used.

### Construction of *S. typhimurium* mutants

The deletion mutant of *ompR* was created by one-step gene inactivation [33]. Entire open reading frame except 40 nucleotides at the beginning and at the end of the gene were replaced by a chloramphenicol resistance marker. Approximately 300 ng of processed PCR product amplified from pKD3 or pKD4 was electroporated into *S. typhimurium* UMR1 containing pKD46. Recovered colonies were purified at least twice on LB medium containing appropriate antibiotics.

Mutant alleles were combined by phage transduction using phage P22 HT105/1 *int-201* whereby the resistance marker of the parent strain was cut out using pCP20 [34]. Transductants were colony purified twice on LB agar plates containing 10 mM EGTA and appropriate antibiotics. All constructed mutants were verified by PCR with control primers located in genes flanking the targeted open reading frame. All quadruple and pentapole mutants were verified after strain construction.

Site directed mutagenesis to replace the glutamate in the EAL motif of STM4264 by alanine was carried out by scar less site directed mutagenesis [35]. In brief, a chloramphenicol resistance cassette fused to an I-*Sce*I recognition site by homologous recombination replaced the codon for glutamic acid 303 of STM4264. Subsequently, plasmid pWRG99, which encodes I-*Sce*I endonuclease under a tetracycline inducible promoter, aided replacement of the chloramphenicol cassette by a DNA fragment of 80 bp (containing the E303A mutation obtained by annealing primers '4264-303A-mut-scarless forw' and '4264-E303A-mut-scarless Rev') after selection on LB agar plates with IPTG, arabinose and tetracycline [35]. STM4264 harboring the desired mutation was verified by DNA sequencing. Primers used in this study are listed in Additional file 1: Table S2.

### Plasmid construction

Plasmid pBAD30::2123 was constructed by cloning STM2123 into pBAD30 with a C-terminal 6xHis tag. STM2123 was amplified with primers 'STM2123 cloning Forw' and 'STM2123 cloning Rev' harboring restriction sites *Xba*I and *Hind*III and the restricted DNA fragment was ligated into pBAD30.

### Construction of mutant GGDEF/EAL proteins

To generate mutations in GGDEF and EAL domains mutagenic oligonucleotides were designed (listed in Additional file 1). The QuickChange mutagenesis kit (Agilent Technologies) was used according to the manufacturer's protocol. The resulting mutations were confirmed by DNA sequencing.

### Rdar morphotype assay

Five microliters of a bacterial suspension in water ($OD_{600}$ of 5) from an overnight culture in LB broth were spotted onto LB without salt agar plates supplemented with Congo red (40 μg $ml^{-1}$) and Coomassie brilliant blue (20 μg $ml^{-1}$). Plates were incubated at 28°C for 48 h. Development of the colony morphology and dye binding was analysed over time.

### Protein techniques

For western blot analysis of CsgD and c-di-GMP turnover protein expression, cells were grown on LB agar plates without salt for 24 h at 28°C. 5 mg (wet weight) cells were harvested, resuspended in 200 µl SDS sample buffer, and incubated at 95°C for 10 min. Membrane proteins were resuspended in sample buffer with 8M Urea. The protein content was analysed by Coomassie blue staining (20% methanol, 10% acetic acid, 0.1% Coomassie brilliant blue G) after sodium dodecyl sulfate-polyacrylamide gel electrophoresis (12% resolving gel and 4% stacking gel). Equal amounts of protein were separated and subsequently transferred to a polyvinylidene difluoride membrane (Immobilon P; Millipore). Detection of CsgD was carried out using a polyclonal anti-CsgD peptide antibody (1:5,000) and detection of 6xHis-tagged proteins with monoclonal anti 6xHis antibody as the primary antibodies and goat anti-rabbit/mouse immunoglobulin G conjugated with horseradish peroxidase (1:2,000; Jackson ImmunoResearch Laboratories Inc.) as the secondary antibodies, respectively [22]. Chemiluminescence (Lumi-Light WB substrate; Roche) was recorded using the LAS-1000 system (FUJIFILM) [25, 36]. Strain *S. Typhimurium* MAE52 was used as a positive control, whereas strain MAE50, a *csgD* deletion mutant, was used as negative control. Western blotted membranes were subjected to the Ponceau S staining to confirm equal loading of protein samples where appropriate.

### Beta galactosidase assay

Promoter activity of *csgD* was assayed with different *csgD* promoter constructs [25, 36]. Expression of *adrA* was analyzed with a chromosomal MudJ transcriptional fusion in *adrA* [37]. Strains were grown on LB without salt plates supplemented with appropriate antibiotics and inducer. Samples were collected after growth for 24 h at 28°C. β-galactosidase activity was the read out for promoter activity [38] after adjustment of bacterial suspension to $OD_{600} = 0.4$ for pUGE13 and to 0.1 for pUGE5, pUGE7 and pUGE19. Normalized β-galactosidase activity was calculated using the formula: Miller units = $1000 \{[OD420 - (1.75 \times OD_{550})] / (t \times V \times OD_{600})\}$ with t = reaction time in min; V = volume of cell suspension. All β-galactosidase measurements were done in duplicates using at least three technical replicates. Statistical analysis was performed applying an unpaired t-test with two-tailed p-value (*** is $p < 0.0001$) using Prism 5 (GraphPad Software).

## Results

### Identification of novel GGDEF domain proteins promoting *csgD* expression

Multicellular behavior as expressed by the rdar biofilm morphotype in *S. typhimurium* UMR1 correlates with expression of the response regulator CsgD, a major target of c-di-GMP signalling. As *csgD* expression is not completely abolished in a STM3388 and STM2123 double mutant [20], re-assessment of the effect of the remaining GGDEF proteins identified STM4551 and STM1987 to additionally activate *csgD* mediated rdar morphotype expression. The respective single deletion mutants exhibited a down regulation of the rdar morphotype (Fig. 1a) and CsgD levels (Fig. 1b) and the double mutant had an additive effect (Fig. 1a).

### Additive effect of GGDEF domain proteins on *csgD* expression

A quadruple mutant with deletion of *STM4551, STM1987, STM3388* and *STM2123* (Δ4DGC) exhibited a smooth and white (saw) morphotype on CR agar plates (Fig. 2a) with CsgD expression additionally down regulated (Fig. 2a). Over expression of the GGDEF protein STM4551 from pBAD30 restored rdar morphotype and *csgD* expression in the Δ4DGC mutant of *S. typhimurium* UMR1. Although STM4551 is an established diguanylate cyclase [30, 39], reportedly its catalytic activity is not required to restore most of the phenotypes

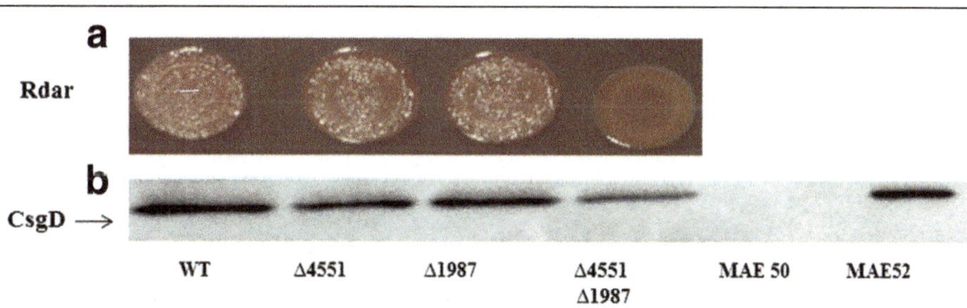


**Fig. 1** Identification of novel GGDEF domain proteins regulating *csgD* expression. **a** Rdar morphotype formation of *S. typhimurium* UMR1 and STM1987 and STM4551 mutants after 48 h of growth on Congo red agar plates. **b** CsgD levels after 24 h of growth at 28°C on LB without salt agar plates. WT is wild type *S. typhimurium* UMR1. MAE50 is a *csgD* deletion mutant of UMR1 (negative control) whereas MAE52 strain is used as a CsgD positive control

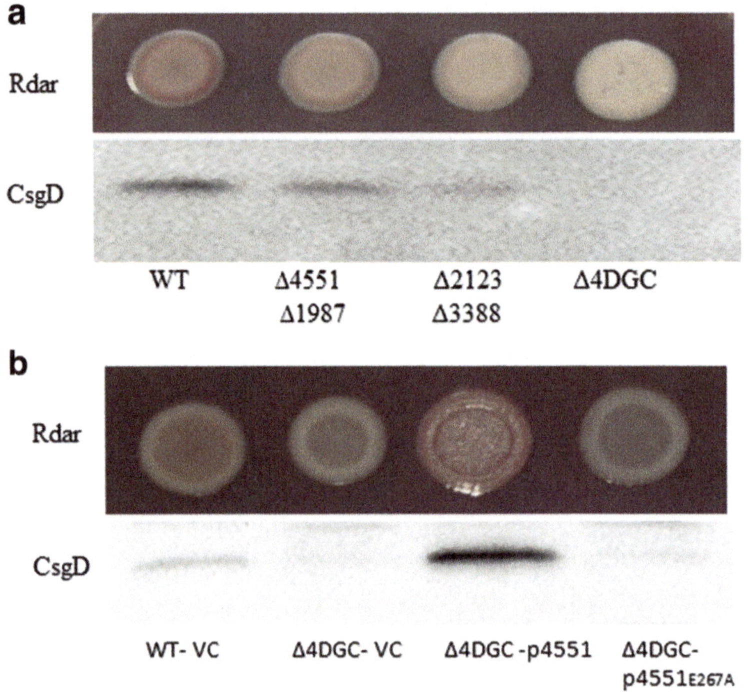


**Fig. 2** Cumulative effect of GGDEF proteins on rdar morphotype and *csgD* expression in *S. typhimurium* UMR1. **a** Rdar morphotype and CsgD levels of double and quadruple mutants of GGDEF proteins after 24 h of growth at 28°C on LB without salt agar plates. **b** Overexpression of the diguanylate cyclase STM4551 in the **Δ**4DGC mutant restored rdar morphotype and *csgD* expression, in contrast to catalytically inactive $STM4551_{E267A}$. Cells were grown for 24 h at 28°C on LB without salt agar plates supplemented with ampicillin (100 μg $ml^{-1}$) and 0.1% L-arabinose. VC = Vector control pBAD30, p4551 = STM4551 cloned in pBAD30, $p4551_{E267A}$ = catalytic mutant $STM4551_{E267A}$ cloned in pBAD30

associated with the deletion of 12 GGDEF domain proteins in *S. enteritidis* including *csgD* expression [39]. However, a catalytic mutant of STM4551 with the GGDEF motif altered to GGAEF did not affect the smooth and white colony morphotype and *csgD* expression, suggesting that the lack of c-di-GMP in Δ4DGC is the only factor mediating the down regulation of rdar morphotype and *csgD* expression.

#### C-di-GMP turnover regulated by GGDEF/EAL domain proteins modulates *csgD* expression

As c-di-GMP is required for *csgD* expression, the contribution of c-di-GMP metabolism to regulate *csgD* expression was investigated for individual GGDEF and EAL proteins. Two GGDEF proteins STM4551 and STM1987 (Fig. 2a) and two GGDEF-EAL proteins STM2123 and STM3388 ([20] and Fig. 2a) promote *csgD* mediated biofilm formation in *S. typhimurium* UMR1. On the other hand, three EAL domain proteins STM4264, STM3611 and STM1827 and the GGDEF-EAL domain protein STM1703 suppress *csgD* expression [31]. Wild type GGDEF domain proteins along with their catalytically inactive variants were overexpressed in the respective chromosomal mutants to assess the impact of catalytic activity. The GGDEF domain protein STM4551 promoted rdar morphotype and *csgD* expression (Fig. 3a) when overexpressed in the respective mutant strain, whereas the catalytically inactive variant $STM4551_{E267A}$, although expressed at the same level as wild type (data not shown), did not affect rdar morphotype and *csgD* expression.

Overexpression of the EAL proteins STM3611 and STM1827 suppressed rdar morphotype and *csgD* expression in the respective chromosomal mutants (Fig. 3b, Additional file 2: Figure S1), whereas the mutants $STM3611_{E136A}$ and $STM1827_{E302A}$ did not exhibit an altered phenotype although mutant proteins were expressed at the same levels as wild type (data not shown). These findings suggest that the action of STM3611 and STM1827 as suppressors of *csgD* expression is due to their c-di-GMP phosphodiesterase activity. A partial effect of the STM3611 protein scaffold on *csgD* expression cannot be excluded at this stage, though.

Expression of the GGDEF-EAL domain protein STM3388 from pBAD30 promotes rdar morphotype formation and *csgD* expression (Fig. 4a) due to its diguanylate cyclase activity. The catalytically inactive GGDEF domain mutant $STM3388_{D342A}$ suppressed rdar morphotype and *csgD* expression slightly, suggesting a minor phosphodiesterase activity. Similarly, expression of

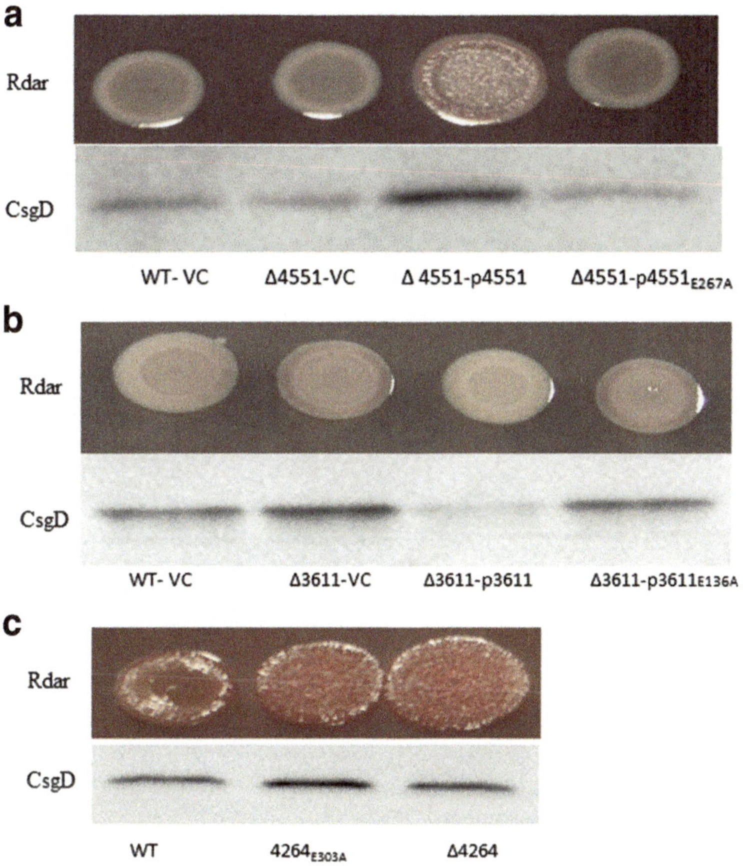


**Fig. 3** Complementation of the rdar morphotype and *csgD* expression phenotypes of GGDEF/EAL mutants of *S. typhimurium* UMR1. Complementation by the GGDEF protein STM4551 (**a**) and the EAL protein STM3611 (**b**) in respective mutants. Catalytically inactive proteins did not restore *csgD* expression and the rdar morphotype. **c** Rdar morphotype formation and CsgD levels of *S. typhimurium* UMR1 upon the chromosomal replacement of the EAL to AAL motif of EAL protein STM4264. Cells were grown for 24 h at 28°C on LB without salt agar plates supplemented with ampicillin (100 μg $ml^{-1}$) and 0.1% L-arabinose

$STM3388_{E467A}$, which contains a mutated EAL motif, slightly promotes rdar morphotype and *csgD* expression. Although protein expression data are missing, these results are consistent with reported apparent time dependent diguanylate cyclase and phosphodiesterase activity of the GGDEF-EAL domain protein STM3388 [20]. A deletion mutant of STM3388 showed enhanced *csgD* expression in early growth phase which was diminished later in the growth phase.

The GGDEF-EAL domain protein STM2123 is a phenotypic diguanylate cyclase [20], consistent with bioinformatic analysis, which predicts a catalytically functional GGDEF domain and a non-functional EAL domain [30]. Over expression of STM2123 enhanced *csgD* expression and c-di-GMP levels [20]. However, reassessment of the cloned STM2123 gene revealed a deletion of 10 amino acids at the C-terminus. Expression of full length STM2123 and its GGDEF mutant $STM2123_{D651A}$ and I site mutant $STM2123_{R640A}$ did not promote rdar morphotype and *csgD* expression at different arabinose concentrations up to 0.1% (Additional file 2: Figure S1, data not shown). In summary, at least two GGDEF proteins STM4551 and STM3388 contribute to *csgD* expression through c-di-GMP turnover. The GGDEF protein STM1987 could not be cloned without mutation and was therefore excluded from the analysis.

The GGDEF-EAL protein STM1703 displays apparent phosphodiesterase activity [31], although bioinformatic analysis predicts a catalytically functional GGDEF and EAL domain. A catalytic mutant in the GGDEF motif, $STM1703_{D313A}$, however, displays down regulation of the rdar morphotype compared to the wild type protein suggesting an active diguanylate cyclase (Fig. 4b). Amino acid exchange of EAL to AAL motif abolished phosphodiesterase activity in all EAL domain proteins examined previously [31]. The $1703_{E437A}$ mutant still down regulates the rdar morphotype and *csgD* expression equally as mutants $STM1703_{E527A}$ and $STM1703_{E613A}$ (data not shown). The catalytic mutants of STM1703 are equally

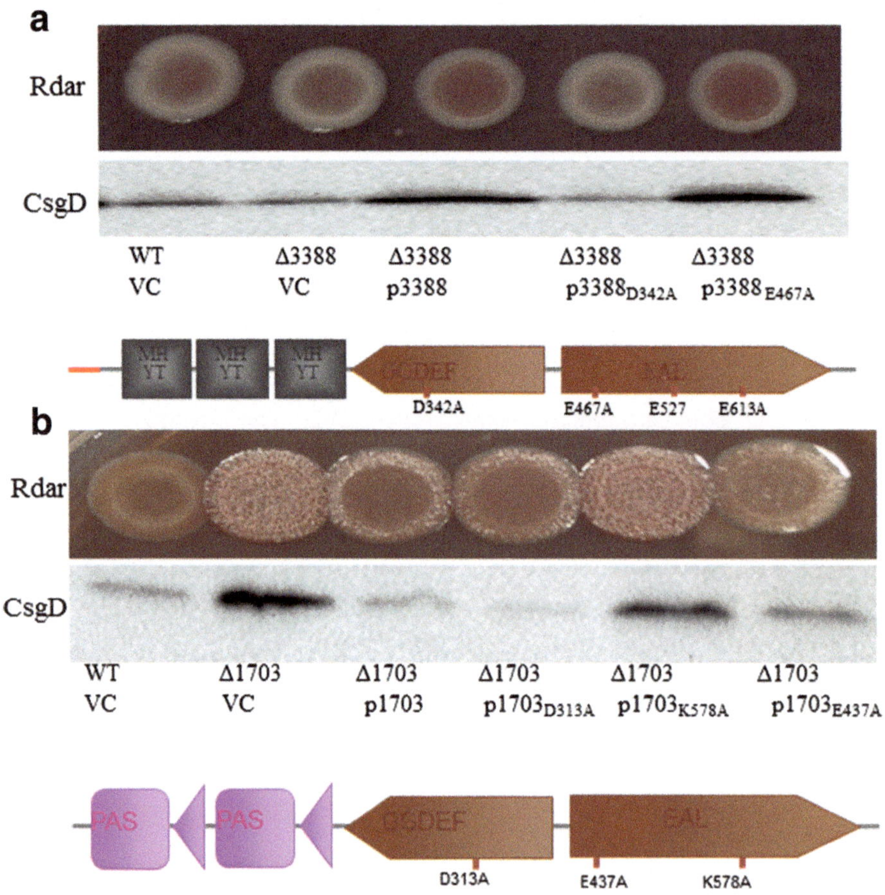


**Fig. 4** The GGDEF-EAL domain protein STM3388 promotes *csgD* expression through its diguanylate cyclase activity, whereas the GGDEF-EAL domain protein STM1703 suppresses *csgD* expression through its phosphodiesterase activity. **a** Rdar morphotype and CsgD levels upon expression of wild type STM3388 and catalytic GGDEF and EAL mutants of STM3388 in STM3388 deletion mutant of *S. typhimurium* UMR1. **b** Rdar morphotype and CsgD levels upon overexpression of STM1703 and catalytic mutants of STM1703 in the *STM1703* deletion background of *S. typhimurium* UMR1. Cells were grown for 24 h at 28°C on LB without salt agar plates supplemented with ampicillin (100 μg $ml^{-1}$) and 0.1% L-arabinose. VC = Vector control pBAD30

expressed as wild type protein except $STM1703_{K578A}$, which did not show functionality (Fig. 4b and data not shown). Interestingly, upon expression of $STM1703_{E437A}$ in the *S. typhimurium* UMR1 background, rdar morphotype downregulation was not observed, while $STM1703_{E527A}$ and $1703_{E613A}$ still displayed downregulation of the rdar morphotype (Additional file 2: Figure S2). These findings suggest a complex role of STM1703 in *csgD* regulation dependent on the enzymatic activity and the protein scaffold.

The EAL domain protein STM4264 could not be cloned without mutation. A scar less single amino acid replacement of the glutamate in the EAL motif led to enhanced rdar morphotype and *csgD* expression (Fig. 3c) similar to the deletion mutant of STM4264. Taken together, these findings indicate that the phosphodiesterase activities of STM4264, STM3611 and STM1827 are required to suppress rdar morphotype formation and *csgD* expression while the situation with respect to STM1703 is more complex. The GGDEF and EAL proteins and respective catalytic mutants are summarized in Additional file 2: Figure S6.

#### Modulation of *csgD* expression by a complex network of GGDEF/EAL domain proteins

Occurrence of multiple diguanylate cyclases and phosphodiesterases dedicated to *csgD* regulation raises the question whether these proteins operate in specific combination i.e.: Do specific phosphodiesterases degrade c-di-GMP synthesized by specific diguanylate cyclases? To identify corresponding diguanylate cyclases and phosphodiesterases, EAL proteins were deleted in the background of GGDEF deletion mutants. If the deleted phosphodiesterase is solely or mainly required for degrading c-di-GMP produced by the deleted diguanylate cyclase, no increase in *csgD* expression is observed. Enhancement of *csgD* expression upon phosphodiesterase deletion in the strain background of deleted diguanylate cyclase indicates no counteraction of the c-di-GMP pool. Deletion of STM3611 in the STM3388 and STM2123 mutants enhanced rdar morphotype and *csgD* expression whereas its deletion in the STM4551 STM1987 double mutant did not have this effect (Fig. 5a) suggesting that STM3611 degrades the c-di-GMP synthesized by the GGDEF domain STM4551 and STM1987. The specificity of STM3611 towards STM1987 and STM4551 is consistent with the role of these proteins in motility regulation [40].

Deletion of the GGDEF-EAL protein STM1703 in the Δ4DGC mutant background invariably showed rdar morphotype formation and *csgD* expression as the STM1703 single mutant (Fig. 5b). Furthermore, combined deletion of STM1703 and the remaining diguanylate cyclases STM1283, STM2672, and AdrA did not diminish *csgD* expression below the level of the STM1703 mutant. This finding leads to the hypothesis that STM1703 degrades c-di-GMP produced from its own GGDEF domain, acting as a diguanylate cyclase locally while it might form a complex with a high affinity c-di-GMP receptor (Fig. 5b).

Deletion of the phosphodiesterase STM4264 in the Δ4DGC mutant enhanced rdar morphotype and CsgD expression to the level of the UMR1 wild type (Fig. 5c) suggesting an additional source of c-di-GMP to contribute to *csgD* expression. Stereomicroscopic visualization of the rdar morphotype of the STM4264 mutant with deletion of the four diguanylate cyclases (Δ4DGC) indicates a 3-D colony architecture different from UMR1 although CsgD expression is at almost similar levels indicating the distinct role of individual c-di-GMP turn

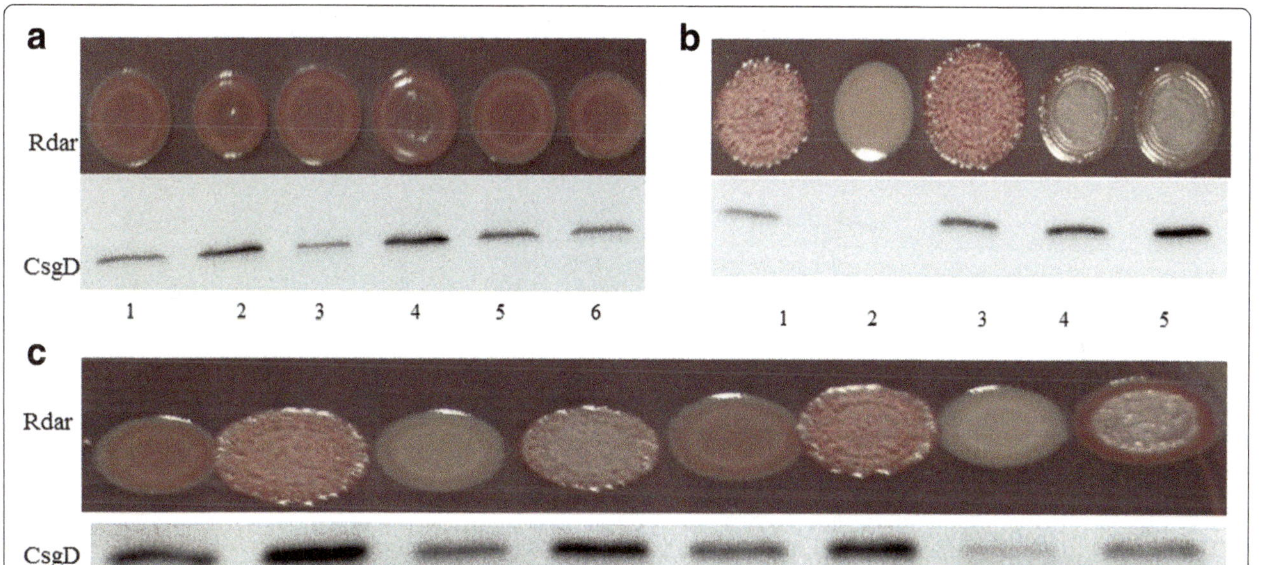


**Fig. 5** Corresponding GGDEF/EAL domain proteins. Rdar morphotype and CsgD levels of *S. typhimurium* UMR1 upon deletion of the EAL domain protein STM3611 (**a**), the GGDEF-EAL protein STM1703 (**b**), and the EAL protein STM4264 (**c**) in different diguanylate cyclase mutant backgrounds. Alteration in rdar morphotype upon *adrA* deletion is due to lack of cellulose synthesis. Stereomicroscopic image of the rdar morphotype formation of the STM4264 mutant upon the deletion of four diguanylate cyclases after 48 h of growth at 28°C on LB without salts plates supplemented with Congo red (**d**) and without Congo red (**e**). Captions (**a**) 1 = UMR1Δ2123, 2 = UMR1Δ3611Δ2123, 3 = UMR1Δ3388, 4 = UMR1Δ3611Δ3388, 5 = UMR1Δ 4551Δ1987, 6 = UMR1Δ3611Δ4551Δ1987. **b** 1 = UMR1Δ1703, 2 = UMR1Δ4551Δ1987Δ2123Δ3388, 3 = UMR1Δ1703Δ4551 Δ1987Δ2123Δ3388, 4 = Δ1703ΔAdrA 5 = Δ1703ΔAdrAΔ1283Δ2672. **c** 1 = UMRI, 2 = UMR1Δ4264, 3 = UMR1 Δ2123Δ3388, 4 = UMR1Δ4264Δ2123Δ3388, 5 = UMR1Δ4551Δ 1987. 6 = UMR1Δ4264Δ4551Δ1987, 7 = UMR1Δ4551Δ1987Δ2123Δ3388, 8 = UMR1Δ4264Δ4551Δ1987Δ2123Δ3388

over proteins in development of the rdar colony (Fig. 5d). Interestingly, 3-D architecture of *S. typhimurium* UMR1 and mutant strains is more pronounced on LB without salt agar plates as compared to LB without salt plates supplemented with Congo red and brilliant blue G (Fig. 5d, e), which indicates a negative effect of these dyes on rdar morphotype formation.

Deletion of STM1827 from single and double mutants of diguanylate cyclases did alter neither rdar morphotype nor *csgD* expression (Additional file 2: Figure S3) indicating that STM1827 contributes to degradation of the global c-di-GMP pool only to some extent.

### C-di-GMP signalling regulates *csgD* expression at multiple levels

We previously proposed transcriptional, posttranscriptional and posttranslational control of *csgD* expression by c-di-GMP [20]. Here, we investigated the target process of c-di-GMP mediated *csgD* expression under physiologically relevant changes in c-di-GMP levels in the Δ4DGC mutant as well as in STM4264 and STM1703 deletion mutants from transcriptional regulation to functionality of CsgD.

Using *csgD* promoter fusions to beta galactosidase that comprise the entire promoter region, UTR and part of the open reading frame from -684 to +441bp (Fig. 6a) indicated statistically significant enhancement of *csgD* promoter activity compared to the single *csgD* deletion background upon deletion of STM1703, whereas deletion of STM4264 tends to increase promoter activity (Fig. 6b). In contrast, the deletion of the 4 DGCs had no effect on *csgD* transcription (Fig. 6c). These results are consistent with previous reports of STM1703 to affect *csgD* transcription [20]. The c-di-GMP pool degraded by STM4264 and produced by 4 DGCs probably affects mainly posttranscriptional events beyond the fusion construct such as mRNA processing and stability.

To narrow down the c-di-GMP responsive regulatory region within the *csgD* upstream region in respect to

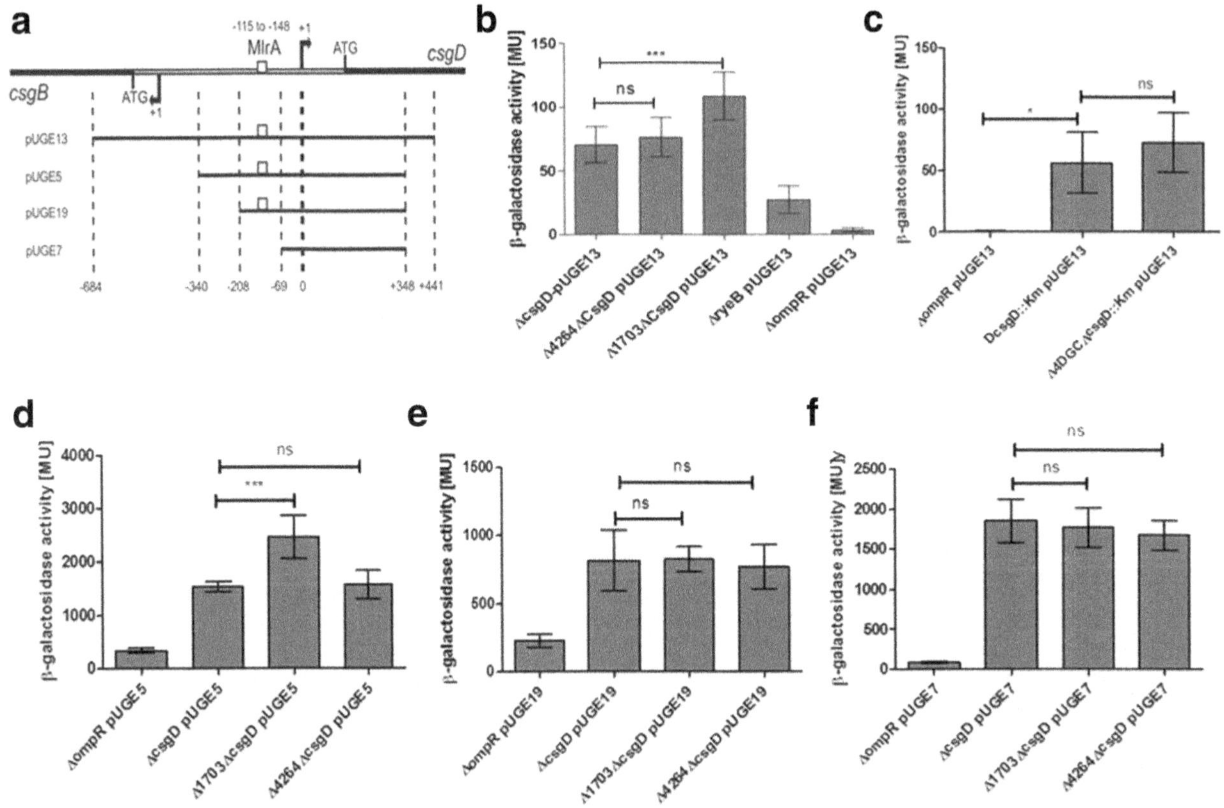


**Fig. 6** Effect of c-di-GMP on *csgD* transcription in the *S. typhimurium csgD* mutant background. **a** Schematic drawing of fusion constructs containing the *csgD* promoter region of different length. Transcriptional activity of the *csgD* promoter region was analyzed in pUGE13 [25, 36] (**b**) upon deletion of STM1703 and STM4264 and (**c**) upon deletion of four diguanylate cyclases (Δ4DGCs) compared to the respective Δ*csgD*::Km background The transcriptional activity was not affected in the Δ4DGC mutant whereas enhanced β-galactosidase activity was observed in Δ*1703* compared to Δ*csgD:101*. **d-f** Identification of the STM1703 regulatory region. As enhanced activity was observed only for pUGE5, but not for other fusions, STM1703 acts via the upstream region between nt −340 and −208. Controls were Δ*ryeB* with partial reduction of *csgD* transcription [28] and Δ*ompR* mutant as negative control. β-galactosidase measurements were done in duplicates using at least three technical replicates. Statistical significance is indicated by $*P < 0.05$ as compared to the respective Δ*csgD* wild type control using unpaired *t*-test (Prism 5, GraphPad Software)

STM1703, ß-galactosidase assays were performed with promoter fusions with subsequent extensions of the *csgD* promoter region (Fig. 6a) and [20, 23]. In summary, the results show that STM1703 acts on *csgD* expression in a distinct region, which could be traced between −340 and −208 upstream to the transcriptional start site of *csgD*.

Investigating the expression of *csgD* from pBAD30 in different GGDEF/EAL mutant backgrounds did not reveal an effect of c-di-GMP (Additional file 2: Figure S5A). Assessment of CsgD functionality, assessed through its effect on *adrA* transcription in different mutant backgrounds did not show an effect of c-di-GMP (Additional file 2: Figure S5B). This finding indicates that the function of CsgD is not dependent on c-di-GMP levels. However, chromosomal over expression of STM4264 suppresses the expression of *csgD* (Fig. 7 and [31]). Cumulatively, these findings suggest that c-di-GMP enhances *csgD* expression by acting on multiple levels. Involvement of GGDEF/EAL domain proteins in regulation of *csgD* expression is summarized in Fig. 8.

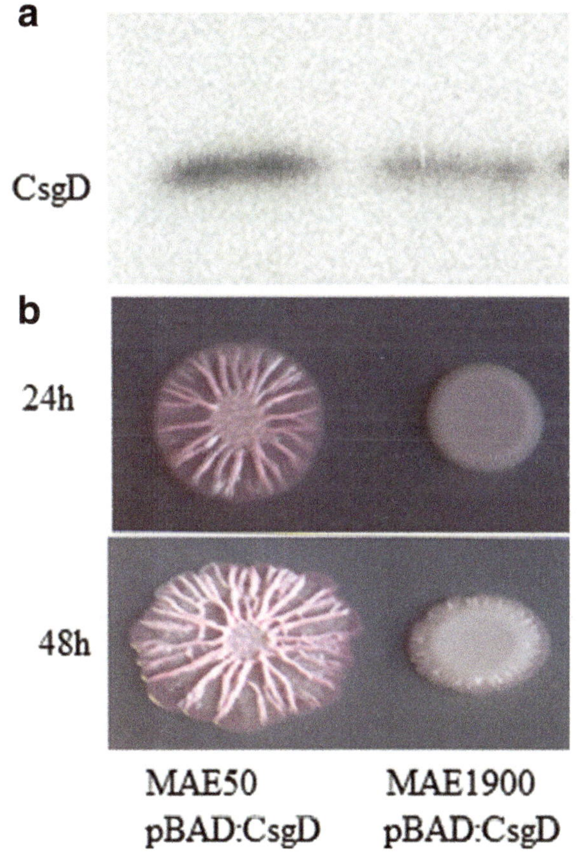


**Fig. 7** Overexpression of the PDE STM4264 from chromosome under *lacUV5* promoter suppresses CsgD production and rdar morphotype formation. CsgD expression (**a**) and rdar morphotype formation (**b**) upon overexpressing *csgD* from pBAD30 in *S. typhimurium* strain MAE1900 where STM4264 is expressed from the constitutive *lacUV5* promoter as compared to the Δ*csgD:101* deletion strain MAE50

## Discussion

Previously, we identified a role for GGDEF/EAL proteins in regulation of *csgD* expression and rdar morphotype formation in *S. typhimurium* [20]. In the present study, we showed that, in most instances, the catalytic activity of GGDEF/EAL domain proteins is required for the regulation of rdar morphotype and *csgD* expression. In addition, networks of corresponding c-di-GMP metabolizing proteins were identified.

GGDEF/EAL domain proteins can regulate biofilm formation through their enzymatic activities, but also, independently, through protein-protein interactions. For example, EAL domain proteins STM1697 and STM1344 are enzymatically inactive, but contribute to *csgD* and rdar morphotype expression by interacting with the $FlhD_2$ subunit of $FlhD_2C_4$, the master regulator of flagella biogenesis in *S. typhimurium* [41, 42]. Similarly, GdpS, a GGDEF domain protein in *Staphylococcus aureus* and *Staphylococcus epidermidis*, does not exhibit diguanylate cyclase activity, but is nevertheless required for biofilm formation and expression of the extracellular matrix polysaccharide PAG (Poly N-acetyl glucosamine) [43, 44]. Moreover, the BLUF-EAL domain protein YcgF of *E. coli* does not exhibit phosphodiesterase activity, but contributes to expression of colonic acid and repression of curli fimbriae [45, 46]. Furthermore, GGDEF/EAL domain proteins regulate rdar formation and *csgD* expression through their enzymatic activities by c-di-GMP turnover with the exception of STM1703. This is in contrast to the previous preposition that *csgD* expression requires the GGDEF domain protein STM4551, but not its catalytic activity [39].

An additional role of the GGDEF proteins STM1987 and STM4551 in *csgD* expression has been identified, those proteins were previously known only to stimulate cellulose production [21].

BphG1, a GGDEF-EAL domain protein of *Rhodobacter sphaeroides*, ScrC of *Vibrio cholerae* and MSDGC1 of *Mycobacterium tuberculosis* are bi-functional GGDEF-EAL domain proteins [47, 48]. In this study, we confirm a bi-functional enzymatic activity of the GGDEF-EAL domain protein STM3388 *in vivo* through construction of mutant proteins.

Occurrence of multiple c-di-GMP metabolizing proteins in bacterial species raises the question of target specificity. Although eight of the GGDEF/EAL domain proteins are found to be involved in the regulation of *csgD* expression, indications for local regulation exists. Particularly, elevated level of *csgD* in the STM1703 mutant could not be restored upon the deletion of several diguanylate cyclases suggesting that the regulation of *csgD* by STM1703 can

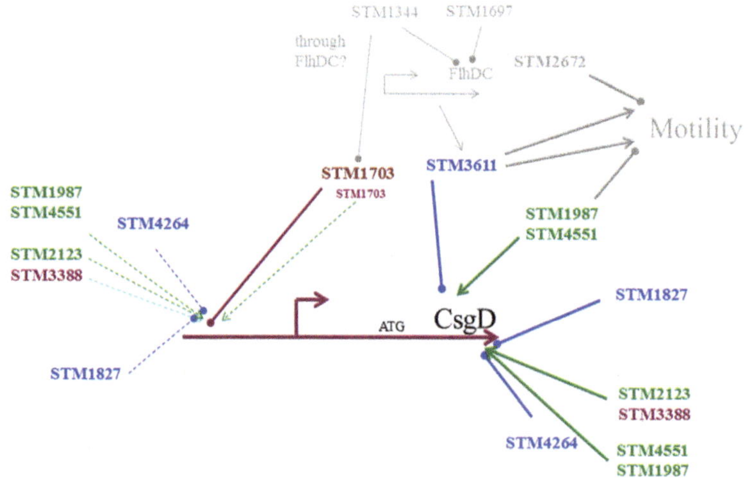


**Fig. 8** Schematic diagram showing the regulatory network of c-di-GMP signaling regulating *csgD* expression. At least three distinct groups of c-di-GMP turnover proteins regulate *csgD* expression. Regulation of *csgD* by STM1703 occurs on the transcriptional level, while the regulatory level by the other groups is unknown and drawn arbitrarily. The STM1987/STM4551/STM3611 group inversely regulates motility [40]. EAL-like proteins STM1344 and STM1697 affect the c-di-GMP signaling network through post-translational inhibition of $FlhD_4C_2$, the regulator of the flagellar cascade and regulation of STM1703 [41, 55]. Green represents a diguanylate cyclase, blue a phosphodiesterase and magenta represents a diguanylate cyclase/ phosphodiesterase; light grey not directly investigated in this work

occur locally. This unconventional behavior of STM1703 led us to hypothesize that STM1703 forms a complex with a high affinity c-di-GMP receptor. Complex formation of GGDEF/EAL domain proteins with c-di-GMP receptors to regulate target processes locally occurs in *E. coli* where the c-di-GMP effector PNPase is physically associated with the diguanylate cyclase DosC and the phosphodiesterase DosP in the RNA degradosome [49]. Recently, the STM1703 homologue in *E. coli* has been proposed to function as a trigger protein, which senses and effectively degrades c-di-GMP produced upstream in the regulatory cascade thereby releasing inhibition of the diguanylate cyclase YdaM and the transcriptional regulator MlrA [50]. However, in *S. typhimurium*, a diguanylate cyclase corresponding to YdaM is not present. Also, the promoter upstream region that mediates STM1703 dependent *csgD* regulation from −208 to −340 does not correspond to the putative MlrA binding site, which we identified to be located at nts −115 to −148. Of note, we were unable to delete *mlrA* in *S. typhimurium* UMR1, but overexpression of MlrA showed the previously reported phenotype of CsgD upregulation [51].

On the other hand, the additive effect of the four diguanylate cyclases on down regulation of rdar morphotype and *csgD* expression in the deletion mutant of the phosphodiesterase STM4264 suggests a global impact of c-di-GMP on *csgD* expression regulated by this phosphodiesterase. This is consistent with c-di-GMP levels to be substantially elevated upon deletion of STM4264, whereas deletion of STM1703 leads only to a marginal increase of c-di-GMP, despite higher *csgD* expression in the STM1703 mutant [49]. Similarly, VpsT, a member of the LuxR-CsgD family in *V. cholerae*, is regulated by a global pool of c-di-GMP assembled by at least five diguanylate cyclases [52].

VpsT is not only regulated by c-di-GMP at multiple levels, but is also able to bind c-di-GMP to efficiently regulate transcription of target genes [53, 54]. In contrast, CsgD from Enterobacteriaceae lacks the c-di-GMP binding motif [29]. The complex regulation of *csgD* expression by c-di-GMP signalling suggests involvement of more than one c-di-GMP effector in modulation of *csgD* expression. Identification of the c-di-GMP receptors and elucidation of molecular mechanisms leading to c-di-GMP mediated regulation of *csgD* expression is an interesting subject for follow-up studies.

## Conclusions

Several GGEDEF/EAL proteins have recently been shown to regulate target processes through protein-protein interactions. Our findings propose that regulation of *csgD* expression and rdar biofilm development by GGDEF/EAL domain proteins is mainly, but not exclusively, mediated through the enzymatic activities of the proteins. The diguanylate cyclase activity of GGDEF domain proteins contributes to promote *csgD* expression. Thereby, the GGDEF proteins STM4551, STM1987, STM3388 and STM2123 have an additive effect on the promotion of *csgD* expression. Moreover, *csgD* is regulated by c-di-GMP signalling at multiple levels. The GGDEF/EAL protein STM 1703 suppresses the transcription of *csgD*, whereas EAL domain STM 4264 suppresses *csgD* by acting on post-transcriptional events.

C-di-GMP signalling has recently emerged as an important intracellular tool to promote biofilm formation in a concerted action in many pathogenic and environmental bacterial species. Our findings extend the understanding of the mechanisms of the regulation of target processes by c-di-GMP signalling in *S. typhimurium*.

## Additional files

**Additional file 1: Table S1.** Strains and plasmids, **Table S2.** Primers and references to supplementary material.

**Additional file 2: Figure S1.** Complementation of rdar morphotype and *csgD* expression by cyclic di-GMP turnover proteins. **Figure S2.** CsgD levels and rdar morphotype formation of *S. typhimurium* UMR1 upon expression of the GGDEF-EAL protein STM1703 and its catalytic mutants. **Figure S3.** STM1827 regulates rdar morphotype and *csgD* expression by degrading the global pools of c-di-GMP. **Figure S4.** Enhanced rdar morphotype in STM4264 and STM1703 mutants is dependent on the transcriptional regulators RpoS and OmpR. **Figure S5.** Effect of c-di-GMP signalling on translation and functionality of CsgD. **Figure S6.** Schematic representation of GGDEF/EAL proteins and mutants used in the study.

### Acknowledgements
We appreciate the assistance of Mohsin Barq in mutant construction during his Master thesis project.

### Funding
Irfan Ahmad has been a recipient of a scholarship from the Higher Education Commission Pakistan. Annika Cimdins has been a recipient of a scholarship from the German Research Foundation (DFG) (CI 239/1-1). This work was funded by the Karolinska Institutet and the Swedish Research Council for Natural Sciences and Engineering.

### Authors' contributions
IA carried out construction of mutant strains and plasmids, participated in western blotting, site directed mutagenesis, phenotypic assays, the design of study, analysis of the data and drafted the manuscript. AC carried out beta-galactosidase assays, rdar morphotype assay and western blotting and participated in the analysis of the data. TB carried out western blotting and rdar morphotype assays. UR designed the study, participated in its coordination and analysis of data. All authors read, revised and approved the final manuscript.

### Competing interests
The authors declare that they have no competing interests.

### Author details
[1]Department of Microbiology, Tumor and Cell Biology, Karolinska Institutet, Stockholm, Sweden. [2]Present Address: Department of Molecular Biology, Umeå University, Umeå, Sweden. [3]Department of Allied Health Sciences, University of Health Sciences, Lahore, Pakistan. [4]Present Address: Department of Biology, Laboratory for Microbiology, Philipps-University Marburg, Marburg, Germany.

### References
1. Hall-Stoodley L, Costerton JW, Stoodley P. Bacterial biofilms: from the natural environment to infectious diseases. Nat Rev Microbiol. 2004;2(2): 95–108.
2. O'Toole G, Kaplan HB, Kolter R. Biofilm formation as microbial development. Annu Rev Microbiol. 2000;54:49–79.
3. Ross P, Weinhouse H, Aloni Y, Michaeli D, Weinberger-Ohana P, Mayer R, Braun S, de Vroom E, van der Marel GA, van Boom JH, et al. Regulation of cellulose synthesis in *Acetobacter xylinum* by cyclic diguanylic acid. Nature. 1987;325:279–81.
4. Jenal U. Cyclic di-guanosine-monophosphate comes of age: a novel secondary messenger involved in modulating cell surface structures in bacteria? Curr Opin Microbiol. 2004;7(2):185–91.
5. Simm R, Morr M, Kader A, Nimtz M, Römling U. GGDEF and EAL domains inversely regulate cyclic di-GMP levels and transition from sessility to motility. Mol Microbiol. 2004;53:1123–34.
6. Römling U, Gomelsky M, Galperin MY. c-di-GMP: The dawning of a novel bacterial signalling system. Mol Microbiol. 2005;57:629–39.
7. Hickman JW, Tifrea DF, Harwood CS. A chemosensory system that regulates biofilm formation through modulation of cyclic diguanylate levels. Proc Natl Acad Sci U S A. 2005;102(40):14422–7.
8. Malone JG, Williams R, Christen M, Jenal U, Spiers AJ, Rainey PB. The structure-function relationship of WspR, a *Pseudomonas fluorescens* response regulator with a GGDEF output domain. Microbiology. 2007;153(Pt 4):980–94.
9. Paul R, Weiser S, Amiot NC, Chan C, Schirmer T, Giese B, Jenal U. Cell cycle-dependent dynamic localization of a bacterial response regulator with a novel di-guanylate cyclase output domain. Genes Dev. 2004;18(6):715–27.
10. Ryjenkov DA, Tarutina M, Moskvin OV, Gomelsky M. Cyclic diguanylate is a ubiquitous signaling molecule in Bacteria: insights into biochemistry of the GGDEF protein domain. J Bacteriol. 2005;187(5):1792–8.
11. Christen M, Christen B, Folcher M, Schauerte A, Jenal U. Identification and characterization of a cyclic di-GMP-specific phosphodiesterase and its allosteric control by GTP. J Biol Chem. 2005;280(35):30829–37.
12. Schmidt AJ, Ryjenkov DA, Gomelsky M. The ubiquitous protein domain EAL is a cyclic diguanylate-specific phosphodiesterase: enzymatically active and inactive EAL domains. J Bacteriol. 2005;187(14):4774–81.
13. Ryan RP, Fouhy Y, Lucey JF, Crossman LC, Spiro S, He YW, Zhang LH, Heeb S, Camara M, Williams P, et al. Cell-cell signaling in *Xanthomonas campestris* involves an HD-GYP domain protein that functions in cyclic di-GMP turnover. Proc Natl Acad Sci U S A. 2006;103:6712–7.
14. Galperin MY, Nikolskaya AN, Koonin EV. Novel domains of the prokaryotic two-component signal transduction systems. FEMS Microbiol Lett. 2001; 203(1):11–21.
15. Galperin MY. Bacterial signal transduction network in a genomic perspective. Environ Microbiol. 2004;6(6):552–67.
16. Römling U, Galperin MY, Gomelsky M. Cyclic di-GMP: the First 25 Years of a Universal Bacterial Second Messenger. Microbiol Mol Biol Rev. 2013;77(1):1–52.
17. Jonas K, Melefors O, Römling U. Regulation of c-di-GMP metabolism in biofilms. Future Microbiol. 2009;4:341–58.
18. Römling U. Great times for small molecules: c-di-AMP, a second messenger candidate in Bacteria and Archaea. Sci Signal. 2008;1(33):e39.
19. Römling U, Amikam D. Cyclic di-GMP as a second messenger. Curr Opin Microbiol. 2006;9(2):218–28.
20. Kader A, Simm R, Gerstel U, Morr M, Römling U. Hierarchical involvement of various GGDEF domain proteins in rdar morphotype development of *Salmonella enterica* serovar Typhimurium. Mol Microbiol. 2006;60:602–16.
21. Garcia B, Latasa C, Solano C, Portillo FG, Gamazo C, Lasa I. Role of the GGDEF protein family in *Salmonella* cellulose biosynthesis and biofilm formation. Mol Microbiol. 2004;54(1):264–77.
22. Römling U, Rohde M, Olsen A, Normark S, Reinköster J. AgfD, the checkpoint of multicellular and aggregative behaviour in *Salmonella typhimurium* regulates at least two independent pathways. Mol Microbiol. 2000;36(1):10–23.
23. Gerstel U, Römling U. The csgD promoter, -control unit for biofilm formation in Salmonella typhimurium-. Res Microbiol. 2003;154:659–67.
24. Gerstel U, Römling U. Oxygen tension and nutrient starvation are major signals that regulate agfD promoter activity and expression of the multicellular morphotype in Salmonella typhimurium. Environ Microbiol. 2001;3(10):638–48.
25. Gerstel U, Park C, Römling U. Complex regulation of csgD promoter activity by global regulatory proteins. Mol Microbiol. 2003;49(3):639–54.

26. Boehm A, Vogel J. The csgD mRNA as a hub for signal integration via multiple small RNAs. Mol Microbiol. 2012;84(1):1–5.
27. Holmqvist E, Reimegård J, Steerk M, Grantcharova N, Römling U, Wagner EGH. Two antisense RNAs target the transcriptional regulator CsgD to inhibit curli synthesis. Embo J. 2010;29:1840–50.
28. Monteiro C, Papenfort K, Hentrich K, Ahmad I, Le Guyon S, Reimann R, Grantcharova N, Romling U. Hfq and Hfq-dependent small RNAs are major contributors to multicellular development in Salmonella enterica serovar Typhimurium. RNA Biol. 2012;9(4):489–502.
29. Zakikhany K, Harrington CR, Nimtz M, Hinton JC, Römling U. Unphosphorylated CsgD controls biofilm formation in *Salmonella enterica* serovar Typhimurium. Mol Microbiol. 2010;77(3):771–86.
30. Ahmad I, Lamprokostopoulou A, Le Guyon S, Streck E, Peters V, Barthel M, Hardt W-D, Römling U. Complex c-di-GMP signaling networks mediate the transition between virulence properties and biofilm formation in *Salmonella enterica* serovar Typhimurium. PLoS One. 2011;6:e28351.
31. Simm R, Lusch A, Kader A, Andersson M, Römling U. Role of EAL-containing proteins in multicellular behavior of *Salmonella enterica* serovar Typhimurium. J Bacteriol. 2007;189(9):3613–23.
32. Morgan JL, McNamara JT, Zimmer J. Mechanism of activation of bacterial cellulose synthase by cyclic di-GMP. Nat Struct Mol Biol. 2014;21(5):489–96.
33. Datsenko KA, Wanner BL. One-step inactivation of chromosomal genes in Escherichia coli K-12 using PCR products. Proc Natl Acad Sci U S A. 2000; 97(12):6640–5.
34. Schmieger H. Phage P22-mutants with increased or decreased transduction abilities. Mol Gen Genet. 1972;119(1):75–88.
35. Blank K, Hensel M, Gerlach RG. Rapid and highly efficient method for scarless mutagenesis within the Salmonella enterica chromosome. PLoS One. 2011;6(1):e15763.
36. Gerstel U, Kolb A, Römling U. Regulatory components at the csgD promoter – additional roles for OmpR and integration host factor and role of the 5′ untranslated region. FEMS Microbiol Lett. 2006;261(1):109–17.
37. Zogaj X, Nimtz M, Rohde M, Bokranz W, Römling U. The multicellular morphotypes of *Salmonella typhimurium* and *Escherichia coli* produce cellulose as the second component of the extracellular matrix. Mol Microbiol. 2001;39(6):1452–63.
38. Miller J. Experiments in Molecular Genetics. Cold Spring Harbor: Cold Spring Harbor Laboratory Press; 1972.
39. Solano C, Garcia B, Latasa C, Toledo-Arana A, Zorraquino V, Valle J, Casals J, Pedroso E, Lasa I. Genetic reductionist approach for dissecting individual roles of GGDEF proteins within the c-di-GMP signaling network in *Salmonella*. Proc Natl Acad Sci U S A. 2009;106(19):7997–8002.
40. Le Guyon S, Simm R, Rhen M, Römling U. Dissecting the c-di-GMP signaling network regulating motility in Salmonella enterica serovar Typhimurium. Environ Microbiol. 2014. n/a-n/a.
41. Simm R, Remminghorst U, Ahmad I, Zakikhany K, Römling U. A Role for the EAL-Like Protein STM1344 in Regulation of CsgD Expression and Motility in Salmonella enterica Serovar Typhimurium. J Bacteriol. 2009;191(12):3928–37.
42. Ahmad I, Wigren E, Le Guyon S, Vekkeli S, Blanka A, el Mouali Y, Anwar N, Chuah ML, Lünsdorf H, Frank R, et al. The EAL-like protein STM1697 regulates virulence phenotypes, motility and biofilm formation in Salmonella typhimurium. Mol Microbiol. 2013;90(6):1216–32.
43. Shang F, Xue T, Sun H, Xing L, Zhang S, Yang Z, Zhang L, Sun B. The *Staphylococcus aureus* GGDEF domain-containing protein, GdpS, influences protein A gene expression in a cyclic diguanylic acid-independent manner. Infect Immun. 2009;77(7):2849–56.
44. Holland LM, O'Donnell ST, Ryjenkov DA, Gomelsky L, Slater SR, Fey PD, Gomelsky M, O'Gara JP. A staphylococcal GGDEF domain protein regulates biofilm formation independently of c-di-GMP. J Bacteriol. 2008;190(15):5178–89.
45. Tschowri N, Lindenberg S, Hengge R. Molecular function and potential evolution of the biofilm-modulating blue light-signalling pathway of *Escherichia coli*. Mol Microbiol. 2012;85(5):893–906.
46. Tschowri N, Busse S, Hengge R. The BLUF-EAL protein YcgF acts as a direct anti-repressor in a blue-light response of *Escherichia coli*. Genes Dev. 2009; 23(4):522–34.
47. Ferreira RB, Antunes LC, Greenberg EP, McCarter LL. *Vibrio parahaemolyticus* ScrC modulates cyclic dimeric GMP regulation of gene expression relevant to growth on surfaces. J Bacteriol. 2008;190(3):851–60.
48. Bharati BK, Sharma IM, Kasetty S, Kumar M, Mukherjee R, Chatterji D. A full length bifunctional protein involved in c-di-GMP turnover is required for long term survival under nutrient starvation in *Mycobacterium smegmatis*. Microbiology. 2012;158(Pt 6):1415–27.
49. Tuckerman JR, Gonzalez G, Gilles-Gonzalez MA. Cyclic di-GMP activation of polynucleotide phosphorylase signal-dependent RNA processing. J Mol Biol. 2011;407(5):633–9.
50. Lindenberg S, Klauck G, Pesavento C, Klauck E, Hengge R. The EAL domain protein YciR acts as a trigger enzyme in a c-di-GMP signalling cascade in E. coli biofilm control. EMBO J. 2013;32(14):2001–14.
51. Brown PK, Dozois CM, Nickerson CA, Zuppardo A, Terlonge J, Curtiss 3rd R. MlrA, a novel regulator of curli (AgF) and extracellular matrix synthesis by Escherichia coli and Salmonella enterica serovar Typhimurium. Mol Microbiol. 2001;41(2):349–63.
52. Beyhan S, Odell LS, Yildiz FH. Identification and characterization of cyclic diguanylate signaling systems controlling rugosity in *Vibrio cholerae*. J Bacteriol. 2008;190(22):7392–405.
53. Srivastava D, Harris RC, Waters CM. Integration of cyclic di-GMP and quorum sensing in the control of *vpsT* and *aphA* in *Vibrio cholerae*. J Bacteriol. 2011; 193(22):6331–41.
54. Shikuma NJ, Fong JC, Yildiz FH. Cellular levels and binding of c-di-GMP control subcellular localization and activity of the *Vibrio cholerae* transcriptional regulator VpsT. PLoS Pathog. 2012;8(5):e1002719.
55. Wada T, Morizane T, Abo T, Tominaga A, Inoue-Tanaka K, Kutsukake K. An EAL-domain protein YdiV acts as an anti-FlhD4C2 factor responsible for nutritional control of the flagellar regulon in *Salmonella enterica* serovar Typhimurium. J Bacteriol. 2011;193(7):1600–11.

# 16

# Selection of *Streptomyces* against soil borne fungal pathogens by a standardized dual culture assay and evaluation of their effects on seed germination and plant growth

Andrea Kunova*†, Maria Bonaldi†, Marco Saracchi, Cristina Pizzatti, Xiaoyulong Chen and Paolo Cortesi

**Abstract**

**Background:** In the search for new natural resources for crop protection, streptomycetes are gaining interest in agriculture as plant growth promoting bacteria and/or biological control agents. Because of their peculiar life cycle, in which the production of secondary metabolites is synchronized with the development of aerial hyphae and sporulation, the commonly used methods to screen for bacterial antagonists need to be adapted.

**Results:** The dual culture assay was standardized in terms of inoculation timing of *Streptomyces* antagonist and pathogen, and growth rate of different fungal pathogens. In case of fast-growing fungi, inoculation of the antagonist 2 or 3 days prior to the pathogen resulted in significantly stronger inhibition of mycelium growth. One hundred and thirty *Streptomyces* strains were evaluated against six destructive soil borne pathogens. The activity of strains varied from broad-spectrum to highly specific inhibition of individual pathogens. All strains inhibited at least one tested pathogen. Three strains, which combined the largest broad-spectrum with the highest inhibition activity, were selected for further characterization with four vegetable species. All of them were able to colonize seed surface of all tested vegetable crops. They mostly improved radicle and hypocotyl growth in vitro, although no statistically significant enhancement of biomass weight was observed in vivo. Occasionally, transient negative effects on germination and plant growth were observed.

**Conclusions:** The adapted dual culture assay allowed us to compare the inhibition of individual *Streptomyces* strains against six fungal soil borne pathogens. The best selected strains were able to colonize the four vegetable crops and have a potential to be developed into biocontrol products. Although they occasionally negatively influenced plant growth, these effects did not persist during the further development. Additional in vivo studies are needed to confirm their potential as biological control or plant growth promoting agents.

**Keywords:** Dual culture assay, Antagonism, Antibiosis, Biological control, Seed colonization, Plant growth promotion

## Background

Modern agricultural systems aim to ensure sustainable crop production and thus they drive the search for new natural resources to find environmentally friendly solutions for crop protection and yield increase. As a part of this research, streptomycetes, a group of Gram-positive, mycelium-forming bacteria are gaining interest in agriculture as plant growth promoting bacteria (PGPB) and/or biological control agents (BCAs). The genus *Streptomyces* comprises more than 500 species ubiquitous in nature, which can be free-living in soil or symbionts of eukaryotic organisms, ranging from fungi to plants, insects and marine animals (reviewed in [1]). Few *Streptomyces* species are pathogens of tuberous and taproot crops causing economically important diseases [2], and very little is known about the effects of other species on

* Correspondence: andrea.kunova@unimi.it
†Equal contributors
Department of Food, Environmental and Nutritional Sciences (DeFENS), Università degli Studi di Milano, via Giovanni Celoria, 2, 20133 Milan, Italy

plant growth and health. Some streptomycetes are able to colonize plant rhizosphere and live as endophytes in the roots of numerous crops [3–6]. Occasional reports demonstrated that *Streptomyces* species were able to suppress plant pathogens, and some of them promoted plant growth [7, 8]. They protected plant roots by inhibiting fungal pathogens through the production of lytic enzymes and antifungal compounds [9–13]. Local and systemic activation of plant defense systems through induced or acquired resistance by *Streptomyces* was described, in which they primed the defense pathways by inducing low-level expression of systemic acquired resistance and jasmonate/ethylene pathways [14–16]. Furthermore, plant growth promotion was observed through the auxin or siderophore production. Positive effects of *Streptomyces* application on seed germination and on root and shoot growth were reported [17–23]. These observations, the ability of *Streptomyces* to produce a huge variety of antimicrobial secondary products [24] and their presence in soil make them promising candidates for the selection and development as plant protection products and biofertilizers.

The selection of biological control agents usually starts with an in vitro screening of a collection of strains against selected pathogens by a dual culture assay, in which the candidate BCA is co-cultivated with the pathogen on agar medium and its antagonistic activity is quantified in terms of inhibition of pathogen's mycelium growth [25]. The production of many secondary metabolites in *Streptomyces* is induced upon the development of aerial hyphae and the beginning of sporulation [26–28]. However, this developmental peculiarity was not always considered in previous studies, where the inoculation timing of the *Streptomyces* strains varied from inoculating *Streptomyces* 7 days before the pathogen to the co-inoculation at the same day [29–32]. Moreover, the growth rate of different pathogens was also not taken into consideration. Therefore, there is a strong need for a standardized methodology that would consider the particular life cycle of *Streptomyces* antagonists and growth rate of different fungal pathogens. This will help to reduce the risk of producing biased results when screening a large pool of streptomycetes against a variety of fungal pathogens. Similarly to the biocontrol studies, the selection of PGP strains involves in vitro screening for compounds such as phytohormones and siderophores, followed by in vivo evaluation of their activity [32–34]. Diverse *Streptomyces* spp. were shown to produce PGP related metabolites and enhance plant growth through phosphate solubilization, iron chelation, and auxin production [22, 35–37]. Indeed, positive effects of streptomycete application on seed germination, root growth, as well as hypocotyl development were observed [17–21, 23].

Biological control and PGP are two different but linked components associated with beneficial plant-microbe interactions [38], and can share the same molecular mechanisms. For instance, some fungal pathogens require iron for their pathogenicity [39]. On the other hand, siderophores produced by beneficial rhizobacteria are responsible for scavenging ferric iron from the environment and may directly inhibit the pathogen growth by iron competition [40]. At the same time, they make the iron available for plant growth and may serve as inducers of plant systemic resistance [34, 41, 42]. Sometimes, negative effects of streptomycetes on plant growth were also observed, e.g., some *Streptomyces* strains adversely influenced the growth of monocotyledonous and dicotyledonous plants [43, 44]. Therefore, evaluating effects of streptomycetes on plant growth in absence of pathogen is an important step in developing *Streptomyces*-based biocontrol products.

The objectives of this work were, 1) to standardize the dual culture assay considering the inoculation timing of *Streptomyces* and the pathogen growth rate, 2) to evaluate the standardized method by screening the antagonism of 130 root endophytic streptomycetes against six common and destructive soil borne fungal pathogens of vegetable crops, and 3) to evaluate the effects of three most active streptomycete strains on seed germination and growth of four vegetable crops.

## Results

### Mycelium growth rate of selected soil borne fungal pathogens

The growth curves of the six soil borne fungal pathogens were determined on CZY medium: *Sclerotinia sclerotiorum* FW361, *Rhizoctonia solani* FW408, *Fusarium oxysporum* f.sp. *lactucae* L74, *Pythium ultimum* FW407, *Phytophthora* sp. FW409 and *Thielaviopsis basicola* FW406 (Fig. 1). Based on their growth rate, they were divided into three groups: 1. fast-growing, 2. medium-growing, and 3. slow-growing. The group 1 comprised *S. sclerotiorum* FW361 and *R. solani* FW408; their colonies reached the edges of the 90 mm Petri plates 4 days after the inoculation. *F. oxysporum* f.sp. *lactucae* L74 and *P. ultimum* FW407 grew slightly slower: 4 days after inoculation, their mean radial growth was 17.5 and 18.1 mm, respectively and were classified as group 2. The group 3 included *Phytophthora* sp. FW409 and *T. basicola* FW406, which grew very slowly; 4 days after the inoculation their mean radial growth was 10.6 and 12.2 mm, respectively.

### Optimization of the dual culture assay

The dual culture assay was optimized to be able to compare the inhibition activity of 130 *Streptomyces* strains against six soil borne fungal pathogens belonging to the

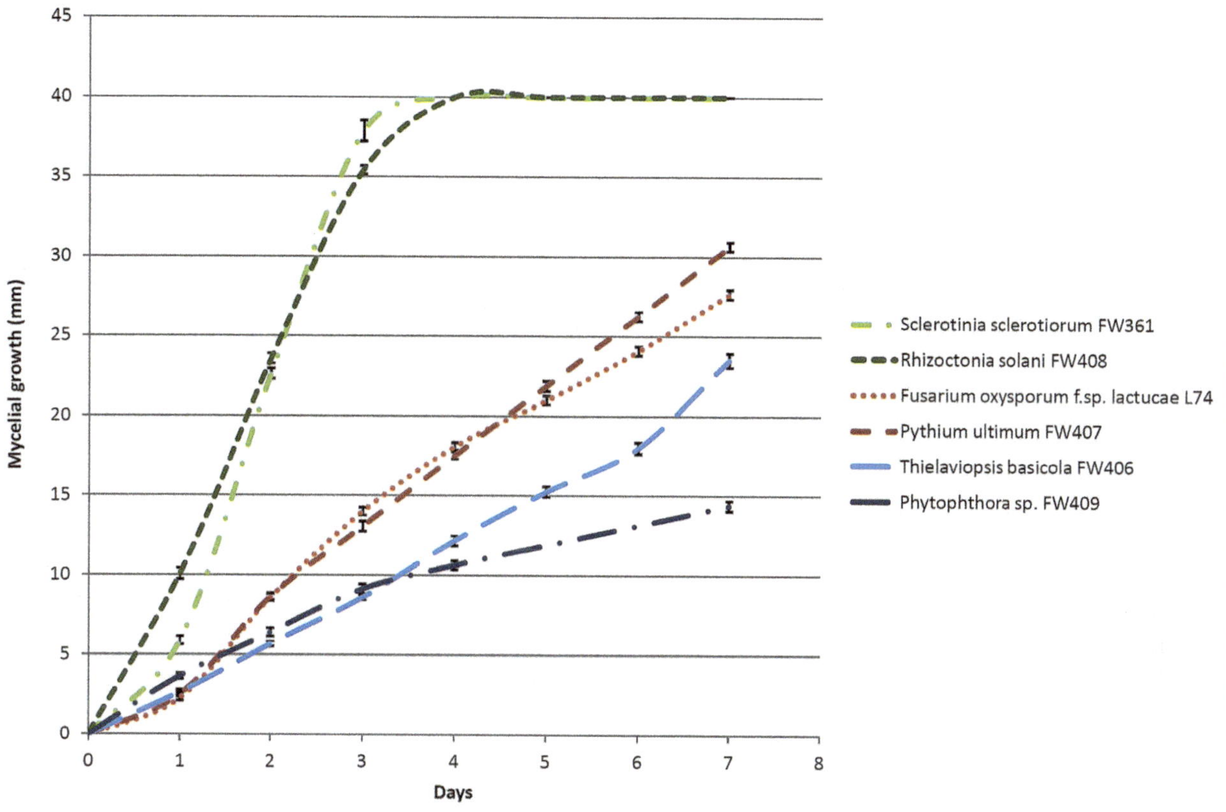


**Fig. 1** Mycelial growth of six tested soil borne fungal pathogens. The pathogens were divided into three groups based on their growth rate: fast-growing (*green*), medium-growing (*red*) and slow-growing (*blue*). The error bars represent the standard error

**Table 1** Inhibition of pathogen mycelium growth based on timing of *Streptomyces* inoculation

| | | Day of *Streptomyces* inoculation[a] | Mycelium growth inhibition (%) | |
|---|---|---|---|---|
| | | | *Sclerotinia sclerotiorum* FW361 | *Fusarium oxysporum* f. sp. *lactucae* L74 |
| *Streptomyces* strain | CMJ57I | 0 | 58.43[b] b[c] | 56.06 a |
| | | −1 | 67.12 b | 56.06 a |
| | | −2 | 77.35 a | 58.82 a |
| | | −3 | 80.99 a | 56.06 a |
| | CX14W | 0 | 63.50 b | 59.41 a |
| | | −1 | 71.61 ab | 55.71 a |
| | | −2 | 74.65 a | 55.88 a |
| | | −3 | 74.57 a | 60.29 a |

[a] Day 0 refers to the co-inoculation of the pathogen and *Streptomyces*, Days −1, −2 and −3 correspond to inoculation of *Streptomyces* 1, 2 or 3 days before the pathogen
[b] mean value of mycelium growth inhibition
[c] Tukey *post-hoc* test; mean values in a column with the same letters are not significantly different ($P = 0.05$)

three groups based on their mycelium growth rate. We standardized the inoculation timing of pathogen and *Streptomyces* and the distance between the microorganisms. First, the inoculation timing of the two microorganisms was investigated. Two *Streptomyces* reference strains (*S. rochei* CMJ57I and *S. anulatus* CX14W) were inoculated 3, 2 or 1 day before or at the same day as *S. sclerotiorum* FW361 (group 1) or *F. oxysporum* f.sp. *lactucae* L74 (group 2) and the inhibition of pathogen mycelium growth was assessed 3 and 7 days after pathogen inoculation, respectively. In case of *S. sclerotiorum* FW361, when *Streptomyces* was inoculated 2 or 3 days before the pathogen, it caused significantly stronger inhibition of mycelium growth, while for *F. oxysporum* f.sp. *lactucae* L74 the inhibition of mycelium growth was not significantly different regardless the timing of streptomycete inoculation (Table 1). Next, to be able to compare the effect of the antagonists on fast- and slow-growing fungi, the distance between the two microorganisms was set as the mean of the 4-day-mycelial growth for the pathogens of the groups 2 and 3; 20 mm and 10 mm respectively; and the antagonist and the pathogen were inoculated the same day. The pathogens of group 1 were inoculated 2 days after the antagonist to allow the production of bio-active compounds by streptomycetes and were placed in the distance corresponding to 2-day-mycelial growth (25 mm). The inhibition of mycelial growth was assessed 3 days after pathogen inoculation for fungi in the group 1, and 6 and 7 days after inoculation for fungi in groups 2 and 3, respectively.

### Selection of streptomycetes based on their antagonistic activity

After the optimization of the dual culture assay, the antagonistic activity of 130 *Streptomyces* strains was evaluated. They showed a wide range of inhibition activity against individual pathogens (Additional file 1: Table S1). The activity against *S. sclerotiorum* FW431 ranged from 0 to 87 %, with *Streptomyces* sp. ALP07R being the most active strain. The activity against *R. solani* FW408 was lower and varied from 0 to 72.4 %, with *Streptomyces* sp. SLF27R being the best strain. *F. oxysporum* f.sp. *lactucae* L74 was the least inhibited pathogen: *Streptomyces* sp. SLF27R, the most active strain, inhibited its growth by 47.9 %; and 15 out of 130 strains did not show any activity against this pathogen. The activity against *P. ultimum* FW407 varied from 0 to 63.64 % and *Streptomyces* sp. CX08W was the best strain. Similarly, *Streptomyces* sp. MR01W was the most active against *T. basicola* FW406 and *Phytophthora* sp. FW409, with inhibition activity of 85.7 and 67.4 %, respectively. The mean rank values were calculated for each *Streptomyces* strain as described in Materials and Methods and the first three strains were chosen for further studies: *S. cyaneus* ZEA17I, *S. anulatus* CMJ58I and *S. albidoflavus* VT111I (Table 2). *S. cyaneus* ZEA17I was the most active strain: it strongly inhibited especially *S. sclerotiorum* FW431 (79.5 %) and *R. solani* FW408 (67.5 %). *S. anulatus* CMJ58I was very active against *T. basicola* FW406 (76.2 %), *P. ultimum* FW407 (57.4 %) and *Phytophthora* sp. FW409 (74.4 %). Also *S. albidoflavus* VT111I resulted very active against *P. ultimum* FW407 (55.9 %), but not against *Phytophthora* sp. FW409 (26.0 %)

### Scanning electron microscope (SEM) studies of streptomycete seed colonization

Prior to evaluating the PGP activity of the three selected *Streptomyces* strains, the colonization of seed surface of lettuce, lamb lettuce, rocket and tomato was examined by SEM. The observations confirmed that seed surfaces of all tested vegetable crops, previously inoculated with streptomycetes, were successfully colonized. The hyphae of *Streptomyces* strains were observed as early as 24 h post inoculation (hpi), and 72 hpi seed surface was abundantly colonized (Fig. 2). Untreated seeds did not show any colonization (data not shown).

### Seed germination

The germination of seeds of cultivated rocket, lamb lettuce, lettuce and tomato inoculated with *S. cyaneus* ZEA17I, *S. anulatus* CMJ58I and *S. albidoflavus* VT111I was assessed both in vitro and in vivo (Table 3). No differences in germination were observed in vitro, while some differences were observed in vivo. The selected strains did not show any negative effects on germination of cultivated rocket and lettuce. However, the germination of lamb lettuce was significantly inhibited after the bacterization with *S. albidoflavus* VT111I (33.0 %) compared to untreated control (64.0 %). Similarly, inhibition of tomato germination was observed after the treatment with *S. cyaneus* ZEA17I (88.3 %) compared to untreated control (96.0 %).

### Seedling root and hypocotyl length

The root and hypocotyl lengths of seedlings grown in vitro varied depending on the combination vegetable crop - streptomycete strain (Table 4). All three strains significantly promoted the growth of radicles and hypocotyls of cultivated rocket and tomato and the radicles of lamb lettuce; e.g. *S. anulatus* CMJ58I augmented radicle growth of cultivated rocket ca. three-times (46.83 mm compared to 15.52 mm in untreated control). Moreover, *S. anulatus* CMJ58I promoted also the growth of lamb lettuce hypocotyls by ca. 30 % and *S. cyaneus* ZEA17I significantly enhanced hypocotyl growth of lettuce (3.51 mm compared to 2.97 mm in untreated control). However, in lamb lettuce, *S. albidoflavus* VT111I and *S. cyaneus* ZEA17I negatively affected the radicle growth by 35 and 46 % respectively; and *S. cyaneus* ZEA17I also significantly reduced hypocotyl length by 18 %.

**Table 2** Mycelial growth inhibition of six fungal soil borne pathogens by *Streptomyces cyaneus* ZEA17I, *S.anulatus* CMJ58I and *S. albidoflavus* VT111I and their respective mean rank positions following dual culture assay

| *Streptomyces* sp. | Mean rank position | Mycelium growth inhibition (%) | | | | | |
|---|---|---|---|---|---|---|---|
| | | *Sclerotinia sclerotiorum* FW361 | *Rhizoctonia solani* FW408 | *Fusarium oxysporum* f.sp. *lattucae* L74 | *Thielaviopsis basicola* FW406 | *Pythium ultimum* FW407 | *Phytophtora* sp. FW409 |
| ZEA17I | 12.0 | 79.49 | 67.54 | 34.06 | 51.79 | 50.74 | 62.00 |
| CMJ58I | 21.8 | 46.79 | 41.75 | 28.99 | 76.19 | 57.35 | 74.42 |
| VT111I | 27.3 | 64.10 | 53.95 | 28.47 | 52.38 | 55.88 | 26.00 |

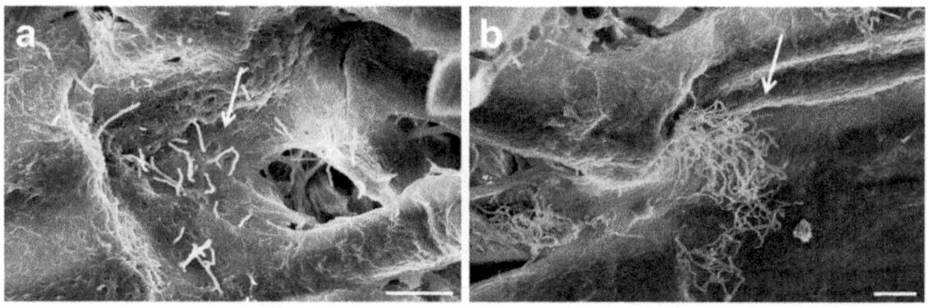


**Fig. 2** Scanning electron microscopy micrographs of tomato seed surface 24 h (**a**) and 72 h (**b**) after inoculation with *Streptomyces albidoflavus* VT111I. Arrows indicate the hyphae and the mycelium of *S. albidoflavus* VT111I. Bar equals to 10 μm

### Biomass dry weight in vivo

The strain-specific effects on plant biomass of four vegetable crops were observed also in vivo in controlled conditions (Table 5). No significant differences were observed among biomass dry weight of any vegetable crop, when the three strains were compared to control. Interestingly, *S. cyaneus* ZEA17I increased the dry weight of 2-week-old tomatoes compared to *S. anulatus* CMJ58I and *S. albidoflavus* VT111I. Instead, in 3-week-old lettuce, *S. anulatus* CMJ58I showed negative effect on biomass compared to *S. cyaneus* ZEA17I and *S. albidoflavus* VT111I.

### Production of secondary metabolites

All three tested strains produced indole-3-acetic acid. The highest amount was produced by *Streptomyces anulatus* CMJ58I (4.25 μg/mL, s.e. 0.063), whereas *S. cyaneus* ZEA17I and *S. albidoflavus* VT111I produced 3.71 μg/mL (s.e. 0.17) and 3.26 μg/mL (s.e. 0.16), respectively. None of the strains produced siderophores.

## Discussion

Over the years, diverse bacteria - such as *Pseudomonas* spp. and *Bacillus* spp. - have been tested as BCAs, mostly studying the antagonism as the mechanism of action to suppress pathogen growth through the production of antifungal compounds [40, 45]. Recently, another group - streptomycetes - are of interest in agriculture as PGPB and BCAs, as they are commonly found in soil and are able to colonize rhizosphere and root tissues [3, 33, 46]. They surely have the potential to act as antagonists against a variety of soil borne pathogens, as they are renowned for the production of bioactive secondary metabolites. However, until now they have been exploited mostly in pharmaceutical industry for the production of antibiotics, antitumorals or immunosuppressives [47, 48].

The most commonly used method for high-throughput screening of antagonistic microorganisms against fungal pathogens in vitro is the dual culture assay [25]. However, streptomycetes are peculiar bacteria characterized by mycelial growth and particular life cycle, where the production of bioactive secondary metabolites is coordinated with the switch to the formation of aerial hyphae and spores [26, 28, 49]. Therefore, the aim of this work was to improve and adapt the dual culture assay for screening the antagonistic activity of

**Table 3** In vitro (a) and in vivo (b) seed germination of cultivated rocket, lamb lettuce, lettuce and tomato untreated and treated with *Streptomyces anulatus* CMJ58I, *S. albidoflavus* VT111I and *S. cyaneus* ZEA17I

| | Days of incubation | Seed treatment | | | |
|---|---|---|---|---|---|
| | | Untreated control | *S. anulatus* CMJ58I | *S. albidoflavus VT111I* | *S. cyaneus* ZEA17I |
| a) In vitro germination (%) | | | | | |
| Cultivated rocket | 6 | 84.3 ± 4.63[a] ns[b] | 80.0 ± 2.00 ns | 85.0 ± 1.73 ns | 74.3 ± 7.22 ns |
| Lamb lettuce | 10 | 88.7 ± 2.96 ns | 90.7 ± 1.76 ns | 87.0 ± 3.61 ns | 85.0 ± 3.06 ns |
| Lettuce | 7 | 79.7 ± 2.85 ns | 78.3 ± 2.40 ns | 81.7 ± 1.86 ns | 84.3 ± 1.86 ns |
| Tomato | 8 | 95.0 ± 1.53 ns | 96.0 ± 1.15 ns | 96.3 ± 2.19 ns | 96.7 ± 1.33 ns |
| b) In vivo germination (%) | | | | | |
| Cultivated rocket | 10 | 65.7 ± 8.09 ns | 80.0 ± 1.15 ns | 78.3 ± 2.19 ns | 58.7 ± 5.36 ns |
| Lamb lettuce | 14 | 64.0 ± 0.58 a | 58.7 ± 4.06 a | 33.0 ± 6.43 b | 44.0 ± 4.04 ab |
| Lettuce | 10 | 76.0 ± 1.00 ns | 71.0 ± 1.15 ns | 71.0 ± 3.06 ns | 70.7 ± 4.37 ns |
| Tomato | 10 | 96.0 ± 1.00 a | 97.7 ± 0.33 a | 92.0 ± 2.52 ab | 88.3 ± 1.33 b |

[a] mean value followed by standard error
[b] Tukey *post-hoc* test; mean values in a row with the same letters are not significantly different ($P = 0.05$); *ns* not significant

**Table 4** Effect of *Streptomyces* sp. on root and hypocotyl length of seedlings of cultivated rocket, lamb lettuce, lettuce and tomato in vitro

| | Length (mm) | Seed treatment | | | |
|---|---|---|---|---|---|
| | | Untreated control | *S. anulatus* CMJ58I | *S. albidoflavus* VT111I | *S. cyaneus* ZEA17I |
| Cultivated rocket | Radicle | 15.52 ± 1.00[a] c[b] | 46.83 ± 2.11 a | 31.11 ± 1.79 b | 29.77 ± 2.05 b |
| | Hypocotyl | 9.34 ± 0.31 b | 11.53 ± 0.36 a | 11.11 ± 0.34 a | 11.75 ± 0.40 a |
| | N | 253 | 239 | 255 | 222 |
| Lamb lettuce | Radicle | 16.79 ± 0.51 b | 26.35 ± 0.74 a | 11.00 ± 0.33 c | 9.08 ± 0.42 c |
| | Hypocotyl | 11.28 ± 0.37 b | 14.55 ± 0.34 a | 10.41 ± 0.30 bc | 9.24 ± 0.37 c |
| | N | 266 | 272 | 261 | 255 |
| Lettuce | Radicle | 16.33 ± 0.68 d | 39.22 ± 1.95 b | 28.31 ± 1.50 c | 58.59 ± 2.10 a |
| | Hypocotyl | 2.97 ± 0.06 b | 2.77 ± 0.05 b | 2.95 ± 0.06 b | 3.51 ± 0.06 a |
| | N | 239 | 235 | 245 | 253 |
| Tomato | Radicle | 54.17 ± 1.09 b | 59.33 ± 1.45 a | 63.96 ± 1.55 a | 63.41 ± 1.00 a |
| | Hypocotyl | 11.52 ± 0.26 c | 13.41 ± 0.33 b | 14.69 ± 0.32 a | 14.79 ± 0.29 a |
| | N | 285 | 288 | 289 | 290 |

[a] The mean value followed by standard error
[b] Tukey *post-hoc* test; mean values in a row with the same letters are not significantly different ($P = 0.05$); *ns* not significant

streptomycetes. We particularly considered the coordination of antibiotic production with the *Streptomyces* developmental program, similarly to Schrey et al. [28], who used the onset of *Streptomyces* sporulation as the time for fungus inoculation in co-culture. Moreover, we put the emphasis on the variable mycelium growth rate of the studied pathogens, so that we would be able to select a pool of strains showing a broad spectrum antagonistic activity against multiple fungal pathogens.

In the first step, we grouped the pathogens based on their mycelium growth rate into three groups and we tested the importance of pathogen growth rate and inoculation timing of both microorganisms in the dual culture assay. In the case of fast-growing fungi such as *S. sclerotiorum* we observed a significantly higher inhibition when *Streptomyces* was inoculated 2–3 days before the pathogen, which was not confirmed for slower growing fungi. This observation is in line with Schrey et al. [28] and confirms that *Streptomyces* start to produce the bioactive secondary metabolites later during their life cycle, and that, especially in the case of fast-growing pathogens, they need to be applied prior the pathogen inoculation, as we observed previously in vivo, where *Streptomyces* application 7 days before lettuce sowing significantly reduced *Sclerotinia* drop [9, 50].

The modified dual culture assay was applied to a collection of 130 streptomycetes and allowed us to compare the inhibition activity of individual *Streptomyces* strains

**Table 5** Effect of *Streptomyces* sp. on plant dry weight of cultivated rocket, lamb lettuce, lettuce and tomato after 2 (a) and 3 (b) weeks of growth in vivo

| Dry weight (mg) | Seed treatment | | | |
|---|---|---|---|---|
| | Untreated control | *S. anulatus* CMJ58I | *S. albidoflavus* VT111I | *S. cyaneus* ZEA17I |
| a) 2 weeks | | | | |
| Cultivated rocket | 5.13 ± 0.31[a] ns[b] | 5.42 ± 0.06 ns | 4.73 ± 0.24 ns | 5.58 ± 0.14 ns |
| Lamb lettuce | 3.91 ± 0.23 ns | 4.6 ± 0.22 ns | 3.93 ± 0.26 ns | 4.65 ± 0.26 ns |
| Lettuce | 3.32 ± .25 ns | 3.16 ± 0.12 ns | 3.14 ± 0.18 ns | 3.66 ± 0.16 ns |
| Tomato | 5.58 ± 0.08 ab | 5.35 ± 0.21 b | 4.95 ± 0.48 b | 6.64 ± 0.26 a |
| b) 3 weeks | | | | |
| Cultivated rocket | 101.88 ± 22.71 ns | 65.95 ± 6.79 ns | 88.28 ± 4.24 ns | 70.80 ± 4.34 ns |
| Lamb lettuce | 33.69 ± 1.39 ns | 31.08 ± 1.49 ns | 35.44 ± 2.27 ns | 28.15 ± 2.62 ns |
| Lettuce | 76.02 ± 4.48 ab | 57.55 ± 4.21 b | 79.37 ± 5.77 a | 77.62 ± 5.91 a |
| Tomato | 43.62 ± 6.35 ns | 48.38 ± 5.15 ns | 56.47 ± 4.31 ns | 45.58 ± 6.93 ns |

[a] The mean value followed by standard error. At 2 weeks (a) the mean value was calculated as the dry weight of 50 plantlets in three replicates, except for lamb lettuce, where 30 plantlets were collected. At 3 weeks (b) the mean value was calculated as the average dry weight of 20 individually grown plantlets
[b] Tukey *post-hoc* test; mean values in a row with the same letters are not significantly different ($P = 0.05$); *ns* not significant

against six fungal soil borne pathogens. We selected three potential wide-spectrum bio-active strains, *S. anulatus* CMJ58I, *S. albidoflavus* VT111I and *S. cyaneus* ZEA17I, showing strong in vitro activity against all six pathogens. The results of the dual culture assay emphasized the antagonist-pathogen specific activity of individual *Streptomyces* strains. These results indicate that diverse streptomycetes use different antagonistic mechanisms and produce different bioactive molecules involved in the antagonism against different pathogens.

In order to develop strains showing biocontrol potential into commercial products, they need to be able to interact with the host to exert their activity in vivo [18]. Therefore, we studied the three selected streptomycetes for their ability to colonize various vegetable crops. The observations of seed surface by scanning electron microscopy revealed that all the three strains were able to grow on all tested hosts as early as 24 hai, as previously observed by Coombs and Franco [51] on wheat seeds inoculated with a *Streptomyces* sp. strain EN27. Moreover, *S. cyaneus* ZEA17I was observed on the surface and inside of 2-week-old lettuce roots by confocal and SEM microscopy [50]. Therefore, *S. anulatus* CMJ58I, *S. albidoflavus* VT111I and *S. cyaneus* ZEA17I are valuable candidates as beneficial bacteria for plant growth promotion and disease prevention.

It is also important that the potential BCAs do not affect negatively the germination and growth of the host plant. Among the variety of secondary metabolites, *Streptomyces* often produce IAA, which can improve the plant growth by stimulating cell elongation and root growth [18]. All the three tested strains produced IAA at high concentrations compared to previous studies [52], and the amount of IAA might be further increased in certain conditions [53]. Some authors observed improvement of plant growth by IAA-producing streptomycetes, for instance *Streptomyces* sp. CMUH009 promoted maize seed germination by 20 % compared to control seeds treated with sterile water [54]. Similarly, the three strains, *S. anulatus* CMJ58I, *S. albidoflavus* VT111I and *S. cyaneus* ZEA17I, showed species-specific differences on germination and plant growth of cultivated rocket, lamb lettuce, lettuce and tomato. Although we did not detect significant differences on seed germination in vitro, inhibition of lamb lettuce and tomato germination was detected in vivo after the treatment with *S. albidoflavus* VT111I and *S. cyaneus* ZEA17I. This might have been due to the production of some secondary metabolites, as these two strains also reduced the growth of lamb lettuce radicles. Another possibility is that the complex interactions among the antagonist, the natural microflora and the host plant might have transiently negatively affected seed germination. Indeed, negative effects have been sometimes observed after application of BCAs on the germination and plant growth in the absence of the pathogen [55]. However, this negative effect might be mitigated during the growth of the host plant, as we did not see any reduction in the biomass dry weight after bacterization by the three antagonists after 2 weeks of growth..

## Conclusion

The standardized dual culture assay proposed in this work, in which the distance between the streptomycete and pathogen was set as a function of the mycelial growth of fungal pathogens, enabled us to compare the inhibition activity of candidate antagonistic bacteria against multiple pathogens showing different mycelial growth. The selected strains - *S. anulatus* CMJ58I, *S. albidoflavus* VT111I and *S. cyaneus* ZEA17I - showed species-specific effects on plant germination and growth. They only rarely negatively influenced plant growth, and these effects did not persist during the further development. Therefore, they are valuable candidates for the development as BCAs. However, further in vivo studies are needed to evaluate their potential as biological control or plant growth promoting agents.

## Methods

### *Streptomyces* strains

The strains were part of a collection of endophytic streptomycetes maintained in the laboratory of Plant Pathology at the Department of Food, Environmental and Nutritional Sciences (DeFENS), University of Milan, Italy [3]. Strains were grown on Czapek-Yeast Extract medium (CZY: 35 g/L Czapek Dox broth, Difco Laboratories, USA; 2 g/L yeast extract, Difco Laboratories, USA; 15 g/L agar, Applichem, Germany) for 3 weeks at 24 °C. Spores were collected in 10 % sterile glycerol + 0.01 % tween20 and filtered through two layers of sterile gauze. The concentration was determined and the spore suspension was stored at −20 °C.

### Pathogen strains

The fungi and oomycetes used in this work are representatives of the main soil borne pathogens of horticultural crops. *Sclerotinia sclerotiorum* FW361, *Fusarium oxysporum* f.sp. *lactucae* L74, *Thielaviopsis basicola* FW406 and *Pythium ultimum* FW407 belong to a collection maintained in the laboratory of Plant Pathology at DeFENS, University of Milan, Italy. Two additional species, *Rhizoctonia solani* FW408 and *Phytophthora* sp. FW409, were kindly provided by Dr. Andrea Minuto (Centro di Sperimentazione e Assistenza Agricola, Albenga, Italy). The pathogens were maintained at 20 °C on Malt-Extract Agar medium (MEA: 30 g/L malt extract, Difco Laboratories, USA; 15 g/L agar, Applichem, Germany), whereas *Phytophthora* sp. FW409 on V8 medium (200 mL/L V8 Vegetable juice, Campbell food, Belgium; 2 g/L $CaCO_3$, Carlo Erba, Italy; 15 g/L agar, Applichem, Germany).

### Pathogen mycelium growth curves

The pathogen agar-mycelium disc (6 mm diameter), taken from the edge of an actively growing fungal colony, was inoculated upside down in the center of the Petri plate containing CZY or Potato Dextrose Agar medium (PDA: 39 g/L Potato Dextrose Agar, Difco Laboratories, USA) and plates were incubated at 24 °C in the dark. Four perpendicularly radial mycelium measurements were taken from the edge of the disc daily for 7 days in three replicates and the mean daily mycelial growth was calculated.

### Dual culture assay

The antibiosis assay was performed in Petri plates (90 mm diameter) containing CZY medium for all fungi and oomycetes, except *T. basicola* FW406, for which PDA was used. The pathogen was inoculated as described above in the center of the Petri plate, whereas 10 μL of streptomycete agar-spore suspension (10 μL of $10^8$ CFU/mL in 90 μL 0.01 % agar) were uniformly distributed along a 40 mm line. The two microorganisms were inoculated the same day at 20 mm and 10 mm distance for group 2- and 3-pathogens, respectively. For the group 1-pathogens, the streptomycetes were inoculated first, and the fungus was inoculated in the center of the plate 2 days after the streptomycete inoculation at 25 mm distance. Three replicates were prepared for each strain and plates inoculated only with the pathogen were used as control. Following the inoculation, plates were incubated at 24 °C in dark. The antagonistic activity was expressed as the percentage of mycelium growth inhibition compared to the control. It was calculated by the formula: $(R1-R2)/R1 \times 100$, where R1 and R2 were the mycelial radial growth of the pathogen in the control and in the presence of the antagonist, respectively.

In the preliminary study, the dual culture assays were performed in order to optimize the timing of inoculation of both microorganisms. *Streptomyces rochei* CMJ57I and *S. anulatus* CX14W were used as reference antagonist strains and *S. sclerotiorum* FW361 and *F. oxysporum* f.sp. *lactucae* L74 as representative pathogens. The antagonists were inoculated on CZY 3, 2 or 1 day before or at the same day as the pathogen and the percentage of inhibition was calculated at 3 and 7 days after pathogen inoculation, respectively.

### Selection of streptomycetes based on their antagonistic activity

One hundred and thirty streptomycete strains were randomly chosen from the collection and their antagonistic activity against the six fungal soil borne pathogens was calculated as described above. The inhibition was assessed 3, 6 or 7 days after pathogen inoculation for group 1, 2 and 3, respectively. Subsequently, rank position was attributed to each streptomycete based on its inhibition activity against each pathogen (from 1 for the most active, to 130 for the least active) and for each strain the average of the six rank positions was calculated. The mean rank values were sorted from the smallest to the highest and the first three strains were chosen for the subsequent experiments (Table 2, Additional file 1: Table S1).

### Scanning electron microscope (SEM) studies of *Streptomyces* seed colonization

The ability of three selected streptomycete strains (*Streptomyces anulatus* CMJ58I, *S. albidoflavus* VT111I and *S. cyaneus* ZEA17I) to colonize seed teguments was assessed under SEM (Leo Electron Microscopy, Cambridge, UK). Seeds of cultivated rocket (*Eruca sativa* Mill.; Fratelli Ingegnoli, Italy), lamb lettuce (*Valerianella locusta* L. 'Accent'; Enza Zaden, Italy), lettuce (*Lactuca sativa* L. 'Bionda ricciolina'; Fratelli Ingegnoli, Italy) and tomato (*Solanum lycopersicum* L. 'Marmande'; Fratelli Ingegnoli, Italy) were surface sterilized in 0.7 % sodium hypochlorite for 5 min and rinsed three times in sterile water. Subsequently, they were immersed in 1 mL of streptomycete spore suspension ($10^8$ CFU/mL) and air-dried under the laminar flow hood. Control seeds were treated with sterile water (untreated control). The seeds were then placed on moist filter paper in a Petri dish (20 cm in diameter) and were incubated at 20 °C in dark. Samples were freeze-dried 24, 48 and 72 h after seed inoculation as described previously [56, 57] and observed by SEM.

### Effects of streptomycetes on seed germination and seedling growth in vitro

Three strains, *Streptomyces anulatus* CMJ58I, *S. albidoflavus* VT111I and *S. cyaneus* ZEA17I, were selected based on the results of the dual culture assay. Seeds of cultivated rocket (*Eruca sativa* Mill.), lamb lettuce (*Valerianella locusta* L. 'Accent'), lettuce (*Lactuca sativa* L. 'Bionda ricciolina') and tomato (*Solanum lycopersicum* L. 'Marmande') were surface sterilized as described above, immersed in a sufficient amount of *Streptomyces* spore suspension ($10^8$ CFU/mL) and air-dried under the laminar flow hood. Control seeds were treated with sterile water (untreated control). One hundred seeds were placed on moist filter paper in three replicates and incubated in the growth chamber at 24 °C, 55 % relative humidity and 15 h photoperiod. During the experiment, filter paper was maintained humid adding a suitable volume of sterile water. Germination was assessed at different times based on vegetable species (Table 3) and the mean germination percentage was calculated. At the end of the experiment, which varied according to vegetable species, root and hypocotyl lengths were measured and the means were calculated.

### Effect of streptomycetes on plant biomass dry weight

Seeds of cultivated rocket (*Eruca sativa* Mill.), lamb lettuce (*Valerianella locusta* L. 'Accent'), lettuce (*Lactuca sativa* L. 'Bionda ricciolina') and tomato (*Solanum lycopersicum* L. 'Marmande') were surface sterilized and bacterized with *Streptomyces anulatus* CMJ58I, *S. albidoflavus* VT111I or *S. cyaneus* ZEA17I as described above. Control seeds were treated with sterile water (untreated control). Two experiments were performed. In the first one, 100 seeds in three replicates for every combination plant species - *Streptomyces* were sown in the non-sterilized Irish and Baltic peat based growing substrate (Vigorplant, Italy) in plastic boxes (10 × 10 × 10 cm) and watered with tap water. The plants were grown in the growth chamber at 24 °C, 55 % relative humidity and 15 h photoperiod and watered as necessary with tap water to keep the soil moist. The number of germinated plants was assessed after 10 days of growth for cultivated rocket, lettuce and tomato, and after 14 days for lamb lettuce. After 2 weeks (in case of cultivated rocket, lettuce and tomato) or 3 weeks in case of lamb lettuce, the plants were carefully removed from the boxes and the growing substrate was carefully removed from roots. Fifty plantlets were counted (except lamb lettuce, where 30 plantlets were counted), left to air-dry at room temperature and the biomass dry weight was determined.

In the second experiment, the bacterized seeds of four vegetable crops were placed individually in polystyrene seed trays (48 cm3/cell) watered with tap water. The plants were grown in the growth chamber as described previously. After 3 weeks (except lamb lettuce, which was grown for 4 weeks), the plants were carefully removed from the seed trays and the growing substrate was carefully removed from roots. The plants were left to air-dry at room temperature and the biomass dry weight was determined.

### Production of compounds involved in plant growth promotion activity

#### *Indole-3-acetic acid (IAA)*

Ten μL of *Streptomyces* spore suspension were inoculated in 5 mL of Czapek broth added with 500 μg/mL of L-tryptophan (Sigma-Aldrich, USA). After 10 days of incubation at 26 °C with constant shaking at 125 rpm in the dark, the production of IAA was determined as described previously [58].

#### *Siderophores*

Ten μL of *Streptomyces* agar-spore suspension were inoculated in the centre of a Petri plate (90 mm diameter), containing the Fe-free Czapek agar medium prepared as previously described [3]. After 14 days of incubation at 24 °C in the dark, the *Streptomyces* colony was overlaid by 15 mL of the chrome azurol S agar [59] as proposed by Pérez-Miranda et al. [60]. After 1 day of incubation at room temperature in the dark, the change of color around the colony (from blue to orange) indicated the siderophore production.

### Statistical analyses

The statistical analyses were performed using R software, version R3.0.2 [61]. The percent data of inhibition activity and seed germination were arcsine root-square transformed and were submitted to ANOVA. Similarly, ANOVA was performed for data of root and hypocotyl lengths, and biomass dry weight, followed by a Tukey *post-hoc* test for multiple comparison ($P = 0.05$) using the TukeyC package [62].

**Acknowledgements**

The authors would like to thank to Elio Burrone for his help with dual culture assay and greenhouse experiments.

**Funding**

This research was in part supported by the research program "Dote ricerca applicata" funded by Lombardy Region and Sipcam Italia Spa.

**Authors' contributions**

AK and MB performed the experiments, helped with the statistical analyses and wrote the manuscript. AK participated to the design of experiments. MS performed SEM sample preparation, observations and acquired SEM pictures. CP helped with the in vivo experiments and conducted the statistical analyses. XC helped with the experiments and participated in discussion of experimental design. PC designed the outline of the study and partly supported the research. All authors read and approved the final manuscript.

**Competing interests**

The authors declare that the research was conducted in the absence of any commercial or financial relationships that could be construed as a potential conflict of interest.

**References**


1. Seipke RF, Kaltenpoth M, Hutchings MI. *Streptomyces* as symbionts: an emerging and widespread theme? FEMS Microbiol Rev. 2012;36:862–76.
2. Loria R, Kers J, Joshi M. Evolution of plant pathogenicity in *Streptomyces*. Annu Rev Phytopathol. 2006;44:469–87.
3. Bonaldi M, Chen X, Kunova A, Pizzatti C, Saracchi M, Cortesi P. Colonization of lettuce rhizosphere and roots by tagged *Streptomyces*. Front Microbiol. 2015;6:25.
4. Cao L, Qiu Z, You J, Tan H, Zhou S. Isolation and characterization of endophytic streptomycete antagonists of fusarium wilt pathogen from surface-sterilized banana roots. FEMS Microbiol Lett. 2005;247:147–52. The Oxford University Press.
5. Coombs JT, Franco CMM. Isolation and identification of actinobacteria from surface-sterilized wheat roots. Appl Environ Microbiol. 2003;69:5603–8.
6. Sardi P, Saracchi M, Quaroni S, Petrolini B, Borgonovi GE, Merli S. Isolation of endophytic *Streptomyces* strains from surface-sterilized roots. Appl Environ Microbiol. 1992;58:2691–3.
7. Dombou CL, Salove MKH, Crawford DL, Beaulieu C. Actinomycetes, promising tools to control plant diseases and to promote plant growth. Phytoprotection. 2001;82:85–102.
8. Palaniyandi SA, Yang SH, Zhang L, Suh JW. Effects of actinobacteria on plant disease suppression and growth promotion. Appl Microbiol Biotechnol. 2013;97:9621–36.
9. Bonaldi M, Kunova A, Saracchi M, Sardi P, Cortesi P. Streptomycetes as biological control agents against basal drop. Acta Hortic ISHS. 2014;1044: 313–8.
10. El-Tarabily KA, Soliman MH, Nassar AH, Al-Hassani HA, Sivasithamparam K, McKenna F, et al. Biological control of *Sclerotinia minor* using a chitinolytic bacterium and actinomycetes. Plant Pathol. 2000;49:573–83.
11. Faheem M, Raza W, Zhong W, Nan Z, Shen Q, Xu Y. Evaluation of the biocontrol potential of *Streptomyces goshikiensis* YCXU against *Fusarium oxysporum* f. sp. *niveum*. Biol Control. 2015;81:101–10.
12. Kanini GS, Katsifas EA, Savvides AL, Karagouni AD. *Streptomyces rochei* ACTA1551, an indigenous greek isolate studied as a potential biocontrol agent against *Fusarium oxysporum* f.sp. *lycopersici*. Biomed Res Int. 2013; 2013:10, 387230. doi:10.1155/2013/387230.
13. Xiao K, Kinkel LL, Samac DA. Biological control of *Phytophthora* root rots on alfalfa and soybean with *Streptomyces*. Biol. Control. 2002;23:285–95.
14. Conn VM, Walker AR, Franco CMM. Endophytic actinobacteria induce defense pathways in *Arabidopsis thaliana*. Mol Plant Microbe Interact. 2008; 21:208–18.
15. Lehr NA, Schrey SD, Hampp R, Tarkka MT. Root inoculation with a forest soil streptomycete leads to locally and systemically increased resistance against phytopathogens in Norway spruce. New Phytol. 2008;177:965–76.
16. Kurth F, Mailänder S, Bönn M, Feldhahn L, Herrmann S, Große I, et al. *Streptomyces*-induced resistance against oak powdery mildew involves host plant responses in defense, photosynthesis, and secondary metabolism pathways. Mol Plant Microbe Interact. 2014;27:891–900.
17. Nassar AH, El-Tarabily KA, Sivasithamparam K. Growth promotion of bean (*Phaseolus vulgaris* L.) by a polyamine-producing isolate of *Streptomyces griseoluteus*. Plant Growth Regul. 2003;40:97–106. Kluwer Academic Publishers.
18. El-Tarabily KA. Promotion of tomato (*Lycopersicon esculentum* Mill.) plant growth by rhizosphere competent 1-aminocyclopropane-1-carboxylic acid deaminase-producing streptomycete actinomycetes. Plant Soil. 2008;308: 161–74.
19. Sadeghi A, Karimi E, Dahaji PA, Javid MG, Dalvand Y, Askari H. Plant growth promoting activity of an auxin and siderophore producing isolate of *Streptomyces* under saline soil conditions. World J Microbiol Biotechnol. 2012;28:1503–9.
20. Gopalakrishnan S, Vadlamudi S, Bandikinda P, Sathya A, Vijayabharathi R, Rupela O, et al. Evaluation of *Streptomyces* strains isolated from herbal vermicompost for their plant growth-promotion traits in rice. Microbiol Res. 2014;169:40–8.
21. Duca D, Lorv J, Patten CL, Rose D, Glick BR. Indole-3-acetic acid in plant-microbe interactions. Antonie Van Leeuwenhoek. 2014;106:85–125.
22. Gopalakrishnan S, Srinivas V, Alekhya G, Prakash B. Effect of plant growth-promoting *Streptomyces* sp. on growth promotion and grain yield in chickpea (*Cicer arietinum* L). 3 Biotech. 2015;5:799–806.
23. Tokala RK, Strap JL, Jung CM, Crawford DL, Salove MH, Deobald LA, et al. Novel plant-microbe rhizosphere interaction involving *Streptomyces lydicus* WYEC108 and the Pea plant (*Pisum sativum*). Appl Environ Microbiol. 2002; 68:2161–71.
24. Watve MG, Tickoo R, Jog MM, Bhole BD. How many antibiotics are produced by the genus *Streptomyces*? Arch. Microbiol. 2001;176:386–90.
25. Pliego C, Ramos C, de Vicente A, Cazorla FM. Screening for candidate bacterial biocontrol agents against soilborne fungal plant pathogens. Plant Soil. 2011;340:505–20.
26. Chater KF. *Streptomyces* inside-out: a new perspective on the bacteria that provide us with antibiotics. Philos Trans R Soc Lond B Biol Sci. 2006;361: 761–8.
27. Hopwood DA. The Leeuwenhoek Lecture, 1987: Towards an understanding of gene switching in *Streptomyces*, the basis of sporulation and antibiotic production. Proc R Soc B, Biol Sci. 1988;235:121–38.
28. Schrey SD, Erkenbrack E, Früh E, Fengler S, Hommel K, Horlacher N, et al. Production of fungal and bacterial growth modulating secondary metabolites is widespread among mycorrhiza-associated streptomycetes. BMC Microbiol. 2012;12:164.
29. Boukaew S, Chuenchit S, Petcharat V. Evaluation of *Streptomyces* spp. for biological control of *Sclerotium* root and stem rot and *Ralstonia* wilt of chili pepper. BioControl. 2011;56:365–74.
30. Chamberlain K, Crawford DL. In vitro and in vivo antagonism of pathogenic turfgrass fungi by *Streptomyces hygroscopicus* strains YCED9 and WYE53. J Ind Microbiol Biotechnol. 1999;23:641–6.
31. Trejo-Estrada SR, Sepulveda IR, Crawford DL. In vitro and in vivo antagonism of *Streptomyces violaceusniger* YCED9 against fungal pathogens of turfgrass. World J Microbiol Biotechnol. 1998;14:865–72. Kluwer Academic Publishers.
32. Ji SH, Gururani MA, Chun SC. Isolation and characterization of plant growth promoting endophytic diazotrophic bacteria from Korean rice cultivars. Microbiol Res Elsevier GmbH. 2014;169:83–98.
33. Jog R, Nareshkumar G, Rajkumar S. Plant growth promoting potential and soil enzyme production of the most abundant *Streptomyces* spp. from wheat rhizosphere. J Appl Microbiol. 2012;113:1154–64.
34. Rungin S, Indananda C, Suttiviriya P, Kruasuwan W, Jaemsaeng R, Thamchaipenet A. Plant growth enhancing effects by a siderophore-producing endophytic streptomycete isolated from a Thai jasmine rice plant (*Oryza sativa* L. cv. KDML105). Antonie Van Leeuwenhoek. 2012;102:463–72.
35. Lugtenberg B, Kamilova F. Plant-growth-promoting rhizobacteria. Annu Rev Microbiol Annual Reviews. 2009;63:541–56.
36. Lugtenberg BJ, Dekkers L, Bloemberg GV. Molecular determinants of rhizosphere colonization by *Pseudomonas*. Annu Rev Phytopathol. 2001;39: 461–90.
37. Verma VC, Singh SK, Prakash S. Bio-control and plant growth promotion potential of siderophore producing endophytic *Streptomyces* from *Azadirachta indica* A. Juss J Basic Microbiol. 2011;51:550–6.
38. Berg G. Plant-microbe interactions promoting plant growth and health: Perspectives for controlled use of microorganisms in agriculture. Appl Microbiol Biotechnol. 2009;84:11–8.
39. Expert D, Franza T, Dellagi A. Iron in plant-pathogen interactions. In: Expert D, O'Brian MR, editors. Mol. Asp. Iron Metab. Pathog. Symbiotic Plant–Microbe Assoc. Netherlands: Springer; 2012. p. 7–39.
40. Beneduzi A, Ambrosini A, Passaglia LMP. Plant growth-promoting rhizobacteria (PGPR): Their potential as antagonists and biocontrol agents. Genet Mol Biol. 2012;35:1044–51.
41. Sharma A, Johri B, Sharma A, Glick B. Plant growth-promoting bacterium *Pseudomonas* sp. strain GRP3 influences iron acquisition in mung bean (*Vigna radiata* L. Wilzeck). Soil Biol Biochem. 2003;35:887–94.
42. Audenaert K, Pattery T, Cornelis P, Höfte M. Induction of systemic resistance to *Botrytis cinerea* in tomato by *Pseudomonas aeruginosa* 7NSK2: role of salicylic acid, pyochelin, and pyocyanin. Mol Plant Microbe Interact. 2002;15: 1147–56.
43. Murao S, Hayashi H. Gougerotin, as a plant growth inhibitor, from Streptomyces sp. No. 179. Agric Biol Chem. 1983;47:1135–6.
44. Heisey RM, Putnam AR. Herbicidal effects of geldanamycin and nigericin, antibiotics from *Streptomyces hygroscopicus*. J Nat Prod. 1986;49:859–65.
45. Babalola OO. Beneficial bacteria of agricultural importance. Biotechnol. Lett. 2010;32:1559–70.
46. Cao L, Qiu Z, You J, Tan H, Zhou S. Isolation and characterization of endophytic *Streptomyces* strains from surface-sterilized tomato (*Lycopersicon esculentum*) roots. Lett Appl Microbiol. 2004;39:425–30.
47. Lucas X, Senger C, Erxleben A, Grüning BA, Döring K, Mosch J, et al. StreptomeDB: A resource for natural compounds isolated from *Streptomyces*

species. Nucleic Acids Res. 2013; 41(Database issue):D1130-6. doi: 10.1093/nar/gks1253.41.

48. de Lima Procópio RE, da Silva IR, Martins MK, de Azevedo JL, de Araújo JM. Antibiotics produced by *Streptomyces*. Braz J Infect Dis. 2012;16:466–71.
49. Flärdh K, Buttner MJ. *Streptomyces* morphogenetics: dissecting differentiation in a filamentous bacterium. Nat Rev Microbiol. 2009;7:36–49.
50. Chen X, Pizzatti C, Bonaldi M, Saracchi M, Erlacher A, Kunova A, et al. Biological control of lettuce drop and host plant colonization by rhizospheric and endophytic streptomycetes. Front Microbiol. 2016;7:714.
51. Coombs JT, Franco CMM. Visualization of an endophytic *Streptomyces* species in wheat seed. Appl Environ Microbiol. 2003;69:4260–2.
52. Abd-alla MH, El-Sayed AE-S, Rasmey A-HM. Indole-3-acetic acid (IAA) production by Streptomyces atrovirens isolated from rhizospheric soil in Egypt. J Biol Earth Sci. 2013;3:B182–93.
53. Harikrishnan H, Shanmugaiah V, Balasubramanian N. Optimization for production of Indole acetic acid (IAA) by plant growth promoting Streptomyces sp VSMGT1014 isolated from rice rhizosphere. Int J Curr Microbiol Appl Sci. 2014;3:158–71.
54. Khamna S, Yokota A, Peberdy JF. Indole-3-acetic acid production by *Streptomyces* sp. isolated from some Thai medicinal plant rhizosphere soils. EurAsian J Biosci. 2010;4:23–32.
55. Koch E. Effect of biocontrol agents on plant growth in the absence of pathogens. IOBC/WPRS Bull. 2001;24:81–9.
56. Assante G, Saracchi M, Farina G, Moricca S. Histological studies on the mycoparasitism of *Clamidosporium tenuissimum* on urediospores of *Uromyces appendiculatus*. Mycol Res. 1992;17:170–82.
57. Rocchi F, Quaroni S, Sardi P, Saracchi M. Studies on *Anthostoma decipiens* involved in *Carpinus betulus* decline. J Plant Pathol. 2010;92:637–44.
58. Bano N, Musarrat J. Characterization of a new *Pseudomonas aeruginosa* strain NJ-15 as a potential biocontrol agent. Curr Microbiol. 2003;46:324–8.
59. Schwyn B, Neilands JB. Universal chemical assay for the detection and determination of siderophores. Anal Biochem. 1987;160:47–56.
60. Pérez-Miranda S, Cabirol N, George-Téllez R, Zamudio-Rivera LS, Fernández FJ. O-CAS, a fast and universal method for siderophore detection. J Microbiol Methods. 2007;70:127–31.
61. R_Core_Team. R: A language and environment for statistical computing. Vienna: The R Foundation for Statistical Computing; 2013.
62. Faria JC, Jelihovschi EG, Allaman IB. Conventinal Tukey Test. Ilheus: Universidade Estadual de Santa Cruz; 2014.

# 17

# Rapid detection of carbapenemase-producing *Klebsiella pneumoniae* strains derived from blood cultures by Matrix-Assisted Laser Desorption Ionization-Time of Flight Mass Spectrometry (MALDI-TOF MS)

Christina Sakarikou[1*], Marco Ciotti[2], Camilla Dolfa[1], Silvia Angeletti[3] and Cartesio Favalli[1,2]

**Abstract**

**Background:** Carbapenemase-producing *Enterobacteriaceae* (CPE), particularly carbapenemase-producing *Klebsiella pneumoniae* isolates, are important causative agents of nosocomial infections associated with significant mortality rates mostly in critical wards. The rapid detection and typing of these strains is critical either for surveillance purposes and to prevent outbreaks and optimize antibiotic therapy. In this study, the MALDI-TOF MS method was used to detect rapidly these isolates from blood cultures (BCs) and to obtain proteomic profiles enable to discriminate between carbapenemase-producing and non-carbapenemase-producing strains.

**Results:** Fifty-five *K. pneumoniae* strains were tested. Identification and carbapenemase-production detection assay using Ertapenem were performed both from bacterial pellets extracted directly from BCs flasks and from subcultures of these strains. For all isolates, a complete antimicrobial susceptibility testing and a genotypic characterization were performed.

We found 100% agreement between the carbapenemase-producing profile generated by MALDI TOF MS and that obtained using conventional methods. The assay detected and discriminated different carbapenemase-producing *K. pneumoniae* isolates within 30 min to 3 h after incubation with Ertapenem.

**Conclusions:** MALDI-TOF MS is a promising, rapid and economical method for the detection of carbapenemase-producing *K. pneumoniae* strains that could be successfully introduced into the routine diagnostic workflow of clinical microbiology laboratories.

**Keywords:** Carbapenemases, Detection, *Klebsiella pneumoniae*, Blood culture, MALDI-TOF MS

* Correspondence: cristina_sakarikou@yahoo.it
[1]Department of Experimental Medicine and Surgery, "Tor Vergata" University of Rome, Via Montpellier 1, 00133 Rome, Italy
Full list of author information is available at the end of the article

## Background

In recent years, there was a worldwide spread of carbapenemase-producing *Enterobacteriaceae* (CPE), mainly carbapenemase-producing *Klebsiella pneumoniae* isolates, that has become an increasing health care problem all over the world [1–3]. CPE are important etiological agents of hospital-acquired infections (HAI), which cause significant mortality particularly among patients in intensive care units (UCI) [4]. In *Enterobacteriaceae,* carbapenem resistance, (CRE), is mainly induced by two mechanisms: (i) expression of carbapenemases enzymes able to degrade carbapenems; (ii) an association of decreased outer membrane permeability (e.g porin loss or deficiency of porin expression) with overexpression of β-lactamases possessing very weak carbapenemase activity (e.g extended-spectrum β-lactamases ESBLs or acquired AmpC-type cephalosporinases) [5]. From a public health standpoint, carbapenemase producers are taken into major consideration than carbapenem-resistant isolates non carbapenemase producers because their resistance trait can be transferred [6]. The carbapenemase genes transmission by mobile genetic elements leads to a rapid dissemination and carries additional non-β-lactam resistance determinants responsible for resistance to different groups of antibiotics giving rise to multidrug- and pandrug-resistant (MDR, PDR) isolates [7]. Accordingly, a rapid detection of carbapenemases is essential in order to prevent spreading, detect outbreak and for an effective antibiotic therapy to improve the outcome of these patients. However, detection of carbapenemases in Enterobacteria is challenging especially in the presence of carbapenemase-producing *Klebsiella pneumoniae* strains with low carbapenem MICs [8]. For that purpose, the latest molecular- and biochemical-based techniques (e.g RT-PCR, microarray, Carba NP test) have been shown to be suitable to detect early (3–5 h) carbapenemase-producing Enterobacteria directly from positive blood cultures [9–13]. Recently, it has been shown that MALDI-TOF MS could represent an alternative to detect rapidly CPE strains as well as to identify rapidly the pathogens [14–17]. As mass spectrometry measures mass differences, a MALDI-TOF MS-based assay was performed to detect the hydrolysis of different β-lactam antibiotics after a short co-incubation of the Gram-negative isolate with the specific antibiotic. Therefore, hydrolyzed and non-hydrolyzed substances which differ in their molecular weights could be monitoring [15, 17].

In this study, 55 consecutive non-replicated clinical strains of *K. pneumoniae,* isolated from hospitalized patients with Gram-negative bacteremia were tested. The study focused on carbapenemase-producing *Klebsiella pneumoniae* strains because of their increasing prevalence in Italian hospitals [18]. The identification and the carbapenemase-production detection assay using Ertapenem were performed from both bacterial pellets extracted directly from BCs flasks and from subcultures of these isolated strains.

## Methods

### Specimens collection

A total of 55 consecutive non-replicated clinical strains of *K. pneumoniae,* isolated from blood cultures of hospitalized patients, were collected from July 2014 to July 2015 in two hospitals of Rome, Italy. In particular, 47 were isolated at the Microbiology Laboratory of Polyclinic "Tor Vergata" Foundation and 8 at the Clinical Pathology and Microbiology Laboratory of the University Hospital "Campus Bio-Medico".

The blood specimens were collected in BD BACTEC™ Plus Aerobic/F and BD BACTEC™ Plus Anaerobic/F vials and incubated in a Bactec FX blood culture system (Becton Dickinson, NJ, USA). Clinical and epidemiological data of patients were also recorded. The isolates were collected for surveillance and public health purposes, as done routinely in clinical setting, and no ethical approval was required.

### Sample preparation

At the Microbiology laboratory of Polyclinic "Tor Vergata" Foundation, a total of 170 consecutive positive BCs containing Gram-negative rods were included in this study. After the BC was flagged as positive, Gram staining was performed. In the presence of Gram-negative rods, MALDI-TOF MS identification was performed directly from the bacterial pellet extracted from the BC flasks, using the MALDI Sepsityper kit (Bruker Daltonics, Bremen, Germany), according to the manufacturer's instructions, in an autoflex speed mass spectrometer (Bruker Daltonics, Bremen, Germany). Moreover, an additional bacterial pellet was extracted for the carbapenemase-production detection assay.

For the identification assay, an ethanol-formic acid extraction was performed on the bacterial pellets obtained using the Sepsityper kit (pellets contained $10^7$ to $10^9$ CFU), according to manufacturer's instructions.

In the Clinical Pathology and Microbiology laboratory of the University Hospital "Campus Bio-Medico", only the MALDI-TOF MS identification was performed directly from the bacterial pellet extracted from the BC flaks using the MALDI Sepsityper kit (Bruker Daltonics, Bremen, Germany) as described above, on a Microflex LT mass spectrometer (Bruker Daltonics, Bremen, Germany). However, the carbapenemase-production detection assay for the same isolates was performed only from subcultures of frozen strains at the Microbiology laboratory of Polyclinic "Tor Vergata" Foundation on an autoflex speed mass spectrometer (Bruker Daltonics, Bremen, Germany).

#### MALDI-TOF MS-based carbapenemase-production detection assay

The bacterial pellets obtained using the Sepsityper kit (pellets contained $10^7$ to $10^9$ CFU) were resuspended in 30 µl of Ertapenem solution (1 mg/mL in 1 mM ammonium hydrogen citrate buffer pH 7.1) and incubated for 3 h at 37 °C under agitation for the carbapenemase-production detection assay. Incubated suspensions with Ertapenem solution were centrifuged at 13.000 rpm for 2 min, and 1 µl of the supernatant was spotted in quadruplicate onto a polished steel target plate MTP 384 and air-dried spots were overlaid with 1 µl of HCCA matrix in organic solvent. Data are collected at 30-min intervals during the entire incubation time. All measurements were performed on an autoflex speed mass spectrometer (Bruker Daltonics, Bremen, Germany), within a mass range of 100 to 1.100 Da, while the spectra were manually examined using the FlexAnalysis 3.4 software (Bruker Daltonics, Bremen, Germany). The spectra were acquired in the linear positive mode at a laser frequency of 200 Hz. A method optimized for the low mass range was set up using the following parameters: acquisition range, 100 to 1,000 Da with maximum laser frequency; acceleration voltage, IS1 voltage 19.66 kV; IS2 voltage, 18.51 kV; lens 6.99 kV and extraction delay time, 350 ns. For instrument calibration, an external standard solution consisting of bradykinin 1–7 (757.40 Da), 2 HCCA peaks (CCA_$[M+H]^+$ 190.05 Da; and CCA_$[2\ M+H]^+$ 379.02 Da) and angiotensin (1049, 54) (Bruker Daltonics, Bremen, Germany) (100–1000 Da range) was used. A reference wild-type strain of *K. pneumoniae* was used as a negative control while a KPC-producing *K. pneumoniae* reference strain was used as a positive control and both of them were included in each run.

For quality control of the antibiotic solution and for detection of spontaneous hydrolysis a sample with Ertapenem solution only, without bacteria, was processed as all the other samples.

The obtained spectra were not reported in this section because they are periodically deleted for room reasons from the computer used for the routine laboratory analyses.

#### Subcultures processing

Identification and the carbapenemase-production detection assay were also performed from subcultures of the strains isolated on MacConkey agar, Columbia blood agar and Chocolate PolyViteX plates (BioMèrieux, France) after 18–24 h incubation at 37 °C. After overnight incubation, 30 µl of Ertapenem solution was inoculated separately with a 1 µl-inoculation loop filled with both single bacteria colonies isolated from overnight cultures and control strains and incubated for 3 h at 37 °C under agitation for the carbapenemase-production detection assay. For the identification assay, a small part of an isolate's colony was inoculated on the polished steel target plate MTP 384 and overlaid with 1 µl of HCCA matrix. In the same manner, the respective measurements were performed as described above.

#### Strain characterization

For all *K. pneumoniae* isolates, a complete antimicrobial susceptibility testing was created from isolates cultured onto chocolate PolyViteX agar after overnight incubation at 37 °C using the Vitek 2 automated system (bioMérieux, France) and the AST cards N-202 for a phenotypic characterization. The MIC values were interpreted according to EUCAST breakpoints tables 2014. A genotypic characterization for the resistance genes $bla_{KPC}$, $bla_{NDM}$, $bla_{OXA\text{-}48}$, $bla_{VIM}$, $bla_{IMP}$, $bla_{SHV}$, $bla_{TEM}$, $bla_{CTX\text{-}M}$,was performed by Polymerase Chain Reaction (PCR) (Qiagen, Hilden, Germany), using primers already reported in literature [11, 17]. The DNA was extracted from each isolated strain by EZI Advanced XL automated system (Qiagen, Hilden, Germany).

### Results

#### MALDI-TOF MS identification performance

At the Microbiology laboratory of Polyclinic "Tor Vergata" Foundation, 175 isolates, tested directly from blood culture samples, were correctly identified at the species level by MALDI-TOF MS; there was a 100% agreement with the subculture MALDI-TOF MS-based identification routinely performed in our laboratory. One-hundred and forty-six isolates fell into the *Enterobacteriaceae* family whereas 29 were non fermentative Gram-negatives rods. Among the *Enterobacteriaceae* isolates, 47 were identified as *K. pneumoniae* strains. All identifications exhibited a Biotyper score value of > 2.2. Moreover, the identifications obtained directly from blood culture samples had a score values >2.3 compared with the score obtained from subculture. Data, using the direct method, were obtained earlier respect to the routine method: 15 min after the BC was flagged as positive versus 12 h of the routine method.

#### MALDI-TOF MS-based carbapenemase-production detection assay performance

Data were collected at 30-min intervals during the entire incubation time. Differences in the hydrolysis rate were easily detected. In the presence of sensitive *K. pneumoniae* strains, intact Ertapenem displayed the following peak pattern: 476.5 Da, 498.5 Da, 520.5 Da, corresponding to the molecular peak of Ertapenem, its single sodium adduct and its double sodium adduct, respectively. These peaks were not found in the spectra obtained from carbapenemase-producing strains. Instead, a mass peak was found at 472.5 Da corresponding to the hydrolyzed decarboxylated sodium adduct while another peak was found at 538.5 Da corresponding to the hydrolyzed double sodium adduct [19]. A m/z 450 Da peak was also detectable in all spectra. Ertapenem only was also incubated up to 3 h as quality control and no spontaneous hydrolysis was seen. Moreover, no difference was observed when the carbapenemase-production detection assay was performed either on solid media growth isolates or isolates extracted directly from BCs. Thus, sensitivities and specificities were 100% for both subcultures and blood cultures. Furthermore, the assay revealed and discriminated KPC-producing strains from other carbapenemases after a 30 min incubation with Ertapenem (Fig. 1). Thus, KPC- plus VIM- producing strains were detected after 2 h incubation with Ertapemen while OXA-48-producing strains were detected after 3 h incubation with Ertapenem. However, after 1 h incubation a partial hydrolysis of Ertapenem was observed in the presence of OXA-48 strains as indicated by the loss of its molecular peak in the presence of the peaks of its salt adducts.

Carbapenemase production was detected by MALDI-TOF MS, and these results were compared with those achieved by routine methods, MIC determination and PCR analysis. No discordance was observed in terms of bacteria classification according to resistance profile (Table 1). Of the 55 isolates investigated, 40 were found to be carbapenemase-producing strains and 15 non-carbapenemase-producing strains.

Considering the time to results, the report of the MALDI-TOF-based assay was available after 3 h the BCs were flagged as positive. Instead, 12 to 24 h were necessary using the routine AST, and 5 h by PCR assay. So, the time saved for generating the report ranged from 11 to 21 h comparing MALDI-TOF *vs* routine AST and 2 h comparing MALDI-TOF with PCR assay.

## Discussion

Antibiotic resistance in CPE is a life-threatening condition, so a prompt diagnosis is crucial for improving patient's prognosis, choose the optimal antibiotic therapy and implement surveillance network. In this context, molecular typing helps to define the extent of the outbreak, identify the source of infection and the reservoir [20, 21], while proteomic profiling may represent a valid

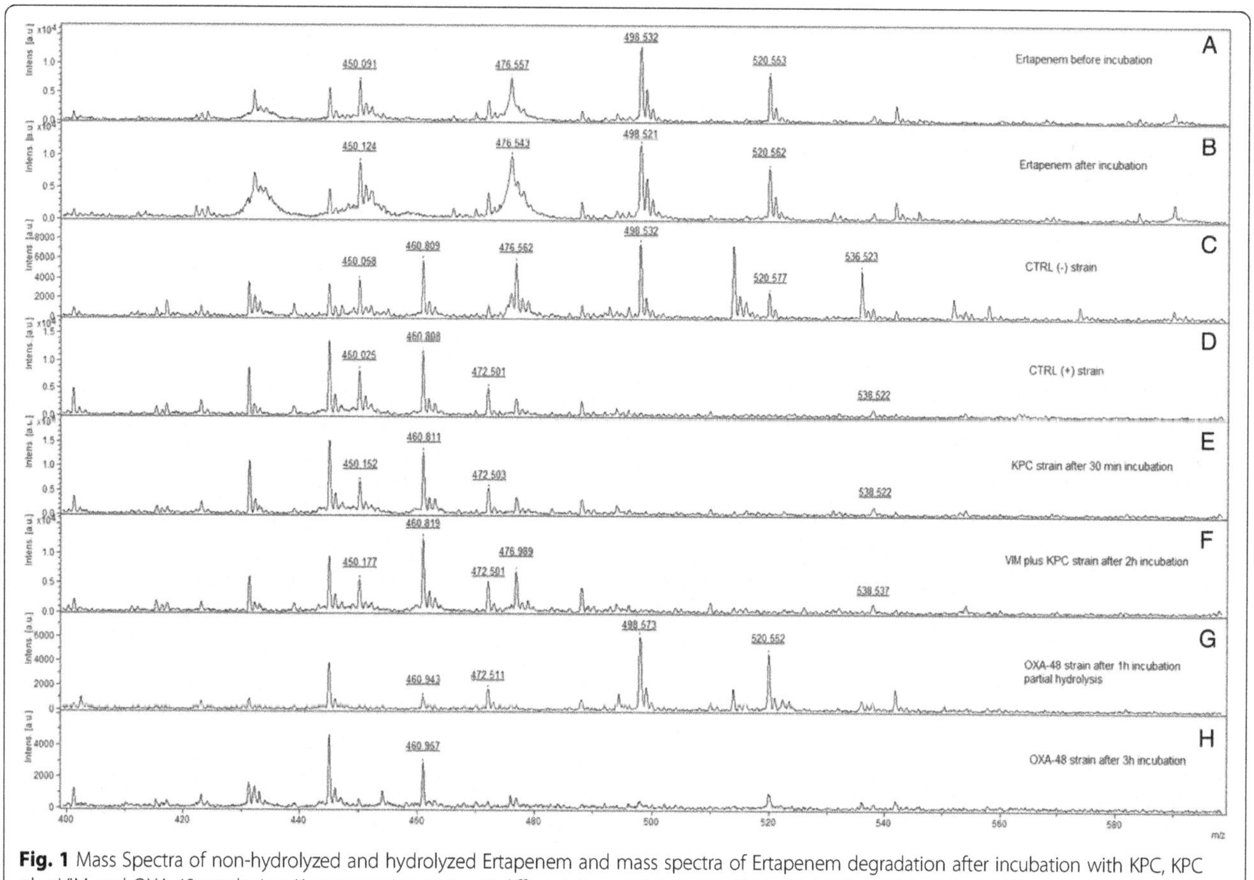


**Fig. 1** Mass Spectra of non-hydrolyzed and hydrolyzed Ertapenem and mass spectra of Ertapenem degradation after incubation with KPC, KPC plus VIM and OXA-48 producing *K. pneumoniae* strains at different time points, respectively. Molecular weight values of Ertapenem and its salt adducts peaks are indicated at the top of the peaks

**Table 1** Description of carbapenemase-producing *K. pneumoniae* strains used in the carbapenemase-production detection assay

| | MIC Ertapenem (mg/l) | MIC Imipenem (mg/l) | MIC Meropenem (mg/l) | Hydrolysis of Ertapenem | Time point of hydrolysis (h) |
|---|---|---|---|---|---|
| 31 *K.pneumoniae* KPC | ≥8 | ≥16 | ≥16 | + | 0.5 h |
| 5 *K.pneumoniae* KPC,VIM | ≥8 | ≥16 | ≥16 | + | 2 h |
| 4 *K.pneumoniae* OXA-48 | ≥8 | ≤2 | ≤2 | + | 3 h |

alternative to biochemical and genome-based identification and susceptibility detection assays.

In this study, we examine the performance of MALDI-TOF MS identification assay and the mass spectrometric carbapenemase-production detection assay directly on positive BC flasks from hospitalized patients with Gram-negative bacteremia. The MALDI-TOF MS technology appeared accurate, cost saving and rapid.

It was investigated the Ertapenem hydrolysis induced by different carbapenemase-producing *K. pneumoniae* strains isolated from positive blood cultures. Using 1 mL of blood culture fluid, the obtained amount of bacteria producing carbapenemase was sufficient to hydrolyze the antibiotic in a time frame ranging from 30 min to 3 h of incubation. The different peak patterns derived from carbapenemase and non-carbapenemase-producing strains allowed the unequivocal identification of these strains. All spectra displayed a mass peak at 450 Da corresponding to the hydrolyzed and decarboxylated forms of Ertapenem. Likely, this peak reflects an unspecific signal derived from a slight background hydrolysis. Similarly, the mass peak at 460 Da detected in all spectra, except those of Ertapenem solution alone, is probably another unspecific signal derived from the blood culture medium [19]. For these reasons, these two peaks were not classified among the spectra originated from BCs (Fig. 1).

A characteristic peak pattern that allowed the discrimination between the carbapenemase-producing and non-carbapenemase-producing plated bacteria was also obtained with the bacteria derived directly from fresh positive BCs. Therefore, it can be envisaged that the growth medium does not affect significantly the activity of the carbapenemase enzymes.

The method presented in this work does not require particularly skilled personnel, and may be used in any laboratory equipped with a MALDI-TOF spectrometer.

In this study, we decided to set up the assay on Ertapenem hydrolysis as it is associated with a characteristic peak pattern easily detected using only the organic matrix HCCA which is used also for bacteria identification. Furthermore, Ertapenem is considered a good candidate for carbapenemase producers detection as reported by Nordmann et al. 2012 [6, 19].

For the 55 K. *pneumoniae* strains studied, no false-negative results were found. This is important, because the aim of the assay is to obtain a rapid information for a timely and effective antibiotic therapy.

Therefore, the test showed a high sensitivity and this is important for an appropriate antibiotic stewardship, as a good sensitivity allows a higher level of confidence in the administration of narrow-spectrum antibiotics in presence of susceptible isolates [17].

Furthermore, carbapenemase-production detection assay was able to detect and discriminate between different carbapenemase-producing *K. pneumoniae* isolates within 30 min to 3 h after incubation with Ertapenem. The rapid detection of carbapenemase-producing bacteria from positive BCs, combined with a proactive antimicrobial stewardship program, is important because the choice of the optimal therapy is crucial for the prognosis of patients with sepsis. With the present protocol, a report regarding the effectiveness of carbapenem therapy is obtainable within a few hours after the BC is flagged positive. The presence/absence of characteristic peak patterns resulting from carbapenemase-producing and non-carbapenemase-producing strains has been used as the cut-off of the assay with maximum sensitivity and specificity.

The MALDI-TOF MS-based assay and the approved routine method provided the same classification results. Furthermore, identification and carbapenemase-production detection testing by MALDI-TOF MS can be performed in parallel, as widely the same, and the additional hands-on time required is moderate.

To evaluate whether potential residues of blood culture fluid could interfere with the reproducibility of the measurements, the analysis was performed in quadruplicate. No interference was observed.

All $bla_{KPC}$ positive isolates showed a peak within the window of 11,109 ± 15 Da that was absent in all $bla_{KPC}$ negative isolates of *K. pneumoniae* (Fig. 2). This peak has been suggested by Lau et al. (2014) as a $bla_{KPC}$-containing pKpQIL plasmid-identifying proteomic marker that could be implemented in the routine screening of the clinical laboratory workflow and may monitoring epidemiological dissemination mapping in real time during any future CRE hospital outbreak [22]. These data suggest a monoclonal spreading of all KPC-producing *K. pneumoniae* isolates in our hospitals, probably correlated to the hyper-epidemic clonal complex (CC) 258 (ST-258 producing KPC-2 or KPC-3, and ST-512 producing

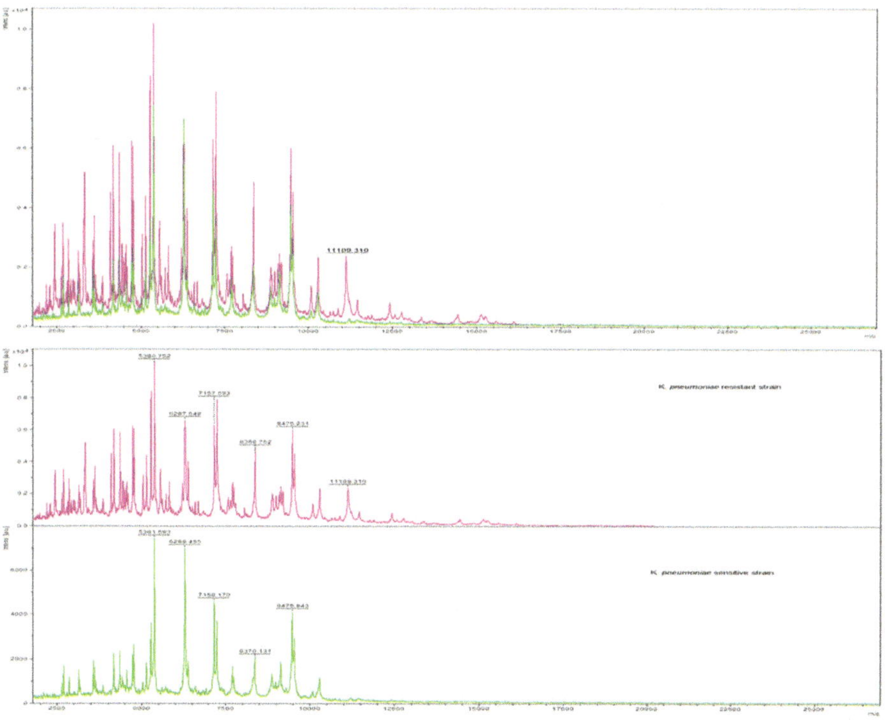


**Fig. 2** Identification of a MALDI-TOF MS peak associated with $bla_{KPC}$. Presence of the 11,109-Da peak in KPC-carrying *K. pneumoniae* isolate mass spectrum (fuchsia) and absence of an 11,109-Da peak in a *K. pneumoniae* non KPC-carrying isolate mass spectrum (*green*), respectively. This specific peak could be a helpful real-time marker in epidemic strain detection in a hospital setting

KPC-3) which was considered as the predominant clone in Italy [18]. The positive detection of ~ 11,109 Da mass peak associated with $bla_{KPC}$-containing pKpQIL plasmid could be a real-time marker of carbapenem-producing strains of great clinical and epidemiological value as suggested by Lau et al. (2014) [22]. It could be easily implemented in the routine diagnostic activity of any microbiology laboratory as part of the organism identification by MALDI-TOF MS without placing additional constraints on laboratory resources. All these findings confirm the capacity of MALDI-TOF MS technology to identify timely a carbapenemase producer causing a bacteremia. Compared with the molecular-based techniques the cost saving is significant and no particular skilled personnel is required [21]. In addition, a proteomic approach unlike a genomic provides information on gene expression without focusing on single genes. Indeed, genomic assays may not reveal novel carbapenemase genes as the range of resistance genes to be detected is predefined.

The MALDI-TOF MS carbapenemase-production detection assay could be used to confirm carbapenemase activity either in routine and reference laboratories and might supplement other rapid and cost-effective tests used in carbapenemase detection such as the phenotypic Carba NP test.

However, these preliminary findings requires further assessment and confirmation, preferably by multicenter validation studies in order to avoid clonal bias as well as analytical error and develop universal guidelines for data interpretation.

## Conclusion

In summary, the presented carbapenemase-producing strain detection by MALDI-TOF MS technology could improve the workflow of BCs positive to *Enterobacteriaceae* strains which show a suspicious susceptibility profile, and enhance the stewardship of carbapenems by revising the resistance surveillance program at different levels.

A final same-day report will be available to clinicians and to hospital epidemiologists improving patient management and treatment; thus, preventing outbreaks and spreading of pathogens in the hospital with beneficial effects on the health care system.

Compared to other alternative approaches, like molecular-genetic methods that remain costly with practical limitations, this method is cost effective, highly sensitivity and easy to perform.

However, a single diagnostic technology cannot be universally applicable in clinical microbiology as the detection of a limited number of specific determinants cannot accomplish the complex issue of antibiotic resistance.

**Abbreviations**

AmpC: Class C β-lactamases; AST: Antimicrobial Susceptibility Test; BC: Blood culture; CC 258: Clonal complex 258; CPE: Carbapenemase-producing *Enterobacteriaceae*; CRE: Carbapenem-resistant *Enterobacteriaceae*; CTX-M: Cefotaximase; ESBLs: extended-spectrum b-lactamases; EUCAST: European Committee on Antimicrobial Susceptibility testing; HAI: Hospital-Acquired infection; HCCA: α-cyano-4-hydroxy cinnamic acid; ICU: Intensive-care unit; IMP: Imipenemase; KPC: *Klebsiella Pneumoniae* Carbapenemase; MALDI-TOF MS: Matrix-assisted laser desorption ionization-time of flight mass spectrometry; MDR: Multidrug resistant; MIC: Minimum inhibitory concentration; NDM: New Delhi Metallo-beta-lactamase; OXA-48: Oxacillinase-48; PCR: Polymerase chain reaction; PDR: Pandrug resistant; SHV: Sulfhydryl variable; ST 258: Sequence type 258; TAT: Turn around time; TEM: A β-lactamase named after a Greek patient Temoneira; VIM: Verona Integron-encoded Metallo-beta-lactamase

**Acknowledgments**

We give special thanks to Dr. Giuseppe Alfieri, Becton Dickinson Italy S.p.A, for kindly providing some of the required reagents for this study.

**Funding**

The study was partially supported by the Department of Experimental Medicine and Surgery, "Tor Vergata" University of Rome, in order to implement the microbiologic characterization of the pharmaco-resistant strains, CPE-CRE strains, as required from the European Surveillance, by a novel technology.

**Authors' contributions**

CS and CF contributed in the conception and design of the study. CS, MC, CD participated in the development of the method and the validation and analyzed the data. SA participated in the microbiologic analysis, and provided strains. CS wrote the manuscript. MC and CF reviewed the manuscript. All authors read and approved the final version of the manuscript.

**Competing interest**

The authors declare that they have no competing interests.

**Author details**

[1]Department of Experimental Medicine and Surgery, "Tor Vergata" University of Rome, Via Montpellier 1, 00133 Rome, Italy. [2]Laboratory of Clinical Microbiology and Virology, Polyclinic "Tor Vergata" Foundation, V.le Oxford 81, 00133 Rome, Italy. [3]Clinical Pathology and Microbiology Laboratory, University Hospital Campus Bio-Medico, Via Alvaro del Portillo 200, 00128 Rome, Italy.

**References**

1. Jeon JH, Lee JH, Lee JJ, Park KS, Karim AM, Lee C-R, Jeong BC, Lee SH. Structural basis for carbapenem-hydrolyzing mechanisms of carbapenemases conferring antibiotic resistance. Int J Mol Sci. 2015;16: 9654–92 [http://dx.doi:10.3390/ijms16059654].
2. Nordmann P, Naas T, Poirel L. Global spread of carbapenemase producing *Enterobacteriaceae*. Emerg Infect Dis. 2011;17:1791–8 [http://dx.doi: 10.3201/eid1710.110655].
3. Tzouvelekis LS, Markogiannakis A, Psichogiou M, Tassios PT, Daikos GL. Carbapenemases in Klebsiella *pneumoniae* and other *Enterobacteriaceae* evolving crisis of global dimensions. Clin Microbiol Rev. 2012;25(4):682 [http://dx.doi:10.1128/CMR.05035-11].
4. Vincent JL, Rello J, Marshall J, Silva E, Anzueto A, Martin CD, Moreno R, Lipman J, Gomersall C, Sakr Y, Reinhar K. International study of the prevalence and outcomes of infection in intensive care units. JAMA. 2009; 302(21):2323–9. [http://dx.doi:10.1001/jama.2009.1754].
5. Nordmann P, Dortet L, Poirel L. Carbapenem resistance in Enterobacteriaceae: here is the storm! Trends Mol Med. 2012;18:263–72 [http://dx.doi: 10.1016/j.molmed.2012.03.003].
6. Nordmann P, Gniadkowski M, Giske CG, Poirel L, Woodford N, Miriagou V, the European Network on Carbapenemases. Identification and screening of carbapenemase-producing Enterobacteriaceae. Clin Microbiol Infect. 2012; 18:432–8. [http://dx.doi: 10.1111/j.1469-0691.2012.03815.x].
7. Kumarasamy KK, Toleman MA, Walsh TR, et al. Emergence of a new antibiotic resistance mechanism in India, Pakistan, and the UK: a molecular, biological, and epidemiological study. Lancet Infect Dis. 2010;10:597–602. [http://dx.doi: 10.1016/S1473-3099(10)70143-2].
8. Cuzon G, Ouanich J, Gondret R, Naas T, Nordmann P. Outbreak of OXA-48-positive carbapenem-resistant *Klebsiella pneumoniae* isolates in France. Antimicrob Agents Chemother. 2011;55:2420–3. [http://dx.doi: 10.1128/AAC.01452-10].
9. Dortet L, Bréchard L, Poirel L, Nordmann P. Rapid detection of carbapenemase-producing Enterobacteriaceae from blood cultures. Clin Microbiol Infect. 2014;20:340–4 [http://dx.doi: 0.1111/1469-0691.12318].
10. Francis RO, Wu F, Della-Latta P, Shi J, Whittier S. Rapid detection of Klebsiella pneumoniae carbapenemase genes in Enterobacteriaceae directly from blood culture bottles by real-time PCR. Am J Clin Pathol. 2012;137: 627–32. [http://dx.doi: 10.1309/AJCP9SNHJG2QGLWU].
11. Bogaerts P, de Castro Rezende R, De Mendonça R, Huang T-D, Denis O, Glupczynski Y. Validation of carbapenemase and extended-spectrum b-lactamase multiplex endpoint PCR assays according to ISO 15189. J Antimicrob Chemother. 2013;68:1576–82 [http://dx.doi:10.1093/jac/dkt065].
12. Avlami A, Bekris S, Ganteris G, et al. Detection of metallo-beta-lactamase genes in clinical specimens by a commercial multiplex PCR system. J Microbiol Methods. 2010;83:185–7 [http://dx.doi: 10.1016/j.mimet.2010.08.014].
13. Naas T, Cuzon G, Bogaerts P, Glupczynski Y, Nordmann P. Evaluation of a DNA microarray (Check-MDR CT102) for rapid detection of TEM, SHV, and CTX-M extended-spectrum beta-lactamases and of KPC, OXA-48, VIM, IMP, and NDM-1 carbapenemases. J Clin Microbiol. 2011;49:1608–13 [http://dx.doi: 10.1128/JCM.02607-10].
14. Burckhardt I, Zimmermann S. Using matrix-assisted laser desorption ionization-time of flight mass spectrometry to detect carbapenem resistance within 1 to 2.5 hours. J Clin Microbiol. 2011;49:3321–4 [http://dx.doi.org/10.1128/JCM.00287-11].
15. Sparbier K, Schubert S, Weller U, Boogen C, Kostrzewa M. Matrix-assisted laser desorption ionization-time of flight mass spectrometry-based functional assay for rapid detection of resistance against beta lactam antibiotics. J Clin Microbiol. 2012;50:927–37 [http://dx.doi.org/10.1128/JCM.05737-11].
16. Hrabák J, Studentová V, Walková R, Zemlicková H, Jakubu V, Chudácková E, Gniadkowski M, Pfeifer Y, Perry JD, Wilkinson K, Bergerová T. Detection of NDM-1, VIM-1, KPC, OXA-48, and OXA-162 carbapenemases by matrix assisted laser desorption ionization-time of flight mass spectrometry. J Clin Microbiol. 2012;50:2441–3. [http://dx.doi.org/10.1128/JCM.01002-12].
17. Jung JS, Popp C, Sparbier K, Lange C, Kostrzewa M, Schubert S. Evaluation of Matrix-Assisted LaserDesorption Ionization-Time of Flight Mass Spectrometry for Rapid Detection of b-Lactam Resistance in *Enterobacteriaceae* Derived from Blood Cultures J. Clin Microbiol. 2014;52: 924–30 [http://dx.doi:10.1128/JCM.02691-13].
18. Giani T, Pini B, Arena F, Conte V, Bracco S, Migliavacca R, et al. Epidemic diffusion of KPC carbapenemase-producing Klebsiella pneumoniae in Italy: results of the first countrywide survey, 15 May to 30 June 2011. Euro Surveill. 2013; 18(22). [http://www.eurosurveillance.org/ViewArticle.asp?ArticleId=20489]
19. Johansson A, Ekelöf J, Giske CG, Sundqvist M. The detection and verification of carbapenemases using ertapenem and Matrix Assisted Laser Desorption Ionization-Time of Flight. BMC Microbiol. 2014;14:89. [http://dx.doi: 10.1186/1471-2180-14-89].
20. Spinali S, van Belkum A, Goering RV, Girard V, Welker M, Van Nuenen M, Pincus DH, Arsac M, Durand G. Microbial typing by matrix-assisted laser desorption ionization–time of flight mass spectrometry: do we need guidance for data interpretation? J Clin Microbiol. 2015;53:760–5. [http://dx.doi:10.1128/JCM.01635-14].
21. Croxatto A, Prod'hom G, Greub G. Applications of MALDI-TOF mass spectrometry in clinical diagnostic microbiology. FEMS Microbiol Rev. 2012; 36:380–407 [http://dx.doi: 10.1111/j.1574-6976.2011.00298.x].
22. Lau AF, Wang H, Weingarten RA, Drake SK, Suffredini AF, Garfield MK, Chen Y, Gucek M, Youn JH, Stock F, Tso H, DeLeo J, Cimino JJ, Frank KM, Dekker JP. A rapid matrix-assisted laser desorption ionization-time of flight mass spectrometry-based method for single-plasmid tracking in an outbreak of carbapenem-resistant Enterobacteriaceae. J Clin Microbiol. 2014;52:2804–12 [http://dx.doi:10.1128/JCM.00694-14].

18

# Effectiveness of ITS and sub-regions as DNA barcode markers for the identification of Basidiomycota (Fungi)

Fernanda Badotti[1], Francislon Silva de Oliveira[2], Cleverson Fernando Garcia[1], Aline Bruna Martins Vaz[3,4], Paula Luize Camargos Fonseca[3], Laila Alves Nahum[2,5], Guilherme Oliveira[2,6] and Aristóteles Góes-Neto[3*]

**Abstract**

**Background:** Fungi are among the most abundant and diverse organisms on Earth. However, a substantial amount of the species diversity, relationships, habitats, and life strategies of these microorganisms remain to be discovered and characterized. One important factor hindering progress is the difficulty in correctly identifying fungi. Morphological and molecular characteristics have been applied in such tasks. Later, DNA barcoding has emerged as a new method for the rapid and reliable identification of species. The nrITS region is considered the universal barcode of Fungi, and the ITS1 and ITS2 sub-regions have been applied as metabarcoding markers. In this study, we performed a large-scale analysis of all the available Basidiomycota sequences from GenBank. We carried out a rigorous trimming of the initial dataset based in methodological principals of DNA Barcoding. Two different approaches (PCI and barcode gap) were used to determine the performance of the complete ITS region and sub-regions.

**Results:** For most of the Basidiomycota genera, the three genomic markers performed similarly, i.e., when one was considered a good marker for the identification of a genus, the others were also; the same results were observed when the performance was insufficient. However, based on barcode gap analyses, we identified genomic markers that had a superior identification performance than the others and genomic markers that were not indicated for the identification of some genera. Notably, neither the complete ITS nor the sub-regions were useful in identifying 11 of the 113 Basidiomycota genera. The complex phylogenetic relationships and the presence of cryptic species in some genera are possible explanations of this limitation and are discussed.

**Conclusions:** Knowledge regarding the efficiency and limitations of the barcode markers that are currently used for the identification of organisms is crucial because it benefits research in many areas. Our study provides information that may guide researchers in choosing the most suitable genomic markers for identifying Basidiomycota species.

**Keywords:** ITS, ITS1, ITS2, Probable correct identification, Barcode gap, Basidiomycota

## Background

Fungi are one of the major eukaryotic lineages that are equivalent in species number to animals but exceed that of plants [1]. Fungi are among the most important organisms in the world because of their vital roles in decomposition, nutrient cycling, and obligate mutualistic symbioses with plants, algae, and cyanobacteria [2]. Fungi also have great economic importance for industrial fermentation, pharmaceutical, and biotechnological industries [3]. They may also cause food spoilage and diseases in plants and animals [4]. The diversity of activities is reflected in the high number of taxa, morphologies, habitats, and life strategies used by this group of organisms. Further studies are necessary to better understand their complex interactions with other organisms and environments.

The phylum Basidiomycota is the second largest of the Fungi kingdom and comprises approximately 30% of all described fungal species [5]. This diverse phylum includes primarily macroscopic but also microscopic fungi, such as mushrooms and basidiomycotan yeasts,

* Correspondence: arigoesneto@icb.ufmg.br
[3]Universidade Federal de Minas Gerais, Departamento de Microbiologia, Av. Antônio Carlos, Belo Horizonte 6627, 31270-901, MG, Brazil
Full list of author information is available at the end of the article

respectively [6, 7]; saprotrophs, such as wood-decaying fungi [8]; pathogens of plants [3] and animals [9, 10]; and mycorrhizal symbionts [11]. Basidiomycota species are grouped into the following subphyla: Agaricomycotina, Pucciniomycotina, and Ustilaginomycotina. The first is the largest subphylum with approximately one-third of all described fungal species [5, 12]. Thus, a substantial amount of the data that is currently available on diversity, distribution, and sequencing has targeted Agaricomycotina, particularly in the orders of Agaricales, Polyporales, and Boletales. This subphylum is primarily composed of wood decayers, litter decomposers, and ectomycorrhizal fungi, as well as pathogens and poisonous, hallucinogenic, or edible species [13].

The identification of fungi at the species level is critical to many research areas, such as health sciences and agriculture, where the determination of causal agents of diseases is central to the definition of the suitable treatment, elucidation of outbreaks, and transmission mechanisms [14, 15]. Furthermore, the understanding of the specific roles of microorganisms in an ecosystem, their abundance, and their community composition in ecological and biodiversity studies can only be attained through their reliable identification [16]. However, discovering and describing all extant fungal species appears challenging. According to the *Dictionary of Fungi*, only approximately 100,000 species have been described thus far [12], and the estimated diversity ranges from 1.5 to 5.1 million [1, 17, 18].

Morphological characteristics are useful for species description; however, they may be limited because many macroscopic structures are produced infrequently and temporarily [19], and many taxa often harbor cryptic species complexes [20]. Molecular tools complementing morphological ones are very promising in identifying species and can be used to rapidly and reliably evaluate biological diversity. These markers have been applied to the identification of fungal species since the 1990s [21, 22]; however, the strategy based on the sequencing of standardized genomic fragments (DNA barcoding) was recognized afterwards [23]. The primary difference between molecular identification tools and the "DNA barcode" approach is that the latter involves the use of a standard DNA region that is specific for a taxonomic group. The use of a segment of the mitochondrial gene encoding the cytochrome c oxidase subunit I (COI) has been proposed for animals [24]. For plants, various loci combinations have been proposed [25]; however, a study conducted by the Consortium for the Barcode of Life (CBOL) Plant Working Group agreed that the combination of sequences of two plastid genes, *mat*K and *rbc*L, is the most promising plant barcode [26]. In 2012, the study conducted by Schoch and colleagues compared six DNA regions as promising universal barcodes for fungi. Mitochondrial COI and other protein-coding nuclear gene regions were excluded as potential markers for various reasons such as difficulties in amplifying DNA and insufficient variability. The nuclear ribosomal RNA internal transcribed spacer (ITS) region exhibited the highest probability of correct identification (PCI) for a wide number of fungal lineages analyzed and the most clearly defined barcode gap [27]. Since then, the ITS region has been accepted as the standard barcode marker for fungi. However, a thorough study of ITS sequences in the International Nucleotide Sequence Database (INSD: GenBank, EMBL and DDBJ) revealed that this region is not equally variable in all groups of fungi [28]. Notably, for some genera of Ascomycota, including *Alternaria* [29], *Aspergillus* [30], *Cladosporium* [31], *Penicillium* [30], and *Fusarium* [32], identification using the ITS barcode has been difficult.

One advantage of using the ITS region as a standard marker is that most fungal species have been identified based on this genomic region. GenBank [33] is the most comprehensive and widely used sequence repository in the field. A database specific for fungal sequences, the UNITE (User-friendly Nordic ITS Ectomycorrhiza Database) has been developed [34]. UNITE aims to unify the fungal taxonomic identification and correct the annotations associated with the taxonomic names to the greatest extent possible. The Barcode of Life Data System - BOLD [35] represents another bioinformatics platform; however, fungi remain underrepresented in it. BOLD supplies tools for the storage, quality warranty, and analysis of specimens and sequences to validate a barcode library. To obtain a barcode status on BOLD, sequences must fulfill some requirements, such as voucher data, collection record, and trace files. In the last few years, the scientific community has observed the rapid improvement of DNA sequencing technologies and the huge volume of data generated. Trimming and identifying this enormous amount of data requires bioinformatics tools, such as automated pipelines and various programs. However, the success of the analysis greatly depends on the correct taxonomic identification of sequences. Specifically, in the case of publicly available fungal ITS sequences, the reliability and technical quality vary significantly [34, 36]. Schoch and colleagues [27] estimate that only approximately 50% of the ITS sequences that are deposited in public databases are annotated at the species level. Moreover, Nilsson and colleagues [37] estimated that more than 10% of these fully identified fungal ITS sequences are incorrectly annotated at the species level. On the other hand, excellent initiatives, such as UNITE and that from NCBI that include a tool which allows flagging a GenBank sequence with type material [38] have emerged to minimize such a problem.

The ITS region comprises two sections (ITS1 and ITS2) that flank the conserved 5.8S region. The identification of multiple species from environmental samples (the DNA metabarcode) requires the use of high-throughput technologies, which may have limitations in sequencing read lengths [39]. For such approaches, only a portion of the ITS region is usually used, the ITS1 or the ITS2. The efficiency of these sub-regions in the identification of species in many fungal lineages has been evaluated, and some authors claim that ITS1 is more variable than ITS2 [28, 40–42]. Others have found opposite results [43] or that both the sub-regions are suitable as metabarcoding markers [44, 45]. In a recent work, Guarnica and colleagues [46] demonstrated that the ITS1 region is not more variable than the ITS2 region for *Cortinarius*. Furthermore, the complete ITS region is highly effective in discriminating among species in this highly sampled genus of Basidiomycota.

In the present study, an extensive comparative analysis based on the probability of correct identification (PCI) and barcode gap analyses was performed using a trimmed dataset composed of all Basidiomycota sequences deposited in GenBank. We evaluated the most widely used genomic markers for Fungi (the complete ITS region and the ITS1 and ITS2 sub-regions) to determine which is the most suitable for the identification of Basidiomycota species. Issues related to the need of additional molecular barcode markers as well as the taxonomic complexities within the subphyla are discussed.

## Methods

### Data acquisition and filtering

In this study, only sequences with complete nuclear ribosomal ITS from permanent collections whose taxonomic identifications were curated by specialists (*voucher* specimens) and deposited in GenBank [33] were used. Taxonomic information regarding the specimens was enriched, when available, from the UNITE database [34]. This step was used after downloading sequences from GenBank and before logical and quality filters were applied. For this enrichment, we firstly downloaded the FASTA sequence files from UNITE, and then we generated a tabular file with the UNITE data, keeping only the access numbers that corresponded to our specimens. Then, we retrieved the information related to sampling area and fungal classification from UNITE. Finally, we used the UNITE information to enrich the GenBank information.

Quality filters removed sequences with one or more IUB/IUPAC ambiguous characters, and logic filters ensured that the sequences were suitable for DNA barcode study in accordance with Barcode of Life recommendations (http://www.barcodeoflife.org/). The first logic filter guaranteed that only sequences identified at the species level were maintained in the database. Therefore, species with inconclusive names ('sp.', 'aff.', 'cf.', and 'uncultured') were removed. FungalITSExtractor [47] was used to guarantee that only sequences with complete ITS regions were maintained in the database. More than 99% of fungal complete ITS sequences deposited in GenBank are shorter than 800 or longer than 400 pb; thus all sequences outside of this interval were excluded from the dataset. The low representativeness together with the potential to distort the multiple sequence alignment justified this filter. Only species with specimens collected from at least three different localities were included to guarantee that only distinct and geographically distant specimens were evaluated and to avoid the possibility of working with genetically identical specimens. The list of all species used to perform the analyses of this study is provided (Additional file 1). All filters were performed using custom scripts written in the Perl programming language and are available upon request. The FungalITSExtractor software was used to identify and extract the ITS, ITS1, and ITS2 regions.

### Data Analysis

The ITS, ITS1, and ITS2 datasets were partitioned in several sub-datasets, each containing sequences belonging to only one genus. Sequences from each sub-dataset were aligned using MUSCLE (version 3.8.31) with default parameters [48]. Distance matrices were generated using an uncorrected p-distance because it is simple and without any biological assumptions [49]. To evaluate the discriminative power of the three genomic markers, the probability of correct identification (PCI) was calculated as the ratio of species successfully identified per total number of species. A species was considered successfully identified if the minimum interspecific distance was larger than its maximum intraspecific distance [50]. Custom Perl scripts were written to calculate the distance matrices and the PCI values. Boxplots were plotted in R language.

Two statistical analyses were performed to graphically represent the data, a scatter plot and a dot plot. The scatter plot aimed to evaluate the correlations between the PCI values for the genomic regions pairwise combinations (ITS versus ITS1, ITS versus ITS2, and ITS1 versus ITS2), and the Spearman correlation coefficient was determined. The dot plot was used to compare the PCI with the barcode gap analyses. For this purpose, the PCI values for the four groups previously defined from the barcode gap analyses (Groups 1 to 4) were represented for each genomic region. All data and graphics were generated using Minitab (Minitab Statistical Software, version 17.3.1, State College, Pennsylvania: Minitab Inc., 2016).

## Results

Our primary dataset was comprised of all complete ITS (ITS1 + 5.8S + ITS2) sequences of Basidiomycota and consisted of 37,699 sequences. The exclusion of sequences without the field 'specimen_voucher' in GenBank file reduced the number to 37,342. Removing sequences with ambiguous nucleotides led to 27,459 sequences, and removing sequences with inconclusive species names resulted in 21,238 sequences. After applying FungalITSExtractor, 19,578 sequences remained. ITS sequences with less than 400 bp and more than 800 bp were also excluded from the dataset, as well as ITS1 and ITS2 sequences less than 100 bp, leaving 19,149 sequences. The last filter was used to ensure that only species with at least three sequences collected from different geographic locations were retained in the dataset. Because most of the sequences did not include information regarding their origin, our final dataset had this number reduced to 7,731 sequences from 112 countries from six continents. This dataset was used to perform all subsequent analyses and represented three subphyla, five classes, 25 orders, 73 families, 211 genera, and 936 species (Additional file 2). This dataset has 167 sequences whose DNA were originated from biological specimens considered as type material. Many sequences from type materials were not included in our dataset only because they did not pass in quality and logic filters.

Although GenBank is known to be the most complete available public database, the amount of sequences is biased in our trimmed dataset as follows: 93.1% (7,197 sequences) belong to species of Agaricomycotina, whereas only 5.7% (442 sequences) come from Pucciniomycotina and 1.2%. (92 sequences) from Ustilaginomycotina. When other taxonomic ranks were analyzed, a similar distribution was observed with the vast majority of species belonging to Agaricomycotina (Fig. 1). Inside the subphyla, the imbalance in the amount of sequences is also enormous. For example, in Agaricomycotina, we found very well represented taxa (such as *Cortinarius*,

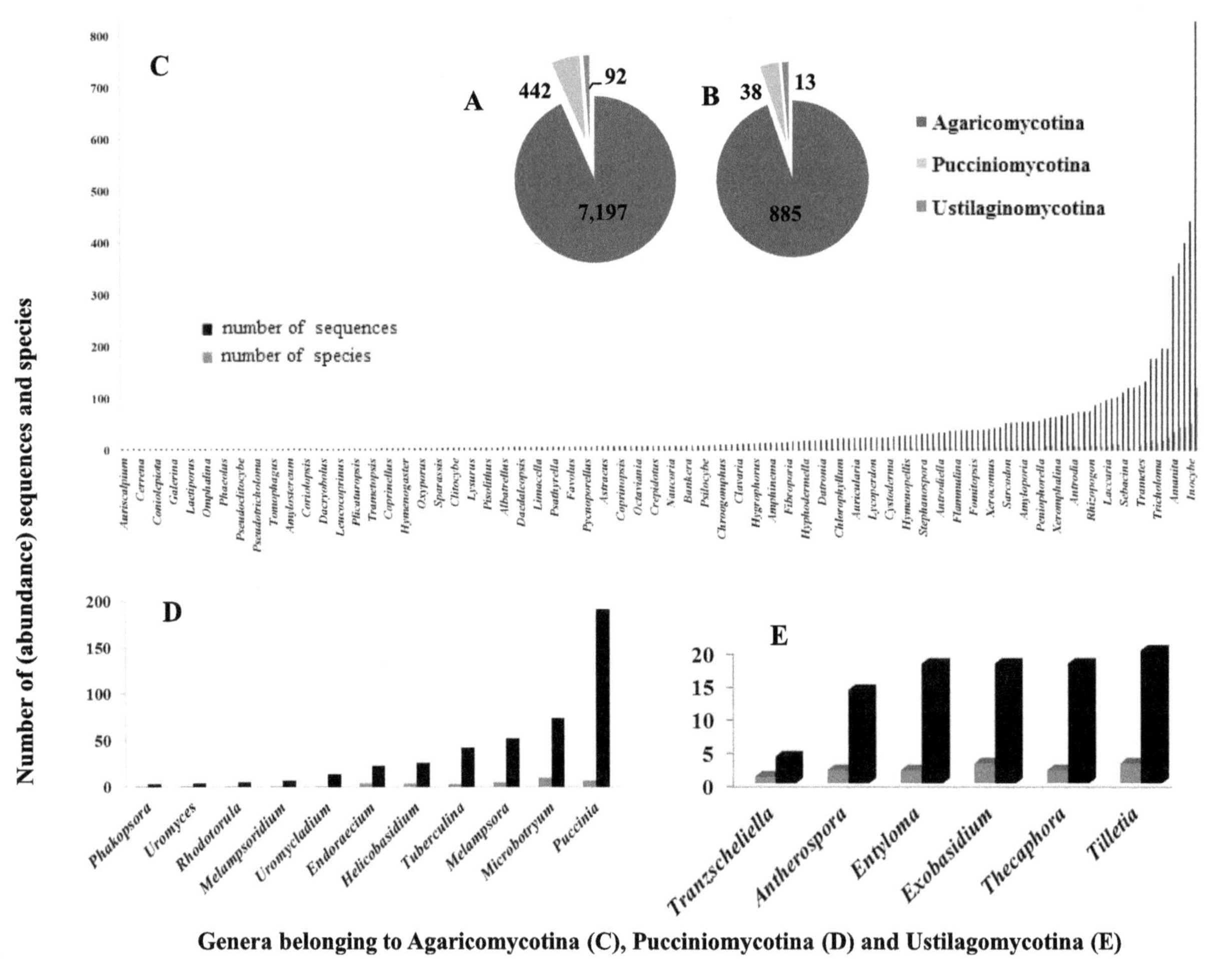


**Fig. 1** Pie charts represent abundance (number) of sequences (**a**) and species (**b**) for the three subphyla represented in the dataset used in this study. The histograms show the number of species and sequences distributed for genera belonging to Agaricomycotina (**c**), Pucciniomycotina (**d**) and Ustilagomycotina (**e**) phylum

with 829 sequences from 124 species) and others that were poorly represented (such as *Auriscalpium*, with only one species represented by three sequences). Most of the genera from the Agaricomycotina dataset were underrepresented; 126 of 194 had 20 or fewer sequences, whereas only 16 genera were represented by more than 100 sequences (Fig. 1 and Additional file 2).

The probability of correct identification (PCI) for the three genomic regions under study was estimated using our trimmed dataset (7,028 sequences from 113 genera). The number of genera analyzed decreased compared with the original dataset (211 genera) because we needed at least two species to estimate intraspecific and interspecific distances. Moreover, the sequences identified as type material are distributed in 27 distinct genera (23.9% of total) (Additional File 3), and only 25 sequences with RefSeq accessions interchangeably with GenBank numbers were identified (Additional File 4). This represented approximately only 0.36% of the sequences that comprised the dataset used to estimate PCI and barcode gap indices.

The mean PCI value for the complete ITS region was 63%, those for the sub-regions were slightly smaller as follows: 59% for ITS1 and 58% for ITS2. For the ITS region, 53.1% of the genera had PCI values higher than the mean, whereas for ITS1 and ITS2, these values were 46% and 48%, respectively (Table 1). The pairwise correlation between the three markers (ITS versus ITS1, ITS versus ITS2 and ITS1 versus ITS2) was estimated considering the PCI values of all genera composing the dataset. The comparisons between complete ITS and the sub-regions showed most of the data on or near the regression line, meaning that most of the PCI values were similar for the genera (Spearman correlation factor for ITS versus ITS1 = 0.8825 and for ITS versus ITS2 = 0.9102). When the sub-regions were associated (ITS1 versus ITS2), the distribution of data had a different profile and a lower correlation was observed (0.8158) (Fig. 2). The pairwise correlation between the genomic regions was carried out at the subphylum level; however, there were no observable patterns at this taxonomic level.

Based on the analysis of the barcode gaps, we assessed and compared the efficiency of the three genomic markers for the identification of Basidiomycota. Thus, we classified the marker performance into the following three distinct categories: *good*, *intermediate*, or *poor*. When a clear barcode gap was present (e.g., *Agaricus*, Fig. 3a), we conventionally stated that the identification was *good*, even if outliers were overlapping. The genomic markers were considered *intermediate* if the whiskers from an intraspecific distance overlap those from an interspecific distance (e.g., *Hebeloma*, Fig. 3b), and *poor* if the boxes overlap or the intraspecific distance values were superior to those of interspecific distance (e.g., *Lactarius*, Fig. 3c). For most of the genera (91.5%) evaluated, the three genomic regions performed similarly, i.e., when the identification is *good* for one region, it is also *good* for the others. The same occurred when the performance was *intermediate* or *poor*. However, for some genera, we found some genomic regions with superior identification performance than others. For instance, the complete ITS had a clearer barcode gap for the genera *Auricularia*, *Flammulina*, *Lentinellus*, *Microbotryum*, *Parasola*, and *Tuberculina* compared with the ITS1 or ITS2 sub-regions. ITS1 performed better than the other regions in the identification of species from the genera *Hygrophorus* and *Stephanospora*, as well as ITS2 for the species belonging to the genera *Amanita*, *Amyloporia*, *Fomitopsis*, *Scleroderma*, and *Strobilurus* (Table 2, Group 2). In some instances, one of the three genetic markers performed worse than the other(s). The ITS1 sub-region is not sufficient to differentiate the species of the genera *Collybia* and *Pleurotus*, and the ITS2 is not a good marker for *Sebacina*, *Hydnellum*, or *Vuilleminia*. Finally, it is important to note that for 11 out of the 113 genera evaluated (*Botyriboletus*, *Clavulina*, *Crepidotus*, *Hohenbuehelia*, *Hydnum*, *Laccaria*, *Lactarius*, *Mucidula*, *Peniophorella*, *Phaeocollybia*, and *Pisolithus*), none of the complete ITS, ITS1 or ITS2 sub-regions could be used to differentiate the species based on the barcode gap analyses (Table 2, Group 4). For a detailed classification of genera considering their barcodes, see Table 2 and Additional File 5, where the boxplots for all analyzed genera are shown.

The results of barcode gap analyses were compared with the PCI values for each genus using a dot plot (Fig. 4). For the genera for which the three genomic markers were classified as *good* in barcode gap analyses (Group 1, Table 2), most of the genera exhibited PCI above the mean value (63%); however, some disagreements were found. Some genera within this group had a PCI equal to zero (*Datronia*, *Hygrocybe*, *Tecaphora*, and *Telephora*) or between 20 and 50% (*Chroogomphus*, *Coprinopsis*, *Lactifluus*, *Melampsora*, *Phellinus*, *Piloderma*, *Puccinia*, *Russula*, *Tilletia*, *Xerocomus*, and *Xeromphalina*) (Fig. 4a). When the group for which one or two genomic regions showing a clearer barcode gap (Group 2, Table 2) was compared with the PCI, most of the genera had a PCI below the mean value (Fig. 4b). When the group for which most of the genomic regions showed an intermediate barcode gap (Group 3, Table 2), only *Lentinus* and *Hyphoderma* had higher PCI than mean value (both for the ITS2 region, Fig. 4c). When the groups for which all three genomic regions were classified as *poor* markers considering the barcode gap (Group 4, Table 2), most of the genera also had a PCI below the mean value (Fig. 4b) with the exception of

**Table 1** Probable Correct Identification (PCI) values (%) for all of the Basidiomycota genera from our trimmed dataset. The PCI values were estimated for the three genomic regions studied, the complete ITS region (ITS1 + 5.8S + ITS2) and the sub-regions ITS1 and ITS2

| Genera | ITS (ITS1 + 5.8S+ ITS2) | ITS1 | ITS2 |
|---|---|---|---|
| *Agaricus* | 100 | 100 | 100 |
| *Alnicola* | 44 | 33 | 33 |
| *Amanita* | 24 | 27 | 27 |
| *Amyloporia* | 25 | 25 | 25 |
| *Antherospora* | 100 | 100 | 100 |
| *Antrodia* | 70 | 70 | 70 |
| *Antrodiella* | 75 | 50 | 75 |
| *Armillaria* | 20 | 20 | 20 |
| *Auricularia* | 80 | 40 | 60 |
| *Boletus* | 32 | 26 | 47 |
| *Butyriboletus* | 67 | 33 | 67 |
| *Calvatia* | 0 | 0 | 0 |
| *Ceriporiopsis* | 100 | 100 | 100 |
| *Chlorophyllum* | 100 | 100 | 100 |
| *Chroogomphus* | 50 | 50 | 50 |
| *Clavaria* | 100 | 100 | 100 |
| *Clavulina* | 33 | 33 | 33 |
| *Collybia* | 50 | 50 | 50 |
| *Coprinopsis* | 50 | 50 | 50 |
| *Cortinarius* | 36 | 45 | 37 |
| *Crepidotus* | 50 | 50 | 50 |
| *Cystoderma* | 50 | 50 | 33 |
| *Cystodermella* | 100 | 100 | 100 |
| *Datronia* | 0 | 50 | 0 |
| *Endoraecium* | 100 | 100 | 100 |
| *Entoloma* | 100 | 86 | 93 |
| *Entyloma* | 100 | 100 | 100 |
| *Exobasidium* | 100 | 100 | 100 |
| *Favolus* | 100 | 100 | 100 |
| *Fibroporia* | 100 | 100 | 67 |
| *Flammulina* | 100 | 50 | 75 |
| *Fomitopsis* | 50 | 25 | 50 |
| *Fuscoporia* | 100 | 100 | 100 |
| *Ganoderma* | 43 | 43 | 43 |
| *Geastrum* | 100 | 67 | 67 |
| *Gloeophyllum* | 100 | 100 | 100 |
| *Gymnopilus* | 100 | 100 | 50 |
| *Gymnopus* | 67 | 67 | 60 |
| *Hebeloma* | 42 | 37 | 32 |
| *Helicobasidium* | 100 | 75 | 100 |
| *Hohenbuehelia* | 0 | 0 | 0 |

**Table 1** Probable Correct Identification (PCI) values (%) for all of the Basidiomycota genera from our trimmed dataset. The PCI values were estimated for the three genomic regions studied, the complete ITS region (ITS1 + 5.8S + ITS2) and the sub-regions ITS1 and ITS2 *(Continued)*

| | | | |
|---|---|---|---|
| *Hydnellum* | 56 | 56 | 56 |
| *Hydnum* | 50 | 50 | 50 |
| *Hygrocybe* | 0 | 0 | 0 |
| *Hygrophorus* | 67 | 67 | 33 |
| *Hymenopellis* | 100 | 100 | 100 |
| *Hyphoderma* | 60 | 60 | 80 |
| *Hyphodermella* | 100 | 100 | 100 |
| *Hypholoma* | 0 | 0 | 0 |
| *Inocybe* | 30 | 19 | 28 |
| *Laccaria* | 0 | 0 | 13 |
| *Lactarius* | 37 | 37 | 41 |
| *Lactifluus* | 50 | 50 | 50 |
| *Lentinellus* | 43 | 43 | 29 |
| *Lentinus* | 57 | 57 | 71 |
| *Lepiota* | 92 | 83 | 75 |
| *Lepista* | 100 | 100 | 100 |
| *Leucoagaricus* | 90 | 100 | 80 |
| *Leucopaxillus* | 100 | 100 | 100 |
| *Lycoperdon* | 100 | 100 | 100 |
| *Lyomyces* | 100 | 100 | 100 |
| *Lyophyllum* | 100 | 100 | 67 |
| *Macrolepiota* | 50 | 50 | 33 |
| *Megacollybia* | 83 | 83 | 33 |
| *Melampsora* | 40 | 40 | 20 |
| *Melanoleuca* | 38 | 38 | 25 |
| *Microbotryum* | 70 | 80 | 50 |
| *Mucidula* | 0 | 0 | 0 |
| *Mycena* | 22 | 22 | 33 |
| *Neofavolus* | 100 | 100 | 100 |
| *Octaviania* | 100 | 100 | 100 |
| *Oligoporus* | 100 | 100 | 100 |
| *Parasola* | 100 | 67 | 67 |
| *Paxillus* | 0 | 0 | 33 |
| *Peniophorella* | 50 | 50 | 50 |
| *Phaeocollybia* | 0 | 0 | 0 |
| *Phanerochaete* | 67 | 67 | 67 |
| *Phellinus* | 100 | 33 | 67 |
| *Phellodon* | 86 | 86 | 71 |
| *Piloderma* | 50 | 50 | 50 |
| *Pisolithus* | 0 | 0 | 0 |
| *Pleurotus* | 29 | 43 | 14 |
| *Pluteus* | 27 | 27 | 23 |

**Table 1** Probable Correct Identification (PCI) values (%) for all of the Basidiomycota genera from our trimmed dataset. The PCI values were estimated for the three genomic regions studied, the complete ITS region (ITS1 + 5.8S + ITS2) and the sub-regions ITS1 and ITS2 *(Continued)*

| | | | |
|---|---|---|---|
| *Polyporus* | 100 | 100 | 86 |
| *Porodaedalea* | 100 | 100 | 100 |
| *Postia* | 50 | 50 | 50 |
| *Psathyrella* | 100 | 100 | 100 |
| *Psilocybe* | 100 | 100 | 100 |
| *Puccinia* | 43 | 43 | 43 |
| *Pycnoporellus* | 100 | 100 | 100 |
| *Ramaria* | 33 | 33 | 33 |
| *Resinicium* | 100 | 100 | 100 |
| *Rhizopogon* | 27 | 27 | 27 |
| *Rhodocollybia* | 100 | 100 | 100 |
| *Rigidoporus* | 100 | 100 | 50 |
| *Russula* | 38 | 27 | 36 |
| *Sarcodon* | 86 | 86 | 86 |
| *Scleroderma* | 33 | 33 | 33 |
| *Sebacina* | 33 | 0 | 33 |
| *Stephanospora* | 80 | 80 | 20 |
| *Strobilurus* | 67 | 67 | 67 |
| *Suillus* | 83 | 83 | 83 |
| *Thecaphora* | 100 | 100 | 100 |
| *Thelephora* | 0 | 0 | 0 |
| *Tilletia* | 100 | 33 | 100 |
| *Tomentella* | 25 | 50 | 25 |
| *Trametes* | 42 | 42 | 42 |
| *Tricholoma* | 43 | 57 | 29 |
| *Tricholomopsis* | 100 | 100 | 100 |
| *Tuberculina* | 67 | 0 | 67 |
| *Vuilleminia* | 33 | 33 | 33 |
| *Xerocomus* | 100 | 50 | 100 |
| *Xeromphalina* | 100 | 100 | 50 |

*Butyriboletus* (with a PCI value above the mean for ITS and ITS2, Fig. 4d).

## Discussion

The accepted DNA barcode for Fungi is the rDNA ITS region [27]. ITS is recognized as a fungal barcode because it is the most sequenced region of fungi and is routinely used for systematics, phylogenetics, and identification [51, 52]. In this study, we downloaded all complete ITS sequences of species belonging to the phylum Basidiomycota from GenBank. Although this is the most complete repository of available ITS sequences, misidentifications or low-quality sequencing have been encountered in this public database [37]. However, some authors think that it is unrealistic that future databases or even a barcode database could be more reliable than GenBank because misidentified sequences would be as common as they are currently and because *vouchers* will not be re-identified by taxonomic experts (for a wide discussion, see [53–55]). To overcome this drawback, logical and quality filters were applied to our original dataset to obtain the most reliable results possible. The restrictiveness in the filtering step aimed to create a high-quality dataset (accurate taxonomic annotation and presence of relevant metadata) that would meet the theoretical assumptions of the biological system of identification via DNA barcode and the principles recommended in BOLD Systems [24, 35].

More than 90% of our trimmed dataset belonged to the subphylum Agaricomycotina. This result is not surprising because it reflects the high diversity of this taxon compared with the other subphyla, which is widely mentioned in the literature [5, 12]. Kirk and colleagues [12] estimated that one-fifth of all known fungal species described belong to the Agaricomycete clade; this diversity is considered to be underestimated because new taxa are continually being described [1, 56]. This discrepancy in the amount of species and sequences from the subphyla may reflect a natural event or may occur due to the specific interests of the scientific community in Agaricomycotina species.

Some criteria have been traditionally used to test the DNA barcoding efficacy to classify and/or identify specimens at the species level, such as similarity measures, tree-based techniques, and identification based on direct sequence comparison [57, 58]. However, all of these approaches present several issues (see [55] for a detailed discussion). Similarity measures are generally used to cluster sequences in "molecular operational taxonomic units"; however, the choice of the threshold value for distinguishing intraspecific and interspecific distances is largely arbitrary [58, 59]. An important and acceptable measure of the efficacy of a genetic marker should reflect the probability of correctly identifying a species. This concept has emerged as the probability of correct identification (PCI) [50, 53, 55, 60]. However, there is no consensus for the definition and calculation of PCI, which currently embraces a broad class of measures. In this work, we assume the concept described by Hollingsworth and colleagues [50] in which the authors considered the "discrimination as successful if the minimum uncorrected interspecific p-distance involving a species was larger than its maximum intraspecific distance" to measure the PCI for each genus included in our dataset. Furthermore, the use of genetic distances enables the observation of the 'barcoding gap', which is possible by plotting the intraspecific and interspecific distances.

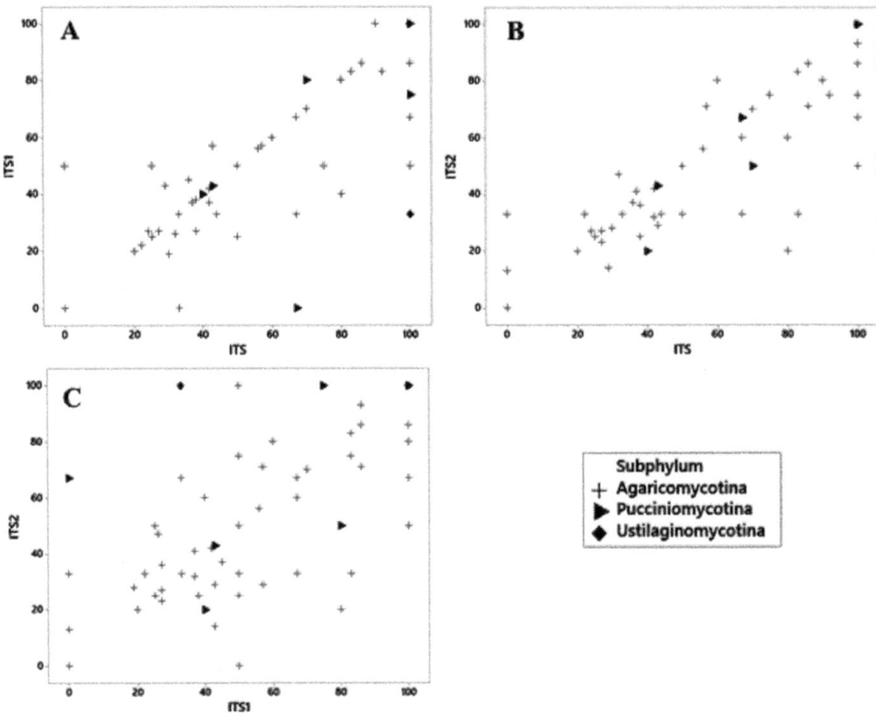


**Fig. 2** Pairwise correlations (**a**, ITS X ITS1, **b**, ITS X ITS2 and **c**, ITS1 X ITS2) between PCI values of all genera from our dataset

Therefore, an ideal barcode marker would reveal intraspecific divergences lower than interspecific divergences [61].

In this study, we aimed to identify the most suitable genomic marker (complete ITS, ITS1 or ITS2) to identify fungal species belonging to Basidiomycota. Our findings, based on PCI and barcode gap analyses, indicated that for most of the genera, the three genomic regions perform similarly, i.e., when one genomic region was considered a good marker (a PCI above the mean value or the presence of a clear barcode gap) the other regions were also; the same was observed when the performance of genomic markers was considered insufficient. When the performance of the genomic markers was individually evaluated, barcode gap analyses provided a more optimistic view than PCI values. Approximately half of the genera exhibited PCI values lower than the mean (63%); however, the three genomic regions were classified as *good* for most of the genera (Table 2) when the barcode gap is taken into account. Accordingly, the comparison between barcode gap and PCI for each genus showed some disagreements. This was primarily observed for

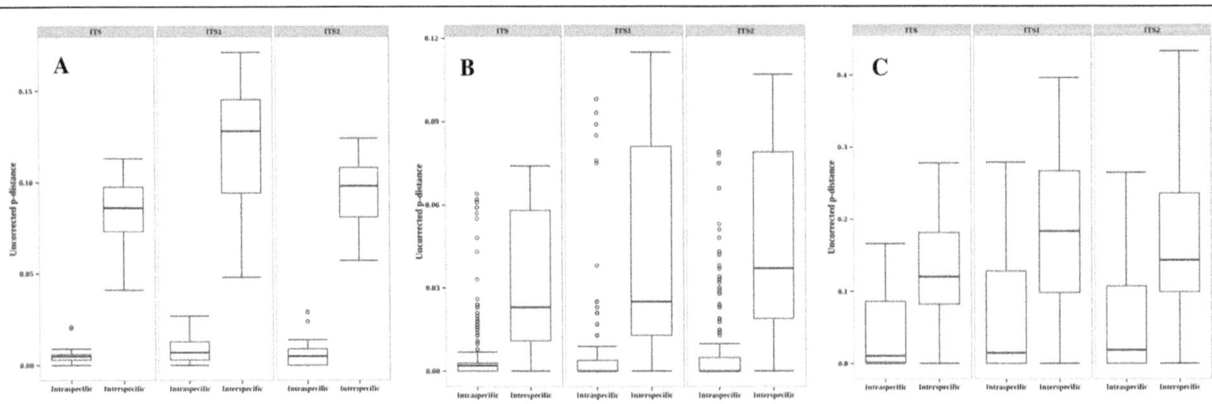


**Fig. 3** Examples of the barcode gap performance classifications used in this study. **a**. Clear barcode gap (identification performance classified as *good*) for the genera *Agaricus*, **b**. *Intermediate* separation between the intra- and interspecific distances for *Hebeloma* and **c**. A *poor* barcode gap for *Lactarius*

**Table 2** Grouping of Basidiomycota genera based on the barcode gap analyses (See Additional file 5 for more details)

| Group 1 | Group 2 | Group 3 | Group 4 |
|---|---|---|---|
| Genera for which all the three genetic regions are *good* markers | Genera for which one or two genetic regions showed a clearer barcode gap and are recommended over the other (s) | Genera for which most of the genetic regions showed *intermediate* barcode gap and their use should be carefully evaluated | Genera for which all three genetic regions are *poor* markers |
| *Agaricus, Antherospora, Antrodia, Ceriporiopsis, Chroogomphus, Clavaria, Coprinopsis, Cystodermella, Datronia, Edoraecium, Entoloma, Entyloma, Exobasidium, Favolus, Fibroporia, Fuscoporia, Geastrum, Gloeophyllum, Helicobasidium, Hygrocybe, Hymenopellis, Hyphodermella, Lactifluus, Lepiota, Lepista, Leucopaxillus, Lycoperdon, Lyomyces, Lyophyllum, Melampsora, Neofavolus, Octaviania, Oligoporus, Phanerochaete, Phellinus, Phellodon, Piloderma, Polyporus, Porodaedalea, Psathyrella, Psilocybe, Puccinia, Pycnoporellus, Resinicium, Rhodocollybia, Russula, Sarcodon, Suillus, Telephora, Thecaphora, Tilletia, Tricholomopsis, Xerocomus, Xeromphalina* | *Amanita* (ITS2), *Amyloporia* (ITS2), *Antrodiella* (ITS and ITS2), *Auricularia* (ITS), *Calvatia* (ITS and ITS1), *Chlorophyllum* (ITS and ITS1), *Flammulina* (ITS), *Fomitopsis* (ITS2), *Ganoderma* (ITS and ITS1), *Gymnopilus* (ITS and ITS1), *Gymnopus* (ITS and ITS1), *Hygrophorus* (ITS1), *Lentinellus* (ITS), *Leucoagaricus* (ITS and ITS1), *Macrolepiota* (ITS and ITS1), *Megacollybia* (ITS and ITS1), *Microbotryum* (ITS), *Parasola* (ITS), *Postia* (ITS and ITS2), *Rigidoporus* (ITS and ITS1), *Scleroderma* (ITS2), *Stephanospora* (ITS1), *Strobilurus* (ITS2), *Tuberculina* (ITS) | *Alnicola, Armillaria, Boletus, Collybia, Cortinarius, Cystoderma, Hebeloma, Hydnellum, Hyphoderma, Hypholoma, Inocybe, Lentinus, Melanoleuca, Mycena, Paxillus, Pleurotus, Pluteus, Ramaria, Rhizopogon, Sebacina, Tomentella, Tricholoma, Trametes, Vulleminia* | *Butyriboletus, Clavulina, Crepidotus, Hohenbuehelia, Hydnum, Laccaria, Lactarius, Mucidula, Peniophorella, Phaeocollybia, Pisolithus* |

some of the genera that showed good identification performance using the barcode gap but had low PCI values (Fig. 4a). The opposite, i.e., high PCI values and poor identification performance via barcode gap, was observed for only one genus, *Botyriboletus* (Fig. 4d).

Initially, the low PCI values found for some genera (such as *Calvatia*, *Datronia*, *Hygrocybe*, *Hohenbuehelia*, *Hypholoma*, *Mucidula*, and *Pisolithus*) could be explained by dataset features, such as the low number of species (genera represented by sequences from only two species) and/or by the high number of outliers, which would have distorted the PCI estimates. Additionally, the taxonomy appears very complex for many of the genera for which the identification performance using ITS and sub-regions were insufficient. Taxonomy issues for two genera (*Hygrocybe* and *Thelephora*) for which PCI values were low and three genera (*Hypholoma*, *Phaeocollybia*, and *Pisolithus*) for which both PCI and

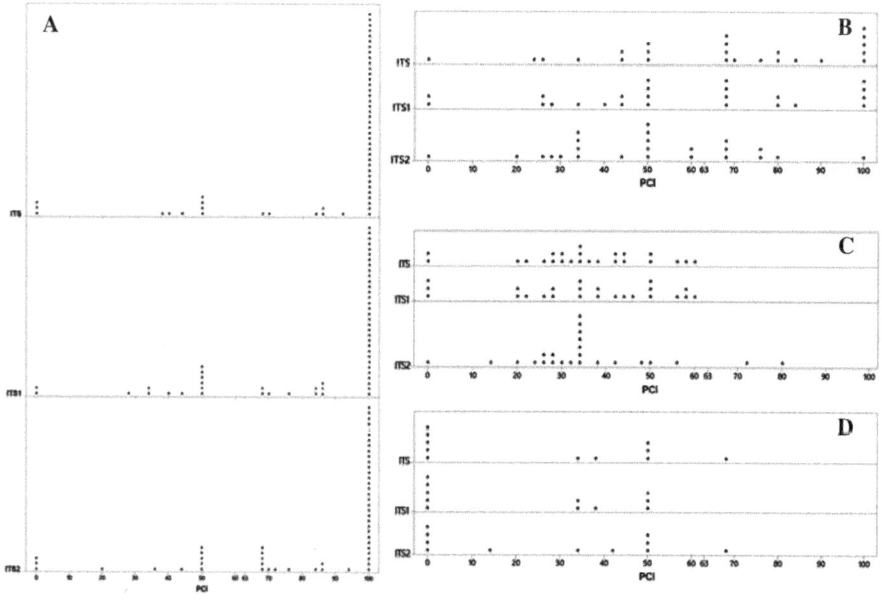


**Fig. 4** PCI values for the genera classified in the Group 1 (**a**), Group 2 (**b**), Group 3 (**c**) and Group 4 (**d**) are represented for the ITS, ITS1 and ITS2 genomic region

barcode gap analyses proved that ITS, ITS1 and ITS2 are not sufficient markers for the identification of species are discussed below based on pertinent literature.

*Hygrocybe* species exhibit extremely high variability in the ITS region, with sequences diverging by more than 25%. Thus, the use of additional DNA barcode markers has been proposed to re-evaluate the taxonomy of this genus [62, 63]. Moreover, significant changes in the classification of *Hygrocybe,* such as its division, are expected [64].

The phylogenetic relationships between and within species of *Thelephora* are also doubtful with ITS. The existence of cryptic species was described, and the importance of integrating morphological and molecular data, as well as employing a meaningful number of samples for the accurate identification is highlighted [65]. *Hypholoma* has been poorly studied. However, a recent study based on the morphological and molecular aspects of *H. cinnabarinum* samples showed that this species is not a member of the genus *Hypholoma* but belongs instead to *Agaricus* [66]. The ecological role of *Phaeocollybia* is uncertain. Smith [67] argues that the genus harbors both saprobes and mycorrhiza formers. Singer [68] considered that members of the genus were not obligatorily ectomycorrhizal, whereas Norvell [69] presented evidence for the consideration of *Phaeocollybia* as a mycorrhizal genus. At the taxonomic level, the complexity remains, as may be exemplified in Norvell [70]. The author proposed the re-evaluation of the genus *Phaeocollybia* by revealing four new agaric species morphologically similar to *Phaeocollybia kauffmanii.* The wide genetic divergence among *Pisolithus* ITS sequences [71–73] indicates significant evolutionary divergence and suggests that this genus encompasses a species complex. This hypothesis was reinforced by Kope and Fortin [74] who separated three groups of *Pisolithus* using incompatibility tests and basidiospore spine morphology.

According to Bickford and colleagues [20], cryptic species are two or more distinct species that are erroneously classified under one species name. Large intraspecific genetic distances associated with morphological and geographical discrete differences have revealed a broad range of cryptic species for many organisms and habitats [75, 76]. Although our knowledge of fungal species remains limited, the presence of cryptic species inside the group is well recognized [20] and was subsequently described for many of the genera covered in this study.

The use of molecular techniques, primarily DNA sequences, generates information to re-evaluate classifications and provides more accurate species delimitations [77]. Currently, the utility of DNA barcoding is evident. However, a universal barcode for the clear identification of all fungal species does not appear feasible, and secondary barcodes for Fungi have already been proposed [78]. In addition to the known limitations of ITS barcodes for some genera of Ascomycota, our results indicated that for some genera of Basidiomycota, such as *Hygrocybe* and *Pisolithus,* additional barcode markers may contribute to a clear elucidation of the complex relationships between and within species. The failure to correctly identify biological species hampers the efforts of the scientific community to conserve, study, or utilize them. Future research in this field should include discovering characteristics that natural selection acts upon [20].

## Conclusions

Progress in many research areas fundamentally depends on the rapid and reliable identification of biological species. Most fungal diversity is unknown, and issues related to the conservation of these organisms are urgent; thus, studies related to species identification are crucial. Knowledge regarding the efficiency and limitations of the barcode markers that are currently used for specific groups of organisms optimize the work of many studies. Therefore, the present study contributes to the rational selection of barcode markers of species belonging to the phylum Basidiomycota.

## Additional files

**Additional file 1:** List of species used in this study, their accession and taxon ID in GenBank and taxonomic affiliations.

**Additional file 2:** Number of species and sequences (specimens) recovered to each genus and their taxonomic affiliations. Data were compiled from our trimmed dataset.

**Additional file 3:** List of genera with sequences originated from type specimens and their PCI values (ITS, ITS1, ITS 2) and groups according to the barcode gap analysis.

**Additional file 4:** List of sequences with RefSeq accessions interchangeably with GenBank numbers.

**Additional file 5:** Barcode gap of all the 113 genera studied for ITS, ITS1 and ITS2 genomic regions by plotting intra- and interspecific distances.

### Abbreviations

BOLD: The barcode of life data system; CBOL: Consortium for the barcode of life; INSD: International nucleotide sequence database; ITS: Nuclear ribosomal RNA internal transcribed spacer region; MUSCLE: Multiple sequence comparison by log-expectation; PCI: Probability of correct identification; UNITE: User-friendly nordic its ectomycorrhiza database

### Acknowledgements

We thank all who contributed directly or indirectly to this work, especially the CNPq (Conselho Nacional de Desenvolvimento Científico e Tecnológico), FIOCRUZ-MG (Fundação Oswaldo Cruz, Minas Gerais), CEFET-MG (Centro Federal de Educação Tecnológica de Minas Gerais), Vale Institute of Technology, and the Graduate Programs of Microbiology and Bioinformatics of the Universidade Federal de Minas Gerais (UFMG).

### Funding

This work was supported by grants from Conselho Nacional de Desenvolvimento Científico e Tecnológico (CNPq, 308148/2013-4 and 564944/2010-6).

### Authors' contributions

FB analyzed the data and drafted the manuscript. FSO wrote the scripts, downloaded and filtered the dataset. CFG and ABMV worked on statistical analyses. PLCF and LN assisted with the data analyses. GO and AGN designed the analyses, analyzed and discussed the data. All authors read and approved the final manuscript.

### Competing interests

The authors declare that they have no competing interests.

### Author details

[1]Centro Federal de Educação Tecnológica de Minas Gerais (CEFET-MG), Departamento de Química, 30.421-169 Belo Horizonte, MG, Brazil. [2]Fundação Oswaldo Cruz (FIOCRUZ), Centro de Pesquisas René Rachou – CPqRR, 30190-002 Belo Horizonte, MG, Brazil. [3]Universidade Federal de Minas Gerais, Departamento de Microbiologia, Av. Antônio Carlos, Belo Horizonte 6627, 31270-901, MG, Brazil. [4]Faculdade de Minas (FAMINAS), 66055-090 Belo Horizonte, MG, Brazil. [5]Faculdade Promove de Tecnologia, 30140-061 Belo Horizonte, MG, Brazil. [6]Instituto Tecnológico Vale, 66055-090 Belém, PA, Brazil.

### References

1. Blackwell M. The fungi: 1, 2, 3 … 5.1 million species? Am J Bot. 2011;98(3): 426–38.
2. Gadd GM. Fungi and their role in the biosphere. In: Jorgensen SE, Brian F, editors. Encyclopedia of ecology. Amsterdam: Elsevier; 2013. p. 1709–17.
3. Lane CR, Beales PA, Hughes KJD. Fungal plant patogens. Wallingford: CAB International; 2012. p. 2012.
4. Fisher MC, Henk DA, Briggs CJ, Brownstein JS, Madoff LC, McCraw SL, Gurr SJ. Emerging fungal threats to animal, plant and ecosystem health. Nature. 2012;484(7393):186–94.
5. Hibbett DS. Major events in the evolution of the Fungi. In: Losos J, editor. Princeton Guide to Evolution, Princeton University Press, vol. 2014. 2014. p. 152–8.
6. Morin E, et al. Genome sequence of the button mushroom *Agaricus bisporus* reveals mechanisms governing adaptation to a humic-rich ecological niche. Proc Natl Acad Sci U S A. 2012;109(43):17501–6.
7. Stajich JE, et al. Insights into evolution of multicellular fungi from the assembled chromosomes of the mushroom *Coprinopsis cinerea* (*Coprinus cinereus*). Proc Natl Acad Sci U S A. 2010;107(26):11889–94.
8. Floudas D, et al. The Paleozoic origin of enzymatic lignin decomposition reconstructed from 31 fungal genomes. Science. 2012;336(6089):1715–9.
9. Brown SM, Campbell LT, Lodge JK. *Cryptococcus neoformans*, a fungus under stress. Curr Opin Microbiol. 2007;10(4):320–5.
10. Dawson Jr TL. *Malassezia globosa* and *restricta*. breakthrough understanding of the etiology and treatment of dandruff and seborrheic dermatitis through whole-genome analysis. J Investig Dermatol Symp Proc. 2007;12(2):15–9.
11. Martin F, et al. The genome of *Laccaria bicolor* provides insights into mycorrhizal symbiosis. Nature. 2008;452:88–92.
12. Kirk PM, Cannon PF, Minter DW, Stalpers JA. Dictionary of the fungi. 10th ed. Wallingford: CABI; 2008.
13. Hibbett DS. A phylogenetic overview of the agaricomycotina. Mycologia. 2006;98(6):917–25.
14. Araujo R. Towards the genotyping of fungi: methods, benefits and challenges. Cur Fung Infect Rep. 2014;8(3):203–10.
15. McNeil M, Roberts AMI, Cockerell V, Mulholland V. Real-time PCR assay for quantification of *tilletia caries* contamination of UK wheat seed. Plant Pathol. 2004;53(6):741–75.
16. Peay KG, Kennedy PG, Bruns TD. Fungal community ecology: a hybrid beast with a molecular master. Bioscience. 2008;58(9):799–810.
17. Hawksworth DL. The fungal dimension of biodiversity: magnitude, significance and conservation. Mycol Res. 1991;95:641–55.
18. Hawksworth DL. The magnitude of fungal diversity: the 1.5 million species estimate revisited. Mycol Res. 2001;105(12):1422–32.
19. Slepecky RA, Starmer WT. Phenotypic plasticity in fungi: a review with observations on *Aureobasidium pullulans*. Mycologia. 2009;101:823–32.
20. Bickford D, Lohman DJ, Sodhi NS, Ng PK, Meier R, Winker K, Ingram KK, Das I. Cryptic species as a window on diversity and conservation. Trends Ecol Evol. 2007;22(3):148–55.
21. White TJ, Bruns T, Lee S, Taylor J. Amplification and direct sequencing of fungal ribosomal RNA genes for phylogenetics. In: Innis MA, Gelfand DH, Sninsky JJ, White TJ, editors. PCR Protocols: a Guide to Methods and Applications. 1990. p. 315–22.
22. Bruns TD, White TJ, Taylor JW. Fungal molecular systematics. Annu Rev Ecol Syst. 1991;22:525–64.
23. Hollingsworth PM. DNA barcoding: potential users. Genom Soc Pol. 2007;3:44–7.
24. Hebert PD, Cywinska A, Ball SL, deWaard JR. Biological identifications through DNA barcodes. Proc Biol Sci. 2003;270(1512):313–21.
25. Pennisi E. Taxonomy. Wanted: a barcode for plants. Science. 2007;318:190–1.
26. Hollingsworth PM, et al. A DNA barcode for land plants. Proc Natl Acad Sci U S A. 2009;106:12794–7.
27. Schoch CL, Seifert KA, Huhndorf S, Robert V, Spouge JL, Levesque CA, Chen W. Fungal barcoding consortium. Nuclear ribosomal internal transcribed spacer (ITS) region as a universal DNA barcode marker for fungi. Proc Natl Acad Sci U S A. 2012;109(16):6241–6.
28. Nilsson RH, Kristiansson E, Ryberg M, et al. Intraspecific ITS variability in the kingdom fungi as expressed in the international sequence databases and its implications for molecular species identification. Evol Bioinforma. 2008;4:193–201.
29. Pryor B, Michailides T. Morphological, pathogenic, and molecular characterization of alternaria isolates associated with alternaria late blight of pistachio. Phytopathology. 2002;92:406–16.
30. Skouboe P, Frisvadm J, Taylor J, Lauritsen D, Boysen M, Rossen L. Phylogenetic analysis of nucleotide sequences from the ITS region of terverticillate *Penicillium* species. Mycol Res. 1999;103:873–81.
31. Schubert K, Groenewald J, Braun U, Dijksterhuis J, Starink M, Hill C, Zalar P, de Hoog G, Crous P. Biodiversity in the *Cladosporium* herbarum complex (Davidiellaceae, Capnodiales), with standardization of methods for *Cladosporium* taxonomy and diagnostics. Stud Mycol. 2007;58:105–56.
32. O'Donnell K, Cigelnik E. Two divergent intragenomic rDNA ITS2 types within a monophyletic lineage of the fungus Fusarium are nonorthologous. Mol Phylogenet Evol. 1997;7:103–16.
33. Benson DA, Clark K, Karsch-Mizrachi I, Lipman DJ, Ostell J, Sayers EW. GenBank. Nucleic Acids Res. 2015;43(Database issue):D30–35.
34. Kõljalg, et al. Towards a unified paradigm for sequence-based identification of fungi. Mol Ecol. 2013;22:5271–7.
35. Ratnasingham S, Hebert PD. A DNA-based registry for all animal species: the barcode index number (BIN) system. PLoS One. 2013;8(7):e66213.
36. Bruns TD, Blackwell M, Edwards I, Taylor AF, Horton T, Zhang N, et al. Preserving accuracy in GenBank. Science. 2008;319(5870):1616.
37. Nilsson RH, Ryberg M, Kristiansson E, Abarenkov K, Larsson KH, Koljalg U. Taxonomic reliability of DNA sequences in public sequence databases: a fungal perspective. PLoS One. 2006;1:e59.
38. Federhen S. Type material in the NCBI taxonomy database. Nucleic Acids Res. 2014;43:D1086–98.
39. Cuadros-Orellana S, Leite LR, Smith A, Medeiros JD, Badotti F, Fonseca PL, Vaz ABM, Oliveira G, Góes-Neto A. Assessment of fungal diversity in the environment using metagenomics: a decade in review. Fung Genom Biol. 2013;3:110.
40. Mullineux T, Hausner G. Evolution of rDNA ITS1 and ITS2 sequences and RNA secondary structures within members of the fungal genera *Grosmannia* and *Leptographium*. Fungal Genet Biol. 2009;46:855–67.
41. Wang XC, Liu C, Huang L, Bengtsson-Palme J, Chen H, Zhang JH, Cai D, Li JQ. ITS1: a DNA barcode better than ITS2 in eukaryotes? Mol Ecol Resour. 2015;15(3):573–86.
42. Ryberg M, Kristiansson E, Sjökvist E, et al. An outlook on the fungal internal transcribed spacer sequences in GenBank and the introduction of a web-based tool for the exploration of fungal diversity. New Phytol. 2009;181:471–7.
43. Bazzicalupo AL, Bálint M, Schmitt I. Comparison of ITS1 and ITS2 rDNA in 454 sequencing of hyperdiverse fungal communities. Fungal Ecol. 2013;6(1): 102–9.

44. Mello A, Napoli C, Murat C, Morin E, Marceddu G, Bonfante P. ITS-1 versus ITS-2 pyrosequencing: a comparison of fungal populations in truffle grounds. Mycologia. 2011;103(6):1184–93.
45. Blaalid R, Kumar S, Nilsson RH, Abarenkov K, Kirk PM, Kauserud H. ITS1 versus ITS2 as DNA metabarcodes for fungi. Mol Ecol Resour. 2013;13(2): 218–24. doi:10.1111/1755-0998.12065.
46. Garnica S, Schön ME, Abarenkov K, et al. Determining threshold values for barcoding fungi: lessons from Cortinarius (Basidiomycota), a highly diverse and widespread ectomycorrhizal genus. FEMS Microbiol Ecol. 2016;92(4):fiw045.
47. Nilsson RH, Bok G, Ryberg M, Kristiansson E, Hallenberg N. A software pipeline for processing and identification of fungal ITS sequences. Source Code Biol Med. 2009;15(4):1.
48. Edgar RC. MUSCLE: multiple sequence alignment with high accuracy and high throughput. Nucleic Acids Res. 2004;32(5):1792–7.
49. Russo CAM, Miyaki CY, Pereira SL. Reconstrução filogenética: Métodos geométricos. In: Matioli SR, editor. Biologia Molecular e Evolução. Ribeirão Preto: Holos Editora; 2012. p. 108–16. 202p.
50. Hollingsworth ML, et al. Selecting barcoding loci for plants: evaluation of seven candidate loci with species-level sampling in three divergent groups of land plants. Mol Ecol Resour. 2009;9(2):439–57.
51. Begerow D, Nilsson H, Unterseher M, Maier W. Current state and perspectives of fungal DNA barcoding and rapid identification procedures. Appl Microbiol Biotechnol. 2010;87(1):99–108.
52. Bellemain E, Carlsen T, Brochmann C, Coissac E, Taberlet P, Kauserud H. ITS as an environmental DNA barcode for fungi: an in silico approach reveals potential PCR biases. BMC Microbiol. 2010;10:189.
53. Spouge JL, Marino-Ramirez L. The practical evaluation of DNA barcode efficacy. Methods Mol Biol. 2012;858:365–77.
54. Seberg O. The future of systematics: Assembling the Tree of Life. Systematist. 2004;23:2–8.
55. Meier R, Shiyang K, Vaidya G, Ng PK. DNA barcoding and taxonomy in Diptera: a tale of high intraspecific variability and low identification success. Syst Biol. 2006;55(5):715–28.
56. Hibbett DS, Ohman A, Glotzer D, Nuhn M, Kirk P, Nilsson RH. Progress in molecular and morphological taxon discovery in Fungi and options for formal classification of environmental sequences. Fungal Biol Rev. 2011; 25(1):38–47.
57. Blaxter M, Mann J, Chapman T, Thomas F, Whitton C, Floyd R, Abebe E. Defining operational taxonomic units using DNA barcode data. Philos Trans R Soc Lond B Biol Sci. 2005;360(1462):1935–43.
58. DeSalle R, Egan MG, Siddall M. The unholy trinity: taxonomy, species delimitation and DNA barcoding. Philos Trans R Soc Lond B Biol Sci. 2005; 360(1462):1905–16.
59. Will KW, Rubinoff D. Myth of the molecule: DNA barcodes for species cannot replace morphology for identification and classification. Cladistics. 2004;20:47–55.
60. Erickson DL, Spouge JL, Resch A, et al. DNA barcoding in land plants: developing standards to quantify and maximize success. Taxon. 2008;3: 1304–16.
61. Meyer CP, Paulay G. DNA barcoding: Error rates based on comprehensive sampling. PLoS Biol. 2005;3:2229–38.
62. Bresinsky A. Beiträge zu einer mykoflora deutschlands (2): Die gattungen hydropus bis hypsizygus mit angaben zur ökologie und verbreitung der arten. Regensburger mykologische schriften band 15. Regensburgische Botanische Gesellschaft: Regensburg; 2008.
63. Babos M, Halász K, Zagyva T, Zöld-Balogh Á, Szegő D, Bratek Z. Preliminary notes on dual relevance of ITS sequences and pigments in *Hygrocybe* taxonomy. Persoonia. 2011;26:99–107. doi:10.3767/003158511X578349.
64. Boertmann D. Fungi of Northern Europe, Volume 1: The Genus Hygrocybe. Oberreifenberg: Koeltz Botanical Books; 2010.
65. Ramírez-López I, Villegas-Ríos M, Salas-Lizana R, Garibay-Orijel R, Alvarez-Manjarrez J. *Thelephora versatilis* and *Thelephora pseudoversatilis*: two new cryptic species with polymorphic basidiomes inhabiting tropical deciduous and sub-perennial forests of the Mexican Pacific coast. Mycologia. 2015; 107(2):346–58.
66. Su M-S, et al. *Hypholoma cinnabarinum* revisited: A contribution to knowledge of *Agaricus* subgenus *Lanagaricus* section *Trisulphurati* (Agaricaceae, Agaricales). Mycoscience. 2014;55(6):449–55.
67. Smith AH. 1957. A contribution toward a monograph of *Phaeocollybia*. Brittonia. 1957. 9:195–217
68. Singer R. The Agaricales in modern taxonomy. 4th ed. Koenigstein: Koeltz Scientific Books; 1986.
69. Norvell LL. The biology and taxonomy of Pacific Northwest species of *Phaeocollybia* Heim (Agaricales, Cortinariaceae) [PhD dissertation]. Seattle, Washinton: University of Washington; 1998. 391 pp.
70. Norvell L. *Phaeocollybia* in western North America. I. The Phaeocollybia kauffmanii complex. Can J Bot. 2000;78:1055–76.
71. Anderson IC, Chambers SM, Cairney JWG. Molecular determination of genetic variation in Pisolithus isolates from a defined region in New South Wales, Australia. New Phytol. 1998;138(1):151–62.
72. Martin F, Delaruelle C, Ivory M. Genetic variability in intergenic spacers of ribosomal DNA in *Pisolithus* isolates associated with pine, eucalyptus and *Afzelia* in Lowland Kenyan forests. New Phytol. 1998;139:341–52.
73. Díez J, Anta B, Manjón JL, Honrubia M. Genetic variability of *Pisolithus* isolates associated with native hosts and exotic eucalyptus in the western Mediterranean region. New Phytol. 2001;149:577–87.
74. Kope HH, Fortin JA. Germination and comparative morphology of basidiospores of *Pisolithus arhizus*. Mycologia. 1990;82:350–7.
75. Hebert PD, Penton EH, Burns JM, Janzen DH, Hallwachs W. Ten species in one: DNA barcoding reveals cryptic species in the neotropical skipper butterfly *Astraptes* fulgerator. Proc Natl Acad Sci U S A. 2004;101(41):14812–7.
76. Grundt HH, Kjolner S, Borgen L, Rieseberg LH, Brochmann C. High biological species diversity in the arctic flora. Proc Natl Acad Sci U S A. 2006;03(4):972–5.
77. Taylor JW, Jacobson DJ, Kroken S, Kasuga T, Geiser DM, Hibbett DS, Fisher MC. Phylogenetic species recognition and species concepts in fungi. Fungal Genet Biol. 2000;31(1):21–32.
78. Stielow JB, Levesque CA, Seifert KA, Meyer W, Iriny L, Smits D, Renfurm R, Verkley GJ, Groenewald M, Chaduli D, et al. One fungus, which genes? Development and assessment of universal primers for potential secondary fungal DNA barcodes. Persoonia. 2015;35:242–63.

19

# Costs and benefits of natural transformation in *Acinetobacter baylyi*

Nils Hülter[1,2†], Vidar Sørum[2†], Kristina Borch-Pedersen[3], Mikkel M. Liljegren[4], Ane L. G. Utnes[2], Raul Primicerio[2], Klaus Harms[2,5*] and Pål J. Johnsen[2*]

**Abstract**

**Background:** Natural transformation enables acquisition of adaptive traits and drives genome evolution in prokaryotes. Yet, the selective forces responsible for the evolution and maintenance of natural transformation remain elusive since taken-up DNA has also been hypothesized to provide benefits such as nutrients or templates for DNA repair to individual cells.

**Results:** We investigated the immediate effects of DNA uptake and recombination on the naturally competent bacterium *Acinetobacter baylyi* in both benign and genotoxic conditions. In head-to-head competition experiments between DNA uptake-proficient and -deficient strains, we observed a fitness benefit of DNA uptake independent of UV stress. This benefit was found with both homologous and heterologous DNA and was independent of recombination. Recombination with taken-up DNA reduced survival of transformed cells with increasing levels of UV-stress through interference with nucleotide excision repair, suggesting that DNA strand breaks occur during recombination attempts with taken-up DNA. Consistent with this, we show that absence of RecBCD and RecFOR recombinational DNA repair pathways strongly decrease natural transformation.

**Conclusions:** Our data show a physiological benefit of DNA uptake unrelated to recombination. In contrast, recombination during transformation is a strand break inducing process that represents a previously unrecognized cost of natural transformation.

**Keywords:** Competence, Natural transformation, Horizontal gene transfer, Bacterial evolution, DNA repair, DprA

## Background

Naturally transformable bacteria take up DNA from the environment upon entering a physiological state called competence [1, 2]. If sufficient homology is present in the taken-up DNA it may be integrated in the recipient genome and in turn combine beneficial alleles in the same genetic background, provide novel traits, and remove deleterious mutations [2–6]. Conversely, natural transformation may also cause harmful changes such as acquisition of deleterious mutations present in external DNA [7]. It is clear that natural transformation shapes bacterial genomes, and the mechanistic insights to DNA uptake and integration is well described for a number of bacterial species, reviewed in [2]. Yet, the selective forces responsible for the evolution and maintenance of natural transformation are still debated [6, 8, 9].

Natural transformation is considered a driving force of bacterial evolution [2, 10, 11], and mathematical models support that natural competence has evolved and is maintained as a mechanism for genome plasticity that lead to increased environmental adaption rates [12, 13]. In contrast, natural transformation has also been hypothesised to support genome stability [14–16] and to remove deleterious mobile genetic elements from chromosomes [17]. Experimental evolution approaches using various bacterial species competent for natural transformation including *Helicobacter pylori*, *Acinetobacter baylyi* and *Streptococcus pneumoniae* have revealed that the impact of natural transformation on rates of adaptive evolution can be beneficial [18, 19], neutral [20], and/or context dependent [21, 22]. Taken together, these reports suggest that the widely accepted idea that natural transformation accelerates bacterial adaptation lacks generality.

* Correspondence: klaus.harms@spdir.net; paal.johnsen@uit.no
†Equal contributors
[2]Department of Pharmacy, Faculty of Health Sciences, UiT The Arctic University of Norway, P.O. Box 6050 Langnes, Tromsø, Norway
Full list of author information is available at the end of the article

Although natural transformation contributes to genome evolution [2], the immediate effects of DNA uptake and recombination in individual cells remain unclear and several non-mutually exclusive hypotheses have been proposed. One hypothesis suggests that taken-up DNA may be used as a source of phosphate, nitrogen, carbon, or nucleotides [8, 23, 24]. Observations that expression of competence genes is controlled by carbon catabolite repression in *Escherichia coli* [25], *Haemophilus influenzae* [25, 26], *Vibrio cholerae* [27] and *Streptococcus gordonii* [28] support this hypothesis. More specific links between intracellular availability of nucleotides, their precursors and natural competence have been found in *H. influenzae* where purine depletion activates the competence activator *sxy* [23], and in *Vibrio cholerae* competence is repressed by exogenous levels of cytidine [29].

An alternative hypothesis suggests that taken-up DNA can be used as template for repair of genomic DNA damages [9, 30]. Four reports by Michod and co-workers provided experimental evidence in favour of this explanation in *Bacillus subtilis* [9, 31–33]. Increased transformation rates were observed when homologous DNA, but not heterologous DNA, was added after exposure of cells to UV-light [33]. However, other reports in *B. subtilis* and *H. influenzae* [34, 35], as well as in *S. pneumoniae* [36, 37] rather suggest that competence induction may be a general stress response. This is further supported by observations that competence-specific transcriptional regulators in *S. peumoniae* and *B. subtilis* control genes whose functions are linked to stress response (reviewed in [2]). The induction of competence by DNA damaging agents in *S. pneumoniae* [38], *Legionella pneumophila* [39] and *H. pylori* [40] also favour the "DNA-repair hypothesis" for the evolution and maintenance of natural transformation.

In the genus *Acinetobacter* competence is not regulated by DNA damages [41, 42] and in *A. baylyi* maximum competence is reached during exponential growth-phase [43]. Under optimal conditions up to 25% of all cells in a competent *A. baylyi* culture are transformed [44]. During transformation competent cells bind extracellular double-stranded DNA [4, 45] and a single-stranded DNA fragment is transported into the cytoplasm [1, 4]. While some bacterial species e.g., *H. influenzae* and species in the family of *Neisseriaceae* selectively take up isogenic DNA [46, 47], other species such as *B. subtilis* and *A. baylyi* transfer DNA of any source into the cytoplasm [44, 48]. Homologous recombination is then initiated by DprA-mediated loading of RecA onto the single-stranded DNA [49, 50] and by subsequent RecA-mediated strand invasion and integration of the donor DNA into the recipient genome. Previous reports on *A. baylyi* indicate that genomic integration frequently cause single-strand breaks in the transforming cell that can turn into double-strand breaks when unrepaired [51, 52].

In this study, we used *A. baylyi* to determine the immediate benefits and costs of DNA-uptake and -integration under both benign and stressful conditions. Our data show that uptake of both homologous and heterologous DNA increased fitness of competent *A. baylyi* cells. The fitness increment was independent of DNA damage and did not involve genomic integration of the acquired DNA. We further show that genomic integration of homologous DNA both reduced transformant fitness when exposed to UV-induced stress, and under benign conditions strongly depended on the RecBCD and RecFOR DNA strand break repair pathways.

## Methods

### Strains, media and DNA

The strains used in this study (Additional file 1: Table S1) are derivatives of *Acinetobacter baylyi* strain JV28 [53] and were constructed using standard molecular biology techniques [54]. Plasmids and primers used in this study are listed in supplementary tables (Additional file 1: Table S2 and Table S3). JV28 is derived from *A. baylyi* ADP1 (GenBank NC_005966) [55]. Strain LCQ2 was constructed by crossing the *trpE*$^+$ wildtype allele into JV28 as described [56]. KOM130 was constructed by replacing the *comB-F* operon of JV28 with the Δ*comB-F::dhfr1* allele [20, 22] by natural transformation with DNA isolated from strain ADP1200Com- [22]. Cotransformation of unwanted markers was excluded phenotypically (TrpE$^-$) or by PCR (*lifO-lipB::aphA3'*). The *dprA* gene (ACIAD0209) of *A. baylyi* was inactivated by replacing an internal 349-bp fragment of *dprA* with an *aacC1* (gentamicin resistance) gene using Splicing by Overlap Extension PCR [57]; see Additional file 2 for details]. The Δ*uvrA* (ACIAD3455), Δ*recF* (ACIAD0003) and Δ*recR* (ACIAD2312) mutations were constructed using the plasmid pGT41 as described [58, 59], see Additional file 2. The Δ*uvrA* strain KOM141 was obtained by replacing *uvrA*$^+$ of JV28 by the Δ*uvrA*::(*nptII sacB*) allele, and subsequent replacement of Δ*uvrA*::(*nptII sacB*) by the Δ*uvrA* allele. The *recO* (ACIAD2578) deletion allele was constructed similarly using a modified pGT41 plasmid (pGT33), and the plasmids containing the Δ*recO*::(*nptII sacB*) and Δ*recO* alleles were provided by J. de Vries (University of Oldenburg, Germany). Similar to Δ*uvrA*, the Δ*recF*, Δ*recR* and Δ*recO* alleles were each crossed into the JV28 strain, giving strains MKD1, MKD2 and KOM82, respectively. The Δ*recR* and Δ*recF* alleles were subsequently inserted into the KOM82 mutant, resulting in strain MKD3 (Δ*recFOR*). Finally, the Δ*recBCD* allele [52] was crossed into each of the MKD3, MKD1, KOM82 and MKD2 mutants, giving the strains MKD6, MKD4, KOM86 and MKD5, respectively. All strains were verified phenotypically (sensitivity to UV irradiation; transformation by 2 kbp donor DNA molecules) and by PCR. The donor DNA plasmid pSBP1 was constructed by inserting a PCR product of the *lifO-lipB::aphA3* insertion allele [60] (amplified with ACIAD3308-f/ACIAD3309-down) of ADP1200Com + $^{KanR}$ [22] into the *Hin*cII site of the pUC19 plasmid vector conferring kanamycin resistance.

Bacterial cultures were incubated at 30 °C in Luria Bertani broth (LB) (BD Difco™, USA). Transformants were scored on LB agar plates supplemented with kanamycin (Sigma-Aldrich, Germany) at 50 mg $l^{-1}$ or 10 mg $l^{-1}$ (plating efficiencies were indistinguishable), and trimethoprim at 250 mg $l^{-1}$.

Genomic and plasmid DNA was isolated using QIAGEN Genomic-tip (QIAGEN, Germany) and the QIAGEN genomic DNA or plasmid DNA purification kit, respectively, according to the manufacturer's instructions.

### Pairwise competitions

Overnight cultures of a transformation-proficient test strain (LCQ2 or NH29) and the DNA uptake-deficient competitor KOM130 were grown in ten ml for 16 h at 30 °C. The cells were pelleted, re-suspended and diluted 100-fold (approximately $10^7$ cells $ml^{-1}$) in PBS [54] pre-warmed to 30 °C. One ml aliquots of each suspension were mixed, transferred to a sterile Petri dish and irradiated with ultraviolet (UV) light using a germicidal lamp. One ml of the irradiated cells was transferred to tubes containing one ml of pre-warmed double-strength LB (2 × LB) with genomic DNA [isogenic (ADP1200Com + $^{KanR}$) or heterologous (salmon sperm); 2 μg $ml^{-1}$] or with DNase 1 (100 μg $ml^{-1}$) and aerated in the dark for 24 h at 30 °C. The cells were plated in appropriate dilution on LB medium with and without trimethoprim (250 mg $l^{-1}$; in some control experiments: 10 mg $l^{-1}$ kanamycin) after two and 24 h. The plates were incubated at 30 °C for 24 h, colonies were counted, and the titers at both time points for the test strain ($T_{t-2}$ and $T_{t-24}$) and the competitor strain ($C_{t-2}$ and $C_{t-24}$) were calculated. The relative fitness $w$ of the test strain was calculated as $w = \ln(T_{t-24}/T_{t-2})/\ln(C_{t-24}/C_{t-2})$.

### UV survival measurements

Overnight cultures of *A. baylyi* were treated as described for the pairwise competitions but without mixing with a competitor and irradiated using a germicidal lamp. Control assays were treated accordingly without irradiation ($UV_0$). One ml of the cells was then mixed with one ml pre warmed 2 × LB, amended with pSBP1 plasmid DNA (0.1 μg $ml^{-1}$; 1800 molecules per cell) linearized with *XmnI* (New England Biolabs) or with homologous genomic DNA from strain ADP1200Com + $^{KanR}$ (2 μg $ml^{-1}$; 50 genome equivalents per cell), and aerated without light for two hours before DNase 1 (100 μg $ml^{-1}$) was added. pSBP1 is pUC19 with a two kbp insert from the *A. baylyi* chromosome with a $Kan^R$ marker (*aphA3*) cloned in the centre of the insert (Additional file 1: Table S1). In some control experiments without DNA, DNase 1 was added at the start. When the Δ*uvrA* strain was used, ten assays (one ml per assay) were irradiated separately and pooled after irradiation of the cells. Cells were plated in appropriate dilution on LB with (survival transformant titer) and without kanamycin (10 mg $l^{-1}$; survival recipient titer) and incubated for 24 h at 30 °C. Colonies were counted, the titers were determined, and the recipient and transformant survival frequencies were calculated as frequencies relative to those at $UV_0$.

### Natural transformation experiments

The preparation of competent cells and the transformation assays were performed as described previously [58]. Briefly, the *A. baylyi* strains were grown in a shaker in LB broth at 30 °C to $1 \times 10^9$ cells $ml^{-1}$ (determined with a Neubauer hemocytometer) and stored at −80 °C as a concentrated stock ($1 \times 10^{10}$ cells $ml^{-1}$ in LB with 10% glycerol) until use. For transformation assays, a freshly thawed bacterial suspension was diluted to $2.5 \times 10^8$ cells $ml^{-1}$ in LB containing 100 ng $ml^{-1}$ donor DNA (isolated from *A. baylyi* KH12), aerated for two hours at 30 °C and plated in appropriate dilution on LB medium (recipient titer) and LB with 10 mg $ml^{-1}$ kanamycin (transformants). The colonies were counted after 16 to 40 h at 30 °C. Transformation frequencies were calculated as transformants per recipient. No kanamycin-resistant colonies were obtained in control experiments without donor DNA.

### Statistical analysis

The relative fitness data were analysed using Two-ways ANOVA, with relative fitness as response variable and competitor strain (two levels), UV-exposure to cells (two levels) and DNA-availability (three levels) as explanatory variables. We also subjected the data to post hoc tests (with Tukey HSD correction) using the same model specifications. Confidence limits (95%) for the group means were obtained using formula based estimation.

The survival data were analysed by regression using a linear Mixed-model controlling for dependencies among the replicates from the same overnight culture used in a series of increasing UV-exposure. In our model, we used relative survival as response and UV-fluence (continuous) and total population or transformants (two levels) as explanatory variables, whereas the overnight culture was included as a random term. The observed trends were linearized by log transforming relative survival and the linearity of the trend was diagnosed, and confirmed, by generalized additive modelling [61]. The transformation frequencies of the set of recombinational repair deficient (Δ*recBCDFOR*) mutants were analysed by one-way ANOVA followed by post hoc tests with TukeyHSD correction. All analyses were performed with the statistical software R [62] and the additional R-packages lme and mgcv.

## Results

### Uptake of exogenous DNA is beneficial regardless of sequence homology, DNA damage and recombination

We investigated the general fitness effects of DNA uptake in mixed culture competition experiments with the transformation-proficient *A. baylyi* wildtype strain (LCQ2) and its non-transformable Δ*comB-F* derivative (Additional file 1: Table S1). The *comB-F* operon encodes genes required for type IV pilus formation, which is essential for transport of free DNA into the periplasm [20, 22, 56] and for spreading growth on semi-solid surfaces (Additional file 1: Fig. S1). To test for differences between benign and DNA damaging conditions, the mixed cultures were treated with or without initial pulses of UV irradiation, and were amended either with homologous DNA, heterologous DNA, or DNase. The UV-dose applied was 216 J $m^{-2}$, which generates one photoproduct every 420 bp on average [63] and killed approximately 98% of the cells. In the absence of DNA, the wildtype strain displayed a relative fitness ($w$) reduction compared with the Δ*comB-F mutant* (Fig. 1a) [$w = 0.88$ (0.83–0.92, 95% CI)]. A mixed culture serial transfer experiment confirmed that the fitness reduction was due to the absence of type IV pilus components in the Δ*comB-F* competitor (Additional file 1: Fig. S2).

The addition of homologous or heterologous DNA increased the fitness of the wildtype compared to the DNA-free experiments (both $P < 0.002$) (Fig. 1a). The fitness effects of added DNA to the competitions were independent of UV-exposure ($P = 0.83$). Together these two results provide a strong argument in favour of a general benefit of DNA uncoupled from recombination.

To further investigate whether the observed fitness increase was independent of recombination, we used a DNA uptake-proficient but transformation-deficient mutant lacking the *dprA* gene [Additional file 1: Table S1] in head-to-head competitions with the Δ*comB-F* mutant. Absence of DprA abolished homologous recombination during natural transformation in *A. baylyi* as reported for *S. pneumoniae* [49] (detection limit $1 \times 10^{-9}$ transformants/recipient). The Δ*dprA* mutant displayed the same cost of type IV pilus formation as the wildtype [$w = 0.86$ (0.82–0.91, 95% CI); Fig. S3]. As seen with the wildtype, addition of both heterologous and homologous DNA increased the fitness of the Δ*dprA* strain relative to the DNA uptake-deficient competitor Δ*comB-F* (both $P < 0.001$) regardless of UV-exposure (Fig. 1b). These data show that under the experimental conditions the DNA uptake machinery is costly to the cell. This cost can be partially compensated by addition of DNA to the competing cultures. However, the benefit of added DNA is unlinked to DNA recombination and thus DNA damage repair. This observation is further supported by the fact that *A. baylyi* takes up DNA from any source [44, 56], suggesting the benefit of competence is independent of DNA sequence homology.

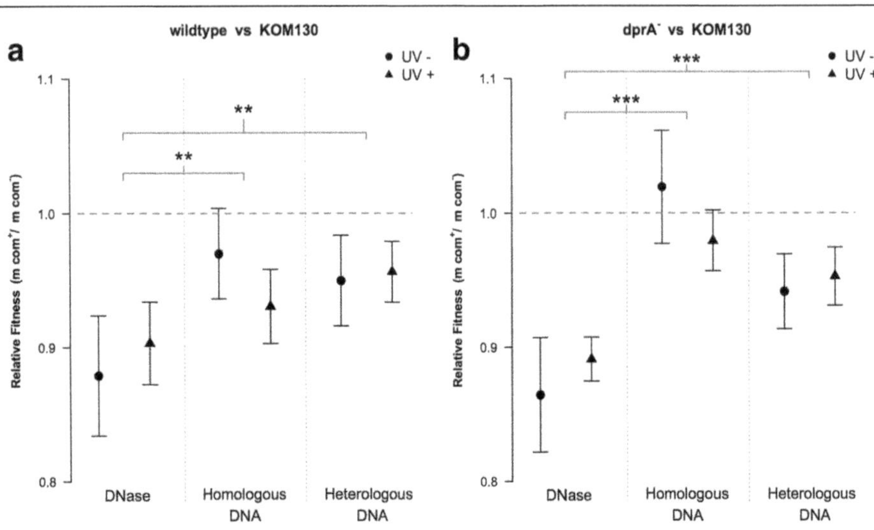


**Fig. 1** Relative fitness of wildtype strain (**a**) and its Δ*dprA* derivative (**b**) grown in head-to-head competitions against the DNA uptake-deficient Δ*comB-F* mutant. The competitions were treated with a combination of an initial UV pulse of 216 J $m^{-2}$ or no UV irradiation with a subsequent amendment of no DNA (DNase), homologous DNA or heterologous DNA. In each group $n = 13$. In the control competitions with no UV and no DNA both the wildtype and its Δ*dprA* derivative displayed a reduced fitness relative to KOM130 of $w = 0.88$ (0.83–0.92, 95% CI) and $w = 0.86$ (0.82–0.91, 95% CI), respectively, due to the biological cost of type IV pilus expression (Additional file 1: Fig. S1). Error bars indicate 95% confidence intervals. Significant differences between the treatment groups (pooled results of both UV- and UV+ competitions) are indicated by asterisks (** $P < 0.001$, *** $P < 0.0001$). A two-way ANOVA of all mean values is provided in (Additional file 1: Table S4)

### Homologous DNA reduces relative survival of transformants after UV-exposure

The competition experiments described above demonstrated a clear fitness increase for DNA-uptake-proficient cells regardless of the cells' ability to use that DNA for recombination under benign or genotoxic conditions. However, the fitness benefit observed might obscure an additional benefit for the transformant fraction of cells in the culture, especially under DNA damaging conditions.

To test that, we investigated the effect of increasing intracellular UV damages in DNA on survival in the transformant fraction compared with the total population in monoculture experiments. We added a short donor DNA substrate (linearized pSBP1; [Additional file 1: Table S1]) containing a chromosomal *A. baylyi* segment (2 kbp) with a kanamycin resistance marker inserted in the center (*lifO-lipB::aphA3*) [22]. Single-locus homologous recombination is largely independent of the chromosomal position [64]. pSBP1 DNA can only be used for recombination at the *lifO-lipB* locus, and recombination results in kanamycin resistance. The donor DNA substrate was added in excess (0.1 μg $ml^{-1}$), and the survival of transformants following UV irradiation from 0 to 216 J $m^{-2}$ was quantified. The results showed an initial benefit of added pSBP1 DNA at low UV doses on survival of the transformant fraction (Fig. 2), similar to what was observed in *H. influenzae* [34]. However, at higher UV-doses (≥144 J $m^{-2}$) that increased the likelihood of UV damages generated in the vicinity of the marker insertion site, this benefit was lost (Fig. 2). The result suggests that recombination with taken-up DNA interfered with repair of UV damages at the recombination locus, which is further supported by a steeper decline in survival of the transformant fraction with increasing UV-doses (Fig. 2 and [Additional file 1: Fig. S4 A]).

We hypothesized that fully homologous DNA as donor DNA would be even more disadvantageous because it can recombine and interfere with UV-damages genome-wide. We tested this by adding genomic *A. baylyi* donor DNA containing the *lifO-lipB::aphA3* marker (2 μg $ml^{-1}$). The donor DNA substrate can undergo phenotypically silent recombinations at any chromosomal locus, and we tracked the integration of the *lifO-lipB::aphA3* marker as a proxy for the genome wide transformation frequency. Transformants survived less than the total population from low to high UV-doses (72–216 J $m^{-2}$) ($P < 0.001$, Fig. 2 and [Additional file 1: Fig. S4 B]). The UV-survival curves of the total populations with pSBP1 or genomic DNA were indistinguishable and these data were pooled (Fig. 2). Control experiments without DNA revealed no difference in kill rates of the total population (Additional file 1: Fig. S4 D).

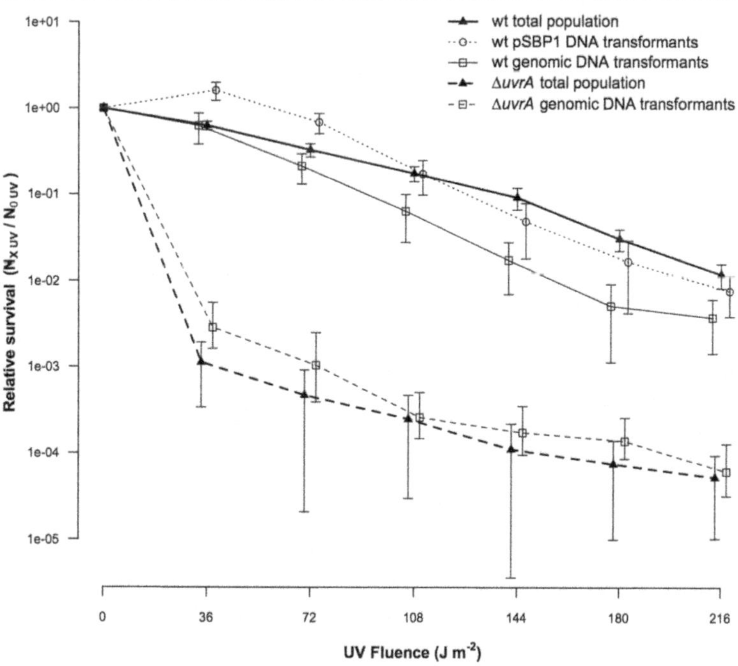


**Fig. 2** Relative survival of the pooled total population of the wildtype (closed triangles and bold solid line, $n = 16$) and the transformant fractions obtained with either pSBP1 DNA (0.1 μg $ml^{-1}$; open circles with dotted line, $n = 10$) or with homologous genomic DNA (2 μg $ml^{-1}$; open squares with solid line, $n = 6$) with increasing levels of UV irradiation. We also determined the relative survival of the total population (closed triangles and bold dashed line, $n = 10$) and the transformant fraction in the *ΔuvrA* mutant obtained with homologous DNA (2 μg $ml^{-1}$; open squares and dashed line, $n = 10$). Error bars denote the 95% confidence intervals. The transformation frequencies without UV irradiation were $(7.7 \pm 1.4) \times 10^{-4}$ (pSBP1 DNA) and $(2.5 \pm 2.8) \times 10^{-2}$ (genomic DNA) for the wildtype; and $(1.9 \pm 1.0) \times 10^{-2}$ (genomic DNA) for the *ΔuvrA* mutant. Mean initial titers (CFU/ml) for the wildtype and the *ΔuvrA* mutant were $1.8 \times 10^7$ and $4.2 \times 10^7$, respectively

UV damages in bacteria can be repaired by nucleotide excision repair (NER) [65]. During NER, the UvrA-directed UvrBC excinuclease [59] cleaves one strand on either side of a UV- lesion in double-stranded DNA. Excision of the incised fragment with the lesion by helicase II is finally followed by fill-in synthesis by DNA polymerase I and sealing of the nick by DNA ligase. With increasing UV-doses, more NER repair sites would remain temporarily unfinished, and gaps and nicks would accumulate. In addition, recombination attempts with taken up DNA will add to the number of single strand breaks, as reported in [52]. We hypothesised that the decrease of transformant survival with increasing UV-doses was caused by interference of NER with transformational recombination (Fig. 3).

To test this hypothesis, we knocked out the key NER gene *uvrA* and investigated relative survival of transformants of the Δ*uvrA* strain with increasing UV-doses as outlined above. As expected, UvrA-deficiency resulted in greatly reduced survival when exposed to UV irradiation (Fig. 2). However, the Δ*uvrA* transformant fraction did not display any detrimental effect with increasing UV doses as observed in the wildtype (Fig. 2), and the kill rates of transformants in the absence of UvrA were the same as that of the total population ($P = 0.79$, [Additional file 1: Fig. S4 C]). This result suggests that in absence of NER, genome-wide recombination with homologous chromosomal DNA provided no additional burden, but also no benefit, to uptake-proficient cells. Cumulatively, NER and recombination with taken-up DNA become a burden to competent cells with increasing levels of DNA damages, due to generation of nicks and gaps that turn into DNA double-strand breaks when unrepaired. Consequently, successful integration of recombination attempts during natural transformation in the wildtype would depend on DNA strand-break repair functions.

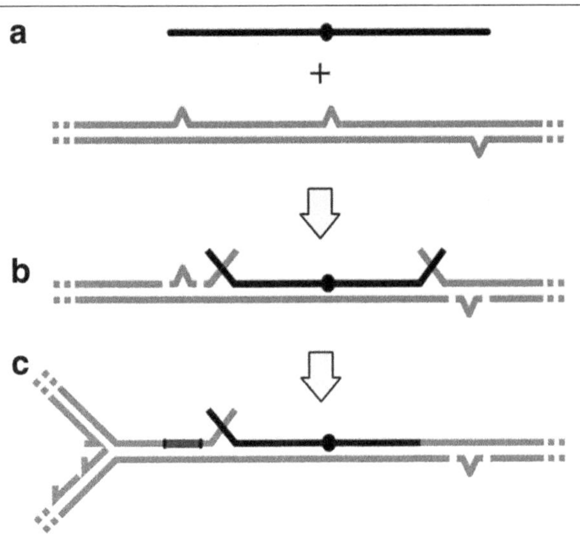


**Fig. 3** Formation of DNA double strand-breaks conferred by nucleotide excision repair during natural transformation. **a**: *Grey lines*: genomic DNA; grey spikes: UV-induced lesions. Black line: taken-up DNA single-strand with genetic marker (*oval*). **b**: Generation of DNA single-strand-breaks by UvrAB-directed, UvrBC-initiated cleavage of the damage-containing strand upstream and downstream of the lesions, and by RecA-mediated strand invasion of the taken-up DNA and cleavage of the displaced DNA strand. **c**: Partial repair of single-strand breaks: 1. by removal of the lesion-containing single-strand fragment (*left lesion*), fill-in synthesis, and covalent ligation (*dark grey*; catalysed by UvrD, DNA polymerase I, and DNA ligase, respectively); and 2. by covalent joining of the invaded taken-up strand with the genomic DNA at one side (downstream of the marker). Single-strand breaks persist when the UV lesion-containing strands are not removed (*right lesion*) and when an invaded DNA strand remains unligated at one end (upstream of the marker). One-sided ligation of the invaded DNA strand is common (see text). Following DNA replication (indicated by a replication fork approaching from the left end), single-strand breaks are converted into potentially lethal double-strand breaks

### Genomic integration of taken-up DNA strongly depends on DNA repair

Two studies on RecA-mediated recombination in *A. baylyi* [51, 52] suggested that both ends of a donor DNA molecule are integrated through two independent ligation events. While one end is covalently joined to the genomic DNA, the opposite end frequently remains temporally unligated, resulting in a single-strand break that can turn into a double-strand break following DNA replication. These reports [51, 52] are in agreement with the observed interaction between NER of UV-damages and recombination on transformant survival which suggest that incomplete recombination attempts during transformation may represent an additional source of DNA damages that rely on DNA strand break repair (Fig. 3). Recombinational DNA strand break repair is initiated in bacteria through two major pathways, the RecBCD and the RecFOR pathways of recombination [66, 67]. DNA single-strand breaks such as nicks or gaps can be repaired through the latter pathway, which requires among others the RecF, RecO and RecR gene products [68]. The RecFOR proteins in concert process gaps and generate RecA-loaded DNA 3′-ends for strand invasion and thus can initiate homologous reciprocal strand transfer and can also restore stalled and broken replication forks [68]. Alternatively, nicks and gaps can turn into double-strand breaks when unprocessed, and these damages are mainly repaired by the RecBCD pathway with the RecBCD heterotrimeric complex as the key repair function [69]. RecBCD processes DNA double-strand ends and generates a recombinogenic RecA-loaded 3′-end that initiates reciprocal strand exchange with a template for repair of the double-strand break [69]. To test the hypothesis that natural transformation depends on DNA strand break repair for successful integration of acquired DNA, we deleted components of the RecFOR pathway and the RecBCD pathway in *A. baylyi* and determined the effects of these deficiencies on natural transformation by homologous chromosomal DNA. The results are shown in Fig. 4. Deletions of *recF*, *recO* and *recR* in combination or alone overall had little or no effect on natural transformation frequencies compared with wildtype, as observed previously in *B. subtilis* and *S. pneumoniae* [70, 71], although in the Δ*recR* mutant it was decreased 3.3-fold ($P = 0.007$). Deletion of the *recBCD* operon, while highly detrimental to viability, did not affect transformation frequency ($P = 0.94$). A recent finding that RecR protein has a role in both RecBCD and RecFOR recombination pathways in mycobacteria [72] supports this observation. In contrast, in a Δ*recBCD* Δ*recFOR* mutant transformation was decreased approximately 10-fold ($P < 0.001$), and this decrease was strongly synergistic (about 5-fold higher than expected from the decreases in frequency of the single mutants, Fig. 4), confirming earlier reports on *H. pylori* [73]. This observation was confirmed using Δ*recF*, Δ*recO* or Δ*recR* mutants of the Δ*recBCD* strain (all *p*-values < 0.001; Fig. 4), with the Δ*recBCD* Δ*recF* double mutant yielding particularly low transformation frequencies (approximately 80-fold decreased compared with wildtype). It is conceivable that in this mutant, the functional RecOR proteins [74] generate recombination intermediates that cannot be resolved in the absence of RecF and RecBCD. An ANOVA table for multiple comparisons of means of the transformation frequencies is provided in (Additional file 1: Table S5). In summary, these results suggest that the majority of natural transformation events depends on DNA strand-break repair and support our hypothesis that taken-up DNA causes damages such as strand-breaks following recombination attempts. It can be hypothesized that amendment of DNA is detrimental to growing cells impaired in recombinational DNA repair. However, comparisons of serial transfer experiments of the Δ*recBCD* Δ*recFOR* mutant with and without DNA suggest that the hypothetical detrimental effect of added DNA on cell survival is masked by the initially observed beneficial effect of DNA (Fig. 1) on cell growth (Additional file 1: Fig. S5).

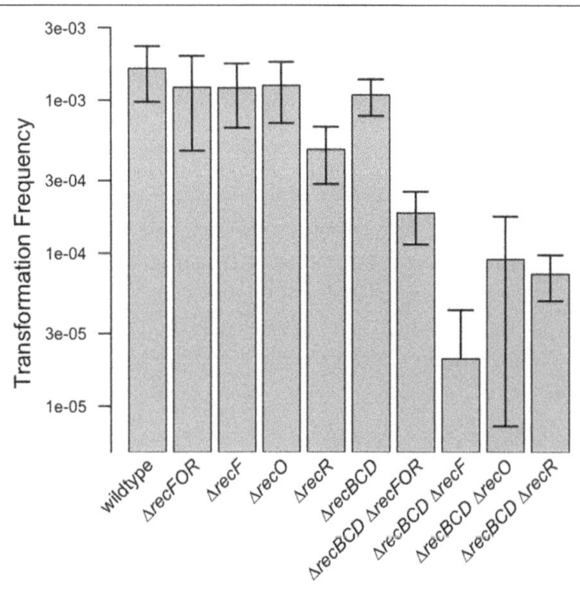


**Fig. 4** Natural transformation frequencies of *A. baylyi* wildtype and DNA recombination-impaired mutants. Transformation frequencies were obtained in liquid transformation experiments with genomic homologous DNA (0.1 μg $ml^{-1}$) containing a kanamycin resistance marker gene and are given as means with 95% confidence intervals. The initial recipient titers are listed in (Additional file 1: Table S4)

## Discussion

In this report we show that the costs of the DNA uptake machinery (*comB-F* gene products) in *A. baylyi* is offset by the amendment of DNA in competition experiments. The beneficial effect of added DNA was independent of UV-exposure to cells, DNA sequence homology, and unlinked to recombination. These results are inconsistent with a role of acquired DNA in strand-break repair as originally proposed by Bernstein and co-workers [30]. The high biological cost of type IV pilus formation (up to 14% reduction in relative fitness) suggests that the ability to take up DNA could rapidly be lost in naturally transformable populations. To this end, reduced transformability have been previously shown in *A. baylyi* during experimental evolution but this is not necessarily dependent on loss of DNA uptake functions since the cost of pilus formation appears to be medium-specific [20, 22]. It is also important to stress that the type IV pilus is involved in motility (Additional file 1: Fig. S1) suggesting that other selective forces may be responsible for the maintenance of central components of the DNA uptake machinery [75].

Early reports favoured the DNA repair hypothesis in *B. subtilis* [9, 31–33], and it has been demonstrated that *B. subtilis* displayed increased transformation frequencies during UV-exposure [33]. However, DNA damage does not induce the competence machinery in *B. subtilis* [35], and more recent reports have made a clear distinction between natural transformation and the physiological state of competence. For example, competence for natural transformation is induced by genotoxic stress in *S. pneumoniae* [38] and it was recently demonstrated that competence development following DNA-damaging stress was beneficial in the presence but not the absence of DNA in *S. pneumoniae* [36]. However, direct evidence demonstrating recombination between incoming donor DNA and damaged areas of the genome was not presented and it is not clear if this benefit was due to other processes, as also pointed out by the authors [36]. Interestingly, competence induction was also beneficial in the presence of non-DNA damaging stress, suggesting a general benefit that extends beyond recombination [36, 37]. Competence induction in

*S. pneumoniae* involves quorum sensing of an extracellular competence-stimulating-peptide (CSP) for a short time during log phase and includes transcriptional regulation by ComX of high number of genes not only limited to DNA uptake and recombination [37]. This is consistent with our finding that taken up DNA increases relative fitness in *A. baylyi*. These results also provide support for a previous report on *H. influenzae*, where an increased survival of the whole population was demonstrated in the presence of both fully and partially homologous DNA [34]. Moreover, we show that the benefit of DNA is also present in an *A. baylyi* strain lacking a functional recombination mediator DprA. Since DprA loads taken-up single-stranded DNA with RecA, and the nucleoprotein filament subsequently initiates recombination [49], the result reported here strongly suggests that the fitness benefit of DNA uptake is not linked to recombination, but rather the uptake of free DNA.

The beneficial effect of DNA uptake does not necessarily exclude a role of the taken-up DNA for repair purposes in the recombining fraction of the competent population. We investigated the effect of exogenous DNA on transformant survival with increasing UV doses. The results revealed that recombination with taken-up DNA is costly. These data provide additional support for the proposed mechanistic model for, and consequences of, integration of DNA into the chromosome following uptake in *A. baylyi* [51, 52]. Kickstein and colleagues provided evidence in favour of separate integration of 5′- and 3′- donor DNA ends during recombination in *A. baylyi*, suggesting frequent occurrence of DNA single-strand breaks. If replication forks reach joints between donor and recipient DNA prior to ligation, a double-strand-break would occur [52]. This suggests that natural transformation in *A. baylyi* strongly depends on DNA strand break repair functions to ensure viability. In our DNA damage experiments we used UV-irradiation to induce intracellular DNA lesions that can lead to single- and double-strand breaks during DNA repair [76]. Consequently, the DNA repair machinery will be saturated due to multiple strand-breaks formed as a response to DNA damage. It is conceivable that the saturation affects initiation of recombinational repair (conferred by RecBCD or by RecFOR), but it is also possible that the repair processes exhaust the pool of cellular RecA protein which is required for initial strand invasion and for repair of strand-breaks due to incomplete insertions of the donor DNA. It can be hypothesized that the repair functions titrate RecA protein away from DprA and thus reduce transformation due to RecA shortage. However, it has been shown that DNA uptake in *A. baylyi* is initiated immediately after exposure to DNA [44], and in our UV experiments the cells were still in lag phase (before initiation of replication) when DNA was added. The integration of acquired DNA into the genome then is likely to cause additional stress in the recipient cells (Fig. 3).

Three lines of evidence presented in this report support this hypothesis. First, we demonstrate in UV-time-kill experiments that the transformant fraction is more susceptible to UV-killing than the total population. Second, UV-exposure reduced transformant survival due to interference between NER and natural transformation (Fig. 2). Third, under benign experimental conditions we demonstrate that the majority of natural transformation events depend on DNA repair functions as demonstrated in *recBCDFOR* mutants (Fig. 4). Taken together, the data presented here and previously [51, 52, 73] strongly suggest that recombination during natural transformation is costly and represents an additional hurdle that the selective forces responsible for the maintenance of natural transformation must overcome. Consequently, our findings are inconsistent with the DNA repair hypothesis for the evolution and maintenance of natural transformation in *A. baylyi* [9, 30].

If DNA uptake is beneficial and unlinked to recombination (which is detrimental), what causes this benefit? Taken-up DNA is rapidly degraded by single-strand-specific DNA exonucleases (RecJ and ExoX) in *A. baylyi* [56]. It is thus possible that the recipient cells can directly access this pool of nucleotides and incorporate them into the DNA biosynthesis pathways as suggested in previous reports [8, 77]. Our results would add to the recent finding that purine depletion induced the *H. influenzae* competence activator *sxy* [23], suggesting that DNA can be utilized as a nutrient or as building blocks in DNA metabolism.

## Conclusions

Our data suggest that the selective pressures responsible for the maintenance of natural transformation are not linked to recombination following DNA uptake. Our findings rather suggest that recombination is a DNA damaging process and that the positive effect of DNA uptake on *A. baylyi* fitness is due to free cytoplasmic DNA where it may be utilized as nutrition or as building blocks for DNA metabolism. In contrast, our data suggest that recombination during transformation is a strand break inducing process that represents a previously unrecognized cost of natural transformation. Our results add new insights to the costs and benefits that play a selective role in the maintenance of natural transformation.

**Abbreviations**

bp: Base pairs; CI: Confidence interval; HSD: Honest significant difference; LB: Luria-Bertani; NER: Nucleotide excision repair; PBS: Phosphate-buffered saline; UV: Ultraviolet

**Acknowledgements**

We are grateful to J. de Vries for providing the pBlue-recO11 and pBlue-recO10 plasmids We thank two anonymous reviewers for constructive criticism and suggestions that improved the manuscript.

**Funding**

This work was supported by the Tromsø Research Foundation (grant number A5870 to P.J.J.), UIT The Arctic University of Norway and the Research Council of Norway (grant number 204263/F20 to P.J.J.). N.H. was partially supported by the European Research Council (grant number 281357). K.H. was supported by the Danish National Research Foundation.
The funding bodies had no role in the writing process, study design and collection, analysis and interpretation of data.

**Authors' contributions**

PJJ and KH conceived the study. NH, VS, KB-P, MML and ALGU conducted the experiments. NH, KH and PJJ supervised the project. NH, VS, KH and PJJ designed the experiments and analysed the data. VS and RP conducted the statistical analyses. VS and NH created the Figures. NH, VS, KH, ALGU and PJJ wrote the manuscript. All authors read and approved the final manuscript.

**Competing interests**

The authors declare that they have no competing interests.

**Author details**

[1]Genomic Microbiology, Institute of Microbiology, Christian-Albrechts-University Kiel, Am Botanischen Garten 11, 24118 Kiel, Germany. [2]Department of Pharmacy, Faculty of Health Sciences, UiT The Arctic University of Norway, P.O. Box 6050 Langnes, Tromsø, Norway. [3]Department of Food Safety and Infection Biology, Faculty of Veterinary Medicine and Biosciences, Norwegian University of Life Sciences, P.O. Box 8146 Dep 0033 Oslo, Norway. [4]Centre for Ecolgical and Evolutionary Synthesis, Faculty of Mathematics and Natural Sciences, University of Oslo, P.O. Box 1066 Blindern0316 Oslo, Norway. [5]Centre for GeoGenetics, Natural History Museum of Denmark, University of Copenhagen, Øster Voldgade 5-7, 1350 Copenhagen K, Denmark.

**References**

1. Chen I, Christie PJ, Dubnau D. The ins and outs of DNA transfer in bacteria. Science. 2005;310(5753):1456–60.
2. Johnston C, Martin B, Fichant G, Polard P, Claverys JP. Bacterial transformation: distribution, shared mechanisms and divergent control. Nat Rev Microbiol. 2014;12(3):181–96.
3. Johnsborg O, Eldholm V, Havarstein LS. Natural genetic transformation: prevalence, mechanisms and function. Res Microbiol. 2007;158(10):767–78.
4. Lorenz MG, Wackernagel W. Bacterial gene transfer by natural genetic transformation in the environment. Microbiol Rev. 1994;58(3):563–602.
5. Thomas CM, Nielsen KM. Mechanisms of, and barriers to, horizontal gene transfer between bacteria. Nat Rev Microbiol. 2005;3(9):711–21.
6. Vos M. Why do bacteria engage in homologous recombination? Trends Microbiol. 2009;17(6):226–32.
7. Redfield RJ. Evolution of bacterial transformation - is sex with dead cells ever better than no sex at all. Genetics. 1988;119(1):213–21.
8. Redfield RJ. Do bacteria have sex? Nat Rev Genet. 2001;2(8):634–9.
9. Michod RE, Wojciechowski MF, Hoelzer MA. DNA-repair and the evolution of transformation in the Bacterium Bacillus-Subtilis. Genetics. 1988;118(1):31–9.
10. Borgeaud S, Metzger LC, Scrignari T, Blokesch M. The type VI secretion system of Vibrio cholerae fosters horizontal gene transfer. Science. 2015; 347(6217):63–7.
11. Croucher NJ, Finkelstein JA, Pelton SI, Mitchell PK, Lee GM, Parkhill J, Bentley SD, Hanage WP, Lipsitch M. Population genomics of post-vaccine changes in pneumococcal epidemiology. Nat Genet. 2013;45(6):656–63.
12. Levin BR, Cornejo OE. The population and evolutionary dynamics of homologous gene recombination in bacterial populations. PLoS Genet. 2009;5(8):e1000601.
13. Moradigaravand D, Engelstadter J. The evolution of natural competence: disentangling costs and benefits of sex in bacteria. Am Nat. 2013;182(4): E112–26.
14. Ambur OH, Engelstadter J, Johnsen PJ, Miller EL, Rozen DE. Steady at the wheel: conservative sex and the benefits of bacterial transformation. Phil Trans R Soc B. 2016;371:20150528.
15. Mell JC, Redfield RJ. Natural competence and the evolution of DNA uptake specificity. J Bacteriol. 2014;196(8):1471–83.
16. Treangen TJ, Ambur OH, Tonjum T, Rocha EP. The impact of the neisserial DNA uptake sequences on genome evolution and stability. Genome Biol. 2008;9(3):R60.
17. Croucher NJ, Mostowy R, Wymant C, Turner P, Bentley SD, Fraser C. Horizontal DNA transfer mechanisms of bacteria as weapons of intragenomic conflict. Plos Biol. 2016;14(3):e1002394.
18. Baltrus DA, Guillemin K, Phillips PC. Natural transformation increases the rate of adaptation in the human pathogen helicobacter pylori. Evolution. 2008; 62(1):39–49.
19. Perron GG, Lee AE, Wang Y, Huang WE, Barraclough TG. Bacterial recombination promotes the evolution of multi-drug-resistance in functionally diverse populations. Proc Biol Sci. 2012;279(1733):1477–84.
20. Bacher JM, Metzgar D, de Crecy-Lagard V. Rapid evolution of diminished transformability in Acinetobacter baylyi. J Bacteriol. 2006;188(24):8534–42.
21. Engelmoer DJ, Donaldson I, Rozen DE. Conservative sex and the benefits of transformation in Streptococcus pneumoniae. PLoS Pathog. 2013;9(11):e1003758.
22. Utnes AL, Sorum V, Hulter N, Primicerio R, Hegstad J, Kloos J, Nielsen KM, Johnsen PJ. Growth phase-specific evolutionary benefits of natural transformation in Acinetobacter baylyi. ISME J. 2015;9(10):2221–31.
23. Sinha S, Mell J, Redfield R. The availability of purine nucleotides regulates natural competence by controlling translation of the competence activator Sxy. Mol Microbiol. 2013;88(6):1106–19.
24. Stewart GJ, Carlson CA. The biology of natural transformation. Annu Rev Microbiol. 1986;40:211–35.
25. Cameron AD, Redfield RJ. Non-canonical CRP sites control competence regulons in Escherichia coli and many other gamma-proteobacteria. Nucleic Acids Res. 2006;34(20):6001–14.
26. Wise EM, Alexander SP, Powers M. Adenosine 3'-5'-cyclic monophosphate as a regulator of bacterial transformation. Proc Natl Acad Sci U S A. 1973; 70(2):471–4.
27. Blokesch M. Chitin colonization, chitin degradation and chitin-induced natural competence of Vibrio cholerae are subject to catabolite repression. Environ Microbiol. 2012;14(8):1898–912.
28. Zheng L, Chen Z, Itzek A, Herzberg MC, Kreth J. CcpA regulates biofilm formation and competence in Streptococcus gordonii. Mol Oral Microbiol. 2012;27(2):83–94.
29. Antonova ES, Bernardy EE. Hammer Natural competence in Vibrio cholerae is controlled by a nucleoside scavenging response that requires CytR-dependent anti-activation. Mol Microbiol. 2012;86(5):1215–31.
30. Bernstein H, Byerly HC, Hopf FA, Michod RE. Origin of sex. J Theor Biol. 1984; 110(3):323–51.
31. Hoelzer MA, Michod RE. DNA-repair and the evolution of transformation in bacillus-subtilis.3. Sex with damaged DNA. Genetics. 1991;128(2):215–23.
32. Michod RE, Wojciechowski MF. DNA-repair and the evolution of transformation.4. DNA-damage increases transformation. J Evol Biol. 1994;7(2):147–75.
33. Wojciechowski MF, Hoelzer MA, Michod RE. DNA-repair and the evolution of transformation in bacillus-subtilis.2. Role of inducible repair. Genetics. 1989;121(3):411–22.
34. Mongold JA. DNA repair and the evolution of transformation in Haemophilus influenzae. Genetics. 1992;132(4):893–8.
35. Redfield RJ. Evolution of natural transformation: testing the DNA repair hypothesis in Bacillus subtilis and Haemophilus influenzae. Genetics. 1993; 133(4):755–61.

36. Engelmoer DJP, Rozen DE. Competence increases survival during stress in streptococcus pneumoniae. Evolution. 2011;65(12):3475–85.
37. Claverys JP, Prudhomme M, Martin B. Induction of competence regulons as a general response to stress in gram-positive bacteria. Annu Rev Microbiol. 2006;60:451–75.
38. Prudhomme M, Attaiech L, Sanchez G, Martin B, Claverys JP. Antibiotic stress induces genetic transformability in the human pathogen Streptococcus pneumoniae. Science. 2006;313(5783):89–92.
39. Charpentier X, Kay E, Schneider D, Shuman HA. Antibiotics and UV radiation induce competence for natural transformation in Legionella pneumophila. J Bacteriol. 2011;193(5):1114–21.
40. Dorer MS, Fero J, Salama NR. DNA damage triggers genetic exchange in Helicobacter pylori. PLoS Pathog. 2010;6(7):e1001026.
41. Hare JM, Ferrell JC, Witkowski TA, Grice AN. Prophage induction and differential RecA and UmuDAb transcriptome regulation in the DNA damage responses of acinetobacter baumannii and acinetobacter baylyi. Plos One. 2014;9(4):e93861.
42. Rauch PJG, Palmen R, Burds AA, Gregg-Jolly LA, van der Zee JR, Hellingwerf KJ. The expression of the Acinetobacter calcoaceticus recA gene increases in response to DNA damage independently of RecA and of development of competence for natural transformation. Microbiol-Uk. 1996;142:1025–32.
43. Palmen R, Vosman B, Kok R, van der Zee JR, Hellingwerf KJ. Characterization of transformation-deficient mutants of acinetobacter-calcoaceticus. Mol Microbiol. 1992;6(13):1747–54.
44. Palmen R, Vosman B, Buijsman P, Breek CK, Hellingwerf KJ. Physiological characterization of natural transformation in Acinetobacter calcoaceticus. J Gen Microbiol. 1993;139(2):295–305.
45. Smith HO, Danner DB, Deich RA. Genetic-transformation. Annu Rev Biochem. 1981;50:41–68.
46. Frye SA, Nilsen M, Tonjum T, Ambur OH. Dialects of the DNA uptake sequence in Neisseriaceae. PLoS Genet. 2013;9(4):e1003458.
47. Mell JC, Hall IM, Redfield RJ. Defining the DNA uptake specificity of naturally competent Haemophilus influenzae cells. Nucleic Acids Res. 2012;40(17):8536–49.
48. Provvedi R, Dubnau D. ComEA is a DNA receptor for transformation of competent Bacillus subtilis. Mol Microbiol. 1999;31(1):271–80.
49. Mortier-Barriere I, Velten M, Dupaigne P, Mirouze N, Pietrement O, McGovern S, Fichant G, Martin B, Noirot P, Le Cam E, et al. A key presynaptic role in transformation for a widespread bacterial protein: DprA conveys incoming ssDNA to RecA. Cell. 2007;130(5):824–36.
50. Quevillon-Cheruel S, Campo N, Mirouze N, Mortier-Barriere I, Brooks MA, Boudes M, Durand D, Soulet AL, Lisboa J, Noirot P, et al. Structure-function analysis of pneumococcal DprA protein reveals that dimerization is crucial for loading RecA recombinase onto DNA during transformation. Proc Natl Acad Sci U S A. 2012;109(37):E2466–75.
51. Harms K, Wackernagel W. The RecBCD and SbcCD DNases suppress homology-facilitated illegitimate recombination during natural transformation of Acinetobacter baylyi. Microbiol-Sgm. 2008;154:2437–45.
52. Kickstein E, Harms K, Wackernagel W. Deletions of recBCD or recD influence genetic transformation differently and are lethal together with a recJ deletion in Acinetobacter baylyi. Microbiology. 2007;153:2259–70.
53. de Vries J, Heine M, Harms K, Wackernagel W. Spread of recombinant DNA by roots and pollen of transgenic potato plants, identified by highly specific biomonitoring using natural transformation of an Acinetobacter sp. Appl Environ Microbiol. 2003;69(8):4455–62.
54. Sambrook J, Fritsch E, Maniatis T. Molecular cloning : a laboratory manual. 2nd ed. Cold Spring Harbor: Cold Spring Harbor Laboratory Press; 1989.
55. Barbe V, Vallenet D, Fonknechten N, Kreimeyer A, Oztas S, Labarre L, Cruveiller S, Robert C, Duprat S, Wincker P, et al. Unique features revealed by the genome sequence of Acinetobacter sp. ADP1, a versatile and naturally transformation competent bacterium. Nucleic Acids Res. 2004;32(19):5766–79.
56. Overballe-Petersen S, Harms K, Orlando LA, Mayar JV, Rasmussen S, Dahl TW, Rosing MT, Poole AM, Sicheritz-Ponten T, Brunak S, et al. Bacterial natural transformation by highly fragmented and damaged DNA. Proc Natl Acad Sci U S A. 2013;110(49):19860–5.
57. Horton RM, Cai ZL, Ho SN, Pease LR. Gene splicing by overlap extension: tailor-made genes using the polymerase chain reaction. Biotechniques. 1990;8(5):528–35.
58. Harms K, Schon V, Kickstein E, Wackernagel W. The RecJ DNase strongly suppresses genomic integration of short but not long foreign DNA fragments by homology-facilitated illegitimate recombination during transformation of Acinetobacter baylyi. Mol Microbiol. 2007;64(3):691–702.
59. Orren DK, Sancar A. The (A)BC excinuclease of Escherichia coli has only the UvrB and UvrC subunits in the incision complex. Proc Natl Acad Sci U S A. 1989;86(14):5237–41.
60. Kok RG, Young DM, Ornston LN. Phenotypic expression of PCR-Generated random mutations in a Pseudomonas putida gene after its introduction into an Acinetobacter chromosome by natural transformation. Appl Environ Microb. 1999;65(4):1675–80.
61. Crawley MJ. Statistical computing. An introduction to data analysis using S-plus. 1st ed. Chichester: Wiley; 2002.
62. R core team. R: A language and environment for statistical computing. 321st ed. Vienna: R Foundation for Statistical Computing; 2015.
63. Courcelle CT, Chow KH, Casey A, Courcelle J. Nascent DNA processing by RecJ favors lesion repair over translesion synthesis at arrested replication forks in Escherichia coli. Proc Natl Acad Sci U S A. 2006;103(24):9154–9.
64. Ray JL, Harms K, Wikmark OG, Starikova I, Johnsen PJ, Nielsen KM. Sexual isolation in Acinetobacter baylyi is locus-specific and varies 10,000-fold over the genome. Genetics. 2009;182(4):1165–81.
65. Van Houten B. Nucleotide excision repair in Escherichia coli. Microbiol Rev. 1990;54(1):18–51.
66. Kowalczykowski SC, Dixon DA, Eggleston AK, Lauder SD, Rehrauer WM. Biochemistry of homologous recombination in Escherichia coli. Microbiol Rev. 1994;58(3):401–65.
67. Kuzminov A. Recombinational repair of DNA damage in Escherichia coli and bacteriophage lambda. MMBR. 1999;63(4):751–813.
68. Morimatsu K, Kowalczykowski SC. RecFOR proteins load RecA protein onto gapped DNA to accelerate DNA strand exchange: a universal step of recombinational repair. Mol Cell. 2003;11(5):1337–47.
69. Dillingham MS, Kowalczykowski SC. RecBCD enzyme and the repair of double-stranded DNA breaks. MMBR. 2008;72(4):642–71.
70. Carrasco B, Fernandez S, Asai K, Ogasawara N, Alonso JC. Effect of the recU suppressors sms and subA on DNA repair and homologous recombination in Bacillus subtilis. Mol Genet Genomics. 2002;266(5):899–906.
71. Johnston C, Mortier-Barriere I, Granadel C, Polard P, Martin B, Claverys JP. RecFOR is not required for pneumococcal transformation but together with XerS for resolution of chromosome dimers frequently formed in the process. PLoS Genet. 2015;11(1):e1004934.
72. Gupta R, Shuman S, Glickman MS. RecF and recr play critical roles in the homologous recombination and single-strand annealing pathways of mycobacteria. J Bacteriol. 2015;197(19):3121–32.
73. Wang G, Lo LF, Maier RJ. The RecRO pathway of DNA recombinational repair in Helicobacter pylori and its role in bacterial survival in the host. DNA Repair (Amst). 2011;10(4):373–9.
74. Sakai A, Cox MM. RecFOR and RecOR as distinct reca loading pathways. J Biol Chem. 2009;284(5):3264–72.
75. Bakkali M. Could DNA, uptake be a side effect of bacterial adhesion and twitching motility? Arch Microbiol. 2013;195(4):279–89.
76. Bonura T, Smith KC. Enzymatic production of deoxyribonucleic acid double-strand breaks after ultraviolet irradiation of Escherichia coli K-12. J Bacteriol. 1975;121(2):511–7.
77. Redfield RJ. Genes for breakfast: the have-your-cake-and-eat-it-too of bacterial transformation. J Hered. 1993;84(5):400–4.

# Permissions

All chapters in this book were first published in BMC MICROBIOLOGY, by BioMed Central; hereby published with permission under the Creative Commons Attribution License or equivalent. Every chapter published in this book has been scrutinized by our experts. Their significance has been extensively debated. The topics covered herein carry significant findings which will fuel the growth of the discipline. They may even be implemented as practical applications or may be referred to as a beginning point for another development.

The contributors of this book come from diverse backgrounds, making this book a truly international effort. This book will bring forth new frontiers with its revolutionizing research information and detailed analysis of the nascent developments around the world.

We would like to thank all the contributing authors for lending their expertise to make the book truly unique. They have played a crucial role in the development of this book. Without their invaluable contributions this book wouldn't have been possible. They have made vital efforts to compile up to date information on the varied aspects of this subject to make this book a valuable addition to the collection of many professionals and students.

This book was conceptualized with the vision of imparting up-to-date information and advanced data in this field. To ensure the same, a matchless editorial board was set up. Every individual on the board went through rigorous rounds of assessment to prove their worth. After which they invested a large part of their time researching and compiling the most relevant data for our readers.

The editorial board has been involved in producing this book since its inception. They have spent rigorous hours researching and exploring the diverse topics which have resulted in the successful publishing of this book. They have passed on their knowledge of decades through this book. To expedite this challenging task, the publisher supported the team at every step. A small team of assistant editors was also appointed to further simplify the editing procedure and attain best results for the readers.

Apart from the editorial board, the designing team has also invested a significant amount of their time in understanding the subject and creating the most relevant covers. They scrutinized every image to scout for the most suitable representation of the subject and create an appropriate cover for the book.

The publishing team has been an ardent support to the editorial, designing and production team. Their endless efforts to recruit the best for this project, has resulted in the accomplishment of this book. They are a veteran in the field of academics and their pool of knowledge is as vast as their experience in printing. Their expertise and guidance has proved useful at every step. Their uncompromising quality standards have made this book an exceptional effort. Their encouragement from time to time has been an inspiration for everyone.

The publisher and the editorial board hope that this book will prove to be a valuable piece of knowledge for researchers, students, practitioners and scholars across the globe.

# List of Contributors

**Haihua Li, Lei Zhang, Longbin Chen, Qi Zhu, Wenjie Wang and Jiayun Qiao**
Tianjin Institute of Animal Husbandry and Veterinary Medicine, Tianjin, China

**Anne-Cécile Normand, Carole Cassagne, Magali Gautier, Stéphane Ranque and Renaud Piarroux**
Laboratoire de Parasitologie-Mycologie, CHU Timone, Université de la Méditerranée, Marseille, France

**Pierre Becker and Marijke Hendrickx**
Service of Mycology and Aerobiology, BCCM/IHEM Fungal Collection, Scientific Institute of Public Health, Brussels, Belgium

**Chao Yu, Nu Wang, Maosen Wu, Fang Tian, Huamin Chen, Fenghuan Yang and Chenyang He**
State Key Laboratory for Biology of Plant Diseases and Insect Pests, Institute of Plant Protection, Chinese Academy of Agricultural Sciences, Beijing 100193, China

**Xiaochen Yuan and Ching-Hong Yang**
Department of Biological Sciences, University of Wisconsin-Milwaukee, Milwaukee, WI 53211, USA

**Jing Li, Yi Zheng, Haiyan Xu, Xiaoxia Xi, Qiangchuan Hou, Shuzhen Feng, Laga Wuri, Yanfei Bian, Zhongjie Yu, Lai-Yu Kwok, Zhihong Sun and Tiansong Sun**
Key Laboratory of Dairy Biotechnology and Engineering, Inner Mongolia Agricultural University, Hohhot 010018, People's Republic of China

**Johannes Schmidpeter, Marlis Dahl, Jörg Hofmann and Christian Koch**
Department of Biology, Division of Biochemistry, Friedrich-Alexander University Erlangen-Nuremberg, Staudtstrasse 5, 91058 Erlangen, Germany

**Shelley M. Horne, Joseph Sayler, Nicholas Scarberry, Meredith Schroeder, Ty Lynnes and Birgit M. Prüß**
Department of Veterinary and Microbiological Sciences, North Dakota State University, Fargo, ND 58103, USA

**A. Sharaf**
Genetic Department, Faculty of Agriculture, Ain Shams University, Cairo 11241, Egypt
Institute of Parasitology, Biology Centre, Czech Academy of Sciences, 37005 České Budějovice, Czechia

**F. Mercati**
Institute of Biosciences and Bioresources (IBBR), National Research Council (CNR) of Italy, 90129 Palermo, Italy

**I. Elmaghraby**
Central Lab. of Organic Agriculture, Agricultural Research Center, Giza 12619, Egypt

**R. M. Elbaz**
Botany and Microbiology Department, Faculty of Science, Helwan University, Ain-Helwan, Cairo 11970, Egypt

**E. M. Marei**
Microbiology Department, Faculty of Agriculture, Ain Shams University, Cairo 11241, Egypt

**Min Liu, Huizhou Liu and Mo Xian**
CAS Key Laboratory of Biobased Materials, Qingdao Institute of Bioenergy
and Bioprocess Technology, Chinese Academy of Sciences, Qingdao 266101, China

**Hailin Chen and Zhe Zhao**
CAS Key Laboratory of Biobased Materials, Qingdao Institute of Bioenergy and Bioprocess Technology, Chinese Academy of Sciences, Qingdao 266101, China
University of Chinese Academy of Sciences, Beijing 100049, China

**Yamei Ding**
Institute of Oceanology, Chinese Academy of Sciences, Qingdao 266071, China

**Guang Zhao**
CAS Key Laboratory of Biobased Materials, Qingdao Institute of Bioenergy and Bioprocess Technology, Chinese Academy of Sciences, Qingdao 266101, China
Randian Technology Company Limited, Tianjin 300457, China

**Dequan Zhang**
Institute of Food Science and Technology, Chinese Academy of Agricultural Sciences/Key Laboratory of Agro-Products Processing, Ministry of Agriculture, Beijing 100193, People's Republic of China

**Wenbo Chen, Honghai Hu, Chunjiang Zhang, Feng Huang and Hong Zhang**
Institute of Food Science and Technology, Chinese Academy of Agricultural Sciences/Key Laboratory of Agro-Products Processing, Ministry of Agriculture, Beijing 100193, People's Republic of China
College of Staple Food Technology, Chinese Academy of Agricultural Sciences, Institute of Food Science and Technology, Harbin 151900, People's Republic of China

**Anna Zanfardino, Adriana Migliardi, Mario Varcamonti and Angela Cordone**
Department of Biology, University of Naples "Federico II", Via Cintia, 80126 Naples, Italy

**Daniele D'Alonzo and Angela Lombardi**
Department of Chemical Sciences, University of Naples "Federico II", Via Cintia, 80126 Naples, Italy

**Ayako Terakawa, Ayane Natsume, Atsushi Okada, Shogo Nishihata and Junko Kuse**
Department of Agrobioscience, Kobe University, 1-1 Rokkodai, Nada, Kobe, Hyogo 657-8501, Japan

**Shinji Takenaka**
Department of Agrobioscience, Kobe University, 1-1 Rokkodai, Nada, Kobe, Hyogo 657-8501, Japan
Organization of Advanced Science and Technology, Kobe University, 1-1 Rokkodai, Nada, Kobe, Hyogo 657-8501, Japan

**Kosei Tanaka**
Organization of Advanced Science and Technology, Kobe University, 1-1 Rokkodai, Nada, Kobe, Hyogo 657-8501, Japan

**Shu Ishikawa**
Department of Science, Technology and Innovation, Kobe University, 1-1 Rokkodai, Nada, Kobe, Hyogo 657-8501, Japan

**Ken-ichi Yoshida**
Organization of Advanced Science and Technology, Kobe University, 1-1 Rokkodai, Nada, Kobe, Hyogo 657-8501, Japan
Department of Science, Technology and Innovation, Kobe University, 1-1 Rokkodai, Nada, Kobe, Hyogo 657-8501, Japan

**Lysiane Duniere, Shanwei Xu, Jin Long, Chijioke Elekwachi, Yuxi Wang, Robert Forster and Tim A. McAllister**
Agriculture and Agri-Food Canada Research Centre, Lethbridge T1J 4B1, AB, Canada

**Kelly Turkington**
Agriculture and Agri-Food Canada (AAFC), Lacombe T4L 1 W1, AB, Canada

**Jing Yan Tang, Jing Ma, Xue Dong Li and Yan Hong Li**
College of Life Science, Capital Normal University, Xisanhuan North Road 105#, Haidian District, Beijing 100048, China

**Luciano Takeshi Kishi, Camila Cesário Fernandes, Wellington Pine Omori, João Carlos Campanharo and Eliana Gertrudes de Macedo Lemos**
Departamento de Tecnologia, Laboratório de Bioquímica de Microrganismos e Planta - LBMP, UNESP - Universidade Estadual Paulista, Faculdade de Ciências Agrárias e Veterinárias, Via de Acesso Prof. Paulo Donato Castellane s/n, 14884-900 Jaboticabal, SP, Brazil

**Annika Cimdins and Ute Römling**
Department of Microbiology, Tumor and Cell Biology, Karolinska Institutet, Stockholm, Sweden

**Irfan Ahmad**
Department of Microbiology, Tumor and Cell Biology, Karolinska Institutet, Stockholm, Sweden
Department of Allied Health Sciences, University of Health Sciences, Lahore, Pakistan
Department of Molecular Biology, Umeå University, Umeå, Sweden

**Timo Beske**
Department of Microbiology, Tumor and Cell Biology, Karolinska Institutet, Stockholm, Sweden
Department of Biology, Laboratory for Microbiology, Philipps-University Marburg, Marburg, Germany

**Andrea Kunova , Maria Bonaldi, Marco Saracchi, Cristina Pizzatti, Xiaoyulong Chen and Paolo Cortesi**
Department of Food, Environmental and Nutritional Sciences (DeFENS), Università degli Studi di Milano, via Giovanni Celoria, 2, 20133 Milan, Italy

**Christina Sakarikou and Camilla Dolfa**
Department of Experimental Medicine and Surgery, "Tor Vergata" University of Rome, Via Montpellier 1, 00133 Rome, Italy

**Cartesio Favalli**
Department of Experimental Medicine and Surgery, "Tor Vergata" University of Rome, Via Montpellier 1, 00133 Rome, Italy
Laboratory of Clinical Microbiology and Virology, Polyclinic "Tor Vergata" Foundation, V.le Oxford 81, 00133 Rome, Italy

**Marco Ciotti**
Laboratory of Clinical Microbiology and Virology, Polyclinic "Tor Vergata" Foundation, V.le Oxford 81, 00133 Rome, Italy

**Silvia Angeletti**
Clinical Pathology and Microbiology Laboratory, University Hospital Campus Bio-Medico, Via Alvaro del Portillo 200, 00128 Rome, Italy

**Fernanda Badotti and Cleverson Fernando Garcia**
Centro Federal de Educação Tecnológica de Minas Gerais (CEFET-MG), Departamento de Química, 30.421-169 Belo Horizonte, MG, Brazil

**Francislon Silva de Oliveira**
Fundação Oswaldo Cruz (FIOCRUZ), Centro de Pesquisas René Rachou - CPqRR, 30190-002 Belo Horizonte, MG, Brazil

**Paula Luize Camargos Fonseca3 and Aristóteles Góes-Neto**
Universidade Federal de Minas Gerais, Departamento de Microbiologia, Av. Antônio Carlos, Belo Horizonte 6627, 31270-901, MG, Brazil

**Aline Bruna Martins Vaz**
Universidade Federal de Minas Gerais, Departamento de Microbiologia, Av. Antônio Carlos, Belo Horizonte 6627, 31270-901, MG, Brazil
Faculdade de Minas (FAMINAS), 66055-090 Belo Horizonte, MG, Brazil

**Laila Alves Nahum**
Fundação Oswaldo Cruz (FIOCRUZ), Centro de Pesquisas René Rachou - CPqRR, 30190-002 Belo Horizonte, MG, Brazil
Faculdade Promove de Tecnologia, 30140-061 Belo Horizonte, MG, Brazil

**Guilherme Oliveira**
Fundação Oswaldo Cruz (FIOCRUZ), Centro de Pesquisas René Rachou - CPqRR, 30190-002 Belo Horizonte, MG, Brazil
Instituto Tecnológico Vale, 66055-090 Belém, PA, Brazil

**Nils Hülter**
Genomic Microbiology, Institute of Microbiology, Christian-Albrechts-University Kiel, Am Botanischen Garten 11, 24118 Kiel, Germany
Department of Pharmacy, Faculty of Health Sciences, UiT The Arctic University of Norway, Tromsø, Norway

**Kristina Borch-Pedersen**
Department of Food Safety and Infection Biology, Faculty of Veterinary Medicine and Biosciences, Norwegian University of Life Sciences, Norway

**Mikkel M. Liljegren**
Centre for Ecolgical and Evolutionary Synthesis, Faculty of Mathematics and Natural Sciences, University of Oslo, Norway

**Ane L. G. Utnes, Vidar Sørum , Pål J. Johnsen and Raul Primicerio**
Department of Pharmacy, Faculty of Health Sciences, UiT The Arctic University of Norway, Tromsø, Norway

**Klaus Harms**
Department of Pharmacy, Faculty of Health Sciences, UiT The Arctic University of Norway, Tromsø, Norway
Centre for GeoGenetics, Natural History Museum of Denmark, University of Copenhagen, Øster Voldgade 5-7, 1350 Copenhagen K, Denmark

# Index